Glutathione in Plant Growth, Development, and Stress Tolerance

Mohammad Anwar Hossain
Mohammad Golam Mostofa
Pedro Diaz-Vivancos • David J. Burritt
Masayuki Fujita • Lam-Son Phan Tran
Editors

Glutathione in Plant Growth, Development, and Stress Tolerance

Springer

Editors
Mohammad Anwar Hossain
Department of Genetics and Plant Breeding
Bangladesh Agricultural University
Mymensingh, Bangladesh

Laboratory of Plant Nutrition and Fertilizers
Graduate School of Agricultural and Life
Sciences, University of Tokyo
Tokyo, Japan

Pedro Diaz-Vivancos
Department of Plant Breeding
CEBAS-CSIC
Murcia, Murcia, Spain

Masayuki Fujita
Department of Applied Biological Science
Kagawa University
Kagawa, Japan

Mohammad Golam Mostofa
Department of Biochemistry and Molecular
Biology
Bangabandhu Sheikh Mujibur Rahman
Agricultural University
Gazipur, Bangladesh

David J. Burritt
Department of Botany
University of Otago
Dunedin, Otago, New Zealand

Lam-Son Phan Tran
RIKEN Center for Sustainable Resource
Science
Signaling Pathway Research Unit
Yokohama, Japan

ISBN 978-3-319-66681-5 ISBN 978-3-319-66682-2 (eBook)
https://doi.org/10.1007/978-3-319-66682-2

Library of Congress Control Number: 2017958440

Printed on acid-free paper

This Springer imprint is published by Springer Nature
The registered company is Springer International Publishing AG
The registered company address is: Gewerbestrasse 11, 6330 Cham, Switzerland

Preface

Glutathione (γ-glutamyl-cysteinyl-glycine) is a ubiquitously distributed sulfur-containing antioxidant molecule that plays key roles in the regulation of plant growth, development, and abiotic and biotic stress tolerance. It is one of the most powerful low-molecular-weight thiols, which rapidly accumulates in plant cells under stress. Recent in-depth studies on glutathione homeostasis (biosynthesis, degradation, compartmentalization, transport, and redox turnover) and the roles of glutathione in cell proliferation and environmental stress tolerance have provided new insights for plant biologists to conduct research aimed at deciphering the mechanisms associated with glutathione-mediated plant growth and stress responses, as well as to develop stress-tolerant crop plants. Glutathione has also been suggested to be a potential regulator of epigenetic modifications, playing important roles in the regulation of genes involved in the responses of plants to changing environments. The dynamic relationship between reduced glutathione (GSH) and reactive oxygen species (ROS) has been well documented, and glutathione has been shown to participate in several cell signaling and metabolic processes, involving the synthesis of protein, the transport of amino acids, DNA repair, the control of cell division, and programmed cell death. Two genes, *gamma-glutamylcysteine synthetase (GSH1)* and *glutathione synthetase (GSH2)*, are involved in GSH synthesis, and genetic manipulation of these genes can modulate cellular glutathione levels. Any fluctuations in cellular GSH and oxidized glutathione (GSSG) levels have profound effects on plant growth and development, as glutathione is associated with the regulation of the cell cycle, redox signaling, enzymatic activities, defense gene expression, systemic acquired resistance, xenobiotic detoxification, and biological nitrogen fixation. Being a major constituent of the glyoxalase system and ascorbate-glutathione cycle, GSH helps to control multiple abiotic and biotic stress signaling pathways through the regulation of ROS and methylglyoxal (MG) levels. In addition, glutathione metabolism has the potential to be genetically or biochemically manipulated to develop stress-tolerant and nutritionally improved crop plants. Although significant progress has been made in investigating the multiple roles of glutathione in abiotic and biotic stress tolerance, many aspects of glutathione-mediated stress responses require additional research.

The main objective of this volume is to explore the diverse roles of glutathione in plants by providing basic, comprehensive, and in-depth molecular information for advanced students, scholars, teachers, and scientists interested in or already engaged in research that involves glutathione. Finally, this book will be a valuable resource for future glutathione-related research and can be considered as a textbook for graduate students and as a reference book for frontline researchers working on glutathione metabolism in relation to plant growth, development, stress responses, and stress tolerance.

As editors of this volume, we are highly thankful to our experienced and well-versed contributors, who cordially accepted our invitation to write their chapters. We would also like to extend our thanks to Dr. Kenneth Teng and the editorial staff of Springer New York, who enabled us to initiate this book project. We believe that the information covered in this book will make a sound contribution to this fascinating area of research.

Mymensingh, Bangladesh	Mohammad Anwar Hossain
Gazipur, Bangladesh	Mohammad Golam Mostofa
Murcia, Spain	Pedro Diaz-Vivancos
Dunedin, New Zealand	David J. Burritt
Kagawa, Japan	Masayuki Fujita
Yokohama, Japan	Lam-Son Phan Tran

Contents

About the Editors

Mohammad Anwar Hossain is a professor in the Department of Genetics and Plant Breeding, Bangladesh Agricultural University, Mymensingh-2202, Bangladesh. He received his B.Sc. in agriculture and M.S. in genetics and plant breeding from Bangladesh Agricultural University, Bangladesh. He also received an M.S. in agriculture from Kagawa University, Japan, in 2008 and a Ph.D. in abiotic stress physiology and molecular biology from Ehime University, Japan, in 2011 through a Monbukagakusho scholarship. In November 2015, he moved to Tokyo University, Japan, as a JSPS postdoctoral researcher to work on isolating low-phosphorus stress-tolerant genes/QTLs from rice. He has over 50 peer-reviewed publications on important aspects of plant physiology and breeding, plant nutrition, plant stress responses and tolerance mechanisms, and exogenous chemical priming-induced abiotic oxidative stress tolerance. He has edited five book volumes, including this one, published by CRC Press, Springer, and Elsevier. He is a professional member of the International Metabolomics Society, Bangladesh Society of Genetics and Plant Breeding, Bangladesh Association for Plant Tissue Culture and Biotechnology, and Seed Science Society of Bangladesh.

Mohammad Golam Mostofa is an associate professor of biochemistry and molecular biology at Bangabandhu Sheikh Mujibur Rahman Agricultural University, Bangladesh. He received his B.Sc. and M.S. in biochemistry and molecular biology from the University of Dhaka, Bangladesh. He also obtained his M.S. in agriculture from Kagawa University, Japan, in 2012 and Ph.D. in plant and environmental sciences from Ehime University, Japan, in 2015 under a Monbukagakusho scholarship. His research focuses on plant physiology and biochemistry, abiotic stress tolerance in plants, oxidative stress responses, and crosstalk between phytohormones and signaling molecules under environmental stresses. His published works deal with the roles of signaling molecules in mitigating abiotic stresses in plants. He has authored over 25 peer-reviewed articles published in international journals. In September 2016, he joined the Signaling Pathway Research Unit at RIKEN Center for Sustainable Resource Science, Japan, as a JSPS postdoctoral researcher to work on the roles of the strigolactone phytohormone and its interactions with other signaling molecules under abiotic stresses in plants.

Pedro Diaz-Vivancos is an associate researcher in the Plant Breeding Department at CEBAS-CSIC (Murcia, Spain). He received his B.Sc. in biological sciences from the University of Murcia and his Ph.D. in 2007 in agricultural and food technology from the University of Cartagena (Spain). He has worked in 2008 and 2009 at Newcastle and Leeds Universities (UK) as a postdoctoral scientist on the role of glutathione in the regulation of cell proliferation and plant development. His main research interests are the study of the role of antioxidative metabolism in plant development and in plant stress responses, including the induction of tolerance against environmental stresses in plum plants by increasing their antioxidant capacity by genetic engineering. He has over 40 peer-reviewed publications and 2 book chapters.

David J. Burritt is an associate professor in the Department of Botany, the University of Otago, Dunedin, New Zealand. He received his B.Sc. and M.Sc. (Hons) in botany and his Ph.D. in plant biotechnology from the University of Canterbury, Christchurch, New Zealand. His research interests include oxidative stress and redox biology, plant-based foods and bioactive molecules, plant breeding and biotechnology, the cryopreservation of germplasm, and the stress biology of plants, animals, and algae. He has over 100 peer-reviewed publications and has edited 2 books for Springer and 1 for Elsevier.

Masayuki Fujita is a professor in the Department of Plant Science, Faculty of Agriculture, Kagawa University, Kagawa, Japan. He received his B.Sc. in chemistry from Shizuoka University, Shizuoka, and his M.Agr. and Ph.D. in plant biochemistry from Nagoya University, Nagoya, Japan. His research interests include physiological, biochemical, and molecular biological responses based on secondary metabolism in plants under biotic (pathogenic fungal infection) and abiotic (salinity, drought, extreme temperatures, and heavy metals) stresses; phytoalexin, cytochrome P450, glutathione *S*-transferase, and phytochelatin; and redox reaction and antioxidants. He has over 150 peer-reviewed publications and has edited 2 books.

Lam-Son Phan Tran is head of the Signaling Pathway Research Unit at RIKEN Center for Sustainable Resource Science, Japan. He obtained his M.Sc. in biotechnology in 1994 and Ph.D. in biological sciences in 1997 from Szent Istvan University, Hungary. After doing his postdoctoral research at the National Food Research Institute (1999–2000) and the Nara Institute of Science and Technology of Japan (2001), in October 2001, he joined the Japan International Research Center for Agricultural Sciences to work on the functional analyses of transcription factors and osmosensors in *Arabidopsis* plants under environmental stresses. In August 2007, he moved to the University of Missouri-Columbia, USA, as a senior research scientist to coordinate a research team working to discover soybean genes to be used for genetic engineering of drought-tolerant soybean plants. His current research interests are the elucidation of the roles of phytohormones and their interactions in abiotic stress responses, as well as translational genomics of legume crops with the aim to enhance crop productivity under adverse environmental conditions. He has published over 110 peer-reviewed papers with more than 80 research and 30 review articles and contributed 8 book chapters to various book editions published by Springer, Wiley-Blackwell, and the American Society of Agronomy, Crop Science Society of America, and Soil Science Society of America. He has also edited nine book volumes, including this one, for Springer and Elsevier.

Chapter 1
Chemistry, Biosynthesis, and Antioxidative Function of Glutathione in Plants

Wilma Sabetta, Annalisa Paradiso, Costantino Paciolla, and Maria Concetta de Pinto

Abstract Glutathione, a tripeptide constituted by glutamate, cysteine, and glycine, is an abundant metabolite that functions as a master regulator of intracellular redox homeostasis. Under optimal conditions, glutathione is mostly present in the reduced form (GSH), with a free thiol group. The link of two molecules of GSH, via a disulfide bond, leads to the formation of glutathione disulfide (GSSG). GSH can be oxidized, directly or indirectly, by reactive oxygen species, working as a scavenger that prevents excessive oxidation of cellular environment. GSH can also react with different thiols to form mixed disulfides. These reversible redox reactions are responsible for many GSH functions. GSH biosynthesis is dependent on the activity of the two ATP-dependent enzymes γ-glutamylcysteine synthetase and glutathione synthetase, encoded, respectively, by the nuclear *GSH1* and *GSH2* genes. The first step of GSH biosynthesis occurs in the plastids, while the second step can take place in both plastids and cytosol. The use of different *gsh1* mutants and *GSH1* overexpressing plants has helped to shed light on the multiple roles of GSH in plant growth, development, and response to changing environment. The maintenance of a high GSH/GSSG ratio is crucial for many physiological functions, and a decrease in this ratio can be utilized as an indicator of oxidative stress. The GSH/GSSG ratio also acts as an important regulator of several mechanisms involved in plant development and in plant stress response. In addition to redox state, also GSH concentration and its subcellular distribution are central factors controlling redox homeostasis and signaling.

Keywords Antioxidant • Cell compartments • Glutathione • Glutathione disulfide • Redox signaling • Redox state

---

W. Sabetta
Spin-off SINAGRI srl, University of Bari "Aldo Moro",
Via Amendola 165/A, I-70126 Bari, Italy

A. Paradiso • C. Paciolla • M.C. de Pinto (✉)
Department of Biology, University of Bari "Aldo Moro", Via Orabona 4, I-70125 Bari, Italy
e-mail: mariaconcetta.depinto@uniba.it

M.A. Hossain et al. (eds.), *Glutathione in Plant Growth, Development, and Stress Tolerance*, https://doi.org/10.1007/978-3-319-66682-2_1

Contents

1 Introduction

Glutathione (GSH) is a ubiquitous low-molecular-weight thiol in eukaryotes (Meister and Anderson 1983). This multifaceted molecule plays a number of key roles in plant biology. GSH is a product of sulfur metabolism, and, being mobile through long-distance transport, it also represents a storage form of reduced sulfur, since it can be remobilized in case of need (Rennenberg 2001). Additionally, GSH, like cysteine, is able to work as an important signal involved in the modulation of sulfate uptake and assimilation (Kopriva and Rennenberg 2004).

GSH plays a crucial role in different phases of plant life cycle; it is involved in embryo and meristem development (Cairns et al. 2006; Vernoux et al. 2000), as well as in pollen germination and in the development of flower primordia (Zechmann et al. 2011; Hatano-Iwasaki and Ogawa 2012; Gulyas et al. 2014). GSH mediates important cellular processes, like cell cycle progression and programmed cell death (Diaz-Vivancos 2010a, b; Kranner et al. 2006). Through the action of glutathione S-transferases (GSTs), GSH is also involved in the detoxification of different toxic compounds, such as xenobiotics, herbicides, and air pollutants (Cummins et al. 2011). Moreover, being the substrate for phytochelatin synthesis, GSH plays a key role in the detoxification of heavy metals (Freeman et al. 2004).

GSH is considered one of the most important cellular antioxidants, since it is able to scavenge directly or indirectly reactive oxygen species (ROS), which are unavoidable by-products of aerobic metabolism. Thus, GSH is a key metabolite in plant responses to biotic and abiotic stresses, in which it serves to remove ROS and, therefore, to limit the extent of oxidative damages (Foyer and Noctor 2005). However, it has been shown that GSH, through interactions with stress hormones, can also be implicated in the strengthening of ROS signals in plants (Han et al. 2013a, b). Therefore, it is conceivable that GSH is not only a simple antioxidant, but, being involved in the control of redox-sensitive proteins, it is able to couple changes in intracellular redox state to development/defense responses of plants, through the ROS-dependent signaling pathways (Foyer and Noctor 2005, 2016; Paciolla et al. 2016).

2 Glutathione Chemistry

GSH is a tripeptide consisting of glutamate, cysteine, and glycine (γ-glutamylcysteinylglycine). The linkage of the γ-carboxyl group of glutamate to the amino group of cysteine renders this bond different from peptide bonds found in proteins and gives stability to the molecule, which cannot be degraded by amino peptidases but requires specific carboxypeptidase and/or γ-glutamyl transpeptidase (Steinkamp and Rennenberg 1985; Wolf et al. 1996; Martin and Slovin 2000; Storozhenko et al. 2002). Some plants also possess GSH homologs in which glycine is substituted by other amino acids, as in the cases of homoglutathione (γ-Glu-Cys-β-Ala) in legumes and hydroxymethylglutathione (γ-Glu-Cys-Ser) in cereals (Klapheck 1988; Klapheck et al. 1992; Meuwly et al. 1993).

Due to the low molecular weight and to the presence of several hydrophilic groups, namely, two carboxylic groups, one amine and one thiol, GSH is a highly water-soluble compound.

The thiol group of the cysteine, being the most important chemically reactive group of GSH, is responsible for the biological and biochemical activity of this tripeptide: it permits redox reactions, as well as reactions of nucleophilic displacement. In the reactions with free radicals, GSH donates hydrogen atoms and produces the thiyl radical, which can also be formed by subtraction of one electron from the thiolate anion by photoionization or metal ions (Wonisch and Schaur 2001). Thiyl radicals, adequately stable and poorly reactive with other hydrogen donors, can dimerize and lead to the formation of glutathione disulfide (GSSG). Apart from GSSG, oxidized forms of glutathione comprise disulfides with other thiols to form "mixed disulfides" and more oxidized forms of the thiol group (Foyer and Noctor 2005). The thiol group of GSH can also act as a nucleophile reacting with a wide spectrum of electrophiles. In this case, it will not lead necessarily to disulfide formation but rather to the formation of GS-conjugate with various compounds. These reactions are important for detoxification of endogenous or xenobiotic compounds (Wang and Ballatori 1998; Dixon and Edwards 2010). Indeed, the GS-conjugates are usually transported by ABCC (subclass C of the ABC transporters) proteins, which are ATP-dependent pumps, to the vacuole, where the amino acids of GSH can be recycled (Martinoia et al. 1993; Lu et al. 1998; Grzam et al. 2006). GSH can also react with nitric oxide (NO), with the formation of nitrosoglutathione (GSNO), a molecule that is receiving increasing consideration for its role as a possible signaling molecule and/or as a NO reservoir (Lindermayr et al. 2005).

The reactions of thiolate-disulfide exchange seem to be due to a nucleophilic displacement on sulfur, in a similar way to those occurring in other nucleophilic displacements (Wonisch and Schaur 2001). The thiol-disulfide exchange reactions of glutathione are important in mediating the reversible oxidation/reduction of the redox-sensitive proteins and therefore play a key role in maintaining cellular redox state (Foyer and Noctor 2005). In addition, GSH also participates in posttranscriptional protein modification through S-glutathionylation, which consists in the formation of a stable mixed disulfide between GSH and a protein thiol. In this way, GSH protects proteins from irreversible modifications that can be induced by oxidation (Noctor et al. 2012).

2.1 Glutathione Oxidation

Chemical oxidation of GSH is strongly dependent on its deprotonation to the thiolate form and, consequently, can be influenced by pH changes. Since the *pKa* of the GSH thiol is about 9.0, approximately only 1% of GSH will be deprotonated in the cytosol (pH 7.2) and a lower percentage of the thiolate will occur in acidic compartments, such as vacuole or apoplast. Consistently, in chloroplasts glutathione's reactivity will be increased in the light, when the photosynthetic electron transport leads to stroma alkalinization (Rahantaniaina et al. 2013 and references therein).

ROS are able to chemically react with GSH, leading to its oxidation (Fig. 1.1). However, the reaction's rate between singlet oxygen and GSH ($2 \times 10^6\ M^{-1}\ s^{-1}$) is lower than that occurring between this reactive species and other antioxidants, such as tocopherols and carotenoids. On the other hand, GSH reacts quickly with hydroxyl radical ($8.1 \times 10^9\ M^{-1}\ s^{-1}$), even if this oxidant reacts very fast with many other metabolites, such as ascorbate (ASC) and sugars that have higher cellular concentrations. The direct reaction of GSH with H_2O_2 is very slow ($0.9\ M^{-1}\ s^{-1}$). Therefore, superoxide, which reacts with GSH at a rate similar to other antioxidants ($7 \times 10^5\ M^{-1}\ s^{-1}$), seems to be the major ROS contributing to un-catalyzed production of GSSG in vivo (Rahantaniaina et al. 2013 and references therein).

Dehydroascorbate (DHA), the stable product of ASC oxidation, is also able to directly oxidize GSH at significant rates (Fig. 1.1), which are higher at pH 8 than at pH 7. In addition, DHA reductases (DHARs) can catalyze the oxidation of GSH to reduce DHA to ASC (Fig. 1.1). In this way, DHARs provide a link between ascorbate and glutathione pools and allow GSH to take part, indirectly, in H_2O_2 reduction that finally relies on electrons derived from NAD(P)H and/or ferredoxin. In the ascorbate-glutathione (ASC-GSH) pathway, ASC peroxidase reduces H_2O_2 to water at the expense of ASC producing monodehydroascorbate; this last is an unstable product that can dismutate to ASC and DHA, which can then be reduced, chemically or by DHARs, by GSH with the simultaneous production of GSSG (Foyer and Halliwell 1976).

GSH oxidation can also depend on the activity of specific peroxidases (Fig. 1.1). Some GSTs, enzymes involved in the formation of a covalent bond between the sulfur atom of GSH and an electrophilic compound, can use GSH to reduce organic hydroperoxides (Wagner et al. 2002; Dixon et al. 2009; Dixon and Edwards 2010; Cummins et al. 2011). A number of GSTs have both conjugase

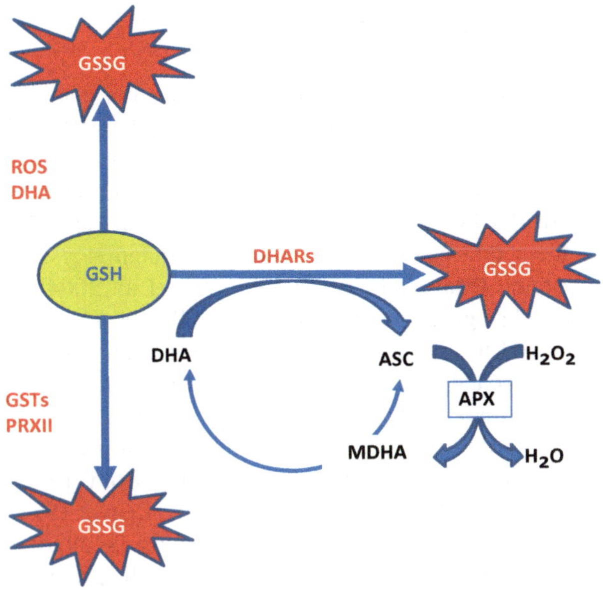

Fig. 1.1 Principal reactions involved in glutathione (GSH) oxidation. GSH can be chemically oxidized by reactive oxygen species (ROS) and dehydroascorbate (DHA). DHA reductases (DHARs) can also oxidizes GSH to regenerate ascorbate (ASC), allowing GSH to take part, indirectly, in H_2O_2 reduction. In this pathway, ASC peroxidase (APX) reduces H_2O_2 to water at the expense of ASC producing monodehydroascorbate (MDHA); this last is an unstable product that can dismutate to ASC and DHA. DHA can then be reduced by DHARs, producing glutathione disulfide (GSSG). Finally, GSH can be also oxidized by the peroxidase activity of glutathione *S*-transferases (GSTs) and type II peroxiredoxins (PRXII). More details are given in the text

and peroxidase activities. The GSTs of the lambda class, which have an active-site cysteine, could generate GSSG by catalyzing the reduction of small molecules or by the deglutathionylation of cysteine residues of proteins (Dixon et al. 2002; Dixon and Edwards 2010).

Several type II peroxiredoxins (PRXs), which are thiol peroxidases that can reduce both H_2O_2 and other organic peroxides, can oxidize GSH, through the action of glutaredoxins (Rouhier2010); on the other hand, GSH peroxidase (GPX), which belongs to the PRX family, contrary to what previously believed, acts as thioredoxin-dependent peroxidase and not as GSH-dependent peroxidase (Iqbal et al. 2006; Navrot et al. 2006).

Other enzymes could be responsible for GSSG formation. For instance, GSNO reductase can produce GSSG from GSH and GSNO (Sakamoto et al. 2002); adenosine phosphosulfate reductase, a key chloroplastic enzyme involved in sulfate reduction, employs GSH as electron donor (Bick et al. 1998); the activity of a plant methionine sulfoxide reductase may need GSH oxidation, which can occur via glutaredoxins (Tarrago et al. 2009).

3 Glutathione Biosynthesis

The biosynthetic pathway of GSH has been characterized in several organisms, and seems to act through a conserved chemical way (Rennenberg 1980; Meister 1988). In higher plants as in animals, GSH biosynthesis occurs in two ATP-dependent steps through the sequential action of γ-glutamylcysteine synthetase (γ-ECS) and GSH synthetase (GS). In the first reaction, γ-ECS, also known as glutamate-cysteine ligase, catalyzes the synthesis of the intermediate γ-glutamylcysteine (γ-EC) from glutamate and cysteine (May and Leaver 1994; Jez et al. 2004; Musgrave et al. 2013). The γ-carboxylate group of glutamate is initially phosphorylated by ATP and subsequently subjected to the nucleophile attack of the amino group of cysteine. The second reaction is catalyzed by GS that, similarly to the γ-ECS, first forms an acylphosphate intermediate at the C-terminal of γ-EC and then, displacing inorganic phosphate, links the amino groups of a glycine to produce GSH (Jez and Cahoon 2004; Herrera et al. 2007; Musgrave et al. 2013). Both enzymes are encoded by single-copy genes, called *GSH1* for γ-ECS (May and Leaver 1994) and *GSH2* for GS (Rawlins et al. 1995; Ullmann et al. 1996), which possess alternate transcription start sites, thus leading to either plastid-targeted or cytosolic proteins (Wachter et al. 2005). Immune-electron microscope analyses in *Arabidopsis* leaves have permitted to sub-localize the γ-ECS to the chloroplasts and the GS both into plastids and cytosol (Hell and Bergman 1988, 1990; Preuss et al. 2014). Activity assays of both enzymes, in cytosol and chloroplast fractions of *Arabidopsis* and wheat leaves, have revealed the 82% of γ-ECS total activity in the chloroplast and the 69% of GS total activity in the cytosol (Noctor et al. 2002a). Thereby, the subcellular localization of these two enzymes in plants makes GSH biosynthesis a compartmentalized process and supports the idea of a specific movement of biosynthetic intermediates between organelles and cytosol. Consistently, most of the neo-synthesized chloroplastic γ-EC moves to the cytosol as substrate of cytosolic GS (Pasternak et al. 2008).

Functional studies on plant lines carrying severe mutations at the *GSH1* and *GSH2* genes have allowed to establish the molecular basis for the comprehension of γ-ECS and GS roles in numerous species. Knockout mutations of these genes result in lethal phenotypes in different eukaryotes, indicating that GSH biosynthesis is essential for cell life (Grant et al. 1996; Kim et al. 2005). In particular, the knockout of the *Arabidopsis* GSH1 gene causes lethality at the embryo stage (Cairns et al. 2006), while *GSH2*-deficient lines show delay in the development, associated with death at the seedling stage (Pasternak et al. 2008). This difference appears to reflect the likely replacement of the missing GSH with the γ-EC intermediate, which accumulates to high levels in *gsh2* mutant plants. Additionally, these plants can partially restore the wild-type phenotype with an increase of the cytosolic γ-ECS activity (Pasternak et al. 2008), thus supporting the evidence of transport of GSH and its biosynthetic intermediates from cytosol to plastids (Noctor et al. 2002a). Less severe mutations, causing reduced GSH levels, have also been obtained by forward genetics approaches. In particular, some *gsh1* mutants are worth mentioning: the root-meristem-less1 (*rml1*) mutant, with only the 5% of wild-type GSH amount, is

Table 1.1 Effects on plant development and response to environmental stress of GSH deficiency in different glutathione biosynthetic mutants

GSH biosynthetic mutants	GSH content	Effects on plant development and response to environmental stress	References
Rml1 (root meristemless1)	3% of wt	Impairment in initiation and maintenance of cell division during post-embryonic root development.	Cheng et al. (1995) Vernoux et al. (2000)
Zir 1 (zinc tolerance induced by iron 1)	15% of wt	Seedlings smaller in size than the wild type. High sensitivity to Zn excess and no Fe-mediated Zn tolerance.	Shanmugam et al. (2012)
Pad2-1 (phytoalexin-deficient 2-1)	22% of wt	Enhanced susceptibility to various pathogens due to low accumulation of antimicrobial defenses and alteration in SA-dependent pathway. Wilting of leaves and downregulation of various stress-responsive genes under combined cold and osmotic stress.	Parisy et al. (2007) Dubreuil-Maurizi et al. (2011) Kumar et al. (2015)
Cad2-1 (cadmium sensitive 2-1)	15–45% of wt	Under Cd stress, deficiency of phytochelatins and consequent heavy metals sensitivity Moderate susceptibility to pathogens.	Howden et al. (1995) Cobbett et al. (1998) Parisy et al. (2007)
Rax1-1 (regulator of APX2 1-1)	20–50% of wt	Constitutive expression of stress-inducible APX2 Under photooxidative stress, altered expression of a wide set of defense-related genes	Ball et al. (2004)

unable to develop root apical meristems (Cheng et al. 1995; Vernoux et al. 2000); less drastic reductions, up to the 50% of total glutathione, do not reveal evident phenotypic aberrations but cause alterations in stress signal transduction and response, such as high sensitivity to cadmium in the *cad2* mutants or to pathogens in the *pad2* mutants (Howden et al. 1995; Cobbett et al. 1998; Parisy et al. 2007). The level of GSH content and the altered response in the development and in the response to environmental changes in different GSH biosynthetic mutants are summarized in Table 1.1.

3.1 Molecular Characteristics of γ-ECS and GS

Numerous studies performed in several organisms have permitted to better understand the structure, mechanism of action, and regulation of the enzymes involved in GSH biosynthesis.

On the basis of sequence analysis from multiple species, three main distinct families of γ-ECSs belonging to the plants and γ-protobacteria group, the α-protobacteria group, and the non-plant eukaryotes group (mammals, yeasts, trypanosomes) have been identified (May and Leaver 1994). Despite sharing a putative distant ancestor, common structural motifs and very similar catalytic mechanism of action, insignificant sequence homology has been found between the three groups (Copley and Dhillon 2002). In particular, the cloning of γ-ECS from different plant species has highlighted the strong sequence diversity with the mammalian and microbial counterparts (Frendo et al. 1999; Wu et al. 2009). Plant γ-ECSs share more structural and functional similarity with yeast γ-ECSs than with the heterodimeric γ-ECSs found in several eukaryotes or with the monomeric γ-ECS of *Escherichia coli* (Seelig et al. 1984; Fraser et al. 2002; Hibi et al. 2004; Biterova and Barycki 2009). In plant γ-ECSs, two magnesium ions in the active site increase the reactivity of the γ-phosphate group of ATP and assist the right orientation of the glutamate γ-carboxylate, thus stabilizing the resulting γ-glutamyl-phosphate intermediate. The cysteine binding site, adjacent to the glutamate binding site, undergoes a significant conformational change upon ligand binding, thus becoming reactive for cysteine attack. A close arginine, in a highly conserved position, usually provides the transition state of this reaction and the final peptide bond formation (Biterova and Barycki 2009).

GSs are members of the ATP-grasp superfamily, consisting of ligases that form amide bonds in peptides after translation (Li et al. 2003; Dinescu et al. 2004). They are usually characterized by the ATP-grasp binding site, with two α-helices and β-sheets, and an active site with high specificity for the substrate. A phylogenetic analysis of members of the GS family from several species allowed to infer the genetic distance and thereby to collocate plant GSs close to yeast GSs, sharing the 40% sequence homology (Wang and Oliver 1997). In contrast to the bacterial GSs that are homo-tetramers, mammalian, plant, and yeast GSs have been identified as homodimers, composed of two identical subunits linked by disulfide bonds (Gushima et al. 1983; Yamaguchi et al. 1993).

The heterologous GS protein, obtained by the overexpression of a cDNA of *Arabidopsis thaliana* GS in bacteria and yeast cells, shares high sequence similarity to *GSH2* products from other species and is characterized by the presence of an extremely conserved glycine-rich domain, close to the carboxy-terminus, typical of the eukaryotic GS family (Rawlins et al. 1995; Ullmann et al. 1996; Wang and Oliver 1997; Galant et al. 2009). A deep structural characterization of the *Arabidopsis* GS revealed the presence, in the active site, of three specific regions, involved in the bonds of ATP, magnesium, and both γ-EC and GSH (Herrera et al. 2007). Experiments of site-directed mutagenesis permitted to identify the role of 15 specific amino acids in GS active site, showing the sensibility of plant GS even to minor amino acid changes in the active site and suggesting that the ATP and the γ-EC binding can enhance reciprocally (Herrera et al. 2007).

As other members of the ATP-grasp family, plant GSs are able to use different amino acids during the tripeptide formation (Skipsey et al. 2005). The replacement of glycine with β-alanine, serine, or glutamate has been observed in several crop species that, unlike animals, can synthesize alternative forms of GSH. Homoglutathione, the most known analog of GSH, characterized by the presence of a β-alanine instead of glycine, is synthesized by homoglutathione synthetases (hGSs). From a structural point of view, plant hGSs are generally quite similar to human, yeast, and *Arabidopsis* GSs, also confirming the invariant amino acidic composition in the binding sites for ATP, Mg^{2+}, and γ-EC (Gogos and Shapiro 2002; Galant et al. 2009). The preference of β-alanine versus glycine is crucially determined by the active site of hGS. The active site of both enzymes is structurally composed by a lid domain, a glycine-rich loop, and an alanine-rich loop. The first two are critical for ATP binding and are responsible for the major conformational changes, while the alanine-rich loop can interact with glycine in the GS structures or with β-alanine in the hGS structure (Gogos and Shapiro 2002; Galant et al. 2009).

3.2 *Regulation of Glutathione Biosynthesis*

Cysteine availability is one of the most important factors affecting GSH biosynthesis. Indeed, a constitutive enhancement in GSH content can be achieved by the supply of exogenous cysteine or by the increase of enzymes involved in cysteine synthesis, which are, in turn, influenced by the availability of reduced sulfur (Buwalda et al. 1988, 1990; Harms et al. 2000; Noji and Saito 2002; Wirtz and Hell 2007). Treatments of leaves and roots of spinach and maize with excess of exogenous cysteine caused a significant GSH increase (Buwalda et al. 1988, 1990; Farago and Brunold 1994). Cysteine concentration presumably regulates GSH biosynthesis independently by the amount of γ-ECS, as shown in leaf discs of both wild-type and transformed poplar lines overexpressing the bacterial γ-ECS (Noctor et al. 1996). The cysteine control on the γ-ECS activity can prevent an excessive GSH production and regulate the cysteine metabolism itself (Buwalda et al. 1988; Rennenberg 1995). In addition to cysteine, glycine and ATP availability can also affect GSH production (Buwalda et al. 1990; Noctor et al. 1997; Ogawa et al. 2004).

Changes in *GSH1* and *GSH2* expression, which can influence γ-ECS and GS levels, represent also a possible way to regulate GSH synthesis. Indeed, heavy metals, jasmonic acid, and oxidative stress seem to activate the expression of both genes, which also respond to high light and other kind of stress (Xiang and Oliver 1998; Sung et al. 2009).

The production of plants overexpressing bacterial *GSH1* and *GSH2* genes has provided important insights into the regulation of GSH biosynthesis and metabolism. Leaves of transformed poplar lines overexpressing the cytosolic bacterial γ-ECS contain higher levels of cysteine, γ-EC, and total glutathione than those measured in leaves of untransformed plants (Noctor et al. 1996; Arisi et al. 1997). A more substantial increase in GSH has been obtained targeting the bacterial γ-ECS to the chloroplast (Noctor et al. 1998). The overexpression of the same bacterial γ-ECS in *Arabidopsis* and Indian mustard leads to a twofold increase of the GSH content, without evident phenotypes or general physiological perturbations (Xiang et al. 2001; Zhu et al. 1999b). A considerably greater increase in GSH content has been obtained by the overexpression into tobacco plants of a bifunctional protein with both γ-ECS and GS activities, isolated from *Streptococcus*, suggesting that GS activity can be limiting when γ-EC is made available (Liedschulte et al. 2010; Noctor et al. 2012). Overexpression of *GSH 2* gene from *Escherichia coli* in the cytosol or in the chloroplast enhances GS activities but leaves unchanged GSH content (Strohm et al. 1995; Foyer et al. 1995; Noctor et al. 1998). On the other hand, the overexpression of the same gene in presence of cadmium, which increases γ-EC intermediate, stimulates GSH biosynthesis (Zhu et al. 1999a). The rate-limiting effect of GS for GSH biosynthesis during stress conditions has been also demonstrated in tobacco plants overexpressing a soybean GS, which confers tolerance to the Fomesafen herbicide (Skipsey et al. 2005).

However, among the two biosynthetic enzymes, γ-ECS is generally thought to play a key role in regulating GSH biosynthesis. The γ-ECS activity is limited by the availability of free cysteine and ATP and is tightly feedback modulated by the end product itself. As in mammals, a feedback inhibition of γ-ECSs by GSH is a useful way to modulate glutathione homeostasis in plants, especially under conditions in which this tripeptide is rapidly consumed (Hell and Bergmann 1990; Noctor et al. 2002a). Moreover, rapid transcriptional activation and post-translational modifications of γ-ECS ensure the strict control of intracellular GSH levels (Hell and Bergmann 1990; May et al. 1998; Noctor et al. 2002a; Jez et al. 2004). A fine post-translational activation of *Arabidopsis* γ-ECS occurs through changes in its redox state. Indeed, the reversible formation of a disulfide bond makes γ-ECS more active (Jez et al. 2004). In this way, under oxidizing conditions γ-ECS is activated in parallel with the increased demand for GSH. As the GSH level increases, the more reduced intracellular environment causes an inactivation of γ-ECS, thus providing an efficient and rapid switch mechanism for the control of GSH biosynthesis (Hicks et al. 2007). The active enzyme in the oxidized status works as a dimer with two intermolecular disulfide bonds located at specific cysteine sites (Cys178-Cys398 and Cys341-Cys356); in a reducing environment, these bonds are disrupt and the enzyme comes back to the less active monomeric form. The first of the two disulfide bonds seems to be essential for the dimer formation, since experiments of site-direct mutagenesis of these cysteine block both the *Arabidopsis* and *Brassica juncea* γ-ECSs in the monomeric form (Hothorn et al. 2006; Hicks et al. 2007). The γ-ECS monomer/dimer transition by disulfide linkages is very common in the plant kingdom and seems to be related to the sub-compartmentalization of GSH biosynthesis in the chloroplast (Gromes et al. 2008).

4 Importance of Glutathione in the Redox Regulation

GSH, being involved directly or indirectly in the removal of ROS, like other numerous metabolites, can work as an antioxidant. However, it is interesting to note that the antioxidant and signaling functions of GSH are interdependent, since both require enzymes such as GSTs and PRXs that reduce H_2O_2 or other organic peroxides through thiol-mediated pathways (Noctor et al. 2012). The uniqueness of GSH as antioxidant and signaling molecule is also due to its high abundance and low redox potential, as well as to an ubiquitous distribution in plant cells.

4.1 Glutathione Redox State

A key feature of the cellular glutathione pool is its high reduction state. The glutathione pool is maintained predominantly in a reduced state by glutathione reductases (GRs), whose activities depend on the key electron carrier, NAD(P)H. GRs are flavoproteins with high affinity for both GSSG and NAD(P)H (Halliwell and Foyer 1978; Edwards et al. 1990). GR activities have been found in chloroplasts, cytosol, mitochondria, and peroxisomes (Foyer and Halliwell 1976; Edwards et al. 1990; Rasmusson and Møller 1990; Jiménez et al. 1997; Stevens et al. 2000; Romero-Puertas et al. 2006). In *Arabidopsis*, two genes encoding dual-targeted GRs have been identified. *GR1* encodes for the cytosolic and peroxisomal GRs and is responsible for the 30–60% of the total leaf enzymatic activity (Marty et al. 2009; Kataya and Reumann 2010; Mhamdi et al. 2010). *GR2* encodes an enzyme that is targeted to plastids and mitochondria (Chew et al. 2003). Loss of function of *GR1* determines only modest GSSG accumulation in leaf tissue, probably due to the supporting GSSG-reducing activity of cytosolic NAD(P)H-thioredoxin (TRX) systems (Marty et al. 2009). On the other hand, mutants for *GR2* are embryo-lethal (Tzafrir et al. 2004).

Under optimal conditions, total tissue glutathione pool is mostly reduced; GSH/GSSG ratios in leaves are usually no less than 20:1 (Noctor et al. 2012). However, this is an average value, and these ratios might be higher or lower depending on the specific considered compartments (Meyer et al. 2007; Queval et al. 2011). Indeed, some compartments, such as the endoplasmic reticulum and vacuoles as well as some cell types, like cells of the quiescent center, or dormant tissues, like seeds, are maintained in a more oxidized state (Hwang et al. 1992; Enyedi et al. 2010; Queval et al. 2011; Kranner and Grill 1996; Kranner et al. 2006).

In absence of stress, the glutathione redox potential, related to $[GSH]^2/[GSSG]$, determined by in vivo studies with the redox-sensitive GFPs (roGFPs), is lower than −300 mV in the cytosol, and similar values have been reported in nuclei (Meyer et al. 2007; Jubany-Mari et al. 2010; Schnaubelt et al. 2015). This low glutathione redox potential suggests that GSSG concentrations should be in the nanomolar range, in contrast with the analyses conducted on the entire tissues, in which, usually, GSSG concentrations are reported in the micromolar range (Noctor et al. 2012 and references therein). It could be possible that the low redox potential reported in the cytosol is due to the sequestration of GSSG in other compartments, such as the endoplasmic reticulum, vacuole, or apoplast, where glutathione reduction capacity is relatively low (Noctor et al. 2012). Under optimal conditions, the preservation of a very low glutathione redox potential in the cytosol could permit the initiation of oxidative signaling, with a quite low accumulation of GSSG, which can be perceived by sensitive thiol proteins, as significant changes in redox potential (Noctor et al. 2012, 2013). It has been suggested that changes in glutathione redox potential of about 50 mV could be sufficient to alter the balance between oxidized and reduced forms in thiol-disulfide status of sensitive TRX-regulated proteins (Setterdahl et al. 2003; Noctor et al. 2012). The glutathione redox potential is a primary component controlling relations between oxidative signals and sensitive protein targets, and it can be affected not only by changes in GSH/GSSG ratio but also by absolute GSH concentration (Meyer et al. 2007). For instance, the cytosolic redox potential, measured by roGFP, is more oxidizing in *GR*-deficient (*gr1*) mutants in which the GSH/GSSG ratio but not the glutathione concentration is decreased, as well as in GSH-deficient *cad2* mutant (Meyer et al. 2007; Marty et al. 2009). An increase in the redox potentials of both cytosol and nuclei has been also shown in wild-type plants in which a buthionine sulphoximine (BSO)-dependent GSH depletion has been obtained (Schnaubelt et al. 2015). However, it is interesting to note that changes in redox potential caused by altered GSH/GSSG ratio or by GSH depletion affect different signaling pathways. For instance, in the *Arabidopsis* catalase-deficient (*cat2*) mutant, in which the GSH/GSSG ratio is altered, GSH has a significant role in the induction of the oxidant-dependent salicylic acid and jasmonic acid signaling pathways (Han et al. 2013a, b). In addition, the *cat2* mutants grown under high light show decreased expression of auxin synthesis genes (Gao et al. 2014). On the other hand, in plants with low GSH levels, such as the *rml1-1* mutant, the profile of stress-responsive salicylic acid and jasmonic acid-dependent genes is not altered, while changes in transcript linked to altered hormone responses occur (Schnaubelt et al. 2015).

Deviation from the high reduced state of glutathione can take place in different conditions where oxidant production occurs. Various biotic or abiotic stresses, affecting the rate of ROS production and/or ROS removal, can change glutathione redox state (Gupta et al. 1991; Vanacker et al. 2000; Gomez et al. 2004). Many studies carried out with plants in which enzymes involved in the H_2O_2 removal had been inhibited show a strict link between high accumulation of H_2O_2 and changes in glutathione redox state (May and Leaver 1993; Willekens et al. 1997; Noctor et al. 2002b; Rizhsky et al. 2002; Queval et al. 2007, 2009; Chaouch et al. 2010). For instance, in *Arabidopsis cat2* mutants the transfer from elevated CO_2 conditions to air causes oxidation of the leaf glutathione pool within hours, which is accompanied in the subsequent days by total glutathione accumulation (Queval et al. 2009). In the same mutants under photorespiratory conditions, a moderate rate of endogenous H_2O_2 production causes a strong decrease in the GSH/GSSG ratio of the whole leaf (Queval et al. 2012).

Under oxidative conditions, GSSG accumulation may be explained as the net result of oxidation processes which overcomes, even if only slightly, the capability of glutathione reduction (Noctor et al. 2013; Fig. 1.2). GR1 activity is necessary under oxidative stress, since it has been shown that in *cat2 gr1* double mutants, deficient in both the major leaf catalase and GR1, accumulation of GSSG is massively increased compared with the parent lines (Mhamdi et al. 2010). However,

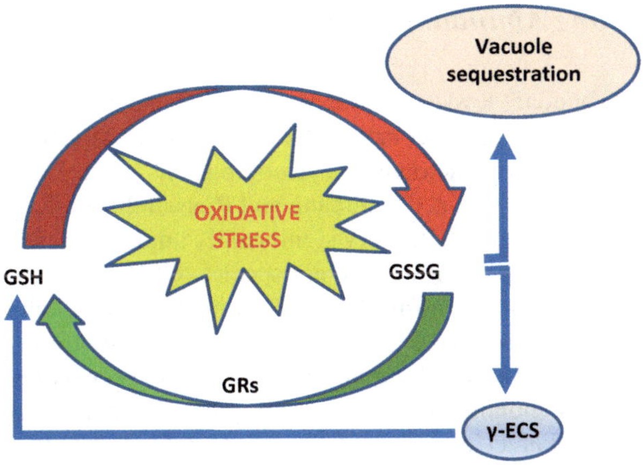

Fig. 1.2 Changes in glutathione content and redox state under oxidative stress. Under stress conditions, the oxidative processes overcome the capability of glutathione reduction, due to glutathione reductases (GR). The increase in glutathione disulfide (GSSG) content stimulates γ-glutamylcysteine synthetase (γ-ECS), leading to an increase of total glutathione pool. On the other hand, the excessive accumulation of GSSG in sensitive subcellular compartments, such as cytosol and nuclei, is avoided by its compartmentalization in vacuoles. More details are given in the text

since GR has a Michaelis constant (Km) for GSSG of 10–50 μM (Smith et al. 1989; Edwards et al. 1990), it is possible that a kinetic limitation of this enzyme occurs (Noctor et al. 2012).

Often under stress conditions, GSSG accumulation in plants is not accompanied by a decrease in GSH, whose levels remain quite constant, rather by an increase in total glutathione pool that seems to be principally due to GSSG accumulation (Smith et al. 1984; Willekens et al. 1997; Mhamdi et al. 2010). This behavior can be explained by new synthesis of GSH and compartmentalization of GSSG (Noctor et al. 2012, 2013; Fig. 1.2).

An increase in GSH neosynthesis can occur during oxidative stress, as a result of activation of cysteine and GSH production at transcriptional and posttranslational levels (Hicks et al. 2007; Gromes et al. 2008; Queval et al. 2009). In addition, a considerable quantity of the GSSG generated by oxidative stress can be accumulated in the vacuole (Queval et al. 2011) by ABCC transporters (Martinoia et al. 1993; Lu et al. 1998). New GSH biosynthesis is necessary for GSSG accumulation and successive sequestration; indeed, introducing a *cad2* mutation in the *cat2* background, GSSG accumulation is inhibited (Han et al. 2013a; Noctor et al. 2013).

The decrease in GSH/GSSG ratio, measured in whole cell extracts, although a merged value of GSH/GSSG ratios in different cell compartments, represents a valuable marker for oxidative stress, because GSSG accumulation in specific compartments, such as the vacuole, is dependent on its increase in other compartments (Noctor et al. 2013).

4.2 Glutathione Abundance and Distribution in Plant Cells

GSH is the most abundant low-molecular-weight thiol in plant tissues, and it generally accumulates to millimolar concentrations (Queval et al. 2011; Koffler et al. 2013). In the *Arabidopsis* gametophyte, cytosol, plastids, nuclei, and mitochondria contain similar amounts of GSH (Zechmann and Russell 2011). On the other hand, in the sporophyte GSH content can vary significantly among different cell compartments. In roots and leaves of different plant species, GSH has been localized with the highest contents in mitochondria, followed by nuclei, peroxisomes, cytosol, and plastids (Zechmann and Müller 2010). In the center of *Arabidopsis* old leaves the calculated GSH concentrations, by quantitative immunoelectron microscopy, vary from 14.8 mM in mitochondria, 6.4 mM in nuclei, 4.5 mM in the cytosol, to 4.4 mM in peroxisomes (Koffler et al. 2013; Zechmann 2014). Concentrations of about 1 mM have been observed in chloroplasts of *Arabidopsis* leaves (Queval et al. 2011; Koffler et al. 2013). On the other hand, very low levels of GSH (0.03–0.08 mM) have been found in vacuoles and, with this technique, GSH is undetectable in the apoplast (Queval et al. 2011; Koffler et al. 2013). Although mitochondria have the highest GSH concentrations, in mesophyll cells of *Arabidopsis* leaves, cytosol and chloroplasts, having the greater volumes, contain 50% and 30%, respectively, of total GSH (Queval et al. 2011).

In the next paragraphs, the role and the importance of GSH and its redox state in different plant cell compartments will be discussed.

4.2.1 Cytosolic Glutathione

The cytosolic glutathione redox potential, as previously said, is negative and quite stable. The control of the redox state in this compartment is favored by the continuous reduction of GSSG by GR and/or by the sequestration of GSSG in the vacuole. Thus, the sequestration keeps GSSG very low in this sensitive compartment and guarantees suitable but not extreme accumulation during oxidative stress (Hartmann et al. 2003). Consistently, different stresses are able to render the cytosolic environment more oxidized (Meyer et al. 2007; Jubany-Mari et al. 2010). The cytosol, although not directly involved in ROS production, plays a key role in the integration of redox signals (Foyer and Noctor 2016; Paciolla et al. 2016). The importance of cytosolic GSH in the signaling events occurring in abiotic and biotic stress response has been confirmed by the use of plants with mutation in the genes coding for proteins required for the transport of γ-EC and GSH across the plastid envelope membranes. These mutants (*clt1clt2clt3*) have an altered partitioning of GSH between plastid and cytosol, with a clear decrease in the cytosolic GSH content. The *clt1clt-2clt3* mutants show enhanced sensitivity to cadmium and to the fungal pathogen *Phytophthora brassicae*, and are not able to activate a correct pathogen defense, defecting in the salicylic acid-dependent expression of pathogenesis-resistance protein (Maughan et al. 2010).

4.2.2 Glutathione in Chloroplasts and Peroxisomes

GSH in chloroplast has an important function in the organelle's protection from possible oxidative damages caused by ROS (Pietrini et al. 2003). Indeed, during stress conditions that induce stomata closure, such as excess light, high salinity, and drought, which induce high ROS production in chloroplasts (Asada 2006; Golan et al. 2006; Pospisil 2012), GSH accumulates not only in the stroma but also in the thylakoid lumen (Heyneke et al. 2013; Zechmann 2014). During oxidative stress occurring in *cat2* mutants, in which the initial increase in H_2O_2 production is extra-chloroplastic, a strong accumulation of GSSG occurs in the chloroplast (Queval et al. 2011). This GSSG accumulation could be dependent on an import from the cytosol or, more probably, on GSH oxidation within the chloroplast (Noctor et al. 2013). However, independently by the mechanism involved, GSSG accumulation in the chloroplasts may influence not only the thiol-dependent reactions in this compartment but also the synthesis pathways contributing to the regulation of total glutathione content (Noctor et al. 2013). On the other hand, insufficient content of GSH in chloroplasts, permitting ROS accumulation, leads to cell death (Doyle et al. 2010). In plants subjected to biotic stress, compartment-specific changes in GSH content can occur. For instance, at the beginning of *Botrytis cinerea* and *Pseudomonas syringae* infections in *Arabidopsis* plants, GSH accumulates in chloroplasts, whereas, at later stages, depletion of GSH in chloroplasts leads to ROS accumulation and progression of disease symptoms (Großkinsky et al. 2012; Simon et al. 2013). The GSH decline in the chloroplast, despite the active GSH synthesis, may be due to its transport in other cellular compartments (Noctor et al. 2002a; Noctor et al. 2013).

Peroxisomal GSH behaves in a similar way to the chloroplastic one. Indeed, during the first phase of pathogen infection, peroxisomes function as GSH accumulators, whereas the GSH decrease in the successive stages contributes to the induction of necrotic lesions (Großkinsky et al. 2012; Simon et al. 2013). In tomato plants infected with *B. cinerea*, the decline in GSH content in peroxisomes has been linked to the pathogen-induced senescence of leaves (Kuźniak and Sklodowska 2001). On the other hand, stress conditions that favor photorespiration and H_2O_2 production in peroxisomes lead to GSH accumulation in this cell compartment (Miller et al. 2010; Hernández et al. 2013).

4.2.3 Mitochondrial Glutathione

Mitochondrial glutathione is strongly reduced (Schwarzlander and Finkemeier 2013) and, as the chloroplastic one, is involved in the direct and indirect removal of ROS, protecting membranes, proteins, as well as DNA (Foyer et al. 2004; Green et al. 2006; Rhoads et al. 2006). However, mitochondrial GSH seems to be necessary for correct plant development. Indeed, the biosynthetic mutants of *Arabidopsis pad2-1*, which show the same level of mitochondrial GSH of the wild-type plants and up to 91% of GSH decrease in other cell compartments (Zechmann et al. 2008), have a normal phenotype when grown in non-stressed conditions (Parisy et al. 2007). On the other hand, the *rml1* mutant, with a 96–98% drop in the GSH levels in all the compartments and with the highest GSH depletion in mitochondria (Zechmann and Müller 2010), forms extremely short roots and small shoots and leaves (Cheng et al. 1995; Vernoux et al. 2000). The importance of mitochondrial GSH for survival has been demonstrated using the *pad2-1* mutants, which under short-term excess light show a huge increase only in mitochondrial GSH (Heyneke et al. 2013). The depletion of mitochondrial GSH, occurring in *Nicotiana tabacum* plants infected with tobacco mosaic virus (TMV) and in *Arabidopsis* plants infected with *B. cinerea*, is accompanied with the development of necrotic lesions (Király et al. 2012; Simon et al. 2013). A drop in total GSH contents and accumulation of GSSG in mitochondria has been also observed in the pathogen-induced senescence of tomato plants (Kuźniak and Sklodowska 2001). Thus, the depletion of mitochondrial GSH seems to promote ROS accumulation and to be responsible for the initiation of programmed cell death (Zechmann 2014).

4.2.4 Nuclear Glutathione

Nuclei of non-stressed leaves, after mitochondria, show the highest concentration of GSH (Koffler et al. 2013), which co-localizes with DNA (Diaz-Vivancos et al. 2010a). Changes in nuclear redox balance of nuclei may cause DNA damages, which could induce mutations and eventually cell death (Diaz-Vivancos et al. 2010b).

When GSH synthesis is impaired, as occurs in the *rml1-1* mutants or in wild-type seedlings treated with BSO, GSH depletion, in the cytosol as well as in nuclei, arrests cell cycle in the roots; the decrease in GSH modulates the expression of genes involved in cell cycle control (Schnaubelt et al. 2015). Accordingly, it has been proposed that high levels of GSH in nuclei during G1 phase represent an essential strategy to permit cell cycle progression (Diaz-Vivancos et al. 2010a, b). In particular, in proliferating cells in the G1 phase, GSH is recruited and sequestered in the nucleus, leading to a strong depletion of the cytoplasmic GSH pool, and a concomitant decrease in transcripts linked to oxidative signaling and stress tolerance (Markovic et al. 2007;Diaz-Vivancos et al. 2010a). The subsequent changes in cytosolic redox state trigger GSH synthesis, with a subsequent increase in the total glutathione pool, which occurs before the disappearance of nuclear envelope. Successively, the cytoplasmic and nuclear GSH pools become in equilibrium (Markovic et al. 2007; Pellny et al. 2009; Diaz-Vivancos et al. 2010a, b). The retention of GSH within the nucleus causes an arrest of cell cycle at the S/G2 phases (Locato et al. 2015).

High levels of GSH in nuclei play important roles in the protection of sensitive nuclear components, such as DNA and proteins, but are also involved in the regulation of the expression of genes involved in the activation of plant defense (Han et al. 2013a, b; García-Giménez et al. 2013). Consistently, an increase in nuclear GSH content is a common event during pathogen attack (Király et al. 2012; Großkinsky et al. 2012; Simon et al. 2013). It is possible that nuclear GSH accumulation, after pathogen infection, can function as a signal to increase total GSH contents (Zechmann 2014). Accordingly, in TMV-infected tobacco plants, as well as in *Arabidopsis* plants infected with *P. syringae* and *B. cinerea*, the increase in nuclear GSH is followed by a strong accumulation of GSH in chloroplasts and cytosol (Király et al. 2012; Großkinsky et al. 2012; Simon et al. 2013).

4.2.5 Glutathione in Other Cell Compartments

Vacuoles, under non-stress conditions, have very low concentrations of GSH (Queval et al. 2011; Koffler et al. 2013), probably for the presence in this compartment of carboxypeptidases involved in its degradation (Steinkamp and Rennenberg 1985; Wolf et al. 1996). However, as discussed above, under oxidative stress, GSSG sequestration in this compartment can function as a protective mechanism involved in the control of cytosolic redox potential (Queval et al. 2011; Noctor et al. 2013). Moreover, vacuoles also function as a sink for GSH conjugates. For instance, cadmium can form complexes with GSH, which are then transported into vacuoles (Van Belleghem et al. 2007). GS-conjugates, successively, could be degraded in this compartment by the action of carboxypeptidase and γ-glutamyl transpeptidase (Steinkamp and Rennenberg 1985; Wolf et al. 1996; Grzam et al. 2007).

In the apoplast in basal conditions, GSH content is very low and sometimes under the level of detection (Vanacker et al. 1998, 2000; Zechmann et al. 2008; Tolin et al. 2013). Also in this case, these low GSH levels can be explained by the presence of the γ-glutamyl transpeptidases, GGT1 and GGT2, located in the cell wall and in the plasma membrane, which degrade GSH (Martin and Slovin 2000; Storozhenko et al. 2002;Ferretti et al. 2009). Consistently, in *Arabidopsis ggt1* mutants the level of apoplastic GSH is similar to the chloroplastic one (Tolin et al. 2013). It has been proposed that GSH content and redox state in the apoplast are involved in sensing and signaling environmental stress (Tolin et al. 2013). For instance, fungal infections in barley plants cause GSH accumulation in apoplast. Moreover, the apoplastic glutathione pool becomes more oxidized during the hypersensitive response (Vanacker et al. 1998, 2000).

Endoplasmic reticulum contains glutathione essentially as GSSG, which is needed in order to create an appropriate environment for disulfide bridges formation and proper folding of proteins (Hwang et al. 1992; Enyedi et al. 2010). In *Arabidopsis gsh2* mutants the accumulation of γ-EC and the low GSH levels have a negative impact on protein folding occurring in this compartment (Au et al. 2012).

5 Conclusions

Over the last decades, many experimental evidences have shown that GSH has a key role as antioxidant and that it is an irreplaceable player in the control of cellular redox state. The maintenance of a high GSH/GSSG ratio is crucial for many physiological functions, and a decrease in this ratio can be utilized as an indicator of oxidative stress. Different reactions could contribute to GSH oxidation during oxidative stress, modifying its redox state (Fig. 1.1). GSH oxidation can occur by chemical reactivity of the thiol group with ROS and DHA or can be catalyzed by specific enzymes. The oxidation of GSH by DHARs makes a link between ASC and GSH pools and allows GSH to take part, indirectly, in H_2O_2 reduction. On the other hand, GSH oxidation by some GSTs and type II PRXs renders the antioxidant and signaling functions of GSH interdependent. Indeed, GSH, by participating in thiol-disulfide exchange, is also involved in the control of ROS-dependent signaling. Consistently, the GSH/GSSG ratio functions as an important regulator of several mechanisms involved in plant development and in plant response to environmental changes (Rahantaniaina et al. 2013; Foyer and Noctor 2016). Even under oxidative stress, the avoidance of an excessive oxidation in various sensitive cell compartments, like cytosol and nuclei, is given by the mutual aid of different mechanisms, among which are de novo GSH synthesis, GSSG reduction, and GSSG sequestration in opportune cell compartments, such as the vacuole (Fig. 1.2). Thus, not only the concentration and the redox state but also the subcellular distribution of GSH are central factors controlling redox homeostasis and signaling, which act as key actors in influencing the outcome of plant responses to environmental changes.

Recently, it has been shown that the depletion of the cytosolic GSH in the *Arabidopsis clt1clt2clt3* triple mutants, which negatively affects biotic stress tolerance (Maughan et al. 2010), has no effect on the decrease in leaf area induced by abiotic stress (Schnaubelt et al. 2013). Given the different responses of plants to the changes in GSH concentration and redox state in different cell compartments, in the near future it will be crucial to understand the specificity of these changes in response to distinct environmental stresses and at different stages of plant development.

References

Arisi ACM, Noctor G, Foyer CH, Jouanin L (1997) Modification of thiol contents in poplars (*Populus tremula* x *P. alba*) overexpressing enzymes involved in glutathione synthesis. Planta 203:362–372

Asada K (2006) Production and scavenging of reactive oxygen species in chloroplasts and their functions. Plant Physiol 141:391–396

Au KKC, Perez-Gomez J, Neto H, Muller C, Meyer AJ, Fricker MD, Moore I (2012) A perturbation in glutathione biosynthesis disrupts endoplasmic reticulum morphology and secretory membrane traffic in *Arabidopsis thaliana*. Plant J 71:881–894

Ball L, Accotto GP, Bechtold U, Creissen G, Funck D, Jimenez A, Kular B, Leyland N, Mejia-Carranza J, Reynolds H, Karpinski S, Mullineaux PM (2004) Evidence for a direct link between glutathione biosynthesis and stress defense gene expression in Arabidopsis. Plant Cell 16:2448–2462

Bick JA, Aslund F, Chen YC, Leustek T (1998) Glutaredoxin function for the carboxyl-terminal domain of the plant-type 5′-adenylylsulfate reductase. Proc Natl Acad Sci U S A 95:8404–8409

Biterova EI, Barycki JJ (2009) Mechanistic details of glutathione biosynthesis revealed by crystal structures of *Saccharomyces cerevisiae* glutamate cysteine ligase. J Biol Chem 284:32700–32708

Buwalda F, Dekok LJ, Stulen I, Kuiper PJC (1988) Cysteine, gamma-glutamyl-cysteine and glutathione contents of spinach leaves as affected by darkness and application of excess sulfur. Physiol Plant 74:663–668

Buwalda F, Stulen I, Dekok LJ, Kuiper PJC (1990) Cysteine, gamma-glutamyl-cysteine and glutathione contents of spinach leaves as affected by darkness and application of excess sulfur II. Glutathione accumulation in detached leaves exposed to H_2S in the absence of light is stimulated by the supply of glycine to the petiole. Physiol Plant 80:196–204

Cairns NG, Pasternak M, Wachter A, Cobbett CS, Meyer AJ (2006) Maturation of Arabidopsis seeds is dependent on glutathione biosynthesis within the embryo. Plant Physiol 141:446–455

Chaouch S, Queval G, Vanderauwera S, Mhamdi A, Vandorpe M, Langlois-Meurinne M, Van Breusegem F, Saindrenan P, Noctor G (2010) Peroxisomal hydrogen peroxide is coupled to biotic defense responses by ISOCHORISMATE SYNTHASE1 in a daylength-related manner. Plant Physiol 153:1692–1705

Cheng JC, Seeley KA, Sung ZR (1995) RML1 and RML2, Arabidopsis genes required for cell-proliferation at the root-tip. Plant Physiol 107(107):365–376

Chew O, Whelan J, Millar AH (2003) Molecular definition of the ascorbate-glutathione cycle in *Arabidopsis* mitochondria reveals dual targeting of antioxidant defenses in plants. J Biol Chem 278:46869–46877

Cobbett CS, May MJ, Howden R, Rolls B (1998) The glutathione-deficient, cadmium-sensitive mutant, cad2-1, of *Arabidopsis thaliana* is deficient in gamma-glutamylcysteine synthetase. Plant J 16:73–78

Copley SD, Dhillon JK (2002) Lateral gene transfer and parallel evolution in the history of glutathione biosynthesis genes. Genome Biol 3:5

Cummins I, Dixon DP, Freitag-Pohl S, Skipsey M, Edwards R (2011) Multiple roles for plant glutathione transferases in xenobiotic detoxification. Drug Metab Rev 43:266–280

Diaz-Vivancos P, Dong YP, Ziegler K, Markovic J, Pallardo FV, Pellny TK, Verrier PJ, Foyer CH (2010a) Recruitment of glutathione into the nucleus during cell proliferation adjusts whole-cell redox homeostasis in *Arabidopsis thaliana* and lowers the oxidative defence shield. Plant J 64:825–838

Diaz-Vivancos P, Wolff T, Markovic J, Pallardo FV, Foyer CH (2010b) A nuclear glutathione cycle within the cell cycle. Biochem J 431:169–178

Dinescu A, Cundari TR, Bhansali VS, Luo JL, Anderson ME (2004) Function of conserved residues of human glutathione synthetase – implications for the ATP-grasp enzymes. J Biol Chem 279:22412–22421

Dixon DP, Edwards R (2010) Glutathione S-transferases. Arabidopsis Book 8:e0131

Dixon DP, Davis BG, Edwards R (2002) Functional divergence in the glutathione transferase superfamily in plants – identification of two classes with putative functions in redox homeostasis in *Arabidopsis thaliana*. J Biol Chem 277:30859–30869

Dixon DP, Hawkins T, Hussey PJ, Edwards R (2009) Enzyme activities and subcellular localization of members of the Arabidopsis glutathione transferase superfamily. J Exp Bot 60:1207–1218

Doyle SM, Diamond M, McCabe PF (2010) Chloroplast and reactive oxygen species involvement in apoptotic-like programmed cell death in Arabidopsis suspension cultures. J Exp Bot 61:473–482

Dubreuil-Maurizi C, Vitecek J, Marty L, Branciard L, Frettinger P, Wendehenne D, Meyer AJ, Mauch F, Poinssot B (2011) Glutathione deficiency of the *Arabidopsis* mutant *pad2*-1 affects oxidative stress-related events, defense gene expression, and the hypersensitive response. Plant Physiol 157:2000–2012

Edwards EA, Rawsthorne S, Mullineaux PM (1990) Subcellular-distribution of multiple forms of glutathione-reductase in leaves of pea (*Pisum sativum* L). Planta 180:278–284

Enyedi B, Varnai P, Geiszt M (2010) Redox state of the endoplasmic reticulum is controlled by Ero1L-alpha and intraluminal calcium. Antioxid Redox Signal 13:721–729

Farago S, Brunold C (1994) Regulation of thiol contents in maize roots by intermediates and effectors of glutathione synthesis. J Plant Physiol 144:433–437

Ferretti M, Destro T, Tosatto SCE, La Rocca N, Rascio N, Masi A (2009) Gamma-glutamyl transferase in the cell wall participates in extracellular glutathione salvage from the root apoplast. New Phytol 181:115–126

Foyer CH, Halliwell B (1976) The presence of glutathione and glutathione reductase in chloroplasts: a proposed role in ascorbic acid metabolism. Planta 133:21–25

Foyer CH, Noctor G (2005) Redox homeostasis and antioxidant signaling: a metabolic interface between stress perception and physiological responses. Plant Cell 17:1866–1875

Foyer CH, Noctor G (2016) Stress-triggered redox signalling: what's in pROSpect? Plant Cell Environ 39:951–964

Foyer CH, Souriau N, Perret S, Lelandais M, Kunert KJ, Pruvost C, Jouanin L (1995) Overexpression of glutathione-reductase but not glutathione synthetase leads to increases in antioxidant capacity and resistance to photoinhibition in poplar trees. Plant Physiol 109:1047–1057

Foyer CH, De Paepe R, Noctor G (2004) Roles of mitochondria in redox signalling and oxidative stress. Acta Physiol Plant 26:86–86

Fraser JA, Saunders RDC, McLellan LI (2002) *Drosophila melanogaster* glutamate-cysteine ligase activity is regulated by a modifier subunit with a mechanism of action similar to that of the mammalian form. J Biol Chem 277:1158–1165

Freeman JL, Persans MW, Nieman K, Albrecht C, Peer W, Pickering IJ, Salt DE (2004) Increased glutathione biosynthesis plays a role in nickel tolerance in Thlaspi nickel hyperaccumulators. Plant Cell 16:2176–2191

Frendo P, Mathieu C, Van de Sype G, Herouart D, Puppo A (1999) Characterisation of a cDNA encoding gamma-glutamylcysteine synthetase in *Medicago truncatula*. Free Radic Res 31:S213–S218

Galant A, Arkus KAJ, Zubieta C, Cahoon RE, Jez JM (2009) Structural basis for evolution of product diversity in soybean glutathione biosynthesis. Plant Cell 21:3450–3458

Gao X, Yuan HM, Hu YQ, Li J, Lu YT (2014) Mutation of Arabidopsis CATALASE2 results in hyponastic leaves by changes of auxin levels. Plant Cell Environ 37:175–188

García-Giménez JL, Markovic J, Dasi F, Queval G, Schnaubelt D, Foyer CH, Pallardo FV (2013) Nuclear glutathione. BBA-Gen Subjects 1830:3304–3316

Gogos A, Shapiro L (2002) Large conformational changes in the catalytic cycle of glutathione synthase. Structure 10:1669–1676

Golan T, Muller-Moule P, Niyogi KK (2006) Photoprotection mutants of *Arabidopsis thaliana* acclimate to high light by increasing photosynthesis and specific antioxidants. Plant Cell Environ 29:879–887

Gomez LD, Vanacker H, Buchner P, Noctor G, Foyer CH (2004) Intercellular distribution of glutathione synthesis in maize leaves and its response to short-term chilling. Plant Physiol 134:1662–1671

Grant CM, MacIver FH, Dawes IW (1996) Glutathione is an essential metabolite required for resistance to oxidative stress in the yeast *Saccharomyces cerevisiae*. Curr Genet 29:511–515

Green RM, Graham M, O'Donovan MR, Chipman JK, Hodges NJ (2006) Subcellular compartmentalization of glutathione: correlations with parameters of oxidative stress related to genotoxicity. Mutagenesis 21:383–390

Gromes R, Hothorn M, Lenherr ED, Rybin V, Scheffzek K, Rausch T (2008) The redox switch of gamma-glutamylcysteine ligase via a reversible monomer-dimer transition is a mechanism unique to plants. Plant J 54:1063–1075

Großkinsky DK, Koffler BE, Roitsch T, Maier R, Zechmann B (2012) Compartment-specific antioxidative defense in arabidopsis against virulent and avirulent *Pseudomonas syringae*. Phytopathology 102:662–673

Grzam A, Tennstedt P, Clemens S, Hell R, Meyer AJ (2006) Vacuolar sequestration of glutathione -conjugates outcompetes a possible degradation of the glutathione moiety by phytochelatin synthase. FEBS Letters 580(27):6384–6390

Grzam A, Martin MN, Hell R, Meyer AJ (2007) Gamma-Glutamyl transpeptidase GGT4 initiates vacuolar degradation of glutathione S-conjugates in *Arabidopsis*. FEBS Lett 581:3131–3138

Gulyas Z, Boldizsar A, Novak A, Szalai G, Pal M, Galiba G, Kocsy G (2014) Central role of the flowering repressor ZCCT2 in the redox control of freezing tolerance and the initial development of flower primordia in wheat. BMC Plant Biol 14:91

Gupta AS, Alscher RG, McCune D (1991) Response of photosynthesis and cellular antioxidants to ozone in populus leaves. Plant Physiol 96:650–655

Gushima H, Miya T, Murata K, Kimura A (1983) Purification and characterization of glutathione synthetase from *Escherichia coli* B. J Appl Biochem 5:210–218

Halliwell B, Foyer CH (1978) Properties and physiological function of a glutathione reductase purified from spinach leaves by affinity chromatography. Planta 139:9–17

Han Y, Chaouch S, Mhamdi A, Queval G, Zechmann B, Noctor G (2013a) Functional analysis of Arabidopsis mutants points to novel roles for glutathione in coupling H_2O_2 to activation of salicylic acid accumulation and signaling. Antioxid Redox Signal 18:2106–2121

Han Y, Mhamdi A, Chaouch S, Noctor G (2013b) Regulation of basal and oxidative stress-triggered jasmonic acid-related gene expression by glutathione. Plant Cell Environ 36:1135–1146

Harms K, von Ballmoos P, Brunold C, Hofgen R, Hesse H (2000) Expression of a bacterial serine acetyltransferase in transgenic potato plants leads to increased levels of cysteine and glutathione. Plant J 22:335–343

Hartmann TN, Fricker MD, Rennenberg H, Meyer AJ (2003) Cell-specific measurement of cytosolic glutathione in poplar leaves. Plant Cell Environ 26:965–975

Hatano-Iwasaki A, Ogawa K (2012) Overexpression of GSH1 gene mimics transcriptional response to low temperature during seed vernalization treatment of Arabidopsis. Plant Cell Physiol 53:1195–1203

Hell R, Bergmann L (1988) Glutathione synthetase in tobacco suspension-cultures – catalytic properties and localization. Physiol Plant 72:70–76

Hell R, Bergmann L (1990) Gamma-Glutamylcysteine synthetase in higher-plants – catalytic properties and subcellular-localization. Planta 180:603–612

Hernández I, Cela J, Alegre L (2013) Antioxidant defenses against drought stress. In: Aroca R (ed) Plant responses to drought stress. Springer, Berlin, pp 231–258

Herrera K, Cahoon RE, Kumaran S, Jez JM (2007) Reaction mechanism of glutathione synthetase from *Arabidopsis thaliana* – site-directed mutagenesis of active residues. J Biol Chem 282:17157–17165

Heyneke E, Luschin-Ebengreuth N, Krajcer I, Wolkinger V, Muller M, Zechmann B (2013) Dynamic compartment specific changes in glutathione and ascorbate levels in Arabidopsis plants exposed to different light intensities. BMC Plant Biol 13:104

Hibi T, Nii H, Nakatsu T, Kimura A, Kato H, Hiratake J, Oda J (2004) Crystal structure of gamma-glutamylcysteine synthetase: insights into the mechanism of catalysis by a key enzyme for glutathione homeostasis. Proc Natl Acad Sci U S A 101:15052–15057

Hicks LM, Cahoon RE, Bonner ER, Rivard RS, Sheffield J, Jez JM (2007) Thiol-based regulation of redox-active glutamate-cysteine ligase from *Arabidopsis thaliana*. Plant Cell 19:2653–2661

Hothorn M, Wachter A, Gromes R, Stuwe T, Rausch T, Scheffzek K (2006) Structural basis for the redox control of plant glutamate cysteine ligase. J Biol Chem 281:27557–27565

Howden R, Andersen CR, Goldsbrough PB, Cobbett CS (1995) A cadmium-sensitive, glutathione-deficient mutant of *Arabidopsis thaliana*. Plant Physiol 107:1067–1073

Hwang C, Sinskey AJ, Lodish HF (1992) Oxidized redox state of glutathione in the endoplasmic-reticulum. Science 257:1496–1502

Iqbal A, Yabuta Y, Takeda T, Nakano Y, Shigeoka S (2006) Hydroperoxide reduction by thioredoxin-specific glutathione peroxidase isoenzymes of *Arabidopsis thaliana*. FEBS J 273:5589–5597

Jez JM, Cahoon RE (2004) Kinetic mechanism of glutathione synthetase from *Arabidopsis thaliana*. J Biol Chem 279:42726–42731

Jez JM, Cahoon RE, Chen SX (2004) *Arabidopsis thaliana* glutamate-cysteine ligase – functional properties, kinetic mechanism, and regulation of activity. J Biol Chem 279:33463–33470

Jiménez A, Hernandez JA, delRio LA, Sevilla F (1997) Evidence for the presence of the ascorbate-glutathione cycle in mitochondria and peroxisomes of pea leaves. Plant Physiol 114:275–284

Jubany-Mari T, Alegre-Batlle L, Jiang K, Feldman LJ (2010) Use of a redox-sensing GFP (c-roGFP1) for real-time monitoring of cytosol redox status in *Arabidopsis thaliana* water-stressed plants. FEBS Lett 584:889–897

Kataya AR, Reumann S (2010) Arabidopsis glutathione reductase 1 is dually targeted to peroxisomes and the cytosol. Plant Signal Behav 5:171–175

Kim RH, Smith PD, Aleyasin H, Hayley S, Mount MP, Pownall S, Wakeham A, You-Ten AJ, Kalia SK, Horne P, Westaway D, Lozano AM, Anisman H, Park DS, Mak TW (2005) Hypersensitivity of DJ-1-deficient mice to 1-methyl-4-phenyl-1,2,3,6-tetrahydropyrindine (MPTP) and oxidative stress. Proc Natl Acad Sci U S A 102:5215–5220

Király L, Kunstler A, Holler K, Fattinger M, Juhasz C, Muller M, Gullner G, Zechmann B (2012) Sulfate supply influences compartment specific glutathione metabolism and confers enhanced resistance to tobacco mosaic virus during a hypersensitive response. Plant Physiol Biochem 59:44–54

Klapheck S (1988) Homoglutathione – isolation, quantification and occurrence in legumes. Physiol Plant 74:727–732

Klapheck S, Chrost B, Starke J, Zimmermann H (1992) Gamma-Glutamylcysteinylserine – a new homolog of glutathione in plants of the family Poaceae. Bot Acta 105:174–179

Koffler BE, Bloem E, Zellnig G, Zechmann B (2013) High resolution imaging of subcellular glutathione concentrations by quantitative immunoelectron microscopy in different leaf areas of *Arabidopsis*. Micron 45:119–128

Kopriva S, Rennenberg H (2004) Control of sulphate assimilation and glutathione synthesis: interaction with N and C metabolism. J Exp Bot 55:1831–1842

Kranner I, Grill D (1996) Significance of thiol-disulfide exchange in resting stages of plant development. Bot Acta 109:8–14

Kranner I, Birtic S, Anderson KM, Pritchard HW (2006) Glutathione half-cell reduction potential: a universal stress marker and modulator of programmed cell death? Free Radic Biol Med 40:2155–2165

Kumar D, Datta R, Hazra S, Sultana A, Mukhopadhyay R, Chattopadhyay S (2015) Transcriptomic profiling of *Arabidopsis thaliana* mutant *pad2*.1 in response to combined cold and osmotic stress. PLoS One 10:e0122690

Kuźniak E, Sklodowska M (2001) Ascorbate, glutathione and related enzymes in chloroplasts of tomato leaves infected by *Botrytis cinerea*. Plant Sci 160:723–731

Li H, Xu HM, Graham DE, White RH (2003) Glutathione synthetase homologs encode alpha-L-glutamate ligases for methanogenic coenzyme F-420 and tetrahydrosarcinapterin biosyntheses. Proc Natl Acad Sci U S A 100:9785–9790

Liedschulte V, Wachter A, An ZG, Rausch T (2010) Exploiting plants for glutathione (GSH) production: uncoupling GSH synthesis from cellular controls results in unprecedented GSH accumulation. Plant Biotechnol J 8:807–820

Lindermayr C, Saalbach G, Durner J (2005) Proteomic identification of S-nitrosylated proteins in Arabidopsis. Plant Physiol 137:921–930

Locato V, Uzal EN, Cimini S, Zonno MC, Evidente A, Micera A, Foyer CH, De Gara L (2015) Low concentrations of the toxin ophiobolin A lead to an arrest of the cell cycle and alter the intracellular partitioning of glutathione between the nuclei and cytoplasm. J Exp Bot 66:2991–3000

Lu YP, Li ZS, Drozdowicz YM, Hortensteiner S, Martinoia E, Rea PA (1998) AtMRP2, an Arabidopsis ATP binding cassette transporter able to transport glutathione S-conjugates and chlorophyll catabolites: functional comparisons with AtMRP1. Plant Cell 10:267–282

Markovic J, Borras C, Ortega A, Sastre J, Vina J, Pallardo FV (2007) Glutathione is recruited into the nucleus in early phases of cell proliferation. J Biol Chem 282:20416–20424

Martin MN, Slovin JP (2000) Purified gamma-glutamyl transpeptidases from tomato exhibit high affinity for glutathione and glutathione S-conjugates. Plant Physiol 122:1417–1426

Martinoia E, Grill E, Tommasini R, Kreuz K, Amrhein N (1993) ATP-dependent glutathione s-conjugate export pump in the vacuolar membrane of plants. Nature 364:247–249

Marty L, Siala W, Schwarzlander M, Fricker MD, Wirtz M, Sweetlove LJ, Meyer Y, Meyer AJ, Reichheld JP, Hell R (2009) The NAD(P)H-dependent thioredoxin system constitutes a functional backup for cytosolic glutathione reductase in *Arabidopsis*. Proc Natl Acad Sci USA 106:9109–9114

Maughan SC, Pasternak M, Cairns N, Kiddle G, Brach T, Jarvis R, Haas F, Nieuwland J, Lim B, Muller C, Salcedo-Sora E, Kruse C, Orsel M, Hell R, Miller AJ, Bray P, Foyer CH, Murray JAH, Meyer AJ, Cobbett CS (2010) Plant homologs of the *Plasmodium falciparum* chloroquine-resistance transporter, PfCRT, are required for glutathione homeostasis and stress responses. Proc Natl Acad Sci U S A 107:2331–2336

May MJ, Leaver CJ (1993) Oxidative stimulation of glutathione synthesis in *Arabidopsis thaliana* suspension-cultures. Plant Physiol 103:621–627

May MJ, Leaver CJ (1994) *Arabidopsis thaliana*gamma-glutamylcysteine synthetase is structurally unrelated to mammalian, yeast, and *Escherichia-Coli* homologs. Proc Natl Acad Sci U S A 91:10059–10063

May MJ, Vernoux T, Sanchez-Fernandez R, Van Montagu M, Inze D (1998) Evidence for post-transcriptional activation of gamma-glutamylcysteine synthetase during plant stress responses. Proc Natl Acad Sci U S A 95:12049–12054

Meister A (1988) Glutathione metabolism and its selective modification. J Biol Chem 263:17205–17208

Meister A, Anderson ME (1983) Glutathione. Annu Rev Biochem 52:711–760

Meuwly P, Thibault P, Rauser WE (1993) Gamma-Glutamylcysteinylglutamic acid – a new homolog of glutathione in maize seedlings exposed to cadmium. FEBS Lett 336:472–476

Meyer AJ, Brach T, Marty L, Kreye S, Rouhier N, Jacquot JP, Hell R (2007) Redox-sensitive GFP in Arabidopsis thaliana is a quantitative biosensor for the redox potential of the cellular glutathione redox buffer. Plant J 52:973–986

Mhamdi A, Hager J, Chaouch S, Queval G, Han Y, Taconnat L, Saindrenan P, Gouia H, Issakidis-Bourguet E, Renou JP, Noctor G (2010) Arabidopsis GLUTATHIONE REDUCTASE1 plays a crucial role in leaf responses to intracellular hydrogen peroxide and in ensuring appropriate gene expression through both salicylic acid and jasmonic acid signaling pathways. Plant Physiol 153:1144–1160

Miller G, Suzuki N, Ciftci-Yilmaz S, Mittler R (2010) Reactive oxygen species homeostasis and signalling during drought and salinity stresses. Plant Cell Environ 33:453–467

Musgrave WB, Yi H, Kline D, Cameron JC, Wignes J, Dey S, Pakrasi HB, Jez JM (2013) Probing the origins of glutathione biosynthesis through biochemical analysis of glutamate-cysteine ligase and glutathione synthetase from a model photosynthetic prokaryote. Biochem J 450:63–72

Navrot N, Collin V, Gualberto J, Gelhaye E, Hirasawa M, Rey P, Knaff DB, Issakidis E, Jacquot JP, Rouhier N (2006) Plant glutathione peroxidases are functional peroxiredoxins distributed in several subcellular compartments and regulated during biotic and abiotic stresses. Plant Physiol 142:1364–1379

Noctor G, Strohm M, Jouanin L, Kunert KJ, Foyer CH, Rennenberg H (1996) Synthesis of glutathione in leaves of transgenic poplar overexpressing gamma-glutamylcysteine synthetase. Plant Physiol 112:1071–1078

Noctor G, Arisi ACM, Jouanin L, Valadier MH, Roux Y, Foyer CH (1997) The role of glycine in determining the rate of glutathione synthesis in poplar. Possible implications for glutathione production during stress. Physiol Plant 100:255–263

Noctor G, Arisi ACM, Jouanin L, Foyer CH (1998) Manipulation of glutathione and amino acid biosynthesis in the chloroplast. Plant Physiol 118:471–482

Noctor G, Gomez L, Vanacker H, Foyer CH (2002a) Interactions between biosynthesis, compartmentation and transport in the control of glutathione homeostasis and signalling. J Exp Bot 53:1283–1304

Noctor G, Veljovic-Jovanovic S, Driscoll S, Novitskaya L, Foyer CH (2002b) Drought and oxidative load in the leaves of C3 plants: a predominant role for photorespiration? Ann Bot 89:841–850

Noctor G, De Paepe R, Foyer CH (2007) Mitochondrial redox biology and homeostasis in plants. Trends Plant Sci 12:125–134

Noctor G, Mhamdi A, Chaouch S, Han Y, Neukermans J, Marquez-Garcia B, Queval G, Foyer CH (2012) Glutathione in plants: an integrated overview. Plant Cell Environ 35:454–484

Noctor G, Mhamdi A, Queval G, Foyer CH (2013) Regulating the redox gatekeeper: vacuolar sequestration puts glutathione disulfide in its place. Plant Physiol 163:665–671
Noji M, Saito K (2002) Molecular and biochemical analysis of serine acetyltransferase and cysteine synthase towards sulfur metabolic engineering in plants. Amino Acids 22:231–243
Ogawa K, Hatano-Iwasaki A, Yanagida M, Iwabuchi M (2004) Level of glutathione is regulated by ATP-dependent ligation of glutamate and cysteine through photosynthesis in *Arabidopsis thaliana*: mechanism of strong interaction of light intensity with flowering. Plant Cell Physiol 45:1–8
Paciolla C, Paradiso A, de Pinto MC (2016) Cellular redox homeostasis as central modulator in plant stress response. In: Gupta DK, Palma JM, Corpas FJ (eds) Redox state as a central regulator of plant-cell stress responses. Springer International Publishing, Cham, pp 1–23
Parisy V, Poinssot B, Owsianowski L, Buchala A, Glazebrook J, Mauch F (2007) Identification of PAD2 as a gamma-glutamylcysteine synthetase highlights the importance of glutathione in disease resistance of Arabidopsis. Plant J 49:159–172
Pasternak M, Lim B, Wirtz M, Hell R, Cobbett CS, Meyer AJ (2008) Restricting glutathione biosynthesis to the cytosol is sufficient for normal plant development. Plant J 53:999–1012
Pellny TK, Locato V, Vivancos PD, Markovic J, De Gara L, Pallardo FV, Foyer CH (2009) Pyridine nucleotide cycling and control of intracellular redox state in relation to poly (ADP-ribose) polymerase activity and nuclear localization of glutathione during exponential growth of Arabidopsis cells in culture. Mol Plant 2:442–456
Pietrini F, Iannelli MA, Pasqualini S, Massacci A (2003) Interaction of cadmium with glutathione and photosynthesis in developing leaves and chloroplasts of *Phragmites australis* (Cav.) Trin. ex steudel. Plant Physiol 133:829–837
Pospisil P (2012) Molecular mechanisms of production and scavenging of reactive oxygen species by photosystem II. BBA-Bioenergetics 1817:218–231
Preuss ML, Cameron JC, Berg RH, Jez JM (2014) Immunolocalization of glutathione biosynthesis enzymes in *Arabidopsis thaliana*. Plant Physiol Biochem 75:9–13
Queval G, Issakidis-Bourguet E, Hoeberichts FA, Vandorpe M, Gakiere B, Vanacker H, Miginiac-Maslow M, Van Breusegem F, Noctor G (2007) Conditional oxidative stress responses in the Arabidopsis photorespiratory mutant cat2 demonstrate that redox state is a key modulator of daylength-dependent gene expression, and define photoperiod as a crucial factor in the regulation of H_2O_2-induced cell death. Plant J 52:640–657
Queval G, Thominet D, Vanacker H, Miginiac-Maslow M, Gakiere B, Noctor G (2009) H_2O_2-activated up-regulation of glutathione in arabidopsis involves induction of genes encoding enzymes involved in cysteine synthesis in the chloroplast. Mol Plant 2:344–356
Queval G, Jaillard D, Zechmann B, Noctor G (2011) Increased intracellular H_2O_2 availability preferentially drives glutathione accumulation in vacuoles and chloroplasts. Plant Cell Environ 34:21–32
Queval G, Neukermans J, Vanderauwera S, Van Breusegem F, Noctor G (2012) Day length is a key regulator of transcriptomic responses to both CO_2 and H_2O_2 in Arabidopsis. Plant Cell Environ 35:374–387
Rahantaniaina MS, Tuzet A, Mhamdi A, Noctor G (2013) Missing links in understanding redox signaling via thiol/disulfide modulation: how is glutathione oxidized in plants? Front Plant Sci 4:477
Rasmusson AG, Møller IM (1990) NADP-utilizing enzymes in the matrix of plant-mitochondria. Plant Physiol 94:1012–1018
Rawlins MR, Leaver CJ, May MJ (1995) Characterization of an *Arabidopsis thaliana* cDNA-encoding glutathione synthetase. FEBS Lett 376:81–86
Rennenberg H (1980) Glutathione metabolism and possible biological roles in higher plants. Phytochemistry 21:2771–2781
Rennenberg H (1995) Processes involved in glutathione metabolism. In: Wallsgrove RM (ed) Amino acids and their derivatives in higher plants. Cambridge University Press, Cambridge, pp 155–171

Rennenberg H (2001) Glutathione: an ancient metabolite with modern tasks. In: Grill D, Tausz M, De Kok LJ (eds) Significance of glutathione to plant adaptation to the environment. Kluwer Academic Publisher, New York, Boston, Dordrecht, London, Moscow, pp 1–11

Rhoads DM, Umbach AL, Subbaiah CC, Siedow JN (2006) Mitochondrial reactive oxygen species. Contribution to oxidative stress and interorganellar signaling. Plant Physiol 141:357–366

Rizhsky L, Hallak-Herr E, Van Breusegem F, Rachmilevitch S, Barr JE, Rodermel S, Inze D, Mittler R (2002) Double antisense plants lacking ascorbate peroxidase and catalase are less sensitive to oxidative stress than single antisense plants lacking ascorbate peroxidase or catalase. Plant J 32:329–342

Romero-Puertas MC, Corpas FJ, Sandalio LM, Leterrier M, Rodriguez-Serrano M, del Rio LA, Palma JM (2006) Glutathione reductase from pea leaves: response to abiotic stress and characterization of the peroxisomal isozyme. New Phytol 170:43–52

Rouhier N (2010) Plant glutaredoxins: pivotal players in redox biology and iron-sulphur centre assembly. New Phytol 186:365–372

Sakamoto A, Ueda M, Morikawa H (2002) Arabidopsis glutathione-dependent formaldehyde dehydrogenase is an S-nitrosoglutathione reductase. FEBS Lett 515:20–24

Schnaubelt D, Schulz P, Hannah MA, Yocgo RE, Foyer CH (2013) A phenomics approach to the analysis of the influence of glutathione on leaf area and abiotic stress tolerance in *Arabidopsis thaliana*. Front Plant Sci 4:416

Schnaubelt D, Queval G, Dong YP, Diaz-Vivancos P, Makgopa ME, Howell G, De Simone A, Bai J, Hannah MA, Foyer CH (2015) Low glutathione regulates gene expression and the redox potentials of the nucleus and cytosol in *Arabidopsis thaliana*. Plant Cell Environ 38:266–279

Schwarzlander M, Finkemeier I (2013) Mitochondrial energy and redox signaling in plants. Antioxid Redox Signal 18:2122–2144

Seelig GF, Simondsen RP, Meister A (1984) Reversible dissociation of gamma-glutamylcysteine synthetase into two subunits. J Biol Chem 259:9345–9347

Setterdahl AT, Chivers PT, Hirasawa M, Lemaire SD, Keryer E, Miginiac-Maslow M, Kim SK, Mason J, Jacquot JP, Longbine CC, de Lamotte-Guery F, Knaff DB (2003) Effect of pH on the oxidation-reduction properties of thioredoxins. Biochemistry 42:14877–14884

Shanmugam V, Tsednee M, Yeh KC (2012) ZINC TOLERANCE INDUCED BY IRON 1 reveals the importance of glutathione in the cross-homeostasis between zinc and iron in *Arabidopsis thaliana*. Plant J 69:1006–1017

Simon UK, Polanschutz LM, Koffler BE, Zechmann B (2013) High resolution imaging of temporal and spatial changes of subcellular ascorbate, glutathione and H_2O_2 distribution during *Botrytis cinerea* infection in Arabidopsis. PLoS One 8:e65811

Skipsey M, Davis BG, Edwards R (2005) Diversification in substrate usage by glutathione synthetases from soya bean (*Glycine max*), wheat (*Triticum aestivum*) and maize (*Zea mays*). Biochem J 391:567–574

Smith IK, Kendall AC, Keys AJ, Turner JC, Lea PJ (1984) Increased levels of glutathione in a catalase-deficient mutant of barley (*Hordeum vulgare* L.) Plant Sci Lett 37:29–33

Smith IK, Vierheller TL, Thorne CA (1989) Properties and functions of glutathione-reductase in plants. Physiol Plant 77:449–456

Steinkamp R, Rennenberg H (1985) Degradation of glutathione in plant-cells – evidence against the participation of a gamma-glutamyltranspeptidase. Zeitschrift Fur Naturforschung C 40:29–33

Stevens RG, Creissen GP, Mullineaux PM (2000) Characterisation of pea cytosolic glutathione reductase expressed in transgenic tobacco. Planta 211:537–545

Storozhenko S, Belles-Boix E, Babiychuk E, Herouart D, Davey MW, Slooten L, Van Montagu M, Inze D, Kushnir S (2002) Gamma-glutamyl transpeptidase in transgenic tobacco plants. Cellular localization, processing, and biochemical properties. Plant Physiol 128:1109–1119

Strohm M, Jouanin L, Kunert KJ, Pruvost C, Polle A, Foyer CH, Rennenberg H (1995) Regulation of glutathione synthesis in leaves of transgenic poplar (*Populus tremula* X *P. alba*) overexpressing glutathione synthetase. Plant J 7:141–145

Sung DY, Kim TH, Komives EA, Mendoza-Cozatl DG, Schroeder JI (2009) ARS5 is a component of the 26S proteasome complex, and negatively regulates thiol biosynthesis and arsenic tolerance in *Arabidopsis*. Plant J 59:802–812

Tarrago L, Laugier E, Zaffagnini M, Marchand C, Le Marechal P, Rouhier N, Lemaire SD, Rey P (2009) Regeneration mechanisms of *Arabidopsis thaliana* methionine sulfoxide reductases B by glutaredoxins and thioredoxins. J Biol Chem 284:18963–18971

Tolin S, Arrigoni G, Trentin AR, Veljovic-Jovanovic S, Pivato M, Zechman B, Masi A (2013) Biochemical and quantitative proteomics investigations in *Arabidopsis* ggt1 mutant leaves reveal a role for the gamma-glutamyl cycle in plant's adaptation to environment. Proteomics 13:2031–2045

Tzafrir I, Pena-Muralla R, Dickerman A, Berg M, Rogers R, Hutchens S, Sweeney TC, McElver J, Aux G, Patton D, Meinke D (2004) Identification of genes required for embryo development in Arabidopsis. Plant Physiol 135:1206–1220

Ullmann P, Gondet L, Potier S, Bach TJ (1996) Cloning of *Arabidopsis thaliana* glutathione synthetase (GSH2) by functional complementation of a yeast gsh2 mutant. Eur J Biochem 236:662–669

Van Belleghem F, Cuypers A, Semane B, Smeets K, Vangronsveld J, d'Haen J, Valcke R (2007) Subcellular localization of cadmium in roots and leaves of *Arabidopsis thaliana*. New Phytol 173:495–508

Vanacker H, Carver TLW, Foyer CH (1998) Pathogen-induced changes in the antioxidant status of the apoplast in barley leaves. Plant Physiol 117:1103–1114

Vanacker H, Carver TLW, Foyer CH (2000) Early H_2O_2 accumulation in mesophyll cells leads to induction of glutathione during the hyper-sensitive response in the barley-powdery mildew interaction. Plant Physiol 123:1289–1300

Vernoux T, Wilson RC, Seeley KA, Reichheld JP, Muroy S, Brown S, Maughan SC, Cobbett CS, Van Montagu M, Inze D, May MJ, Sung ZR (2000) The ROOT MERISTEMLESS1/CADMIUM SENSITIVE2 gene defines a glutathione-dependent pathway involved in initiation and maintenance of cell division during postembryonic root development. Plant Cell 12:97–109

Wachter A, Wolf S, Steininger H, Bogs J, Rausch T (2005) Differential targeting of GSH1 and GSH2 is achieved by multiple transcription initiation: implications for the compartmentation of glutathione biosynthesis in the Brassicaceae. Plant J 41:15–30

Wagner U, Edwards R, Dixon DP, Mauch F (2002) Probing the diversity of the Arabidopsis glutathione S-transferase gene family. Plant Mol Biol 49:515–532

Wang W, Ballatori N (1998) Endogenous glutathione conjugates: occurrence and biological functions. Pharmacol Rev 50:335–355

Wang CL, Oliver DJ (1997) Glutathione synthetase: similarities of the proteins from *Schizosaccharomyces pombe* and *Arabidopsis thaliana*. Biochem J 326:563–566

Willekens H, Chamnongpol S, Davey M, Schraudner M, Langebartels C, VanMontagu M, Inze D, VanCamp W (1997) Catalase is a sink for H_2O_2 and is indispensable for stress defence in C-3 plants. EMBO J 16:4806–4816

Wirtz M, Hell R (2007) Dominant-negative modification reveals the regulatory function of the multimeric cysteine synthase protein complex in transgenic tobacco. Plant Cell 19:625–639

Wolf AE, Dietz KJ, Schroder P (1996) Degradation of glutathione S-conjugates by a carboxypeptidase in the plant vacuole. FEBS Lett 384:31–34

Wonisch W, Schaur RJ (2001) Chemistry of glutathione. In: Grill D, Tausz M, De Kok LJ (eds) Significance of glutathione to plant adaptation to the environment. Kluwer Academic Publisher, New York, Boston, Dordrecht, London, Moscow, pp 13–26

Wu JM, Qu T, Chen SY, Zhao ZG, An LZ (2009) Molecular cloning and characterization of a gamma-glutamylcysteine synthetase gene from *Chorispora bungeana*. Protoplasma 235:27–36

Xiang CB, Oliver DJ (1998) Glutathione metabolic genes coordinately respond to heavy metals and jasmonic acid in Arabidopsis. Plant Cell 10:1539–1550

Xiang CB, Werner BL, Christensen EM, Oliver DJ (2001) The biological functions of glutathione revisited in Arabidopsis transgenic plants with altered glutathione levels. Plant Physiol 126:564–574

Yamaguchi H, Kato H, Hata Y, Nishioka T, Kimura A, Oda J, Katsube Y (1993) Three-dimensional structure of the glutathione synthetase from *Escherichia coli* B at 2.0 A resolution. J Mol Biol 229:1083–1100

Zechmann B (2014) Compartment-specific importance of glutathione during abiotic and biotic stress. Front Plant Sci 5:566

Zechmann B, Muller M (2010) Subcellular compartmentation of glutathione in dicotyledonous plants. Protoplasma 246:15–24

Zechmann B, Russell SD (2011) Subcellular distribution of glutathione in the gametophyte. Plant Signal Behav 6:1259–1262

Zechmann B, Mauch F, Sticher L, Muller M (2008) Subcellular immunocytochemical analysis detects the highest concentrations of glutathione in mitochondria and not in plastids. J Exp Bot 59:4017–4027

Zechmann B, Koffler BE, Russell SD (2011) Glutathione synthesis is essential for pollen germination in vitro. BMC Plant Biol 11:54

Zhu YL, Pilon-Smits EAH, Jouanin L, Terry N (1999a) Overexpression of glutathione synthetase in Indian mustard enhances cadmium accumulation and tolerance. Plant Physiol 119:73–79

Zhu YL, Pilon-Smits EAH, Tarun AS, Weber SU, Jouanin L, Terry N (1999b) Cadmium tolerance and accumulation in Indian mustard is enhanced by overexpressing gamma-glutamylcysteine synthetase. Plant Physiol 121:1169–1177

Chapter 2
Synthesis and Roles of Glutathione and Homoglutathione in the Nitrogen-Fixing Symbiosis

Eric Boncompagni, Geneviève Alloing, Karine Mandon, and Pierre Frendo

Abstract Glutathione (GSH) is a major antioxidant molecule in plants. It is involved in regulating plant development and responses to abiotic and biotic environment changes. In leguminous plants, a GSH homolog, homoglutathione is also found. Most legumes can develop a symbiotic interaction with soil bacteria of the rhizobium family under nitrogen deficiency. This symbiosis allows the reduction of atmospheric nitrogen by the bacteria in plant organs called root nodules. In this chapter, we summarize studies that describe the synthesis and the roles of GSH and hGSH in the nitrogen-fixing symbiosis.

Keywords Nitrogen-fixing symbiosis • Rhizobia • Legumes • Glutathione • Redox

Contents

1 Introduction

Glutathione (GSH) is a tripeptide (γ-glutamyl-cysteinyl glycine) present in a broad range of organisms, from bacteria to humans. It is synthesized in a two-step process. In the first step, γ-glutamyl-cysteine ligase (γGCL), also called γ-glutamylcysteine synthetase (γECS), catalyzes the ATP-dependent formation of γ-glutamyl-cysteine

E. Boncompagni • G. Alloing • K. Mandon • P. Frendo (✉)
INRA, Université Côte d'Azur, CNRS, UMR 1355-7254,
Institut Sophia Agrobiotech, 06900 Sophia Antipolis, France
e-mail: frendo@unice.fr

M.A. Hossain et al. (eds.), *Glutathione in Plant Growth, Development, and Stress Tolerance*, https://doi.org/10.1007/978-3-319-66682-2_2

Fig. 2.1 Synthesis and location of glutathione and homoglutathione in a plant cell. Glutathione/homoglutathione ((h)GSH) are synthesized in a two-step pathway involving γ-glutamylcysteine ligase/synthetase (γECS) and glutathione/homoglutathione synthetase ((h)GSHS). The redox state of (h)GSH is regulated by glutathione reductase (GR) that catalyzes the reduction of glutathione/homoglutathione disulfide ((h)GSSG) to (h)GSH

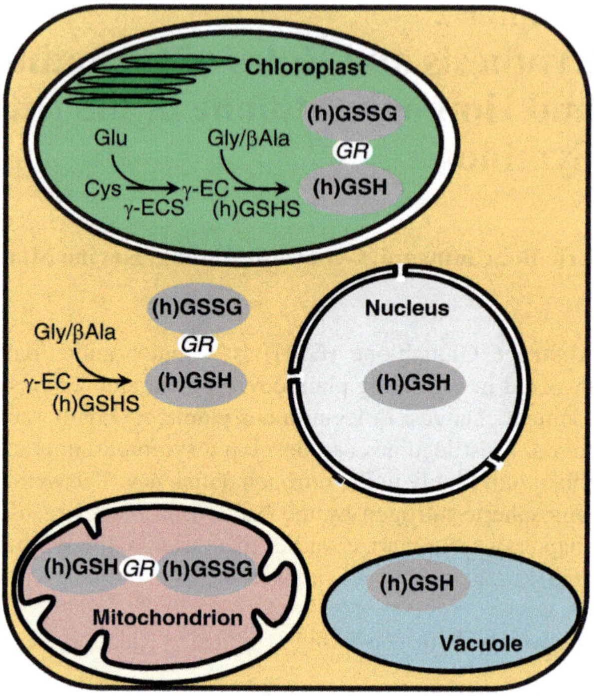

(γEC) from glutamate and cysteine. The primary sequence of γECS is not conserved in these different groups of organisms suggesting an acquisition of this activity at different times during the evolution (Copley and Dhillon 2002). In the second step, glutathione synthetase (GSHS) catalyzes the ATP-dependent formation of GSH from γEC and glycine. The primary sequences of GSHS also differ between eukaryotes and prokaryotes (Copley and Dhillon 2002). Cysteine is the direct precursor of GSH. In this context, sulfur assimilation plays a major role in GSH formation. Taken up from the soil, sulfate is activated to adenosine-5-phosphosulfate prior to its reduction to sulfite and then sulfide, and its incorporation into cysteine. Cysteine, the key metabolite in the synthesis of sulfur-containing compounds in plants and the major pool of sulfur, is stored in GSH when it is not integrated in proteins (Hell and Wirtz 2011). In plants, the GSH synthesis pathway is located in both the plastids and the cytosol (Fig. 2.1). γECS is encoded by a nuclear gene (*GSH1*) and is targeted to the plastids (May and Leaver 1994; Wachter et al. 2005). GSHS is also encoded by a nuclear gene (*GSH2*) and is found in both the plastids and the cytosol (Rawlins et al. 1995; Wachter et al. 2005). GSH accumulates to millimolar (mM) concentrations within cells and multiple GSH homologs have been detected in plants. One of the most frequently observed is homoglutathione (hGSH), which replaces or is present in addition to GSH in the large and diverse plant family Leguminosae (Carnegie 1963; Klapheck 1988). Its synthesis requires a specific homoglutathione synthetase (hGSHS), encoded by a gene derived from the *GSHS* gene by gene duplication (Galant et al. 2011).

The biological functions of GSH relate principally to reversible redox reactions of the cysteine sulfur group, resulting in the coexistence of a reduced state (GSH) and an oxidized state (GSSG), in which two GSH molecules are linked via a disulfide bond. The cellular GSH pool is mostly reduced under optimal conditions. The redox status of GSH is kept high by glutathione reductase (GR), which uses NADPH as reducing power. Typically, the GSH/GSSG ratio is well over 1000 with the GSH concentration close to mM (Queval et al. 2011). In plants, GSH along with ascorbate are the major cellular redox buffers (Rouhier et al. 2008a). GSH is abundant in the plant cell cytosol, chloroplasts, mitochondria, and nucleus. GSH can react with protein cysteine residues to form mixed disulfides via a glutathionylation process. Protein glutathionylation has been extensively investigated in animals (Mieyal and Chock 2012; Xiong et al. 2011; Mieyal et al. 2008; Dalle-Donne et al. 2009), but much less is known about this process in plants (Dixon et al. 2005; Zaffagnini et al. 2012). Glutaredoxins (GRX), which couple GSH redox potential to changes in protein thiol-disulfide status, are involved in the deglutathionylation process and in the regeneration of multiple enzymes, such as peroxiredoxins and methionine sulfoxide reductases (Rouhier et al. 2008b; Rouhier 2010; Lillig and Berndt 2012). GSH may also react with numerous endogenous and xenobiotic electrophilic compounds, via glutathione-S-transferases (Cummins et al. 2011; Dixon et al. 2010). Finally, GSH also protects plants against heavy metals, through the formation of phytochelatins (PC), which are GSH polymers. PC are synthesized by phytochelatin synthase, which uses GSH as a substrate (Clemens 2006).

GSH plays a crucial role in plant development and the adaptation of plants to the environment. Analyses of the phenotypes of *Arabidopsis thaliana* GSH-deficient mutants have shown that GSH is involved in embryo and meristem development (Vernoux et al. 2000; Cairns et al. 2006; Schippers et al. 2016). GSH has also been shown to be involved in light signaling, in studies of the *Arabidopsis rax1* mutant, which has only half the normal level of GSH in its leaves and displays constitutive expression of the photo-oxidative stress-inducible ascorbate peroxidase 2 (Ball et al. 2004). However, the role of GSH is not restricted to the regulation of the plant growth and adaptation to the abiotic environment. This molecule is also involved in the responses of plant to the biotic environment (Frendo et al. 2013).

GSH is required for optimal plant defense against pathogenic microorganisms. The *Arabidopsis* phytoalexin-deficient (*pad2*) mutant, which is deficient in *GSH1* (Parisy et al. 2007), displays impaired production of defense proteins, salicylic acid, and enhanced susceptibility to the pathogens *Pseudomonas syringae*, *Phytophthora porri*, and *Botrytis cinerea* (Roetschi et al. 2001; Ferrari et al. 2003). This association between GSH content and plant defense has also been demonstrated in other GSH1-deficient mutants, *cad2-1* and *rax1-1*, which are less resistant than wild-type plants to avirulent strains of *P. syringae* (Ball et al. 2004). The thiol-disulfide redox status is clearly involved in the regulation of a major regulatory protein, NPR1 (nonexpressor of pathogenesis-related protein 1), involved in plant defense reactions (Spoel et al. 2010). NPR1 must be converted from its oligomeric form to a monomer in order to be translocated from the cytosol to the nucleus, and this requires the reduction of the disulfide bonds of the oligomeric form. This conversion occurs in vitro in a buffer GSH/GSSG at physiological concentration (Mou et al. 2003). GSH is also involved in plant defense against insects. Resistance to the generalist insect *Spodoptera littoralis* is compromised in the *Arabidopsis* mutant *pad2*, because the two major indole and aliphatic glucosinolates of *Arabidopsis* produced in response to insect feeding accumulate to a much lesser extent in this mutant than in wild-type plants (Schlaeppi et al. 2008). GSH also appears to be involved in the plant response to obligate parasites. GSH content increases during interactions between root-knot nematodes and *Medicago truncatula* (Baldacci-Cresp et al. 2012). Similarly, GSH accumulation in the roots of members of the Brassicaceae is a positive marker of clubroot disease, induced by the biotrophic pathogen *Plasmodiophora brassicae* (Wagner et al. 2012). Finally, GSH plays a key role in the nitrogen-fixing symbiotic interaction.

Plants are often involved into symbiotic interactions with fungi and bacteria in order to enhance access to various nutrients. Symbiotic interactions with fungi are very old associations that help plants to more easily access water, nitrogen, phosphorous, iron, and others nutrients (Behie and Bidochka 2014). In exchange, fungi receive photosynthates from the plant. In a similar way, leguminous plants, including clover, alfalfa, bean, soybean, and peas, establish a symbiotic interaction with soil proteobacteria rhizobia that enables them to fix atmospheric N_2 when soil nitrogen levels are insufficient for optimal plant growth (Patriarca et al. 2004; Desbrosses and Stougaard 2011). Symbiotic nitrogen fixation is performed in root nodules, with nodules that are spherical classified as determinate and nodules that are cylindrical and often branched according to their mode of development classified as indeterminate (Fig. 2.2). Legumes with determinate nodules are mainly tropical and subtropical plants such as phaseoloid legumes (including *Glycine*, *Phaseolus*, and *Vigna*) and *Lotus japonicus* in the tribe Loteae. The meristem of determinate nodules is transiently functioning, leading to spherical nodule. In plants developing indeterminate nodules, such as pea, alfalfa, and *M. truncatula*, the nodule meristem is persistent giving an elongated shape to the nodule (Fig. 2.2 b and e; Timmers et al. 1999; Xiao et al. 2014).

Nitrogen-fixing symbiosis (NFS) involves multiple and complex processes (Vernie et al. 2015). A complex molecular cross-talk between the plant and bacteria leads to the entry of the bacteria into plant tissues (Oldroyd 2013). Flavonoids, which are present in plant exudates, attract bacteria and trigger the production of nodulation factors (NF) by the bacteria. The recognition of such NF by the plant initiates many of the processes associated with NFS.

Using the two model legumes for symbiotic interactions, *M. truncatula* and *L. japonicus*, important progress has been achieved toward characterization of the initial recognition between both partners and root nodule development (Suzaki et al. 2015). Signal exchanges between both partners induce the formation of the characteristic shepherd's crook in root hair tips (Fig. 2.2c) and elicit cell division in roots inner cortex, promoting the initiation of nodule meristems (Fig. 2.2c) (Oldroyd and Downie 2008). To get from the outside to the inside of the plant, rhizobia grow and

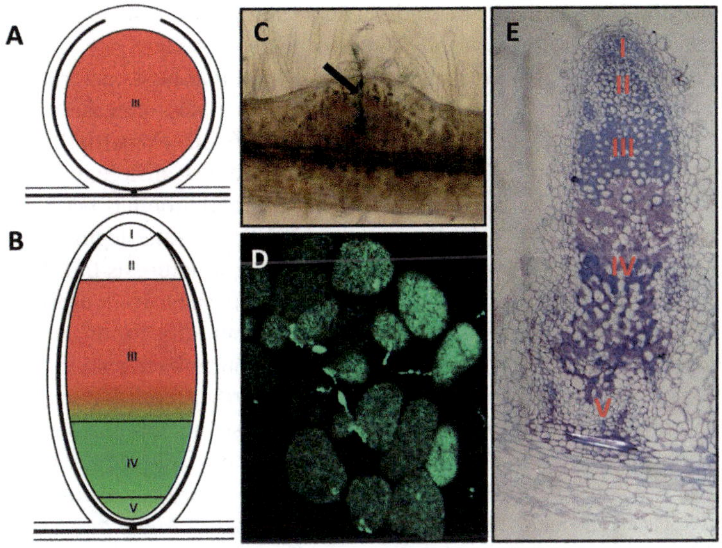

Fig. 2.2 The different steps of the root nodule formation and root nodule structure. Structure of (**a**) determinate and, (**b**) indeterminate nodules with the meristem zone I, the infection zone II, the nitrogen-fixing zone III, the senescence zone IV, and the saprophytic zone V. (**c**) Development of root nodules with the infection thread in blue (*black arrow*). (**d**) Magnification of the infected zone II stained with SYTO 9 DNA staining molecules, stars showing infection thread between infected cells filled with symbiosomes. (**e**) Toluidine blue staining of a thin section of an indeterminate root nodule with the different zones

divide in tubules called infection threads. Bacteria progress toward cortical root cells and finally are released into the cytoplasm of plant cells forming a new organelle-like compartment, the symbiosome inside infected cells (Fig. 2.2d) (Wang and Dong 2011). Intercellular bacterial penetration of root cells, using crack entry followed by the formation of an infection pocket, is an alternative mechanism allowing the infection of the plant cells (Boogerd and van Rossum 1997; Oldroyd and Downie 2004). The above interaction finally leads to the formation of a functional nodule. Although the two model legumes differ in nodule morphology, no major difference has been seen in the molecular signaling of nodule development. *L. japonicus* symbiosis leads to a determinate round shaped nodule and all stages of development within the nodule appear successively. *M. truncatula* nodules, as an example of an indeterminate nodule, are highly structured organs with distinct zones corresponding to different stages of development. At 4 weeks postinoculation (wpi), all the nodule zones are visible (Fig. 2.2b and e). The apical meristematic region (zone I) ensures indeterminate growth of the nodule. Newly formed cells issued from the meristematic zone are infected by *S. meliloti* (zone II). Host plant cells and microsymbionts (interzone II–III) undergo a differentiation process to form the N_2-fixing symbiotic cells (zone III). At later stage, a senescence zone (zone

IV) in which the plant cells are degraded and a saprophytic zone (zone V) in which free bacteria live as parallel saprophytic partners also appear.

In order to become fully functional symbiotic cell that host bacteria, the nodule meristematic cells undergo a differentiation process. Cells increase in size, up to 80-fold larger, and in DNA content reaching 32–64 C (Vinardell et al. 2003; Kondorosi and Kondorosi 2004). In parallel, bacteria also differentiate, becoming fully functional bacteroids. Bacteroid terminal differentiation has been found to be dependent on the plant host (Kondorosi et al. 2013). In some legumes, the bacteroids retain the same morphology and the same DNA structure as free-living bacteria and are able to divide freely outside of nodule tissues. In contrast, bacteroid terminal differentiation is associated with extreme modifications, with bacterial DNA endoreduplication and five- to tenfold extension in bacteroid size. This irreversible transformation of bacteroids prevents any further reproduction.

The comparison of the transcriptomes of plant hosts inducing bacterial terminal differentiation (*M. truncatula*) or not (*L. japonicum*) has allowed the identification of a large family of several hundred legume nodule-specific cysteine-rich (NCR) peptides. NCRs have been found in *M. truncatula* and in *Medicago sativa*, *Pisum sativum*, *Vicia faba*, and *Astragalus sinicus*. All of these plant species host bacteroids undergoing terminal differentiation in their nodules (Van de Velde et al. 2010). In *M. truncatula*, the NCR peptides are involved in terminal bacteroid differentiation (Czernic et al. 2015).

In the nitrogen-fixing zone (zone III), fully differentiated bacteroids express nitrogenase and reduce atmospheric N_2 into ammonia (NH_4^+). NH_4^+ is exported to the plant in return for carbohydrates, which feeds the bacteroids (Crespi and Galvez 2000). This symbiotic interaction is time limited, and from four wpi proximal nodule cells initiate a senescence process, which induces a rapid and complete degradation of bacteroids and host plant cells (Zone IV) (Van de Velde et al. 2006).

2 Synthesis and Roles of GSH and hGSH in Plant Partner During the Nodulation Process

2.1 *Synthesis of GSH and hGSH*

Sulfur (S) metabolism, which supplies cysteine, is strongly involved in GSH formation. S is also an essential element for symbiotic nitrogen fixation. Nitrogenase, the bacterial enzyme responsible for the reduction of N_2, is a complex [Fe-S] enzyme (Rees and Howard 2000). The *L. japonicus Sst1* gene, which is expressed in a nodule-specific manner, encodes a sulfate transporter that is essential for NFS (Krusell et al. 2005). The SST1 protein appears to reside on the symbiosome membrane (Wienkoop and Saalbach 2003) and is thought to transport sulfate from the plant cell cytoplasm to the bacteroids (Krusell et al. 2005). Finally, recent data have shown that the active sulfur metabolism is present in nodules, highlighting the function of nodules as important sites for S-assimilation (Kalloniati et al. 2015).

GSH is one of the main antioxidants involved in plant cell redox homeostasis (Rouhier et al. 2015; Considine and Foyer 2014). In legumes, homoglutathione (hGSH; γ-Glutamyl-Cysteinyl-β-Alanine), a structural homolog of GSH, may be present in addition to GSH or replace it (Frendo et al. 2001; Matamoros et al. 2003; Colville et al. 2015). GSH and hGSH are present at higher concentration in nodules than in roots and in a similar range in leaves (Frendo et al. 2001; Matamoros et al. 1999; Colville et al. 2015). An electron microscopy study of pea nodules using a glutathione-specific antibody showed that both reduced and oxidized GSH are present in the bacteroids, mitochondria, cytosol, and nuclei of infected cells (Matamoros et al. 2013). Analyses of gold particle number in the different cell compartments suggest that total GSH increases in the bacteroids, cytosol, and nuclei of mature nodules compared to young ones.

The synthesis of GSH and hGSH is performed in a two-step pathway, which involves γ-glutamylcysteine synthetase (γECS) in the first step and glutathione synthetase (GSHS) or homoglutathione synthetase (hGSHS) in the second step (Fig. 2.1). The production of GSH or hGSH is linked to the substrate specificity of GSHS and hGSHS. Whereas GSHS presents a high specificity for glycine, hGSHS may accept both glycine and β-Alanine as substrates with a much higher affinity for β-Alanine (Frendo et al. 2001; Galant et al. 2009). The enzyme specificity is strongly linked to two continuous amino acid residues present in the active site, leucine and proline in hGSHS and two alanines in GSHS (Frendo et al. 2001; Galant et al. 2009). GSHS and hGSHS share high homology (~70% amino acid identity) and the two genes are located in tandem on the same chromosome in the model legumes *M. truncatula* (Frendo et al. 2001) and *L. japonicus* (Matamoros et al. 2003). These findings suggest that the *hGSHS* gene derives from the *GSHS* gene by a duplication event occurred after the divergence between the Fabales, Solanales, and Brassicales (Frendo et al. 2001). *GSHS* and *hGSHS* are differentially regulated in plant organs, and their expression is correlated to the GSH and hGSH content (Frendo et al. 1999; Matamoros et al. 1999; Matamoros et al. 2003). *M. truncatula* produces exclusively GSH in the leaves and both GSH and hGSH in the roots and nodules. In contrast, *L. japonicus* produces almost exclusively hGSH in the roots and leaves and more GSH than hGSH in the nodules. The expression of *γECS*, *GSHS*, and *hGSH* is regulated during the nodulation process in *M. truncatula* (Table 2.1). During the last years, transcriptomic analyses of legumes have been particularly developed. The data integrated in bioinformatics gateways represent a powerful tool to perform genome-, transcriptome-, and proteome-wide analyses. In particular, the Legoo site (https://www.legoo.org/) allows the combination of multiple portal dedicated to omics analyses of *M. truncatula*. In parallel, an approach based upon RNA-seq coupled to laser microdissection has been used to analyze gene expression in *M. truncatula* indeterminate nodules, where spatial zonation corresponds to successive developmental stages (Roux et al. 2014). Data were obtained simultaneously for both the plant and the bacteria, with a comparison between five nodule regions, from the apical meristem to the nitrogen-fixation zone. RNA-seq data were also produced from whole (non-laser dissected) nitrogen-starved root systems and mature nodules, allowing differential analyses of gene expression between roots and nodules. This

Table 2.1 Expression of plant thiol-based redox genes in *M. truncatula* nodules. Gene accession numbers are indicated in the figure. Gene annotation is based on candidate orthologues and interprodomain signature. The different columns correspond to the nodule zones: meristematic zone (zone 1), distal infection zone (zone 2 d), proximal infection zone (zone 2 p), infection/fixation interzone (iz), and nitrogen-fixing zone (zone 3). The numbers in the different columns correspond to total reads ribominus. All RNA-seq read values were normalized (Roux et al. 2014). The total reads are reported from the symbimics bioinformatics site. The full organs are nitrogen starved roots and 10-day-old nodules. The grey, yellow, and brown colors correspond to 0–20%, 21–40%, and more than 40% of total expression, respectively. The percentages of expression are reported from the symbimics bioinformatics site

	Full organs		Dissected nodules				
Gene family	**Root**	**Nodule**	**zone 1**	**zone 2 d**	**zone 2 p**	**interzone**	**zone 3**
Glutathione metabolism							
γecs Mt0009_10600	3013	2626	273	329	414	529	657
gshs Mt0036_00123	219	782	123	150	170	386	235
hgshs Mt0036_00122	1328	741	101	66	33	71	108
gr cytosolic Mt0007_00036	4153	3458	203	192	352	494	541
gr plastidial Mt0004_00926	543	656	119	91	100	63	82
Glutathione peroxidases							
Mt0004_11112	2166	27	0.3	0.5	0.7	1.3	0.8
Mt0009_00569	3295	4793	166	173	76	201	291
Mt0009_10967	3255	3275	298	528	620	664	837
Peroxiredoxines							
Mt0004_00346	138	115	100	82	103	89	54
Mt0024_10234	8586	4544	2348	1537	2056	1204	677
Mt0031_10289	1076	467	1104	718	233	534	441
Mt0042_00028	469	498	436	370	256	222	236
Mt0068_10080	1360	1006	465	346	514	767	481
Sulfiredoxin							
Mt0003_11470	218	114	16	20	9	40	39
Thioredoxin							
trx m1 Mt0006_10007	23	64	3	0.2	nd	nd	2
trx m2 Mt0019_00013	213	326	149	136	253	323	273
trx m3 Mt0003_11549	82	280	132	171	276	582	658
Glutaredoxins							
Mt0003_11235	968	1834	62	36	1,2	nd	0,3
Mt0010_10266	23	52	24	36	22	147	63
Mt0037_00004	252	3840	31	329	563	629	583
Glutathione Stransferases							
Mt0001_10040	634	1192	98	100	121	120	160
Mt0051_10130	3732	4868	494	438	493	887	1130
Mt0077_10011	678	1229	43	1023	1912	143	1057

enormous amount of data could be analyzed through the symbimics informatics server (https://iant.toulouse.inra.fr/symbimics/). In silico analyses of *γECS*, *GSHS*, and *hGSHS* expression in different plant organs (https://www.legoo.org/) and in the different nodule zones (https://iant.toulouse.inra.fr/symbimics/) show that expression of the genes is modulated in the different nodule zones during the development of the nodule. Multiple sequences appear to be homologous to chloroplast-targeted *γECS* (Mt0009_10600). The other homologous transcripts are called GCS2 (Mt0005_00269, Mt0005_00270, Mt0005_00271, and Mt0005 00272) and are present at low or undetected levels of expression in roots and nodules. It will be important in the future to perform biochemical characterizations of these proteins in order to ensure the activities of these γECS-like proteins. One interesting feature of the expression pattern of genes involved in GSH and hGSH synthesis in nodule is the upregulation of *γECS* in the nodule nitrogen-fixing zone compared to nodule meristematic zone (Table 2.1). Similarly, *GSHS* expression is also upregulated in the nodule nitrogen-fixing zone, but the highest transcript accumulation is detected in the nodule interzone in which the plant cells complete their differentiation before nitrogen fixation. In contrast to *GSHS*, *hGSHS* expression is maximal in the meristematic zone and in the nitrogen-fixing zone. Similarly to *GSHS* and *γECS*, the cytosolic isoform of *glutathione reductase* (*GR*) is strongly upregulated in the nitrogen-fixing zone, suggesting that an increased reducing system is needed to maintain GSH in an appropriate reduction state (Table 2.1). In contrast, expression of the plastid isoform of GR is not significantly modified in the different nodule zones. The transcription efficiency of *γECS*, *GSHS*, and *hGSHS* in nodules has been analyzed by the detection of β-glucuronidase activity from promoter/*β-glucuronidase* fusions expressed in transgenic *M. truncatula* nodules (El Msehli et al. 2011). *γECS* transcription efficiency seems to be stronger in the nodule meristematic zone and the infection zone than in the nitrogen-fixing zone. In contrast, *GSHS* transcription is stronger in the nodule infection and nitrogen-fixing zones. Finally, *hGSHS* transcription is higher in the meristematic zone than in the other zones but seems to be generally lower than the *GSHS* transcription. These data are consistent with the in silico analyses and suggest that GSH synthesis pathway is more active in nodule than that of hGSH.

2.2 *Roles of GSH and hGSH in the Plant Partner During the Nodulation Process*

The involvement of (h)GSH in nodule formation was tested by using pharmacological and genetic inhibition of (h)GSH synthesis. The application of a specific inhibitor of *γECS* (buthionine sulfoximine, BSO) or the expression of *(h)GSHS* in antisense orientation caused depletion of (h)GSH in *M. truncatula* roots (Frendo et al. 2005). The deficiency of (h)GSH synthesis led to a significant decrease in the number of nascent nodules and in the expression of some early nodulin genes (Frendo et al. 2005). In contrast, the number of infection threads is not affected by (h)GSH deficiency. A similar reduction in the number of nodules was observed in peanut plants treated with BSO (Bianucci et al. 2008). These results, along with the role of GSH in the proper functioning of root and shoot apical meristems (Vernoux et al. 2000; Reichheld et al. 2007; Rouhier et al. 2015), suggest that (h)GSH is required for the initiation and maintenance of the nodule meristem. Transcriptomic analysis of (h)GSH-depleted plants during early nodulation showed that genes involved in meristem formation are significantly downregulated in these plants compared to inoculated control plants. Moreover, genes involved in plant defense and particularly salicylic acid (SA)-regulated genes are upregulated in (h)GSH-depleted roots (Pucciariello et al. 2009). These results suggest that SA is at least partially involved in the nodulation phenotype caused by (h)GSH deficiency.

Nitrogen-fixing efficiency is positively correlated with the concentration of (h) GSH in nodules during nodule development (Dalton et al. 1993). Moreover, the reduction in nitrogen-fixing efficiency correlates with a decrease in (h)GSH levels during nodule aging (Evans et al. 1999; Groten et al. 2005) and stress-induced senescence (Escuredo et al. 1996; Gogorcena et al. 1997; Matamoros et al. 1999; Marino et al. 2007; Naya et al. 2007). These results strongly suggest that (h)GSH is important for nodule functioning. To test this hypothesis, genetic approaches using the nodule nitrogen-fixing zone-specific nodule cysteine-rich (NCR001) promoter were employed to increase and to reduce the *γECS* expression and thus the (h)GSH content specifically in the nodule nitrogen-fixing zone (El Msehli et al. 2011). Downregulation of the *γECS* expression by RNA interference results in the reduction in (h)GSH content, and in a significantly lower N_2 fixation efficiency associated with a significant reduction in root nodule length. This lower N_2 fixation efficiency is correlated with the lower expression of *leghemoglobin* and *thioredoxin S1* genes, two genes involved in nodule functioning (Ott et al. 2005) and bacteroid differentiation (Ribeiro et al. 2017). Conversely, overexpression of *γECS* results in an elevated (h)GSH content, which was associated with enhanced N_2 fixation and significantly higher expression of the *sucrose synthase-1* and *leghemoglobin* genes, two genes involved in nodule functioning (Baier et al. 2007). All these data show the importance of (h)GSH in nodule development and functioning.

2.3 Roles of Redoxins and Glutathione-S-Transferases During the Nodulation Process

GSH is involved in the functioning of multiple enzymes in plants. Among them, the glutaredoxins and glutathione-S-transferase are the main protein families that directly use GSH. In addition, glutathione peroxidases, thioredoxins, peroxiredoxins and sulfiredoxins are actively involved in the thiol-based redox-signaling network (Fig. 2.3).

Glutathione peroxidases (Gpx) are involved in the reduction of lipid hydroperoxides to their corresponding alcohols and in the reduction of hydrogen peroxide to water (Brigelius-Flohe and Maiorino 2013). Plant Gpx appear to be related to peroxiredoxins, even though their primary sequences are similar to animal glutaredoxins (Passaia and Margis-Pinheiro 2015). Six Gpx genes have been identified in the *L. japonicus* genome. The different isoforms have cytoplasmic, plastidial, and mitochondrial localizations in plant cells (Ramos et al. 2009) and two isoforms,

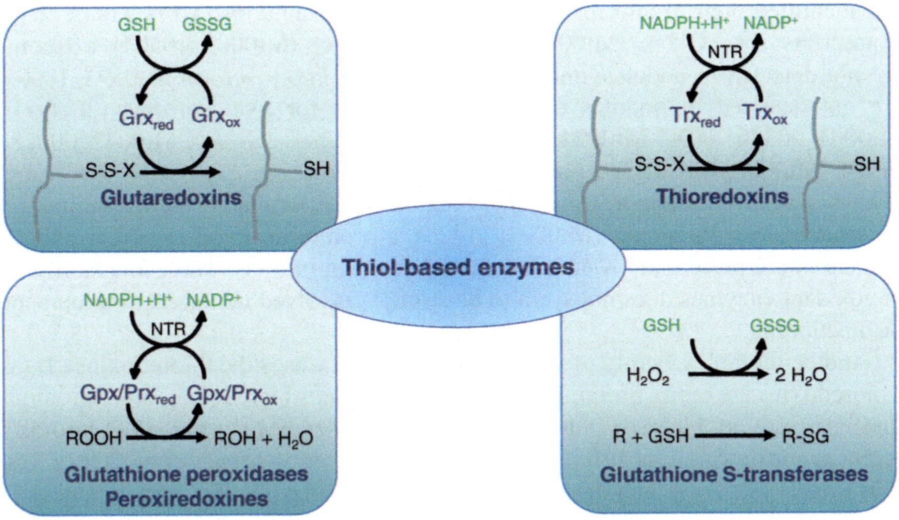

Fig. 2.3 Role of the main thiol-based redox enzymes. The substrates and the major sources of reducing power used by the different enzymes are indicated in the figure

LjGpx1 and LjGpx3 (Matamoros et al. 2015), are highly expressed in nodules. The *LjGpx1* and *LjGpx3* genes encode thioredoxin-dependent phospholipid hydroperoxidases and are differentially regulated in response to nitric oxide (NO) and plant hormones. The enzymes are highly expressed in the infected zone, but *LjGpx3* mRNA is also detected in the cortex and vascular bundles. LjGpx1 is localized to the plastids and nuclei and LjGpx3 to the cytosol and endoplasmic reticulum. Based on yeast complementation experiments, both enzymes protect against oxidative stress, salt stress, and membrane damage, suggesting that both LjGpxs prevent lipid peroxidation and other oxidative processes at different subcellular sites within vascular and infected cells. In *M. truncatula*, in silico analysis of the symbimics database identified seven transcripts showing significant identity with Gpx (Table 2.1). Among them, Mt0004_11112 accumulates less in nodules than in roots, with an 80-fold reduction. Interestingly, Mt0009_00569 and Mt0009_10967 accumulate more in the nitrogen-fixing zone, suggesting a need for a more active antioxidant defense potentially linked to the protection from lipid peroxidation.

Another similar protein class involved in the reduction of hydrogen peroxide is the peroxiredoxins (Prx). In plants, Prxs are grouped into four classes (PrxQ, PrxII, 2-CPrx, and 1-CPrx) that differ in their catalytic mechanisms and subcellular locations (Sevilla et al. 2015). The *L. japonicus* genome encodes nine Prxs that are predicted to be localized in chloroplasts (PrxQ, 2C-PrxA, 2C-PrxB, and PrxIIE), mitochondria (PrxIIF), the cytosol (PrxIIB and 1C-Prx), and the nucleus (1C-Prx). Western blot analysis has shown that nodules contain PrxIIB and PrxIIF (Tovar-Méndez et al. 2011). In *M. truncatula*, in silico analysis of the symbimics database has identified eight transcripts showing significant similarity to *Prx* (Table 2.1). Transcripts for 1C-Prx (Mt0001_01787) and PrxQ (Mt0001_10504) isoforms are not detected in nodules, and transcripts for the PrxQ isoform Mt0393_10001 are not detected in nodules or roots. Transcripts for five isoforms (2C-PrxB Mt0004_00346; TPX1 Mt0024_10234, Mt0031_10289, 2C-PRX Mt0042_00028, PrxIIF Mt0068_10080) are present in nodules. However, expression levels are always similar or lower to transcript levels in roots. Similarly, the transcripts of *M. truncatula* sulfiredoxin (Mt0003_11470), a protein involved in peroxiredoxin reduction, are present at lower levels in nodules than in roots. Thus, this family of antioxidant enzymes does not seem to be strongly involved in nodule development or functioning.

Another complex family of disulfide oxidoreductases is the thioredoxins (Trx). Thioredoxins are redox proteins that act as antioxidants by reducing other proteins through cysteine thiol-disulfide exchange. These enzymes are classified into different groups located in different cellular compartments (Meyer et al. 2012). Trx f, Trx m, Trx x, Trx y, and Trx z are localized in the plastids, Trx h in the cytosol, and Trx o in the mitochondria. Recent data suggests that the nucleus is also enriched in specific Trx isoforms (Delorme-Hinoux et al. 2015). In legumes, the Trx protein family has been analyzed in detail in two model plants, *M. truncatula* (Alkhalfioui et al. 2008) and *L. japonicus* (Tovar-Mendez et al. 2011). In nodules, the different isoforms are all expressed but at different levels. In *M. truncatula*, in silico analysis of the symbimics database shows that *Trx m* transcripts (Mt0006_10007, Mt0019_00013, Mt0003_11549) corresponding to plastidial isoforms accumulate more in nodules than in roots, suggesting that plastidial metabolism is modified in nodules compared to roots. In accordance, a more active plastid starch production has been suggested as genes encoding putative plastidial *phosphoglucomutase* and *phosphoglucose isomerase* are upregulated in *L. japonicus* nitrogen-fixing nodules (Flemetakis et al. 2006). Moreover, modification of plastidial metabolism has been also observed at the proteomic level in nodule (Molesini et al. 2014). The Trx h and Trx o isoforms are maintained in a reduced form by cytosolic and mitochondrial NADPH-dependent thioredoxin reductases (NTRs) A and B, whereas in the plastids this function is probably performed by the ferredoxin-Trx reductases (Tovar-Mendez et al. 2011). Nodule plastids also contain low levels of a singular NTR protein. NTRC is characterized by the presence of a Trx domain that enables the enzyme to directly reduce 2C-Prxs with high catalytic efficiency (Pulido et al. 2010), or BAS1, a plastidial peroxiredoxin (Alkhalfioui et al. 2007). The crucial role of Trx in nodules was demonstrated by the finding that in soybean, RNAi-induced suppression of a Trx isoform induces severe impairment of nodule formation and development (Lee et al. 2005). Thioredoxin affinity chromatography followed by mass spectrometry showed that Nodulin-35, a subunit of uricase, is a target of this Trx (Du et al. 2010). *M. truncatula* also contains specific Trx isoforms, Trx s, which are associated with symbiosis (Alkhalfioui et al. 2008). Recently, one of these isoforms, Trx s1, was shown to be targeted to the symbiosomes. Trx s1 interacts with NCR247 and NCR335 and increases the cytotoxic effects of NCR335 on *S. meliloti*. Trx s silencing impairs bacteroid growth and endoreduplication, two features of terminal bacteroid differentiation, and the ectopic expression of Trx s1 in *S. meliloti* partially complements the silencing phenotype. Thus, Trx s1 is targeted to the bacterial endosymbiont, where it controls NCR activity and bacteroid terminal differentiation (Ribeiro et al. 2017).

Glutaredoxins (Grx) are small redox enzymes that use GSH as a source of reducing potential. Grx are oxidized by substrates and reduced nonenzymatically by GSH. Beside their function in antioxidant defense, bacterial and plant Grx were shown to bind to iron-sulfur clusters and to deliver these clusters to enzymes (Rouhier et al. 2008b), but information on plant Grx in legume nodules is limited. Proteomic analyses identified two Grx (GrxC2 and GrxC4) in *L. japonicus* nodules (Tovar-Mendez et al. 2011), while in *M. truncatula* a search for "glutaredoxin" in the symbimics database identified 84 transcripts showing homology with *Grx*. Among them, 12 genes show significantly higher expression in nodules than in roots. *GrxS5* (Mt0003_11235) is significantly upregulated in the meristematic zone, while a putative *Grx* (Mt0010_10266) is significantly and specifically upregulated in the nodule interzone (Table 2.1). Finally, Mt0037_00004 is strongly upregulated in the infection zone, the interzone, and the nitrogen-fixing zone, suggesting an important role in cells that harbor bacteria.

Finally, glutathione S-transferases (GSTs) are ubiquitous enzymes that catalyze the conjugation of toxic xenobiotics and oxidatively produced compounds to reduced glutathione, which facilitates their metabolism, sequestration, or removal (Labrou et al. 2015). Soybean root nodules contain at least 14 forms of GST, with GST9 being most prevalent. Purified, recombinant GSTs were shown to have wide-ranging kinetic properties, suggesting that this suite of GSTs could provide the physiological flexibility to deal with numerous stresses. Levels of GST9 increased with aging, suggesting a role related to senescence. Downregulation of GST9 by RNA interference led to a decrease in nitrogenase (acetylene reduction) activity and an increase in oxidatively damaged proteins (Dalton et al. 2009). In *M. truncatula*, a search for "glutathione S-transferase" in the symbimics database identified 19 transcripts having homology with GST. Among them, four are expressed at higher levels in nodules compared to roots (Table 2.1). Glutathione S-transferase zeta class (Mt0001_10040) is significantly upregulated in the nitrogen-fixing zone compared to the meristematic zone. Similarly, glutathione S-transferase PHI 9 (Mt0051_10130) is significantly upregulated in the interzone and in the nitrogen-fixing zone. Finally, glutathione S-transferase TAU 19 (Mt0077_10011) is significantly upregulated in the infection zone, in the interzone, and in the nitrogen-fixing zone, suggesting a role in the perception of the bacteria.

Recently, omics methods have allowed important advances in the characterization of genomes and transcriptomes. Proteomics analyses have also allowed the characterization of numerous proteins involved in the nodulation process. However, the importance and the roles of the different enzyme isoforms still have to be defined in order to provide a clear view of their contribution to the nodulation process.

3 The Critical Role of GSH in the Bacterial Partner During the Nodulation Process

In bacteria, GSH has been shown to be involved in adaptation to various stresses, notably high redox and acidic conditions. Rhizobia in the soil frequently face stressful environmental conditions, such as variations in temperature and pH or nutrient starvation. Variability in edaphic factors, in addition to more specific stresses including salinity or heavy metal contamination, affects both the persistence of rhizobia in the soil and the initiation of symbiosis. During plant infection and bacteroid differentiation, rhizobia also have to cope with various stressful conditions. They encounter reactive oxygen species (ROS) and reactive nitrogen species (RNS) produced by the host during all steps of the symbiotic process (Puppo et al. 2013). They must adapt to differentiation into bacteroids, whether it is a terminal process or not. Inside the infected host cells, bacteroids are enclosed within a peribacteroid membrane to form a symbiosome, which is an acidic and lytic compartment (Pierre et al. 2013), and they deal with low oxygen concentrations in the functional nitrogen-fixing zone III. In this respect, genes involved in GSH synthesis and function are crucial for the development and sustainability of the symbiosis (Table 2.2).

3.1 *Importance of the Bacterial GSH Synthesis in Symbiosis*

Control of cellular redox homeostasis is a determining factor during the different steps of the interaction. GSH is the major low molecular weight thiol in Gram-negative bacteria and may play a pivotal role in fine-tuning the cellular redox balance of rhizobia, particularly during symbiosis. Studies with bacterial mutants impaired in GSH synthesis established the importance of GSH pathway in the symbiotic performance of various rhizobia.

In most bacteria, the two steps of GSH biosynthesis are catalyzed by γECS and GSHS, encoded by the *gshA* and *gshB* genes, respectively. A *S. meliloti gshA* mutant is unable to grow in minimal medium, whereas a *gshB* mutant is only slightly affected, showing that GSH is essential for growth and can be partially substituted by γECS (Harrison et al. 2005). Both mutants exhibited higher catalase activity than the wild type, indicating that the two strains experience oxidative stress. Inoculation of *M. sativa* with the *gshA* mutant gave rise to no nodules, while inactivation of *gshB* triggered a delayed nodulation phenotype coupled with a 75% reduction in the nitrogen-fixation capacity of bacteroids. This phenotype was linked to abnormal development and early senescence of nodules. In particular, amyloplast-rich cells, which are characteristic of the IZ, could be observed directly adjacent to the senescence zone IV in nodules infected with the *gshB* mutant (Harrison et al. 2005). These results showed the key role of the *S. meliloti* GSH pool in the interaction with the host plant.

Table 2.2 Functional characterization of genes involved in GSH pathway in rhizobia

GSH-related genes	Rhizobium species	Mutant phenotype, free-living	Mutant phenotype, symbiosis	references
Glutathione synthesis				
gshA (γ-glutamylcysteine synthetase)	*Sinorhizobium meliloti*	No growth in minimal medium; under oxidative stress	No nodulation	Harrison et al. (2005)
	Bradyrhizobium sp. SEMIA 6144	Reduced growth in minimal medium	Impaired nodulation competitiveness	Sobrevals et al. (2006)
	Bradyrhizobium japonicum		Decreased nitrogenase activity	Sobrevals et al. (2006)
gshB (glutathione synthase)	*S. meliloti*	Reduced growth in minimal medium; under oxidative stress	Delayed nodulation; decreased nitrogen fixation	Harrison et al. (2005)
	Rhizobium etli	Gln uptake inhibited	Impaired nodulation competitiveness; early nodule senescence	Tate et al. (2012)
	Rhizobium tropici	Increased sensitivity to acidity and various environmental stresses	Impaired nodulation competitiveness; early nodule senescence	Riccillo et al. (2000) Muglia et al. (2008)
Glutathione reductase				
gor	*R. etli*	Gln uptake inhibited	Impaired nodulation competitiveness; early nodule senescence	Tate et al. (2012)
Glutaredoxins				
grx1	*S. meliloti*	Reduced growth in minimal medium; increased sensitivity to H_2O_2	Impaired bacteroid differentiation	Benyamina et al. (2013)
grx2	*S. meliloti*	Reduced growth in minimal medium	Decreased nitrogenase activity	Benyamina et al. (2013)
grx3	*S. meliloti*	Idem WT	Idem WT	Benyamina et al. (2013)
Glutathione S-transferase				
gstA	*Rhizobium leguminosarum*	Idem WT	Idem WT	Tawfiq Alkafaf et al. (1997)

The importance of GSH in symbiosis efficiency is correlated with a high level of *gshA*, *gshB*, and glutathione reductase (GR) gene (*gor*) expression in bacteroids from different zones of the *M. truncatula* nodules (Roux et al. 2014; Table 2.3). The expression level of the three genes is, however, reduced as compared with that in free-living *S. meliloti* (Capela et al. 2006; Table 2.3). Similar transcriptomic analyses of bacteroids also revealed decreased expression of *gshA*, *gshB*, and *gor* genes in *Bradyrhizobium japonicum* (Cuklina et al. 2016; Pessi et al. 2007), while genes in *Rhizobium leguminosarum* and *Azorhizobium caulinodans* bacteroids are transcribed at the same rate (Tsukada et al. 2009; Karunakaran et al. 2009). In *S. meliloti*, the expression of *gshB* is modulated by the alternative sigma factors RpoH1 and RpoH2 (Schluter et al. 2013) and by the LysR-type transcriptional regulator LsrB (Lu et al. 2013). Both *rpoH1* and *lsrB* mutations trigger the formation of ineffective nodules displaying early senescence, with an increased ROS content in zone III of the nodules infected by the *lsrB* mutant (Luo et al. 2005; Mitsui et al. 2004). The RpoH1-regulated genes are involved in heat shock, acidic pH, and antioxidant responses, suggesting a protecting role against environmental stresses encountered by the bacteria in free-living state and within the host plant (de Lucena et al. 2010; Ono et al. 2001).

Deleterious effects of GSH deficiency were subsequently observed in various rhizobium-legume interactions. During the symbiosis between *Phaseolus vulgaris* and *Rhizobium tropici*, a *gshB* mutant is significantly affected in its ability to compete for nodule occupancy, and plants infected with a *gshB* mutant showed a reduction in the dry weight of aerial plant parts. Nodules induced by the *gshB* mutant present a premature senescent pattern, containing very few bacteroids with signs of degradation, and a central zone that was filled with starch granules and devoid of bacteroids. The early senescence of nodules is correlated to enhanced levels of superoxide accumulation. In parallel, an increasing *gshB* expression is observed at the late stage of development of wild-type nodules (Riccillo et al. 2000; Muglia et al. 2008). These data suggest that the bacterial GSH could contribute to the persistence of N_2-fixing bacteroids by protecting them against ROS (Muglia et al. 2008). *gshB* expression may respond to increasing ROS concentration due to functioning of the nodule and/or to other environmental factors such as acidic conditions, which were shown to induce gene expression in free-living *R. tropici* cells. Given the acid pH of the peribacteroid space of nitrogen-fixing bean nodules, the acidic environment may also contribute to the activation of *gshB* expression in *R. tropici* bacteroids.

Infection of *P. vulgaris* with a *Rhizobium etli gshB* mutant is coupled to delayed nodulation and to the formation of undeveloped and ineffective nodules, exhibiting a low nitrogen-fixation efficiency (23.7% of that of WT) and clear signs of early senescence. Expression of the *gshB* gene depends on a transcriptional repressor responding to heat shock and oxidative environment suggesting that, as in other bacteria, genes involved in GSH synthesis contribute to the maintenance of ROS homeostasis in functional nodules (Vercruysse et al. 2011). In addition, *gshB* and *gor* mutants are unable to use Gln as a sole source of carbon, energy, and nitrogen. The mutants were shown to be affected with respect to Gln uptake, suggesting a

Table 2.3 Expression of bacterial genes of the GSH metabolic pathway in *M. truncatula* nodules. Gene accession numbers are indicated on the left side. Gene name and annotation are based on candidate orthologues and interprodomain signature

	Full organs	Dissected nodules					
Gene Family	Nodule vs Free-living	Total reads	zone 1	zone 2 d	zone 2 p	interzone	zone 3
Glutathione metabolism							
γecs SMc00825	-1.1	3820,0	16.2	21.9	29.9	16.5	15.6
gshs SMc00419	-1,0	6433.2	8,0	9.2	24.3	52.5	6,0
gr SMc00154	-1.8	4477.1	19.2	12.1	27.6	24,0	17.2
Glutaredoxins							
grx1 SMc02443	1.6	9123,0	6.8	6.5	19.6	15.8	51.3
grx2 SMc00538	-0.4	10138,0	24.1	24,0	21.3	22.1	8.5
grx3 SMa0280	-0.4	1571.4	17.0	18.0	22.1	21.8	21.1
Glutathione S-Transferases							
gst1 SMc00036	-1.5	670.0	7.7	14.7	21.2	40.2	16.2
gst2 SMc00097	0.3	439.6	5.4	11.3	13.4	11.5	58.3
gst3 SMc00383	-0.8	216.2	28.9	6.8	27.7	24.6	12.0
gst4 SMc00407	-0.9	3933.3	23.7	17.8	21.0	23.3	14.2
gst5 SMc01238	1.2	541.2	28.3	24,0	10.1	12.9	24.7
gst6 SMc01443	1.2	1283.9	14.1	12.6	14.8	23.2	35.4
gst7 SMc02390	0.7	1969.0	12.9	10.3	33.4	27.1	16.3
gst8 SMc03882	-0.8	1593.2	16.9	19.7	31.6	26.6	5.3
gst9 SMc04141	-0.4	163.5	30.0	21.7	9.5	9.6	29.2
gst10 SMc04321	1.7	802.3	16.6	13.9	18.3	18.0	33.2
gst11 SMc00916	-1.1	2709.9	12.0	10.9	25.2	41.6	10.3
gst12 SMb21449	3.7	1340.4	5.7	4.9	3.1	9.8	76.6
gst13 SMa2115	0.1	127.6	20.4	17.4	4.4	17.4	40.3
gst15 SMb20005	1.2	82.2	13.5	1.7	4.8	8.2	71.7
gst17 SMa1497	0.2	747.1	10.3	5.7	8.2	15.0	60.7
SMb20420	-1.7	544.2	34.1	23.9	8.5	12.7	20.8
S-Nitrosoglutathione Reductases							
SMc01270	-0.7	1662.7	23.3	16.7	27.3	28.2	4.5
SMa2113	-1.4	424.2	17.9	12.2	18.6	14.5	36.8
SMb20170	-0.6	1821.3	45.9	23.2	6.7	4.8	19.4

Nodule/free-living expression ratio numbers correspond to $\log_2$ ratios (nodule/free-living bacteria) from whole-genome microarrays analysis using *M. truncatula* nodules collected 18 days postinoculation and *S. meliloti* cultures in exponential phase (Capela et al. 2006).
The values corresponding to dissected nodules are from the left to the right: total reads from laser-capture microdissection and their distribution in each zone (%), as reported by Roux and coll. (Roux et al. 2014)

GSH-glutamine metabolic relationship in the bacteria. These data reveal a potential role of GSH in the regulation of amino acid uptake and hence assimilation in bacteroids (Tate et al. 2012).

To date one rhizobial-legume interaction, the interaction between *Bradyrhizobium* sp. SEMIA 6144 and peanut (*Arachis hypogaea* L.), was shown to give rise to effective nodules independently of the bacterial GSH pool (Sobrevals et al. 2006). In the free-living state, however, growth of the *gshA* mutant in minimal medium is severely affected. Moreover, high osmolarity, H_2O_2, or acidic stress conditions strongly increase the growth deficiency (Sobrevals et al. 2006). This growth defect cannot be rescued by external GSH, underlying that GSH uptake is not enough to restore the phenotype in the tested conditions. In the wild-type strain, the endogenous level of GSH increases under conditions of stress, which strengthens the role of GSH in stress protection. During the interaction with peanut, the *Bradyrhizobium* sp. SEMIA 6144 *gshA* mutant produced a reduced number of nodules per plant and was affected in its capacity to compete during the nodulation process, although it induced efficient nodules (Sobrevals et al. 2006). In the same study, it was reported that a *B. japonicum gshA* mutant developed nodules with a strong nitrogen-fixation deficiency during symbiotic interaction with soybean, reflecting difference between bacteria and plant species.

Overall, the bacterial GSH biosynthesis pathway is mostly essential for the proper development of root nodules. This means that GSH from the host plant cannot replace GSH of bacterial origin, particularly in *S. meliloti* and *R. etli* where exogenously provided GSH could compensate for GSH deficiency in growing cultures (Harrison et al. 2005; Tate et al. 2012). It is very likely that the weakly permeable PBM membrane prevents the uptake of GSH into bacteroids (Udvardi and Day 1997). Similarly, GSH biosynthesis contributes to the virulence of pathogenic bacteria, even in bacteria such as *Listeria monocytogenes* where GSH is also imported from the host (Reniere et al. 2015). Altogether, these results highlight the critical role of GSH in bacteroids, most likely via the involvement of GSH-dependent enzymes. This finding was confirmed by molecular studies on *S. meliloti* Grx.

3.2 Role of *S. meliloti* Glutaredoxins in Symbiosis

An in silico analysis of the *S. meliloti* genome led to the identification of three open reading frames that potentially encode Grx from different classes, SmGRX1, SmGRX2, and SmGRX3. SmGRX1 contains the CGYC redox active site of the classical dithiol class I Grxs and SmGRX2 the CGFS redox active site of monothiol class II Grxs. SmGRX3 carries two domains, an N-terminal Grx domain with a CPYG active site and a C-terminal domain with a methylamine utilization protein (MauE) motif, consisting of five putative transmembrane domains (Benyamina et al. 2013). Both classes I and II are ubiquitous in bacteria. For example, in *Escherichia coli* three dithiol Grx and one monothiol Grx have been characterized. Most of the rhizobia sequenced genomes contain one copy of each gene class, or sometimes two, as is the case of *Bradyrhizobium* sp. BTAi1 (two class I Grxs) and in the tropical legume symbiont *A. caulinodans* (two class I and two class II Grx). In contrast, SmGRX3 orthologues are found only in cyanobacteria and some proteobacteria, more particularly in *S. meliloti* and *Rhizobium* sp. NGR324 (Rhizobase: http://genome.annotation.jp/rhizobase).

Both *Smgrx1* and *Smgrx3* are induced during symbiosis as compared to the free-living state, while *Smgrx2* is similarly expressed under both conditions (Capela et al. 2006). RNA-seq data, obtained from microdissected nodule zones, indicate that *Smgrx1* and *Smgrx2* are maximally and minimally expressed in zone III, respectively, and that *Smgrx3* expression is similar in the different zones (Table 2.3). The symbiotic upregulation of *Smgrx1* might be mediated by NCR peptides and/or acidic environment. Indeed, in culture both NCR treatment and pH downshift in an RpoH1-dependent manner enhanced the expression of *Smgrx1* (initially named *grxC*) (de Lucena et al. 2010; Tiricz et al. 2013). In parallel with these different expression patterns, biochemical and genetic analyses point to distinct functions for the three proteins (Benyamina et al. 2013). SmGrx1 and SmGrx3 recombinant proteins reduce the mixed disulfide bond formed between GSH and 2-hydroethyldisulfide whereas SmGrx2 does not, showing that SmGrx2 is not able to perform deglutathionylation. The SmGRX3 mutant does not display growth or symbiotic defects, and so far its biological role remains unknown. Conversely, both SmGRX1 and SmGRX2 inactivation impair the growth of free-living bacteria and the nitrogen-fixation capacity of bacteroids. The SmGRX1 mutant presents an increased sensitivity to oxidative stress caused by H_2O_2, which is associated with a higher level of glutathionylated proteins under non-stress conditions, and hence SmGRX1 has a key role in protein deglutahionylation. During the interaction with *M. truncatula*, the Sm*GRX1* mutant induced abortive, spherical nodules where bacteroids did not undergo visible differentiation. Moreover, the expression of zone III marker genes is severely reduced in SmGRX1 nodules. These data suggest that redox control, via protein glutathionylation, is involved in bacteroid differentiation. It is possible that Grx1 interacts with NCRs to regulate their activity, as shown for *M. truncatula* Trx s1 (see Sect. 2.3) and for Trx with human defensins (Schroeder et al. 2011).

SmGrx2 inactivation in free-living bacteria results in decreased activities of the Fe/S cluster containing enzymes succinate deshydrogenase and aconitase, suggesting that SmGRX2 participates in the assembly of the Fe/S cluster (Benyamina et al. 2013). In addition, the expression of iron-regulated genes, which are directly controlled by the Rhizobium iron responsive A (RirA) transcriptional regulator, is enhanced in the SmGRX2 mutant, as well as the total intracellular iron content. Thus, SmGRX2 plays a crucial role in the regulation of iron homeostasis, either through a role in the Fe/S cluster assembly machinery or through the regulation of RirA. During the interaction between *S. meliloti* and *M. truncatula*, *grx2* inactivation affects the nodulation process, nodule development, and nitrogen-fixation efficiency. Nodules induced by the SmGRX2 mutant are elongated, yet smaller than WT nodules, they contain differentiated bacteroids, and zone III-specific genes are fully expressed. The nitrogen-fixation deficiency of SmGRX2 bacteroids could result from a direct effect on nitrogenase activity, which involves the interaction of two major components, the iron (Fe) protein containing a Fe/S cluster and the MoFe protein containing Fe/S and Fe/S/Mo clusters. More generally, Grx2 may be required to compensate for the oxidative loss of Fe/S clusters caused by ROS in nodules. The finding that a mutant in *sufT*, involved in Fe/S cluster metabolism, also has a lowered nitrogen-fixation capacity (Sasaki et al. 2016) strengthens this theory.

Overall, SmGRX1 and smGRX2 play critical but distinct roles in the interaction of *S. meliloti* with the host plant, in protein degluthationylation and iron homeostasis, respectively. In *B. japonicum* similarly to *S. meliloti*, both class I and II *grxs* genes are expressed in mature nodules, suggesting a role during the interaction with soybean (Pessi et al. 2007). Likewise, the arbuscular mycorrhizal fungus *Rhizophagus irregularis* possesses two dithiol GRX that might contribute to oxidative stress protection during the in planta phase and two monothiol GRXs that might be involved in the regulation of iron metabolism (Tamayo et al. 2016). To date the roles of Grxs have been poorly analyzed during plant microbe interaction, but their importance during this process will surely be emphasized in the future.

3.3 The Role of Bacterial Glutathione S-Transferases in Symbiosis

Bacterial GSTs are specialized for the detoxification of harmful endobiotics and xenobiotics (Allocati et al. 2009). Rhizobia as a free-living bacterium in the rhizosphere and during interaction with the host plant are potentially exposed to these toxic molecules. The genomes of Gram-negative bacteria and especially rhizobia contain multiple *gst* genes of widely divergent sequences and unknown function. The genomes of *S. meliloti*, *R. leguminosarum bv viciae*, and *B. japonicum* possess 16, 24, and 26 genes encoding putative GSTs, respectively, as described in the Rhizobase.

Several biochemical analyses provided evidence for an involvement of rhizobial GST in detoxification processes, more particularly in cadmium (Cd) detoxification. Cd toxicity is based on its ability to induce oxidative stress by increasing cellular ROS levels and because of its high affinity toward sulfhydryl groups (SH); thus it can inactivate metabolical important enzymes (Bruins et al. 2000). In this context, symbiotic microorganisms may improve plant growth under Cd exposure through absorption and accumulation of the metal as well as detoxification mechanisms (Ghnaya et al. 2015; Teng et al. 2015). In *R. leguminosarum*, GST isoforms of strains with distinct tolerances to Cd were purified and their activity investigated. The relationship between chelation efficiency and enzymatic activity of GSTs has been demonstrated, supporting the hypothesis that GSTs are involved in the formation of GSH-Cd complexes and in tolerance to Cd (Corticeiro et al. 2013). An increase in GST activity was observed in *B. japonicum* exposed to Cd, supporting a role in counteracting metal-induced cellular damage (Bianucci et al. 2013). Nevertheless, not all rhizobial GSTs appear to be efficient against Cd. For example, in diverse strains of *Bradyrhizobium* sp. the total GST activity was inhibited by Cd exposure (Bianucci et al. 2012), and none of the *S. meliloti gst* genes were upregulated upon Cd exposure (Rossbach et al. 2008). Finally, two GSTs with a high affinity to herbicides have also been characterized in *R. leguminosarum* (Faraone et al. 2003).

Some studies have highlighted a putative role of GST in symbiosis. A comparative genomic analysis of rhizobia and non-rhizobia identified two *gst* genes in *S. meliloti* (*gst3* and *gst10*) among the genes overrepresented in symbiotic rhizobia, suggesting that they could contribute to adaptation to symbiosis (Amadou et al. 2008). The expression of *gst10*, as well as four other *gst* genes, is enhanced during symbiosis, two of them (*gst12* and *gst15*) being more specifically expressed in zone III (Table 2.3). In *S. meliloti*, two potential *gst* genes (*gst1*, *gst2*) are located next to *lsrA*, a LysR-type regulator-encoding gene that probably regulates their expression and is crucial for symbiosis (Luo et al. 2005). According to in silico data, these genes are expressed at the late stages of nodule development (Table 2.3). Finally, the expression of a *gst* gene is more than 100-fold upregulated in a *B. japonicum* bacteroid mutant that transiently increases nodule numbers in soybean, possibly resulting from redox changes in the mutant cells (Ohkama-Ohtsu et al. 2016). Functional analyses will be helpful to better understand the contribution of GST with respect to the efficiency of symbiosis.

3.4 The Role of Other GSH-Dependent Enzymes: Glutathione Peroxidases and S-Nitrosoglutathione Reductases

ROS and RNS play crucial roles in the rhizobium-legume symbiosis (Puppo et al. 2013; Ribeiro et al. 2015). GSH reacts with NO to form S-nitrosoglutathione (GSNO), which can then transfer its NO group to other cellular thiols to form S-nitrosothiols. GSNO has been proposed to be a significant player in NO regulatory mechanisms, particularly in S-nitrosylation of proteins (Broniowska et al. 2013). There is growing evidence that GSNO reductase and GSH-dependent glutathione peroxidases (Gpx), which are involved in the degradation of GSNO and the reduction of organic hydroperoxides, respectively, contribute to ROS/RNS detoxication and signaling in microbial systems (Liu et al. 2001; Laver et al. 2013; Boronat et al. 2014). However, studies regarding their role in rhizobia are particularly scarce. The genomes of *S. meliloti* and *B. japonicum* harbor three and one putative GSNO reductase-encoding genes, respectively, whereas a potential Gpx-encoding gene is only present in *B. japonicum* (Rhizobase). Gpx and GSNO reductase genes are downregulated in *B. japonicum* bacteroids, and a low level of GSNO reductase transcripts has been detected in different zones of *S. meliloti*-infected nodules (Table 2.3). Gpx and GSNO reductase activities have been observed in cytosolic extracts of *B. japonicum* and *S. meliloti* (Maiti et al. 2012; Bianucci et al. 2013). The apparent contradiction between the detection of Gpx activity in *S. meliloti* and the absence of a *gpx* gene in the genome of the bacterium may be explained by the existence of some GST(s) displaying a GSH-dependent peroxidase activity (Allocati et al. 2009). Gpx activity was shown to be induced upon Cd exposure and GSNO reductase activity to increase under nitrosative stress, suggesting that these enzymes contribute to antioxidant defenses (Maiti et al. 2012; Bianucci et al. 2013). The functional role of these genes during symbiosis should be further explored to clearly define their importance.

4 Conclusion and Future Perspectives

Over the last years, many advances have been made on the characterization of enzymes involved in thiol biosynthetic pathway and in regulatory network in legume model plants, *M. truncatula* and *L. japonicus*. The importance of thiols and glutaredoxins in the bacteria has also been demonstrated. However, the importance of the majority of the proteins involved in the thiol regulatory network is still to be determined. The generation and the screening of bacterial/plant mutants will be very helpful to establish the function of individual proteins and metabolites in the legume-rhizobial symbiosis.

Our knowledge on the functions of the thiol regulatory network in legume nodules is also at its infancy. It will be necessary to identify the nodule targets of the different thiol-related enzymes and to assess their roles during rhizobial infection and nitrogen-fixing symbiosis. Redox-dependent posttranslational modifications constitute an adaptive mechanism to changing conditions. In nodule, these changing conditions include the biotic interactions between the plant and the bacteria, the modification in cell metabolism from the meristematic cell to the nitrogen-fixing cell, and the abiotic environment modifications which impact the carbon-fixing metabolism and more generally the plant physiology. The development of redox proteomics will allow the large-scale identification of proteins that are oxidized, nitrosylated, or glutathionylated in response to specific stimuli to analyze the redox modifications involved in nodule metabolism regulation.

Acknowledgment Research from our laboratory is supported a grant from ANR (program STAYPINK). This work was supported by the French Government (National Research Agency, ANR) through the "Investments for the Future" LABEX SIGNALIFE: program reference # ANR-11-LABX-0028-01.

References

Alkhalfioui F, Renard M, Montrichard F (2007) Unique properties of NADP-thioredoxin reductase C in legumes. J Exp Bot 58:969–978

Alkhalfioui F, Renard M, Frendo P, Keichinger C, Meyer Y, Gelhaye E, Hirasawa M, Knaff DB, Ritzenthaler C, Montrichard F (2008) A novel type of thioredoxin dedicated to symbiosis in legumes. Plant Physiol 148:424–435

Allocati N, Federici L, Masulli M, Di Ilio C (2009) Glutathione transferases in bacteria. FEBS J 276:58–75

Amadou C, Pascal G, Mangenot S, Glew M, Bontemps C, Capela D, Carrere S, Cruveiller S, Dossat C, Lajus A, Marchetti M, Poinsot V, Rouy Z, Servin B, Saad M, Schenowitz C, Barbe V, Batut J, Medigue C, Masson-Boivin C (2008) Genome sequence of the beta-rhizobium *Cupriavidus taiwanensis* and comparative genomics of rhizobia. Genome Res 18:1472–1483

Baier MC, Barsch A, Kuster H, Hohnjec N (2007) Antisense repression of the *Medicago truncatula* nodule-enhanced sucrose synthase leads to a handicapped nitrogen fixation mirrored by specific alterations in the symbiotic transcriptome and metabolome. Plant Physiol 145:1600–1618

Baldacci-Cresp F, Chang C, Maucourt M, Deborde C, Hopkins J, Lecomte P, Bernillon S, Brouquisse R, Moing A, Abad P, Herouart D, Puppo A, Favery B, Frendo P (2012) (Homo) glutathione deficiency impairs root-knot nematode development in *Medicago truncatula*. PLoS Pathog 8:e1002471

Ball L, Accotto GP, Bechtold U, Creissen G, Funck D, Jimenez A, Kular B, Leyland N, Mejia-Carranza J, Reynolds H, Karpinski S, Mullineaux PM (2004) Evidence for a direct link between glutathione biosynthesis and stress defense gene expression in *Arabidopsis*. Plant Cell 16:2448–2462

Behie SW, Bidochka MJ (2014) Nutrient transfer in plant-fungal symbioses. Trends Plant Sci 19:734–740

Benyamina SM, Baldacci-Cresp F, Couturier J, Chibani K, Hopkins J, Bekki A, De Lajudie P, Rouhier N, Jacquot JP, Alloing G, Puppo A, Frendo P (2013) Two *Sinorhizobium meliloti* glutaredoxins regulate iron metabolism and symbiotic bacteroid differentiation. Environ Microbiol 15:795–810

Bianucci E, Tordable Mdel C, Fabra A, Castro S (2008) Importance of glutathione in the nodulation process of peanut (*Arachis hypogaea*). Physiol Plant 134:342–347

Bianucci E, Furlan A, Rivadeneira J, Sobrino-Plata J, Carpena-Ruiz RO, Tordable Mdel C, Fabra A, Hernandez LE, Castro S (2012) Influence of cadmium on the symbiotic interaction established between peanut (*Arachis hypogaea* L.) and sensitive or tolerant bradyrhizobial strains. J Environ Manag 130:126–134

Bianucci E, Fullana C, Furlan A, Castro S (2013) Antioxidant defense system responses and role of nitrate reductase in the redox balance maintenance in *Bradyrhizobium japonicum* strains exposed to cadmium. Enzym Microb Technol 53:345–350

Boogerd FC, Van Rossum D (1997) Nodulation of groundnut by Bradyrhizobium: a simple infection process by crack entry. FEMS Microbiol Rev 21:5–27

Boronat S, Domenech A, Paulo E, Calvo IA, Garcia-Santamarina S, Garcia P, Encinar Del Dedo J, Barcons A, Serrano E, Carmona M, Hidalgo E (2014) Thiol-based H_2O_2 signalling in microbial systems. Redox Biol 2:395–399

Brigelius-Flohe R, Maiorino M (2013) Glutathione peroxidases. Biochim Biophys Acta 1830:3289–3303

Broniowska KA, Diers AR, Hogg N (2013) S-nitrosoglutathione. Biochim Biophys Acta 1830: 3173–3181

Bruins MR, Kapil S, Oehme FW (2000) Microbial resistance to metals in the environment. Ecotoxicol Environ Saf 45:198–207

Cairns NG, Pasternak M, Wachter A, Cobbett CS, Meyer AJ (2006) Maturation of *Arabidopsis* seeds is dependent on glutathione biosynthesis within the embryo. Plant Physiol 141:446–455

Capela D, Filipe C, Bobik C, Batut J, Bruand C (2006) *Sinorhizobium meliloti* differentiation during symbiosis with alfalfa: a transcriptomic dissection. Mol Plant-Microbe Interact 19:363–372

Carnegie PR (1963) Structure and properties of a homologue of glutathione. Biochem J 89:471–478

Clemens S (2006) Evolution and function of phytochelatin synthases. J Plant Physiol 163:319–332

Colville L, Saez CM, Lewis GP, Kranner I (2015) The distribution of glutathione and homoglutathione in leaf, root and seed tissue of 73 species across the three sub-families of the Leguminosae. Phytochemistry 115:175–183

Considine MJ, Foyer CH (2014) Redox regulation of plant development. Antioxid Redox Signal 21:1305–1326

Copley SD, Dhillon JK (2002) Lateral gene transfer and parallel evolution in the history of glutathione biosynthesis genes. Genome Biol 3:research0025

Corticeiro S, Freitas R, Figueira E (2013) The role of GSTs in the tolerance of *Rhizobium leguminosarum* to cadmium. Biometals 26:879–886

Crespi M, Galvez S (2000) Molecular mechanisms in root nodule development. J Plant Growth Regul 19:155–166

Cuklina J, Hahn J, Imakaev M, Omasits U, Forstner KU, Ljubimov N, Goebel M, Pessi G, Fischer HM, Ahrens CH, Gelfand MS, Evguenieva-Hackenberg E (2016) Genome-wide transcription start site mapping of *Bradyrhizobium japonicum* grown free-living or in symbiosis – a rich resource to identify new transcripts, proteins and to study gene regulation. BMC Genomics 17:302

Cummins I, Dixon DP, Freitag-Pohl S, Skipsey M, Edwards R (2011) Multiple roles for plant glutathione transferases in xenobiotic detoxification. Drug Metab Rev 43:266–280

Czernic P, Gully D, Cartieaux F, Moulin L, Guefrachi I, Patrel D, Pierre O, Fardoux J, Chaintreuil C, Nguyen P, Gressent F, Da Silva C, Poulain J, Wincker P, Rofidal V, Hem S, Barriere Q, Arrighi JF, Mergaert P, Giraud E (2015) Convergent evolution of endosymbiont differentiation in dalbergioid and inverted repeat-lacking clade legumes mediated by nodule-specific cysteine-rich peptides. Plant Physiol 169:1254–1265

Dalle-Donne I, Rossi R, Colombo G, Giustarini D, Milzani A (2009) Protein S-glutathionylation: a regulatory device from bacteria to humans. Trends Biochem Sci 34:85–96

Dalton DA, Baird LM, Langeberg L, Taugher CY, Anyan WR, Vance CP, Sarath G (1993) Subcellular localization of oxygen defense enzymes in soybean (*Glycine max* [L.] Merr.) root nodules. Plant Physiol 102:481–489

Dalton DA, Boniface C, Turner Z, Lindhal A, Kim HJ, Jelinek L, Govindarajulu M, Finger RE, Taylor CG (2009) Physiological roles of glutathione S-transferases in soybean root nodules. Plant Physiol 150:521–530

De Lucena DK, Puhler A, Weidner S (2010) The role of sigma factor RpoH1 in the pH stress response of *Sinorhizobium meliloti*. BMC Microbiol 10:265

Delorme-Hinoux V, Bangash SA, Meyer AJ, Reichheld JP (2015) Nuclear thiol redox systems in plants. Plant Sci 243:84–95

Desbrosses GJ, Stougaard J (2011) Root nodulation: a paradigm for how plant-microbe symbiosis influences host developmental pathways. Cell Host Microbe 10:348–358

Dixon DP, Skipsey M, Grundy NM, Edwards R (2005) Stress-induced protein S-glutathionylation in *Arabidopsis*. Plant Physiol 138:2233–2244

Dixon DP, Skipsey M, Edwards R (2010) Roles for glutathione transferases in plant secondary metabolism. Phytochemistry 71:338–350

Du H, Kim S, Nam KH, Lee MS, Son O, Lee SH, Cheon CI (2010) Identification of uricase as a potential target of plant thioredoxin: implication in the regulation of nodule development. Biochem Biophys Res Commun 397:22–26

El Msehli S, Lambert A, Baldacci-Cresp F, Hopkins J, Boncompagni E, Smiti SA, Herouart D, Frendo P (2011) Crucial role of (homo)glutathione in nitrogen fixation in *Medicago truncatula* nodules. New Phytol 192:496–506

Escuredo PR, Minchin FR, Gogorcena Y, Iturbe-Ormaetxe I, Klucas RV, Becana M (1996) Involvement of activated oxygen in nitrate-induced senescence of pea root nodules. Plant Physiol 110:1187–1195

Evans PJ, Gallesi D, Mathieu C, Hernandez MJ, De Felipe M, Halliwell B, Puppo A (1999) Oxidative stress occurs during soybean nodule senescence. Planta 208:73–79

Faraone A, Petrucci M, Paludi D, Aceto A, Dainelli B (2003) Purification and characterisation of two GST's forms from *Rhizobium leguminosarum* with a high affinity to herbicides. Int J Immunopathol Pharmacol 16:55–60

Ferrari S, Plotnikova JM, De Lorenzo G, Ausubel FM (2003) *Arabidopsis* local resistance to *Botrytis cinerea* involves salicylic acid and camalexin and requires EDS4 and PAD2, but not SID2, EDS5 or PAD4. Plant J 35:193–205

Flemetakis E, Efrose RC, Ott T, Stedel C, Aivalakis G, Udvardi MK, Katinakis P (2006) Spatial and temporal organization of sucrose metabolism in Lotus japonicus nitrogen-fixing nodules suggests a role for the elusive alkaline/neutral invertase. Plant Mol Biol 62:53–69

Frendo P, Mathieu C, Van De Sype G, Hérouart D, Puppo A (1999) Characterisation of a cDNA encoding γ-glutamylcysteine synthetase in *Medicago truncatula*. Free Radic Res 31:213–218

Frendo P, Jimenez MJ, Mathieu C, Duret L, Gallesi D, Van De Sype G, Herouart D, Puppo A (2001) A *Medicago truncatula* homoglutathione synthetase is derived from glutathione synthetase by gene duplication. Plant Physiol 126:1706–1715

Frendo P, Harrison J, Norman C, Hernandez Jimenez MJ, Van De Sype G, Gilabert A, Puppo A (2005) Glutathione and homoglutathione play a critical role in the nodulation process of *Medicago truncatula*. Mol Plant-Microbe Interact 18:254–259

Frendo P, Baldacci-Cresp F, Benyamina SM, Puppo A (2013) Glutathione and plant response to the biotic environment. Free Radic Biol Med 65:724–730

Galant A, Arkus KA, Zubieta C, Cahoon RE, Jez JM (2009) Structural basis for evolution of product diversity in soybean glutathione biosynthesis. Plant Cell 21:3450–3458

Galant A, Preuss M, Cameron J, Jez J (2011) Plant glutathione biosynthesis: diversity in biochemical regulation and reaction products. Front Plant Sci 2:45

Ghnaya T, Mnassri M, Ghabriche R, Wali M, Poschenrieder C, Lutts S, Abdelly C (2015) Nodulation by *Sinorhizobium meliloti* originated from a mining soil alleviates Cd toxicity and increases Cd-phytoextraction in *Medicago sativa* L. Front Plant Sci 6:863

Gogorcena Y, Gordon AJ, Escuredo PR, Minchin FR, Witty JF, Moran JF, Becana M (1997) N_2 fixation, carbon metabolism and oxidative damage in nodules of dark stressed common bean plants. Plant Physiol 113:1193–1201

Groten K, Vanacker H, Dutilleul C, Bastian F, Bernard S, Carzaniga R, Foyer CH (2005) The roles of redox processes in pea nodule development and senescence. Plant Cell Environ 28:1293–1304

Harrison J, Jamet A, Muglia CI, Van De Sype G, Aguilar OM, Puppo A, Frendo P (2005) Glutathione plays a fundamental role in growth and symbiotic capacity of *Sinorhizobium meliloti*. J Bacteriol 187:168–174

Hell R, Wirtz M (2011) Molecular biology, biochemistry and cellular physiology of cysteine metabolism in Arabidopsis thaliana. Arabidopsis Book 9:e0154

Kalloniati C, Krompas P, Karalias G, Udvardi MK, Rennenberg H, Herschbach C, Flemetakis E (2015) Nitrogen-fixing nodules are an important source of reduced sulfur, which triggers global changes in sulfur metabolism in *Lotus japonicus*. Plant Cell 27:2384–2400

Karunakaran R, Ramachandran VK, Seaman JC, East AK, Mouhsine B, Mauchline TH, Prell J, Skeffington A, Poole PS (2009) Transcriptomic analysis of *Rhizobium leguminosarum* biovar viciae in symbiosis with host plants *Pisum sativum* and *Vicia cracca*. J Bacteriol 191:4002–4014

Klapheck S (1988) Homoglutathione: isolation, quantification and occurrence in legumes. Physiol Plant 74:727–732

Kondorosi E, Kondorosi A (2004) Endoreduplication and activation of the anaphase-promoting complex during symbiotic cell development. FEBS Lett 567:152–157

Kondorosi E, Mergaert P, Kereszt A (2013) A paradigm for endosymbiotic life: cell differentiation of Rhizobium bacteria provoked by host plant factors. Annu Rev Microbiol 67:611–628

Krusell L, Krause K, Ott T, Desbrosses G, Kramer U, Sato S, Nakamura Y, Tabata S, James EK, Sandal N, Stougaard J, Kawaguchi M, Miyamoto A, Suganuma N, Udvardi MK (2005) The sulfate transporter SST1 is crucial for symbiotic nitrogen fixation in *Lotus japonicus* root nodules. Plant Cell 17:1625–1636

Labrou NE, Papageorgiou AC, Pavli O, Flemetakis E (2015) Plant GSTome: structure and functional role in xenome network and plant stress response. Curr Opin Biotechnol 32:186–194

Laver JR, Mclean S, Bowman LA, Harrison LJ, Read RC, Poole RK (2013) Nitrosothiols in bacterial pathogens and pathogenesis. Antioxid Redox Signal 18:309–322

Lee MY, Shin KH, Kim YK, Suh JY, Gu YY, Kim MR, Hur YS, Son O, Kim JS, Song E, Lee MS, Nam KH, Hwang KH, Sung MK, Kim HJ, Chun JY, Park M, Ahn TI, Hong CB, Lee SH, Park HJ, Park JS, Verma DP, Cheon CI (2005) Induction of thioredoxin is required for nodule development to reduce reactive oxygen species levels in soybean roots. Plant Physiol 139:1881–1889

Lillig CH, Berndt C (2012) Glutaredoxins in thiol/disulfide exchange. Antioxid Redox Signal 18:1654–1665

Liu L, Hausladen A, Zeng M, Que L, Heitman J, Stamler JS (2001) A metabolic enzyme for S-nitrosothiol conserved from bacteria to humans. Nature 410:490–494

Lu D, Tang G, Wang D, Luo L (2013) The *Sinorhizobium meliloti* LysR family transcriptional factor LsrB is involved in regulation of glutathione synthesis. Acta Biochim Biophys Sin Shangai 45:882–888

Luo L, Yao SY, Becker A, Ruberg S, Yu GQ, Zhu JB, Cheng HP (2005) Two new *Sinorhizobium meliloti* LysR-type transcriptional regulators required for nodulation. J Bacteriol 187:4562–4572
Maiti D, Sarkar TS, Ghosh S (2012) Detection of S-nitrosothiol and nitrosylated proteins in *Arachis hypogaea* functional nodule: response of the nitrogen fixing symbiont. PLoS One 7:e45526
Marino D, Frendo P, Ladrera R, Zabalza A, Puppo A, Arrese-Igor C, Gonzalez EM (2007) Nitrogen fixation control under drought stress. Localized or systemic? Plant Physiol 143:1968–1974
Matamoros MA, Moran JF, Iturbe-Ormaetxe I, Rubio MC, Becana M (1999) Glutathione and homoglutathione synthesis in legume root nodules. Plant Physiol 121:879–888
Matamoros MA, Clemente MR, Sato S, Asamizu E, Tabata S, Ramos J, Moran JF, Stiller J, Gresshoff PM, Becana M (2003) Molecular analysis of the pathway for the synthesis of thiol tripeptides in the model legume *Lotus japonicus*. Mol Plant-Microbe Interact 16:1039–1046
Matamoros MA, Fernandez-Garcia N, Wienkoop S, Loscos J, Saiz A, Becana M (2013) Mitochondria are an early target of oxidative modifications in senescing legume nodules. New Phytol 197:873–885
Matamoros MA, Saiz A, Penuelas M, Bustos-Sanmamed P, Mulet JM, Barja MV, Rouhier N, Moore M, James EK, Dietz KJ, Becana M (2015) Function of glutathione peroxidases in legume root nodules. J Exp Bot 66:2979–2990
May MJ, Leaver CJ (1994) *Arabidopsis thaliana* γ-glutamylcysteine synthetase is structurally unrelated to mammalian, yeast, and *E. coli* homologs. Proc Natl Acad Sci U S A 91:10059–10063
Meyer Y, Belin C, Delorme-Hinoux V, Reichheld JP, Riondet C (2012) Thioredoxin and glutaredoxin systems in plants: molecular mechanisms, crosstalks, and functional significance. Antioxid Redox Signal 17:1124–1160
Mieyal JJ, Chock P (2012) B. Posttranslational modification of cysteine in redox signaling and oxidative stress: focus on s-glutathionylation. Antioxid Redox Signal 16:471–475
Mieyal JJ, Gallogly MM, Qanungo S, Sabens EA, Shelton MD (2008) Molecular mechanisms and clinical implications of reversible protein S-glutathionylation. Antioxid Redox Signal 10:1941–1988
Mitsui H, Sato T, Sato Y, Ito N, Minamisawa K (2004) *Sinorhizobium meliloti* RpoH1 is required for effective nitrogen-fixing symbiosis with alfalfa. Mol Gen Genomics 271:416–425
Molesini B, Cecconi D, Pii Y, Pandolfini T (2014) Local and systemic proteomic changes in *Medicago truncatula* at an early phase of *Sinorhizobium meliloti* infection. J Proteome Res 13:408–421
Mou Z, Fan W, Dong X (2003) Inducers of plant systemic acquired resistance regulate NPR1 function through redox changes. Cell 113:935–944
Muglia C, Comai G, Spegazzini E, Riccillo PM, Aguilar OM (2008) Glutathione produced by *Rhizobium tropici* is important to prevent early senescence in common bean nodules. FEMS Microbiol Lett 23:191–198
Naya L, Ladrera R, Ramos J, Gonzalez EM, Arrese-Igor C, Minchin FR, Becana M (2007) The response of carbon metabolism and antioxidant defenses of alfalfa nodules to drought stress and to the subsequent recovery of plants. Plant Physiol 144:1104–1114
Ohkama-Ohtsu N, Honma H, Nakagome M, Nagata M, Yamaya-Ito H, Sano Y, Hiraoka N, Ikemi T, Suzuki A, Okazaki S, Minamisawa K, Yokoyama T (2016) Growth rate of and gene expression in *Bradyrhizobium diazoefficiens* USDA110 due to a mutation in blr7984, a TetR family transcriptional regulator gene. Microbes Environ 31:249–259
Oldroyd GE (2013) Speak, friend, and enter: signalling systems that promote beneficial symbiotic associations in plants. Nat Rev Microbiol 11:252–263
Oldroyd GE, Downie JA (2004) Calcium, kinases and nodulation signalling in legumes. Nat Rev Mol Cell Biol 5:566–576
Oldroyd GE, Downie JA (2008) Coordinating nodule morphogenesis with rhizobial infection in legumes. Annu Rev Plant Biol 59:519–546
Ono Y, Mitsui H, Sato T, Minamisawa K (2001) Two RpoH homologs responsible for the expression of heat shock protein genes in *Sinorhizobium meliloti*. Mol Gen Genet 264:902–912

Ott T, Van Dongen JT, Gunther C, Krusell L, Desbrosses G, Vigeolas H, Bock V, Czechowski T, Geigenberger P, Udvardi MK (2005) Symbiotic leghemoglobins are crucial for nitrogen fixation in legume root nodules but not for general plant growth and development. Curr Biol 15:531–535

Parisy V, Poinssot B, Owsianowski L, Buchala A, Glazebrook J, Mauch F (2007) Identification of PAD2 as a gamma-glutamylcysteine synthetase highlights the importance of glutathione in disease resistance of *Arabidopsis*. Plant J 49:159–172

Passaia G, Margis-Pinheiro M (2015) Glutathione peroxidases as redox sensor proteins in plant cells. Plant Sci 234:22–26

Patriarca EJ, Tate R, Ferraioli S, Iaccarino M (2004) Organogenesis of legume root nodules. Int Rev Cytol 234:201–262

Pessi G, Ahrens CH, Rehrauer H, Lindemann A, Hauser F, Fischer HM, Hennecke H (2007) Genome-wide transcript analysis of *Bradyrhizobium japonicum* bacteroids in soybean root nodules. Mol Plant-Microbe Interact 20:1353–1363

Pierre O, Engler G, Hopkins J, Brau F, Boncompagni E, Herouart D (2013) Peribacteroid space acidification: a marker of mature bacteroid functioning in *Medicago truncatula* nodules. Plant Cell Environ 36:2059–2070

Pucciariello C, Innocenti G, Van De Velde W, Lambert A, Hopkins J, Clement M, Ponchet M, Pauly N, Goormachtig S, Holsters M, Puppo A, Frendo P (2009) (Homo)glutathione depletion modulates host gene expression during the symbiotic interaction between *Medicago truncatula* and *Sinorhizobium meliloti*. Plant Physiol 151:1186–1196

Pulido P, Spinola MC, Kirchsteiger K, Guinea M, Pascual MB, Sahrawy M, Sandalio LM, Dietz KJ, Gonzalez M, Cejudo FJ (2010) Functional analysis of the pathways for 2-Cys peroxiredoxin reduction in *Arabidopsis thaliana* chloroplasts. J Exp Bot 61:4043–4054

Puppo A, Pauly N, Boscari A, Mandon K, Brouquisse R (2013) Hydrogen peroxide and nitric oxide: key regulators of the legume-Rhizobium and mycorrhizal symbioses. Antioxid Redox Signal 18:2202–2219

Queval G, Jaillard D, Zechmann B, Noctor G (2011) Increased intracellular H(2)O(2) availability preferentially drives glutathione accumulation in vacuoles and chloroplasts. Plant Cell Environ 34:21–32

Ramos J, Matamoros MA, Naya L, James EK, Rouhier N, Sato S, Tabata S, Becana M (2009) The glutathione peroxidase gene family of *Lotus japonicus*: characterization of genomic clones, expression analyses and immunolocalization in legumes. New Phytol 181:103–114

Rawlins MR, Leaver CJ, May MJ (1995) Characterisation of an *Arabidopsis thaliana* cDNA encoding glutathione synthetase. FEBS Lett 376:81–86

Rees DC, Howard JB (2000) Nitrogenase: standing at the crossroads. Curr Opin Chem Biol 4:559–566

Reichheld JP, Khafif M, Riondet C, Droux M, Bonnard G, Meyer Y (2007) Inactivation of thioredoxin reductases reveals a complex interplay between thioredoxin and glutathione pathways in *Arabidopsis* development. Plant Cell 19:1851–1865

Reniere ML, Whiteley AT, Hamilton KL, John SM, Lauer P, Brennan RG, Portnoy DA (2015) Glutathione activates virulence gene expression of an intracellular pathogen. Nature 517:170–173

Ribeiro CW, Alloing G, Mandon K, Frendo P (2015) Redox regulation of differentiation in symbiotic nitrogen fixation. Biochim Biophys Acta 1850:1469–1478

Ribeiro CW, Baldacci-Cresp F, Pierre O, Larousse M, Benyamina S, Lambert A, Hopkins J, Castella C, Cazareth J, Alloing G, Boncompagni E, Couturier J, Mergaert P, Gamas P, Rouhier N, Montrichard F, Frendo P (2017) Regulation of differentiation of nitrogen-fixing bacteria by microsymbiont targeting of plant Thioredoxin s1. Curr Biol 27:250–256

Riccillo PM, Muglia CI, De Bruijn FJ, Roe AJ, Booth IR, Aguilar OM (2000) Glutathione is involved in environmental stress responses in *Rhizobium tropici*, including acid tolerance. J Bacteriol 182:1748–1753

Roetschi A, Si-Ammour A, Belbahri L, Mauch F, Mauch-Mani B (2001) Characterization of an *Arabidopsis-Phytophthora* pathosystem: resistance requires a functional PAD2 gene and is independent of salicylic acid, ethylene and jasmonic acid signalling. Plant J 28:293–305

Rossbach S, Mai DJ, Carter EL, Sauviac L, Capela D, Bruand C, De Bruijn FJ (2008) Response of *Sinorhizobium meliloti* to elevated concentrations of cadmium and zinc. Appl Environ Microbiol 74:4218–4221

Rouhier N (2010) Plant glutaredoxins: pivotal players in redox biology and iron-sulphur centre assembly. New Phytol 186:365–372

Rouhier N, Koh CS, Gelhaye E, Corbier C, Favier F, Didierjean C, Jacquot JP (2008a) Redox based anti-oxidant systems in plants: biochemical and structural analyses. Biochim Biophys Acta 1780:1249–1260

Rouhier N, Lemaire SD, Jacquot JP (2008b) The role of glutathione in photosynthetic organisms: emerging functions for glutaredoxins and glutathionylation. Annu Rev Plant Biol 59:143–166

Rouhier N, Cerveau D, Couturier J, Reichheld JP, Rey P (2015) Involvement of thiol-based mechanisms in plant development. Biochim Biophys Acta 1850:1479–1496

Roux B, Rodde N, Jardinaud MF, Timmers T, Sauviac L, Cottret L, Carrère S, Sallet E, Courcelle E, Moreau S, Debellé F, Capela D, De Carvalho-Niebel F, Gouzy J, Bruand C, Gamas P (2014) An integrated analysis of plant and bacterial gene expression in symbiotic root nodules using laser-capture microdissection coupled to RNA sequencing. Plant J 77:817–837

Sasaki S, Minamisawa K, Mitsui H (2016) A *Sinorhizobium meliloti* RpoH-regulated gene is involved in iron-sulfur protein metabolism and effective plant symbiosis under intrinsic iron limitation. J Bacteriol 198:2297–2306

Schippers JH, Foyer CH, Van Dongen JT (2016) Redox regulation in shoot growth, SAM maintenance and flowering. Curr Opin Plant Biol 29:121–128

Schlaeppi K, Bodenhausen N, Buchala A, Mauch F, Reymond P (2008) The glutathione-deficient mutant pad2-1 accumulates lower amounts of glucosinolates and is more susceptible to the insect herbivore *Spodoptera littoralis*. Plant J 55:774–786

Schluter JP, Reinkensmeier J, Barnett MJ, Lang C, Krol E, Giegerich R, Long SR, Becker A (2013) Global mapping of transcription start sites and promoter motifs in the symbiotic alphaproteobacterium *Sinorhizobium meliloti* 1021. BMC Genomics 14:156

Schroeder BO, Wu Z, Nuding S, Groscurth S, Marcinowski M, Beisner J, Buchner J, Schaller M, Stange EF, Wehkamp J (2011) Reduction of disulphide bonds unmasks potent antimicrobial activity of human beta-defensin 1. Nature 469:419–423

Sevilla F, Camejo D, Ortiz-Espin A, Calderon A, Lazaro JJ, Jimenez A (2015) The thioredoxin/peroxiredoxin/sulfiredoxin system: current overview on its redox function in plants and regulation by reactive oxygen and nitrogen species. J Exp Bot 66:2945–2955

Sobrevals L, Muller P, Fabra A, Castro S (2006) Role of glutathione in the growth of *Bradyrhizobium sp.* (peanut microsymbiont) under different environmental stresses and in symbiosis with the host plant. Can J Microbiol 52:609–616

Spoel SH, Tada Y, Loake GJ (2010) Post-translational protein modification as a tool for transcription reprogramming. New Phytol 186:333–339

Suzaki T, Yoro E, Kawaguchi M (2015) Leguminous plants: inventors of root nodules to accommodate symbiotic bacteria. Int Rev Cell Mol Biol 316:111–158

Tamayo E, Benabdellah K, Ferrol N (2016) Characterization of three new Glutaredoxin genes in the arbuscular mycorrhizal fungus *Rhizophagus irregularis*: putative role of RiGRX4 and RiGRX5 in iron homeostasis. PLoS One 11:e0149606

Tate R, Cermola M, Riccio A, Diez-Roux G, Patriarca EJ (2012) Glutathione is required by *Rhizobium etli* for glutamine utilization and symbiotic effectiveness. Mol Plant-Microbe Interact 25:331–340

Tawfiq Alkafaf NK, Yeoman KH, Wexler M, Hussain H, Johnston AW (1997) Analysis of a *Rhizobium leguminosarum* gene encoding a protein homologous to glutathione S-transferases. Microbiology 143:813–822

Teng Y, Wang X, Li L, Li Z, Luo Y (2015) Rhizobia and their bio-partners as novel drivers for functional remediation in contaminated soils. Front Plant Sci 6:32

Timmers AC, Auriac MC, Truchet G (1999) Refined analysis of early symbiotic steps of the *Rhizobium-Medicago* interaction in relationship with microtubular cytoskeleton rearrangements. Development 126:3617–3628

Tiricz H, Szucs A, Farkas A, Pap B, Lima RM, Maroti G, Kondozozi E, Kereszt A (2013) Antimicrobial nodule-specific cysteine-rich peptides induce membrane depolarization-associated changes in the transcriptome of *Sinorhizobium meliloti*. Appl Environ Microbiol 79:6737–6746

Tovar-Mendez A, Matamoros MA, Bustos-Sanmamed P, Dietz KJ, Cejudo FJ, Rouhier N, Sato S, Tabata S, Becana M (2011) Peroxiredoxins and NADPH-dependent thioredoxin systems in the model legume *Lotus japonicus*. Plant Physiol 156:1535–1547

Tsukada S, Aono T, Akiba N, Lee KB, Liu CT, Toyazaki H, Oyaizu H (2009) Comparative genome-wide transcriptional profiling of *Azorhizobium caulinodans* ORS571 grown under free-living and symbiotic conditions. Appl Environ Microbiol 75:5037–5046

Udvardi MK, Day DA (1997) Metabolite transport across symbiotic membranes of legume nodules. Annu Rev Plant Physiol Plant Mol Biol 48:493–523

Van De Velde W, Guerra JC, De Keyser A, De Rycke R, Rombauts S, Maunoury N, Mergaert P, Kondorosi E, Holsters M, Goormachtig S (2006) Aging in legume symbiosis. A molecular view on nodule senescence in *Medicago truncatula*. Plant Physiol 141:711–720

Van De Velde W, Zehirov G, Szatmari A, Debreczeny M, Ishihara H, Kevei Z, Farkas A, Mikulass K, Nagy A, Tiricz H, Satiat-Jeunemaitre B, Alunni B, Bourge M, Kucho K, Abe M, Kereszt A, Maroti G, Uchiumi T, Kondorosi E, Mergaert P (2010) Plant peptides govern terminal differentiation of bacteria in symbiosis. Science 327:1122–1126

Vercruysse M, Fauvart M, Beullens S, Braeken K, Cloots L, Engelen K, Marchal K, Michiels J (2011) A comparative transcriptome analysis of *Rhizobium etli* bacteroids: specific gene expression during symbiotic nongrowth. Mol Plant-Microbe Interact 24:1553–1561

Vernie T, Kim J, Frances L, Ding Y, Sun J, Guan D, Niebel A, Gifford ML, De Carvalho-Niebel F, Oldroyd GE (2015) The NIN transcription factor coordinates diverse nodulation programs in different tissues of the *Medicago truncatula* root. Plant Cell 27:3410–3424

Vernoux T, Wilson RC, Seeley KA, Reichheld JP, Muroy S, Brown S, Maughan SC, Cobbett CS, Van Montagu M, Inze D, May MJ, Sung ZR (2000) The ROOT MERISTEMLESS1/CADMIUM SENSITIVE2 gene defines a glutathione-dependent pathway involved in initiation and maintenance of cell division during postembryonic root development. Plant Cell 12:97–110

Vinardell JM, Fedorova E, Cebolla A, Kevei Z, Horvath G, Kelemen Z, Tarayre S, Roudier F, Mergaert P, Kondorosi A, Kondorosi E (2003) Endoreduplication mediated by the anaphase-promoting complex activator CCS52A is required for symbiotic cell differentiation in *Medicago truncatula* nodules. Plant Cell 15:2093–2105

Wachter A, Wolf S, Steininger H, Bogs J, Rausch T (2005) Differential targeting of GSH1 and GSH2 is achieved by multiple transcription initiation: implications for the compartmentation of glutathione biosynthesis in the *Brassicaceae*. Plant J 41:15–30

Wagner G, Charton S, Lariagon C, Laperche A, Lugan R, Hopkins J, Frendo P, Bouchereau A, Delourme R, Gravot A, Manzanares-Dauleux MJ (2012) Metabotyping: a new approach to investigate rapeseed (*Brassica napus L.*) genetic diversity in the metabolic response to clubroot infection. Mol Plant-Microbe Interact 25:1478–1491

Wang D, Dong X (2011) A highway for war and peace: the secretory pathway in plant-microbe interactions. Mol Plant 4:581–587

Wienkoop S, Saalbach G (2003) Proteome analysis. Novel proteins identified at the peribacteroid membrane from *Lotus japonicus* root nodules. Plant Physiol 131:1080–1090

Xiao TT, Schilderink S, Moling S, Deinum EE, Kondorosi E, Franssen H, Kulikova O, Niebel A, Bisseling T (2014) Fate map of *Medicago truncatula* root nodules. Development 141:3517–3528

Xiong Y, Uys JD, Tew KD, Townsend DM (2011) S-glutathionylation: from molecular mechanisms to health outcomes. Antioxid Redox Signal 15:233–270

Zaffagnini M, Bedhomme M, Marchand CH, Morisse S, Trost P, Lemaire SD (2012) Redox regulation in photosynthetic organisms: focus on glutathionylation. Antioxid Redox Signal 16:567–586

Chapter 3
Involvement of Thiol-Based Mechanisms in Plant Growth, Development, and Stress Tolerance

Marta Gietler and Małgorzata Nykiel

Abstract Thiol-based mechanisms of plant growth regulation and stress response rely on cellular redox potential, depending mostly on glutathione content ($E_{GSSG/2GSH}$). Thiol (SH) groups play various roles in the cell, with their redox state affecting the activity and structure of many enzymes, receptors, and transcription factors. The oxidation of -SH to the sulfinic ($R\text{-}S0_2H$) and sulfonic ($R\text{-}S0_3H$) acid may cause an irreversible enzyme inactivation and intermolecular protein cross-linking. The reversible oxidation of protein cysteine residues with the rise of formation of radical thiol, sulfenic acid (R-SOH), and S-nitrosothiol (SNT) is often an intermediate step to the formation of a mixed disulfide. Therefore, the most probable modification of these single reactive sulfhydryls is S-thiolation, resulting in formation of the mixed disulfide with low-molecular-weight cellular thiols such as glutathione. This modification is metabolically labile as it is evidenced by rapid "dethiolation" by several reductive processes. Intensity of revocation of Cys residues to reduced state is strongly based on glutaredoxin (Grx) and thioredoxin (Trx) activities, which are part of antioxidative system, regulating thiol-disulfide homeostasis in plant cells. Thus, the dynamic modification of proteins by S-thiolation/dethiolation represents one of the more important adaptive functions by reprogramming metabolism and protecting protein synthesis against irreversible oxidation. Alternatively, it may serve a regulatory role analogous to other posttranslational modifications such as protein phosphorylation. Thus, it can modulate cellular life cycle processes (division, differentiation, programmed cell death), energy metabolism, protein folding and degradation, pathogen resistance, and many others.

Keywords Thiols • Glutathione • S-thiolation • Dethiolation • Oxidative stress • Plant development

M. Gietler (✉) • M. Nykiel
Department of Biochemistry, Warsaw University of Life Sciences,
Nowoursynowska 159, 02-776 Warsaw, Poland
e-mail: marta_gietler@sggw.pl

M.A. Hossain et al. (eds.), *Glutathione in Plant Growth, Development, and Stress Tolerance*, https://doi.org/10.1007/978-3-319-66682-2_3

Contents

1 Introduction

Reactive oxygen species (ROS) and the cellular thiol redox state ($E_{thiol\text{-}disulfide/2thiol}$) are crucial mediators of multiple cell processes like growth, cell proliferation and differentiation, and stress tolerance. With respect to proteins, the thiol group of cysteinyl side chains is susceptible to a number of oxidative modifications, for instance, the formation of inter- or intramolecular disulfides between protein thiols or between protein thiols and low-molecular-weight thiols such as glutathione, the oxidation to sulfenic (P-SOH), sulfinic (P-SO_2H), and sulfonic (P-SO_3H) acid and S-nitrosylation (P-S-NO). Thiol-based mechanisms protect proteins against irreversible reactive oxygen species-related modifications. However, these modifications may also change the function of various proteins containing cysteines residue, protect catalytic centers of enzymes, and participate in fine-tuning of protein activity (Zaffagnini et al. 2012a; Kuźniak et al. 2013). To a great extent, the redox state of these cysteinyl residues is controlled by peroxiredoxins, sulfiredoxins, glutaredoxins, and thioredoxins systems, based on enzymes with disulfide bond cleavage activity (Jung and Thomas 1996). However, S-thiolation/dethiolation process is also dependent on ratio of small reduced thiols to mixed disulfides; therefore, it may be regulated by changes in thiol equilibrium (Klatt and Lamas 2000).

Here, we discuss the evidence supporting the view that thiol-based mechanisms are important in the control of the plant development and adaptive response to changing environment.

2 Role of Thiols in Biological Systems

Thiol compounds protect cells against oxidative stress and electrophilic xenobiotics. The most important nonprotein thiols in living organisms are the tripeptide glutathione (γ-L-Glu-L-Cys-Gly), amino acids like cysteine, and γ-glutamyl-cysteine.

Glutathione (GSH) is the most abundant low-molecular-weight (LMW) thiol which is present at millimolar concentration in nearly all eukaryotic cells. The most important biological functions of GSH are detoxification of free radicals, electrophiles xenobiotics, and maintenance of normal structure and function of proteins. Detoxification of xenobiotics in plant starts with conjugation to glutathione in the cytoplasm, followed by the transport of the conjugates into the vacuole. This pathway is used to counter the toxic effects of some herbicides and environmental pollutants and overlaps or parallels the pathway of the biosynthesis of anthocyanins. The process of connecting xenobiotics with glutathione is catalyzed by the enzymes glutathione-S-transferases (GSTs, EC 2.5.1.18). GSTs can also be involved in transport of flavonoids, detoxification of ROS, programmed cell death, signaling through flavonoids, and in the fumarate synthesis (Dixon et al. 2002).

The thiol moiety of GSH enables this molecule to act as a scavenger reacting with reactive oxygen species (ROS) and reactive nitrogen species (RNS) during oxidative stress. Shift in the GSH to GSSG (its oxidized form) ratio toward GSSG in response to increased intracellular ROS availability is part of the signaling pathway leading to programmed cell death (Kranner et al. 2010). The ratio of reduced to oxidized glutathione is constant and characteristic value for a particular development stage, as well as for specific genotype (Zagorchev et al. 2012). However, preferable parameter of glutathione redox state is the glutathione half-cell reduction potential ($E_{GSSG/2GSH}$), which takes into consideration molar concentration of GSH and GSSG, and is crucial for plant tolerance to unfavorable environmental conditions (Kranner et al. 2006).

Glutathione is also crucial for cell proliferation process. Sequestering of GSH in nucleus takes place in early stages of the cell proliferation cycle; however, it also leads to deficit of cytoplasmic glutathione pool. Glutathione present in nucleus affects transcriptions of genes related to stress tolerance, cell division, redox regulation, auxin signaling, as well as to regulation of transcription factors. It results in increase of ROS abundance in cytoplasm and changes in cell redox homeostasis in favor of the decrease of oxidative signaling and the increase of GSH synthesis. Moreover, it is possible that high level of nucleus GSH also contributes to preservation of the cellular redox environment memory in plant cells after proliferation (Diaz-Vivancos et al. 2010, 2015; Schnaubelt et al. 2015).

The oligomerization of GSH to produce phytochelatins (PCs) is also induced during stress conditions. Its main task is the chelation of potentially deleterious metal ions. The resulting complexes are sequestered in the vacuole, where the toxic effects of metals are reduced (Zagorchev et al. 2013). During the stress, not only glutathione but also other LMW thiols including γ-glutamyl-cysteine and cysteine are responsible for maintaining redox homeostasis; hence not only $E_{GSSG/2GSH}$ but the whole LMW thiol-disulfide-based redox environment described as $E_{thiol\text{-}disulfide/2thiol}$ is of paramount importance, especially upon the signaling mechanism (Birtić et al. 2011; Suzuki et al. 2012). Cysteine, thiol amino acid, is a factor reducing agent, the substrate in the biosynthesis of proteins and glutathione (GSH), and a precursor of reactive sulfur sulfone. Cys is also a sulfur donor for the synthesis of methionine, iron-sulfur clusters, vitamins, lipoic acid, and coenzyme A (Hell and Wirtz 2011). In response to oxidative stress, the level of Cys increases; however, its overaccumulation can be toxic. Due to the high reactivity of -SH group, Cys is often irreversibly oxidized to different by-products. Moreover, Cys to a certain degree can act as a chelator of heavy metals ions, forming Cys-metal ion complexes, which may trigger the Fenton reaction, leading to the production of toxic •OH radicals (Bashir et al. 2013). Cys may be involved in intra- and intermolecular disulfide bond formation, which is of paramount importance during protein folding, or in their reduced form they can act as reactive group in catalytic activity sites. Alternation in intramolecular disulfide bond formation may lead to protein misfolding, which in turn may cause loss of their biological activity and formation of aggregates (Trivedi et al. 2009). Also between two free Cys amino acids disulfide bond can be formed. It was proven that the redox potential of the CysS/2Cys couple is a marker associated with human diseases, a part of antioxidative system in parasites, and probably a component of stress response in plants (Zagorchev et al. 2013). Cysteine residues are also located on the surface in various types of proteins such as antibodies, receptors, hormones, or enzymes like thioredoxins (Trx) and glutaredoxins (Grx). Both Trx and Grx are involved in protection of protein against further oxidation and regulation of their activity. Thiol groups often play a catalytic or structural role, and their modifications have a major impact on the protein functions (Bykova and Rampitsch 2013) and metabolic regulation (Møller et al. 2011).

3 Oxidation Reactions of Protein -SH Groups

Cysteine is a redox-sensitive amino acid residue in proteins and is mostly rapidly oxidized. The reversible oxidation of the SH groups of Cys residues in proteins, or the formation of thiyl radical, sulfenic acid (R-SOH), and S-nitrosothiol (SNT), is often an intermediate step in the formation of a mixed disulfides. In contrast, the oxidation of the -SH to sulfinic acid ($R\text{-}SO_2H$) and sulfonic acid ($R\text{-}SO_3H$) is an irreversible process involved in loss of the biological function of the proteins. The reversible oxidation of protein thiols is one of the control mechanisms for plant growth and development.

3.1 Generation of Thiyl Radicals

Cysteine thiyl radicals (CysS•) are important intermediates in the one-electron oxidation of cysteine (Cys) and one-electron reduction of protein disulfide bonds (reactions I and II):

I. CysSH + OX —> CysS• + $OX^{\bullet -}$+H
II. CysSSCys + e^- + H^+ —> CysS• + CysSH

CysS• are highly reactive; therefore they can easily interact with other thiols with thiyls radicals and form dimmers, or react with oxygen and its derivatives, which leads to the formation of other reactive molecules (Obinger et al. 1997).

The CysS• show low sensitivity to oxygen; however, they are highly reactive toward a large range of double bonds. CysS• belongs to the reactive sulfur species (RSS) (Giles et al. 2001, 2002; Jacob and Anwar 2008) and are engaged in several physiological processes. It was proven that sulfur-centered molecules are involved in the regulation of ethylene-induced stomatal closure (Hou et al. 2013) and reaction to salinity stress, and together with hydrogen peroxide (H_2O_2) they are taking part in regulation of the plasma membrane Na^+/H^+ antiporter system (Li et al. 2014). Thiyl radical formation in vivo is carried out by various mechanisms. The most important is the one-electron oxidation of thiols. The process can be carried out by several distinct oxidants, i.e., hydroxyl (•OH), carbonate ($\bullet CO_3^-$), nitrogen dioxide ($\bullet NO_2$), peroxyl and phenoxyl radicals, oxo-metal complexes (compound I and II of some haem peroxidases), and superoxide radical ($O_2^{\bullet -}$) (Trujillo et al. 2016). Formation of thiyl radicals by reaction with hydroxyl can be highly selective, due to the fact that cysteine residues are able to bind Fe^{2+} and Cu^+. Metal-binded cysteine interaction with H_2O_2 leads to •OH creation in Fenton reaction. Newly created hydroxyl reacts with cysteine at the metal binding site, which leads to controlled thiyls formation (Giles et al. 2003).

The second mechanism by which thiyl radicals of cysteine or related compounds are formed in vivo is the reaction with haem peroxidases. Haem peroxidases (EC 1.11.1.7) are known to have low specificity to the substrates. They can catalyze the oxidation of various organic and inorganic compounds such as phenols, arylamines, halides, and thiols and with most combinations of peroxidases and substrates such oxidations have been found to proceed univalently. It was proven that horseradish peroxidase, lactoperoxidase, and myeloperoxidase (Svensson 1993) promoted thiol oxidation by H_2O_2-independent reaction (Obinger et al. 1997).

The third interesting pathway of thiyl radical formulation is reaction of phenoxyl radicals with intra- or intermolecular electron transfer mechanisms. During this process, initial tyrosyl radical is repaired by a nearby cysteine residue, which leads to CysS• formation (Bhattacharjee et al. 2007; Zhang et al. 2008). Moreover, there is also a pathway during which thiols can react with free radicals derived from peroxynitrite. This process leads to the formation of thiyls and oxygen consumption (Quijano et al. 1997).

3.2 S-Nitrosylation Reactions

S-nitrosylation is a posttranslational modification of proteins involving the covalent attachment of NO (or higher oxide) to cysteine residues and leading to formation of S-nitrosothiol (-SNO). S-nitrosylation can result from three main pathways: oxidative or radical-mediated pathway and metal catalyzed S-nitrosothiols (RSNO) formation.

In the oxidative pathway, N_2O_3 is generated from protonated nitrite at very low pH or by the auto-oxidation of NO• with the presence of O_2. Then N_2O_3 donates NO group to nucleophilic protein thiols (RS^-) to form RSNO:

I. $2NO_2 + 2H^+ \rightarrow 2HNO_2 \rightarrow N_2O_3 + H_2O$
$N_2O_3 + RS^- \rightarrow RSNO + NO_2^-$
II. $\bullet NO + ½ O_2 \rightarrow \bullet NO_2 + \bullet NO \rightarrow N_2O_3$
$N_2O_3 + RS^- \rightarrow RSNO + NO_2^-$

During radical-mediated pathway, thiyl radicals can directly interact with NO• and form RSNO, or in the presence of NO•, thiolate anions (RS^-) can act as proton donors for peroxynitrite which results in the formation of RSNO. Moreover, NO• can form an intermediate radical with thiols (RSH). The intermediate can be further oxidized to RSNO:

I. $\bullet NO + \bullet RS \rightarrow RSNO$
II. $\bullet NO + O_2^- + H^+ \rightarrow OONOH + RS^- \rightarrow RSNO + OOH^-$
III. $\bullet NO + RSH \rightarrow RSN^{\bullet}\text{-}OH + O_2 \rightarrow RSNO + H^+ + O_2^-$

During metal catalyzed RSNO formation, iron can chelate NO^+ and GS^- and act as a NO donor toward RS^- and format RSNO with the release of glutathione (Kuruthukulangarakoola and Lindermayr 2013).

$$Fe^+\left(GS^-\right)_2\left(NO^+\right)_2 + 2\,RS^- \rightarrow 2RSNO + 2GS^- + Fe^{2+}$$

Moreover, not only proteins but also glutathione can be S-nitrosylated in the reaction of N_2O_3 with GSH or NO with glutathione free radical (Broniowska et al. 2013). It was proven that S-nitrosoglutathione (GSNO) content is directly correlated with the level of protein S-nitrosothiols (Corpas et al. 2013; Lindermayr et al. 2005), and it acts as NO reservoir (Leterrier et al. 2012). For years, it was believed that the control of S-nitrosylation process was dependent on the biosynthesis of NO (Stamler et al. 2001). However, more recent studies have shown that S-nitrosylation can also be controlled by the adjustment of the low-molecular-weight SNO by activity of nitrosoglutathione S-reductase (GSNOR) (Leterrier et al. 2011). Studies carried out by Feechan et al. (2005) showed that in *Arabidopsis thaliana* AtGSNOR1 gene is present. It is responsible for the synthesis of GSNOR enzyme, and its sequence is corresponding to the bacterial and yeast GSNOR. Although GSNOR does not reduce S-nitrosylated proteins directly, by limiting the content of GSNO it also regulates transnitrosylation process. GSNOR is not the only enzyme controlling S-nitrosylation. It was found that GSH-dependent formaldehyde dehydrogenase displays a strong reducing activity toward the GSNO (Diaz et al. 2003). However, recent research has focused on the activity of alcohol dehydrogenase belonging to class III of dehydrogenases (alcohol dehydrogenase 3; ADH3), which is one of the main enzymes of GNSOR group. The major ADH3 activity is the formaldehyde detoxification, but it also has ability to catalyze NADH-dependent reduction of GSSG to the GSNO and NH_3, which can in turn be regarded as a mechanism for denitrification (Lee et al. 2008; Staab et al. 2008). A strong resemblance between plant (obtained from tomato *Solanum lycopersicum*) and human ADH3 was observed. The main, but important, difference between these enzymes was the binding pocket anions in the active center of the enzyme, which resulted in a reduction in affinity for the carboxyl groups of hydroxyfatty acids (Kubienová et al. 2013).

3.2.1 The Significance of S-Nitrosylated Protein

S-nitrosylation affects both the activity and the conformation of the modified protein. Due to the reversible nature and high substrate specificity (Palmieri et al. 2010), it is suggested that S-nitrosylation may be involved in short-range signal transduction (Martinez-Ruiz et al. 2013) and multilevel plant stress response (Janssen-Heininger et al. 2008). Changes of protein activity are result of S-nitrosylation-induced conformational changes, oligomerization by disulfide bond formation between monomers, cofactor binding inhibition, and modification of catalytic Cys residues (Lamotte et al. 2014).

Research carried out on the *A. thaliana* protein GapC1, a cytoplasmic isoform of glyceraldehyde-3-phosphate dehydrogenase, suggests that S-nitrosylation may be controlled by the ratio GSH/GSNO. It is an example of a modification alternative to glutathionylation, because the same cysteine residue (Cys149) is modified. The treatment of the extracted GapC1 proteins with GSNO proved that this enzyme was in 90% S-nitrosylated in this conditions and its activity was completely inhibited (Zaffagnini et al. 2013). However, S-nitrosylation not always plays an inhibitory role. For example, S-nitrosylation of ascorbate peroxidase (APX) enhances its activity in *Pisum sativum* (Begara-Morales et al. 2014). On the other hand, APX S-nitrosylation may promote its degradation through the ubiquitin-proteasome pathway in *Nicotiana tabacum* (de Pinto et al. 2013).

It also has been shown that the *A. thaliana* NPR1 transcription factor (activating TGA1—a factor binding motif TGACG), a key regulator of the salicylic acid (SA) signaling, can be activated by S-nitrosylation (Lindermayr et al. 2010). On the other hand, this modification results in a change of protein conformation and in consequence NPR1 inactivation by the oligomer formation in the cytoplasm (Tada et al. 2008). In *A. thaliana* S-nitrosylation of SABP3 results in inhibition of its ability to activate carbonic anhydrase, an important signal transduction factor associated with a SA-dependent pathway (Wang et al. 2009a). This mechanism could be regarded as negative feedback modulating the immune system of plants or the phenomenon of abolishing pathogen-induced hypersensitivity response (Gaupels et al. 2011).

S-nitrosylation of RuBisCO occurs under stress conditions. It results in the removal of RuBisCO carboxylase activity, which proves existence of S-nitrosylation-mediated pathway of sugar metabolism regulation in plants (Abat and Deswal 2009). There is probably S-nitrosylation-based fine regulatory mechanism of Calvin-Benson cycle, because, besides previously described RuBisCO and glyceraldehyde-3-phosphate dehydrogenase, other enzymes such as triosephosphate isomerase, phosphoglycerate kinase, fructose-1,6-bisphosphate aldolase, transketolase, sedoheptulose-1,7-bisphosphatase, ribulose-5-phosphate isomerase, ribose 5-phosphate kinase, malate dehydrogenase and RuBisCO activase are S-nitrosylated. However, its effect on enzyme activity remains unknown (Michelet et al. 2013). Detailed information related to the S-nitrosylation of proteins are attached to Tables 3.1 and 3.2.

Under physiological conditions S-nitrosylation is not a random cysteine residue modification. Among the factors determining the susceptibility of a particular cysteine residue to this posttranslational modification are local pH, the state

Table 3.1 Proteins undergoing S-nitrosylation in plant upon treatment with S-nitrosylation factors

S-nitrosylated proteins in plant under treatment with S-nitrosylation factors		
Protein	Organism	Literature
Glycolysis, gluconeogenesis and the pentose phosphate pathway (energy metabolism of sugars)		
Glyceraldehyde 3-phosphate dehydrogenase C	*Arabidopsis thaliana*	Zaffagnini et al. (2013)
Glyceraldehyde 3-phosphate dehydrogenase A	*Chlamydomonas reinhardtii*	Morisse et al. (2014)
Glycine decarboxylase	*Arabidopsis thaliana*	Palmieri et al. (2010)
Phosphoglycerate kinase	*Solanum tuberosum*	Kato et al. (2013)
Fructose-1,6-bisphosphatase	*Chlamydomonas reinhardtii*	Morisse et al. (2014)
Photosynthesis and photorespiration		
Ribulose-1,5-bisphosphate carboxylase/oxygenase	*Chlamydomonas reinhardtii*	Morisse et al. (2014)
Fructose-1,6-bisphosphate aldolase	*Chlamydomonas reinhardtii*	Morisse et al. (2014)
	Solanum tuberosum	Kato et al. (2013)
Aconitase	*Arabidopsis thaliana*	Lindermayr et al. (2005)
Phosphoribulokinase	*Chlamydomonas reinhardtii*	Morisse et al. (2014)
NADP-dependent malate dehydrogenase	*Chlamydomonas reinhardtii*	Morisse et al. (2014)
Replication and transcription		
MYB domain protein 2	*Arabidopsis thaliana*	Serpa et al. (2007)
Proteolysis and posttranslational modifications of proteins		
TGA1 binding factor	*Arabidopsis thaliana*	Lindermayr et al. (2010)
Non-expressor of PR1 (NPR1)	*Arabidopsis thaliana*	Tada et al. (2008)
Prometacaspase 9	*Arabidopsis thaliana*	Belenghi et al. (2007)
Metacaspase AtMC9	*Arabidopsis thaliana*	Belenghi et al. (2007)
Amino acid metabolism		
S-adenosylmethionine synthetase-1	*Arabidopsis thaliana*	Lindermayr et al. (2006)
Methionine adenosyltransferase 1	*Arabidopsis thaliana*	Lindermayr et al. (2006)
Chaperons		
HSP 90	*Arabidopsis thaliana*	Lindermayr et al. (2005)
Transport and transport ATPases		
Transport inhibitor response 1 (TIR1)	*Arabidopsis thaliana*	Terrile et al. (2012)
ATP synthase	*Arabidopsis thaliana*	Lindermayr et al. (2005)
Cytoskeleton		
Tubulin α	*Arabidopsis thaliana*	Lindermayr et al. (2005)
Tubulin β	*Arabidopsis thaliana*	Lindermayr et al. (2005)
Other or unknown		
Salicylic acid-binding protein 3	*Arabidopsis thaliana*	Wang et al. (2009b)

Table 3.2 Proteins undergoing S-nitrosylation in plant upon stress treatment or under physiological conditions

S-nitrosylated protein in plant under stress treatment or physiological conditions			
Protein	Organism	Literature	Stress
Glycolysis, gluconeogenesis, and the pentose phosphate pathway (energy metabolism of sugars)			
Glyceraldehyde 3-phosphate dehydrogenase B	*Citrus aurantium*	Tanou et al. (2012)	Salinity
Triosephosphate isomerase	*Citrus aurantium*	Tanou et al. (2012)	Salinity
Transketolase	*Arabidopsis thaliana*	Fares et al. (2011)	–
Sedoheptulose-1 7-bisphosphatase	*Citrus aurantium*	Tanou et al. (2012)	Salinity
Ribose-5-phosphate isomerase	*Ozyra sativa*	Lin et al. (2012)	–
Alcohol dehydrogenase	*Arabidopsis thaliana*	Fares et al. (2011)	–
Phosphoglucomutase	*Triticum aestivum*	Gietler et al. (2016)	Dehydration
Beta-amylase	*Triticum aestivum*	Gietler et al. (2016)	Dehydration
UDP-glucose dehydrogenase	*Arabidopsis thaliana*	Fares et al. (2011)	–
Photosynthesis and photorespiration			
Ribulose-1,5-bisphosphate carboxylase/oxygenase	*Brassica juncea*	Abat and Deswal (2009)	Low temperature
Glycolate oxidase	*Pisum sativum*	Ortega-Galisteo et al. (2012)	Cadmium and herbicide
Photosystem I A2 apoprotein	*Arabidopsis thaliana*	Fares et al. (2011)	–
Photosystem II-evolving complex 33	*Citrus aurantium*	Tanou et al. (2009)	Salinity
RuBisCO activase	*Ozyra sativa*	Lin et al. (2012)	–
Replication and transcription			
N5′-nucleotidase	*Triticum aestivum*	Gietler et al. (2016)	Dehydration
ATP-dependent RNA helicase DDX11	*Triticum aestivum*	Gietler et al. (2016)	Dehydration
Serine/arginine-rich splicing factor RS31	*Triticum aestivum*	Gietler et al. (2016)	Dehydration
Maturase K	*Triticum aestivum*	Gietler et al. (2016)	Dehydration
Glycine-rich RNA-binding protein blt801	*Triticum aestivum*	Gietler et al. (2016)	Dehydration
DNA topoisomerase II	*Kalanchoe pinnata*	Abat and Deswal (2008)	–

(continued)

Table 3.2 (continued)

S-nitrosylated protein in plant under stress treatment or physiological conditions			
Protein	Organism	Literature	Stress
Nucleotide metabolism			
Uridylate kinase	*Triticum aestivum*	Gietler et al. (2016)	Dehydration
Translation			
Nonsymbiotic hemoglobin AHb1	*Arabidopsis thaliana*	Perazzolli et al. (2004)	Hypoxic stress
EF1-a elongation factor	*Arabidopsis thaliana*	Fares et al. (2011)	–
50S ribosomal protein L27	*Triticum aestivum*	Gietler et al. (2016)	Dehydration
50S ribosomal protein L16	*Triticum aestivum*	Gietler et al. (2016)	Dehydration
EF-Tu elongation factor precursor	*Arabidopsis thaliana*	Romero-Puertas et al. (2008)	*Pseudomonas syringae*
Proteolysis and posttranslational modifications of proteins			
ATP-dependent Clp protease	*Triticum aestivum*	Gietler et al. (2016)	Dehydration
Calcium-dependent protein kinase isoform 11	*Triticum aestivum*	Gietler et al. (2016)	Dehydration
26S protease regulatory subunit 7-like isoform X1	*Triticum aestivum*	Gietler et al. (2016)	Dehydration
Amino acid metabolism			
Glycine dehydrogenase	*Arabidopsis thaliana*	Perazzolli et al. (2004)	Hypoxic stress
3-isopropylmalate dehydratase	*Arabidopsis thaliana*	Fares et al. (2011)	–
Chaperons			
Chaperone protein dnaJ 11	*Triticum aestivum*	Gietler et al. (2016)	Dehydration
Calnexin 1	*Arabidopsis thaliana*	Fares et al. (2011)	–
Redox regulation			
Catalase	*Pisum sativum*	Ortega-Galisteo et al. (2012)	Cadmium and herbicide
NADPH oxidase	*Arabidopsis thaliana*	Yun et al. (2011)	*Pseudomonas syringae*
Phytochelatin	*Arabidopsis thaliana*	Elviri et al. (2010)	Cadmium
Germin-like protein	*Citrus aurantium*	Tanou et al. (2009)	Salinity
Ascorbate peroxidase	*Antiaris toxicaria*	Bai et al. (2011)	Desiccation

(continued)

Table 3.2 (continued)

S-nitrosylated protein in plant under stress treatment or physiological conditions			
Protein	Organism	Literature	Stress
Peroxiredoxin II	*Arabidopsis thaliana*	Romero-Puertas et al. (2008)	*Pseudomonas syringae*
Superoxide dismutase	*Citrus aurantium*	Tanou et al. (2009)	Salinity
Cysteine protease Rd21	*Arabidopsis thaliana*	Fares et al. (2011)	–
Cysteine protease inhibitor	*Citrus aurantium*	Tanou et al. (2009)	Salinity
Dehydroascorbate reductase	*Antiaris toxicaria*	Bai et al. (2011)	Desiccation
Thioredoxin-like protein	*Oryza sativa*	Lin et al. (2012)	–
Probable trehalose-phosphate phosphatase 9	*Triticum aestivum*	Gietler et al. (2016)	Dehydration
Glutaredoxin-C10	*Triticum aestivum*	Gietler et al. (2016)	Dehydration
Thioredoxin M2	*Triticum aestivum*	Gietler et al. (2016)	Dehydration
Pathogenesis-related protein P2	*Triticum aestivum*	Gietler et al. (2016)	Dehydration
Leaf-specific thionin BTH6	*Triticum aestivum*	Gietler et al. (2016)	Dehydration
Leaf-specific thionin DB4	*Triticum aestivum*	Gietler et al. (2016)	Dehydration
Cyanate hydratase	*Triticum aestivum*	Gietler et al. (2016)	Dehydration
Bifunctional monodehydroascorbate reductase and carbonic anhydrase nectarin-3	*Triticum aestivum*	Gietler et al. (2016)	Dehydration
Hsp70-Hsp90 organizing protein	*Triticum aestivum*	Gietler et al. (2016)	Dehydration
Glutathione reductase	*Antiaris toxicaria*	Bai et al. (2011)	Desiccation
Transport and transport ATPases			
ATPase subunit	*Arabidopsis thaliana*	Fares et al. (2011)	–
H^+-ATPase	*Citrus aurantium*	Tanou et al. (2009)	Salinity
Phosphate transporter 3;1	*Arabidopsis thaliana*	Fares et al. (2011)	–
Adenylate translocator	*Arabidopsis thaliana*	Fares et al. (2011)	–
Mitochondrial uncoupling protein 2	*Triticum aestivum*	Gietler et al. (2016)	Dehydration

(continued)

Table 3.2 (continued)

S-nitrosylated protein in plant under stress treatment or physiological conditions			
Protein	Organism	Literature	Stress
Succinate dehydrogenase [ubiquinone] iron-sulfur subunit 1	*Triticum aestivum*	Gietler et al. (2016)	Dehydration
ATP synthase	*Triticum aestivum*	Gietler et al. (2016)	Dehydration
Voltage-dependent anion channel	*Citrus aurantium*	Tanou et al. (2009)	Salinity
Cytoskeleton			
Rho-like GTP-binding protein	*Triticum aestivum*	Gietler et al. (2016)	Dehydration
Actin-related protein 2/3 complex	*Triticum aestivum*	Gietler et al. (2016)	Dehydration
Actin	*Arabidopsis thaliana*	Rodriguez-Serrano et al. (2014)	Herbicide
Lipid metabolism			
Phospholipase A1-II delta	*Triticum aestivum*	Gietler et al. (2016)	Dehydration
Lipoxygenase 1	*Triticum aestivum*	Gietler et al. (2016)	Dehydration
Other or unknown			
CP12	*Ozyra sativa*	Lin et al. (2012)	–
Oxygen-evolving enhancer proteins (OEE1)	*Arabidopsis thaliana*	Romero-Puertas et al. (2008)	*Pseudomonas syringae*
Oxygen-evolving enhancer proteins (OEE2)	*Triticum aestivum*	Gietler et al. (2016)	Dehydration
Inorganic pyrophosphatase	*Arabidopsis thaliana*	Fares et al. (2011)	–
FGGY carbohydrate kinase domain-containing protein	*Triticum aestivum*	Gietler et al. (2016)	Dehydration
Tetraacyldisaccharide 4-kinase family protein	*Triticum aestivum*	Gietler et al. (2016)	Dehydration
Protein BPS1	*Triticum aestivum*	Gietler et al. (2016)	Dehydration
R-linalool synthase QH5	*Triticum aestivum*	Gietler et al. (2016)	Dehydration
Rac-like GTP-binding protein ARAC2	*Triticum aestivum*	Gietler et al. (2016)	Dehydration
Kinesin-like protein KLP1	*Triticum aestivum*	Gietler et al. (2016)	Dehydration
Phosphatidylinositol/ceramide inositolphosphotransferase 3	*Triticum aestivum*	Gietler et al. (2016)	Dehydration
Alpha-1,4-glucan-protein synthase [UDP-forming]	*Triticum aestivum*	Gietler et al. (2016)	Dehydration

of redox potential, the presence of metal ions (Hess et al. 2001)—particularly calcium (Lai et al. 2001)—and electrostatic interactions affecting pKa thiol groups, as well as the hydrophobicity (Hess et al. 2005). However, recent studies suggest that protein S-nitrosylation of specific cysteine residue is not determined only by the immediate environment but by the theme of acid-alkaline amino acid residues situated about 8 Å away from a modified SH groups. Arrangement of acidic and alkaline residues influences the distribution of the charge over the whole surface of the protein and thus its ability to interact with other molecules (Marino and Gladyshev 2010).

Research carried out by Gietler et al. (2016) on wheat seedlings with different drought tolerance proved that profiles of S-nitrosylated proteins were altered in response to dehydration. Protein profiles of stressed seedlings were, however, very similar regardless of their ability to maintain water deficit, which may lead to a conclusion that both types of seedlings respond similarly to dehydration. In tolerant wheat seedlings, protein which changed their abundance at least twofold were those involved mainly in stress response and signal transduction (29%), storage metabolism (16%), and nucleic acid and protein metabolism (25%). In sensitive wheat seedlings upon severe dehydration, the abundance of protein involved in stress response and signal transduction (35%), energy metabolism (20%), and nucleic acid and protein metabolism (30%) was changed (Gietler et al. 2016).

In tolerant wheat seedlings, the abundance of approximately 75% of S-nitrosylated proteins decreased upon dehydration. Changes in S-nitrosylation concerned mainly the proteins related to nucleic acid, protein metabolism, and signal transduction (Fig. 3.1). The most important seem to be changes in serine/arginine-rich splicing factor, a part of the spliceosome (Reddy 2001), and 5′-nucleotidase, a purine and cytokinin metabolism enzyme (Chen and Kristopeit 1981) which may influence gene expression during abiotic stress (Gietler et al. 2016). Moreover, in tolerant seedlings there was an increase of S-nitrosylated oxygen-evolving enhancer protein 2 (OEE2) and calcium-dependent protein kinase isoform 11 (CDPK 11) (Gietler et al. 2016). OEE2 is involved in photoinhibition which abundance increases in response to stress (Pérez-Bueno et al. 2004), however, S-nitrosylation of this protein may affect its activity. CDPK initiate calcium-dependent signaling processes in response to salinity, drought, and cold and their overexpression enhanced plant stress tolerance (Asano et al. 2012).

3.3 Irreversible Oxidation of -SH Groups

Protein-SH groups can undergo both reversible and irreversible modifications. The bridge between those two modifications is sulfenylation (SOH). Sulfenic acid residue is created at the lowest oxidation state induced by the ROS, and it is considered as a critical intermediate in redox-signaling (Roos and Messens 2011). It is highly reactive and unstable modification therefore SOH can react with other thiols and create mixed disulfides or can be further oxidized to sulfinic (SO_2H) and then

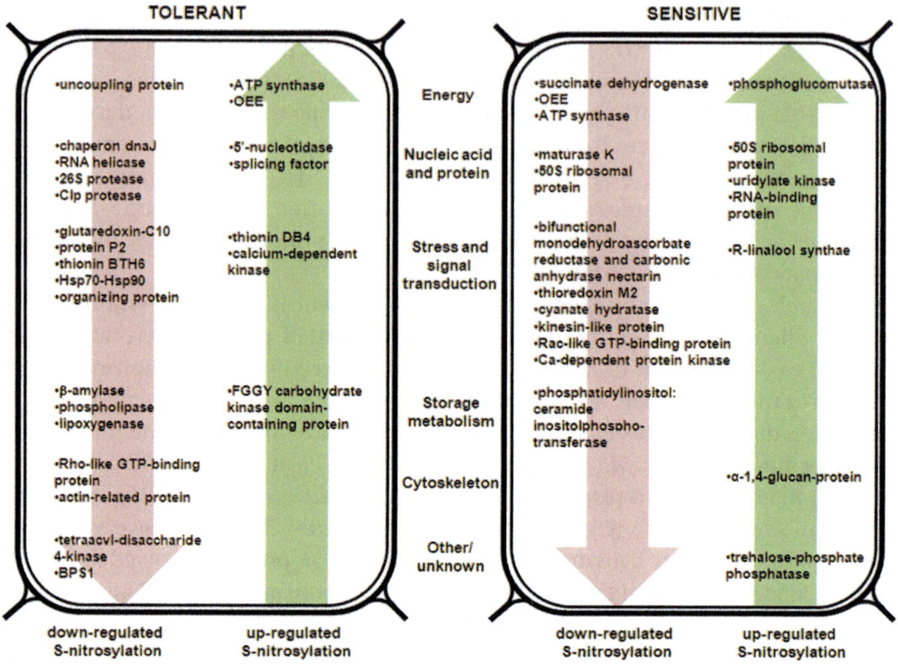

Fig. 3.1 S-nitrosylated up- and downregulated proteins in wheat seedlings (*Triticum aestivum* L.) tolerant and sensitive to dehydration

sulfonic (SO_3H) acids. These modifications have higher oxidation state (+2 and +4, respectively) and are considered irreversible (Chung et al. 2013).

$$\begin{array}{llll} \text{Protein} - \text{SH} \leftrightarrow & \text{Protein} - \text{SOH} \rightarrow & \text{Protein} - \text{SO}_2\text{H} \rightarrow & \text{Protein} - \text{SO}_3\text{H} \\ & \text{Sulfenic acid} & \text{Sulfinic acid} & \text{Sulfonic acid} \end{array}$$

However, there are some papers indicating the possibility of sulfinic acids reduction by specific enzymes. This ATP-dependent mechanism occurs in Eukaryota, where cysteine sulfinic acid residue in peroxiredoxins can be reduced by sulfiredoxin. Sulfiredoxin is highly specific and does not act as a reducing agent for any other molecule than 2-Cys peroxiredoxin (Biteau et al. 2003; Woo et al. 2005; Lowther and Haynes 2011) and mitochondrial PrxIIF (Iglesias-Baena et al. 2011). Nevertheless, apart from this special case there are no evidences of possibility of reduction of cysteine residues after their oxidation to +2 or +4 state; thus oxidation to sulfinic and sulfonic acid is still considered an irreversible process that causes protein damage (Biteau et al. 2003; Woo et al. 2005; Rey et al. 2007; Lowther and Haynes 2011). Both sulfinylation and sulfonylation lead to alternation of protein activity and its increased susceptibility to formation of aggregates, as well as to degradation (Roos and Messens 2011).

4 Protein Modifications Based on S-Thiolation Pathway

The formation and reduction of disulfide bonds are largely dependent on the availability of donors and electron acceptors, which determines the potential redox environment. So, changes in the ratio of thiols/disulfides, or modifications in the redox potential in the cell or plasma, have a significant impact on the structure and function of proteins. Protein thiols in reactions with LMW thiols may form mixed disulfides in process called the S-thiolation that can occur via two different mechanisms: (1) the thiol-disulfide exchange and (2) as a result of the reaction of -SH groups of reversibly oxidized proteins with low-molecular-weight thiol compounds. Protein S-thiolation is now considered a redox posttranslational mechanism protecting protein thiols from irreversible oxidation. Thus, S-thiolation can influence protein function which leads to regulation of the signaling and metabolic pathways (Kuźniak et al. 2013).

4.1 Thiol-Disulfide Exchange

Thiol-disulfide exchange can be performed by direct substitution or by thiol oxidation (Nagy 2013). In the pathway of direct substitution there is a one-step reaction which consists of nucleophile attack of thiol (R_1S^-) on central sulfur of the disulfide (R_2SSR_3) resulting in a trisulfide-like transition state structure. This structure is disintegrated with release of R_2SH and R_1SSR_3 (Bach et al. 2008, Nagy 2013):

$$R_1S^- + R_2SSR_3 \rightarrow R_2\overset{\overset{\displaystyle R_1S}{|}}{S}SR_3 \rightarrow R_2S^- + R_1SSR_3$$

There are two types of thiol-disulfide exchange by oxidation, by two-electron or one-electron oxidation pathways. In two-electron pathway, mostly sulfenic acid (RSOH) or sulfenyl (RSX, e.g., RSCl, RSBr, or RSI) reacts with thiols to form disulfide. Sulfenic acid, which is intermediate for this mechanism, is formed by reaction with oxidants, hypohalous acids (HOX) or hypothiocyanite ($OSCN^-$) (Skaff et al. 2009; Nagy 2013). RSX may be formed by reaction of thiols with appropriate HOX. Moreover, before disulfide bond creation, RSX can be transformed to sulfenic acid by hydrolysis; therefore those two sub-pathways are partially connected (Nagy and Winterbourn 2010; Nagy 2013):

I. $R_1SX + R_2SH \rightarrow R_1SSR_2 + HX$
II. $R_1SOH + R_2SH \rightarrow R_1SSR_2 + H_2O$

During one-electron oxidation mechanism, thiyl radicals are generated from Cys residues (CysS•), as CysS• radicals with unpaired electron can react with each other to form a disulfide:

I. $R_1S\bullet + R_2S\bullet \rightarrow R_1SSR_2$
II. $R_1S\bullet + R_2SH \rightarrow R_1SSR_2\bullet^- + H^+$
$R_1SSR_2\bullet + O_2 \rightarrow R_1SSR_2 + O_2\bullet^-$

However, as thiyl radicals are highly reactive, they can also attack thiols, which are more common in physiological conditions, and create disulfide radical anion. Disulfide radical anion reacts with oxygen and forms disulfide and superoxide in diffusion-controlled reaction (Winterbourn and Metodiewa 1999; Nagy and Winterbourn 2010; Nagy 2013).

4.2 Reaction of the Reversible Oxidized Protein -SH Groups with Low-Molecular-Weight Thiol Compounds

It was observed that the S-thiolation of proteins may be initiated by reactive oxygen species (ROS). This phenomenon occurs without apparent increase in the level of GSSG in cells. In this mechanism, the first oxidant reacts directly with the -SH groups of proteins to temporarily form reversibly oxidized cysteine residues, or thiyl radicals proteins, S-nitrosothiols, or sulfenated acids. Then these reactive forms of thiols can react with LMW thiols, leading to the formation of mixed disulfides. Among nonprotein LMW thiols the best described beside glutathione are cysteine, thiocysteine, cystathionine, homocysteine, glutathione disulfide, cysteinyl-glycine, γ-glutamylcysteine, homoglutathione, hydroxymethyl-glutathione, lipoic acid, reduced lipoic acid, cysteamine, phytochelatins, 1-p-menthene-8-thiol, and methanethiol (Pivato et al. 2014).

4.3 Participation of Glutathione in the Reactions of Protein S-Thiolation

S-thiolation with the formation of disulfide bond between glutathione and cysteine residue of protein is called S-glutathionylation. S-glutathionylation can occur by a number of different mainly non-enzymatic pathways, but participation of glutaredoxins (Grx) in the process cannot be excluded. Spontaneous S-glutathionylation can occur only in limited circumstances. During the oxidative stress, the ratio of reduced to oxidized glutathione (GSH/GSSG) decreases. The reaction proceeds according to the equation (where P means protein):

$$P - SH + GSSG \leftrightarrow P - SSG + GSH$$

The equilibrium of the reaction is defined as K_{OX}, i.e., equivalent of S-glutathionylation value to 50% of proteins targeted by this reaction (Zaffagnini et al. 2012a):

$$K_{OX} = \frac{[P-SSG]\times[GSH]}{[P-SH]\times[GSSG]} = \frac{[P-SSG]}{[P-SH]} \times \frac{[GSH]}{[GSSG]}$$

Spontaneous S-glutathionylation by exchange of the thiol groups occurs only for proteins with constant K_{OX} in the range corresponding to the ratio of GSH/GSSG (Gallogly et al. 2009). However, such a mechanism of S-glutathionylation is a slow process, and therefore it is believed that it is probably strongly limited by kinetics of reaction (Zaffagnini et al. 2012a).

S-glutathionylation mechanisms of great in vivo importance are based on prior modifications of cysteine residues, such as the oxidation to sulfenic acids, the formation of cysteinyl free radicals, and S-nitrosylation (Zaffagnini et al. 2012b). Protein thiol groups can react with ROS leading to formation of sulfenic acids. However, sulfenic acid is an unstable intermediate; therefore it can be further converted into irreversible sulfinates and sulfonates (as described in 4.3 section) or react with LMW thiols, for example, glutathione, to form disulfides including glutathione-protein complexes (Reddie and Carroll 2008). Upon oxidative stress the number of S-glutathionylated proteins and their abundance tend to increase (Hill and Bhatnagar 2012). Therefore it was proposed that S-thiolation of target proteins can be a mechanism of sensing amplified level of ROS, especially H_2O_2, which is involved in oxidation of thiol groups (Kuźniak et al. 2013).

5 Enzymatic Regulation of Protein Dethiolation

Enzymatic dethiolation of protein can be catalyzed by several enzymes with disulfide bond cleavage activity (Jung and Thomas 1996). The specificity and efficiency of enzyme activity are highly dependent on the location of disulfide bond in molecule structure (Nordstrand et al. 1999). S-thiolation/dethiolation process is highly dependent on ratio of reduced LMW thiols (i.e., glutathione) to the ratio of mixed disulfides. Therefore enzymes may regulate the process of dethiolation not only by direct reaction with mixed disulfides but also by shifting the equilibrium of thiol ratio. Enzymatic control of thiolation and dethiolation is a highly specific process, which regulates protein function by their posttranslational modification (Klatt and Lamas 2000).

5.1 Peroxiredoxins

Peroxiredoxins (Prx; EC 1.11.1.15) are crucial part of antioxidative system and dithiol/disulfide redox-based regulation system. Prx are present in majority of cell compartments, e.g., nucleus, cytosol, mitochondrion, and chloroplasts (Netto et al. 2007). In plants there are four isoforms of Prx 2-CysPrx, 1-CysPrx, PrxQ, and type

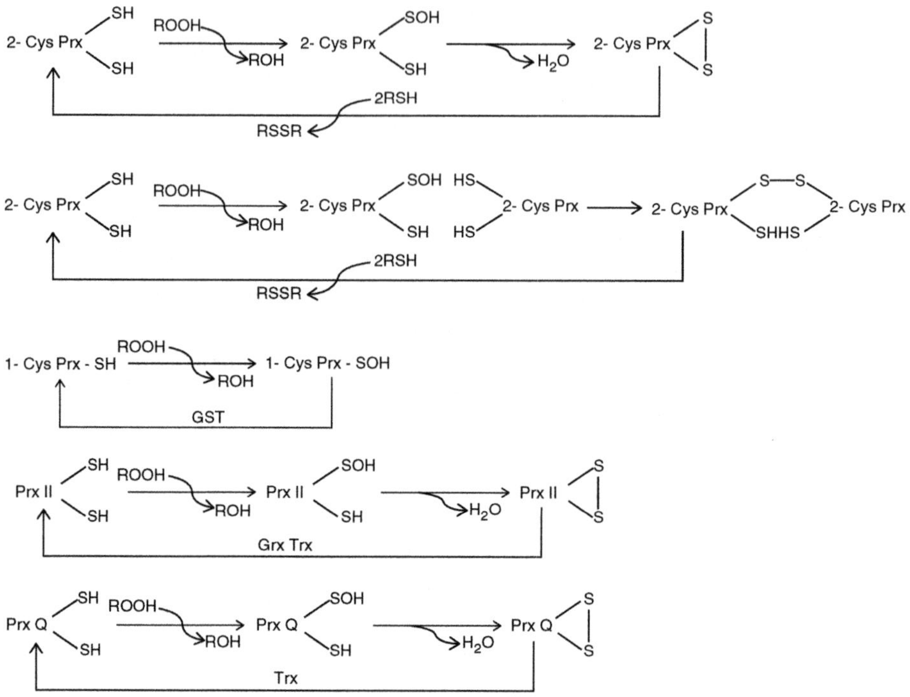

Fig. 3.2 Mechanisms of peroxidative reduction, resolving, and regeneration of different types of peroxiredoxins

II Prx. During their catalytic cycle, Prx are undergoing peroxidative reduction, resolving, and regeneration (Fig. 3.2). In second (resolving) state disulfide bond can be created between two Prx molecules in 2-CysPrx or between two thiol groups in the same molecule (PrxQ and type II Prx) (Bhatt and Tripathi 2011). The most common electron donors for Prx in their regeneration state are thioredoxins, glutaredoxins, cyclophilins, glutathione, and ascorbic acid (Dietz 2011). 2-CysPrx can be inactivated during catalysis by overoxidation to sulfinic acid form (Cys-SO_2H). However, in chloroplast, this normally irreversible modification can by undone by ATP-dependent reduction by sulfiredoxin (Srx), which was proven by studies on *A. thaliana* (Rey et al. 2007; Iglesias-Baena et al. 2010).

Peroxiredoxins, besides their primary function of ROS detoxification, can also act as chaperons. During their chaperon function, Prx form high-molecular-weight protein complexes, which lack peroxidase activity, but they prevent misfolding and aggregation of intracellular macromolecules caused by internal stresses (Chuang et al. 2006; Kim et al. 2009).

Even though Prx do not participate directly in protein dethiolation process, their activity influences level of oxidation of glutathione, thioredixins (Trx), glutaredoxins (Grx), and sulfiredoxins (Srx). During oxidative stress, the activity of Prx increases, and therefore the availability of reduced enzymes, crucial for protein dethiolation, is limited.

5.2 *Sulfiredoxins*

Sulfiredoxins (Srx; EC 1.8.98.2) are strictly eukaryotic enzymes with cysteine residue at C-end. Plant Srx are present in chloroplast, and their enzymatic activity is ATP dependent (Iglesias-Baena et al. 2010). Their main function is restoration of 2-CysPrx after its overoxidation (Rey et al. 2007). In addition, Srx can also be involved in dethiolation (deglutathionylation) of proteins (Findlay et al. 2006). Sulfiredoxins can attack disulfide bond between glutathione and target protein, which results in release of protein in reduced form and glutathionylation of Srx (Iglesias-Baena et al. 2010):

$$\mathrm{PS-SG+Srx-SH \rightarrow PSH+Srx-S-SG}$$

One of the main proteins undergoing deglutathionylation by Srx-dependent mechanism is Prx1 and probably other typical 2-CysPrx. Glutathionylation/deglutathionylation of Prx1 is of paramount importance, because it regulates its chaperon activity. Glutathionylated Prx1 forms dimmers displaying antioxidant activity, whereas deglutathionylated Prx1 tends to form decamers and shows chaperon activity (Chae et al. 2012). Prx is not the only protein undergoing deglutathionylation via pathway associated with Srx activity. It was proven that Srx can dethiolate calcium-binding protein A4, actin, protein-tyrosine phosphatase 1B, and probably many others; therefore it is not substrate specific (Mishra et al. 2015).

5.3 *Glutaredoxins*

Glutaredoxins (Grx) are small enzymes belonging to the oxidoreductases and containing two Cys residues able to form a disulfide bond. Grx can be reduced by glutathione or by thioredoxin reductase (Rouhier et al. 2008). In plants, Grx are involved in oxidative stress response, antioxidative enzyme regeneration, regulation of transcription factors, and deglutathionylation process (Rouhier 2010).

Grx in photosynthetic organisms are assigned into six classes; however, class V and VI are present only in *Cyanobacteria* (Couturier et al. 2009). Grx are participating in deglutathionylation (Fig. 3.3). There are two types of dethiolation catalyzed by Grx: monothiol and dithiol mechanism. Grx facilitate the nucleophilic attack on glutathione being part of PS-SG. Reduced PSH is released, and glutathione is bound to the sulfhydryl group of glutaredoxin to form its glutathionyl form (Herrero and de la Torre-Ruiz 2007; Gao et al. 2010).

In monothiol mechanism, catalyzed by Grx I, Grx-SSG is reduced by reduced glutathione (GSH), which results in formation of GSSG and reduced Grx (Fig. 3.3) (Zaffagnini et al. 2012c; Peltoniemi et al. 2006). In dithiol mechanism two active Cys residues of Grx II are involved in a reaction, and Grx-SSG is reduced by the other Cys residue within the same molecule. The nucleophilic attack of reduced Cys residue on glutathionylated Cys is resulting in release of reduced glutathione, and oxidated Grx, with disulfide bond between cysteines. Grx is then reduced by thioredoxin reductases dependent on NADPH or ferredoxin (Fig. 3.3) (Zaffagnini et al. 2012c).

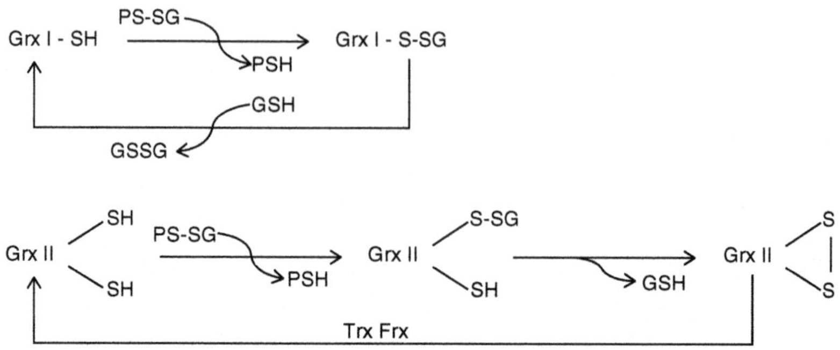

Fig. 3.3 Mechanisms of deglutathionylation by different types of glutaredoxins

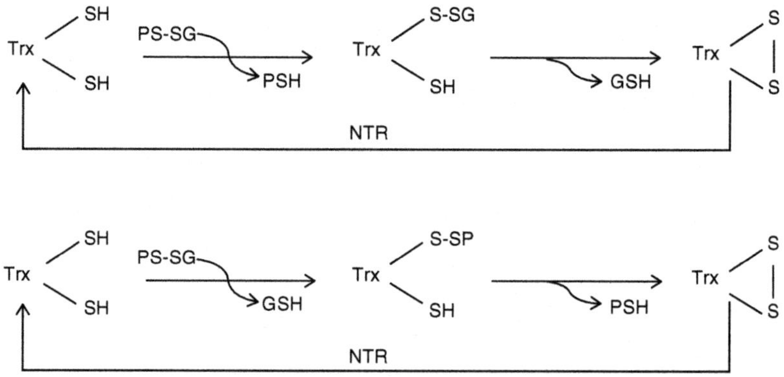

Fig. 3.4 Mechanisms of thioredoxin-dependent deglutathionylation

5.4 Thioredoxins

Thioredoxins (Trx) are small enzymes with molecular weight between 10 and 12 kDa. They have numerous functions based on catalysis of redox reactions. In the active site of Trx, there is a central Cys residue, responsible for determination of enzymatic activity (Fig. 3.4). Trx with reduced Cys residue attacks oxidized protein, which leads to formation of disulfide bond in Trx and protein reduction (Serrato et al. 2013). Trx have been classified into several groups in plants. Most of Trx classes (f, m, x, y, and z) are localized to plant chloroplasts, whereas class h and o are located in the cytosol and mitochondria, respectively. However, apart from typical Trx, there are also Trx-like proteins with domain specific to Trx (Meyer et al. 2012).

Most of Trx have very conservative motif WC[G/P]PC, which allows disulfide bond formation within active site of molecule (Jacquot et al. 2002). Typically both Cys residues are involved in reduction of disulfide bridges of target proteins, which includes dethiolation. During dethiolation, Trx is at first reduced by electron donors specific for the compartment. There are few mechanisms of Trx reduction based on ferrodoxin, based on NADPH, or based on glutathione/glutaredoxins system. Reduced Trx can attack mixed disulfides in target proteins and lead to their dethiolation (Gelhaye et al. 2005).

6 A Regulatory Mechanism of S-Thiolation in Cell Signaling and Physiological Processes

S-thiolation, a redox modification of cysteine residues by formation of disulfides between protein and LMW thiols, is engaged in alternation of protein function, as well as signaling and regulation of cell metabolism (Kuźniak et al. 2013). The most common types of S-thiolation are S-cysteinylation (protein bonded with cysteine), S-cysteaminylation (protein bonded with γ-glutamylcysteine), and S-glutathionylation (protein bonded with glutathione). However, due to the fact that glutathione is by far the most frequent LMW thiol in a plant cell, S-glutathionylation is far more frequent (Spadaro et al. 2010). Protein S-glutathionylation is involved in signal-transduction cascades resulting in the expression of the cellular response to oxidative stress. In yeast, S-glutathionylation of one of the "housekeeping" enzyme GAPDH isoform is crucial for normal cell growth during oxidative stress and may prevent this enzyme against irreversible oxidation leading to its total inhibition (Grant et al. 1999). Such mechanism may also exist in plants, because S-glutathionylation of GAPDH isoforms A and C was confirmed in *A. thaliana* and photosynthetic organism such as *Chlamydomonas reinhardtii* (Zaffagnini et al. 2012b; Morisse et al. 2014).

Another pathway of cell signaling regulation is the modulation of the cellular phosphorylation state by S-glutathionylation of tyrosine phosphatases. The enzymes regulating protein phosphorylation are subjected to redox regulation. Oxidation of protein tyrosine phosphatase (PTP) by ROS leads to its irreversible inactivation, due to the modification of thiol group of Cys215 to sulfenic acid. S-glutathiolation of this PTP residue may protect the enzyme and be easily reversed by glutaredoxins (Barrett et al. 1999). Phosphorylation of tyrosine residue in proteins by PTP is a mechanism of signal transduction, involved in such physiological processes as growth, differentiation, metabolism, cell cycle progression, and cytoskeletal function (Klatt and Lamas 2000). Although this mechanism was described mainly in mammals, S-glutathionylation of protein tyrosine phosphatase has been described in *Glycine max* (Dixon et al. 2005a).

It is postulated that redox-regulated protein S-glutathionylation can control protein degradation via the ubiquitin-proteasome pathway (Klatt and Lamas 2000). During oxidative stress, inactivation of protein ubiquitination is observed; however, restoration of redox homeostasis leads to rapid recovery of this pathway. Studies on neuronal cells during oxidative conditions revealed that ubiquitination is inhibited due to the creation of GSH-protein mixed disulfides in ubiquitin-activating, ubiquitin-carrier, and ubiquitin-ligating enzymes. Therefore, S-glutathiolation probably protects the repair and signaling functions of ubiquitinating enzymes during oxidative stress (Figueiredo-Pereira et al. 1998). The same regulation is probable in plants. In response to stress, the S-glutathionylation of ubiquitin-like protein-NEDD8-like protein RUB3 and membrane-anchored ubiquitin-fold protein 2 was observed in *Triticum aestivum* L. (Gietler et al. 2016). Modification of this proteins, as well as S-thiolation of proteasome 20S observed in *A. thaliana* (Dixon et al. 2005b), suggests that a similar mechanism is present in higher plants.

S-glutathionylation fulfills certain criteria to consider it regulatory mechanism under physiological conditions. It occurs on a specific cysteine residues and changes protein function. S-glutathionylation occurs in chemically and kinetically competent manner with target proteins at their natural abundance. Moreover, there is a correlation between magnitude of physiological-response change and extent of thiol modification. Finally, there is a rapid and efficient mechanism for reversing the S-glutathionylation reaction (Mieyal and Chock 2012).

Even though that S-glutathionylation is considered a modification closely related to stress response, there are proofs of its occurrence under physiological conditions (Dalle-Donne et al. 2009). S-glutathionylation has been suggested as regulatory mechanism for sugar metabolism, which is fundamental for plant growth and development. A number of enzymes involved in reductive and oxidative pentose phosphate pathways including trisphosphate isomerase, fructose-1,6-bisphosphate aldolase, glyceraldehydes-3-phosphate dehydrogenase, enolase and others are S-glutathionylated (Henmi et al. 2007). Results of S-glutathionylation-dependent activation/inhibition are strictly protein dependent, thus S-glutathionylation abolishes the activity of enolase and 6-phosphogluconolactonase, but on the other hand it activates apoptosis related signal transduction kinase 1 and GST (Zagorchev et al. 2013).The best described enzyme regulated by Cys modification is fructose-1,6-bisphosphate aldolase. Several Calvin cycle enzymes were regulated by reduction through Trx; however, recently it was discovered that activity of fructose-1,6-bisphosphate aldolase is controlled by both Trx reduction and S-glutathionylation. In *A. thaliana* chloroplasts, reduction by Trx inhibited enzyme activity and S-glutathionylation acted as the enzyme activator by inhibition of the interaction of Trx with the target site of fructose-1,6-bisphosphate aldolase (Matsumoto and Ogawa 2008). Moreover S-glutathionylation can also control plant development and photosynthesis by regulation of nitrogen fixation via modification of key enzyme of GS-GOGAT cycle, glutamine synthetase (Dixon et al. 2005b).

7 An Antioxidant Defense Mechanism of S-Thiolation in Response to Environmental Stresses

Protein S-thiolation has been proposed to serve an antioxidant function by preventing the irreversible oxidation of cysteine residues to higher oxidation states (e.g., sulfenic acid) following exposure to ROS.

Large-scale studies focused on identification of protein with redox-reactive Cys residues that are susceptible to S-glutathionylation under oxidative stress were performed in animals (Baty et al. 2002), bacteria (Leichert and Jakob 2004), and plants (Lee et al. 2004). However, many studies on proteome are limited to artificial oxidative stress (Table 3.3) rather than to its induction by biotic and abiotic stress (Table 3.4).

Table 3.3 Proteins undergoing S-glutathionylation in plant upon treatment with S-glutathionylation factors

S-glutathionylated proteins in plants treated with S-glutathionylation factors		
Protein	Organism	Literature
Glycolysis, gluconeogenesis, and the pentose phosphate pathway (energy metabolism of sugars)		
Glyceraldehyde 3-phosphate dehydrogenase C	*Arabidopsis thaliana*	Bedhomme et al. (2012)
Glyceraldehyde 3-phosphate dehydrogenase A	*Chlamydomonas reinhardtii*	Zaffagnini et al. (2012b)
Glycine decarboxylase	*Arabidopsis thaliana*	Palmieri et al. (2010)
Phosphoglycerate kinase	*Chlamydomonas reinhardtii*	Zaffagnini et al. (2012b)
Triosephosphate isomerase	*Arabidopsis thaliana*	Ito et al. (2003)
Putative plastidic aldolase	*Arabidopsis thaliana*	Ito et al. (2003)
Fructose-1,6-bisphosphatase	*Chlamydomonas reinhardtii*	Zaffagnini et al. (2012b)
Transketolase	*Chlamydomonas reinhardtii*	Zaffagnini et al. (2012b)
Sedoheptulose-1 7-bisphosphatase	*Chlamydomonas reinhardtii*	Zaffagnini et al. (2012b)
Ribose-5-phosphate isomerase	*Chlamydomonas reinhardtii*	Michelet et al. (2008)
Enolase	*Arabidopsis thaliana*	Dixon et al. (2005b)
UDP-glucose pyrophosphorylase	*Arabidopsis thaliana*	Dixon et al. (2005b)
Alcohol dehydrogenase	*Arabidopsis thaliana*	Dixon et al. (2005b)
UDP-glucose dehydrogenase	*Arabidopsis thaliana*	Dixon et al. (2005b)
Transaldolase	*Chlamydomonas reinhardtii*	Michelet et al. (2008)
Photosynthesis and photorespiration		
Ribulose-1,5-bisphosphate carboxylase/oxygenase	*Chlamydomonas reinhardtii*	Zaffagnini et al. (2012b)
Fructose-1,6-bisphosphate aldolase	*Chlamydomonas reinhardtii*	Zaffagnini et al. (2012b)
	Arabidopsis thaliana	Ito et al. (2003)
Phosphoribulokinase	*Chlamydomonas reinhardtii*	Zaffagnini et al. (2012b)
NADP-dependent malate dehydrogenase	*Chlamydomonas reinhardtii*	Zaffagnini et al. (2012b)
Ribulose-1,5-bisphosphate carboxylase/oxygenase activase	*Chlamydomonas reinhardtii*	Zaffagnini et al. (2012b)
Photosystem I reaction center subunit N	*Arabidopsis thaliana*	Lee et al. (2004)
Photosystem I subunit III	*Arabidopsis thaliana*	Lee et al. (2004)
NADP-malic enzyme	*Arabidopsis thaliana*	Dixon et al. (2005b)
Aconitate hydratase-like protein	*Arabidopsis thaliana*	Dixon et al. (2005b)
RuBisCO binding protein subunit α	*Arabidopsis thaliana*	Dixon et al. (2005b)
Replication and transcription		
DNA damage-repair/toleration protein DRT102	*Arabidopsis thaliana*	Dixon et al. (2005b)

(continued)

Table 3.3 (continued)

S-glutathionylated proteins in plants treated with S-glutathionylation factors		
Protein	Organism	Literature
Translation		
RNA-binding protein RB38	*Chlamydomonas reinhardtii*	Michelet et al. (2008)
RNA-binding protein RB60	*Chlamydomonas reinhardtii*	Michelet et al. (2008)
Tu elongation factor	*Chlamydomonas reinhardtii*	Michelet et al. (2008)
Proteolysis and posttranslational modifications of proteins		
TGA1	*Arabidopsis thaliana*	Lindermayr et al. (2010)
Proteasome 20S	*Arabidopsis thaliana*	Dixon et al. (2005b)
Cytosolic aminopeptidase	*Arabidopsis thaliana*	Dixon et al. (2005b)
Leucine aminopeptidase	*Arabidopsis thaliana*	Dixon et al. (2005b)
Proline iminopeptidase	*Arabidopsis thaliana*	Dixon et al. (2005b)
Nitrilase	*Arabidopsis thaliana*	Dixon et al. (2005b)
Protein tyrosine phosphatase	*Glycine max*	Dixon et al. (2005b)
Amino acid metabolism		
S-adenosylmethionine synthetase 1	*Arabidopsis thaliana*	Lee et al. (2004)
Glutathione transferase	*Arabidopsis thaliana*	Dixon et al. (2005b)
Methionine synthase	*Arabidopsis thaliana*	Dixon et al. (2005b)
Glutamine synthetase	*Arabidopsis thaliana*	Dixon et al. (2005b)
Glycine dehydrogenase	*Arabidopsis thaliana*	Dixon et al. (2005b)
Arginase	*Arabidopsis thaliana*	Dixon et al. (2005b)
3-isopropylmalate dehydratase	*Arabidopsis thaliana*	Dixon et al. (2005b)
Beta-ureidopropionase	*Arabidopsis thaliana*	Dixon et al. (2005b)
Aspartate aminotransferase	*Chlamydomonas reinhardtii*	Michelet et al. (2008)
Nucleotide metabolism		
Sulfate adenylyltransferase	*Arabidopsis thaliana*	Dixon et al. (2005b)
Inosine-5′-monophosphate dehydrogenase	*Arabidopsis thaliana*	Dixon et al. (2005b)
Amidophosphoribosyltransferase	*Arabidopsis thaliana*	Dixon et al. (2005b)
Lipid and fatty acid metabolism		
Acetyl-CoA carboxylase	*Arabidopsis thaliana*	Dixon et al. (2005b)
Isocitrate lyase	*Chlamydomonas reinhardtii*	Michelet et al. (2008)
Chaperons		
HSP 60	*Arabidopsis thaliana*	Dixon et al. (2005b)
HSP 70	*Chlamydomonas reinhardtii*	Michelet et al. (2008)
Cpn 20	*Chlamydomonas reinhardtii*	Michelet et al. (2008)
FKBP	*Chlamydomonas reinhardtii*	Michelet et al. (2008)
Redox regulation		
Peroxiredoxin II	*Populus tremula*	Noguera-Mazon et al. (2006)
Violaxanthin de-epoxidase	*Arabidopsis thaliana*	Lee et al. (2004)
Carbonic anhydrase	*Arabidopsis thaliana*	Lee et al. (2004)

(continued)

Table 3.3 (continued)

S-glutathionylated proteins in plants treated with S-glutathionylation factors		
Protein	Organism	Literature
Dehydroascorbate reductase	*Arabidopsis thaliana*	Dixon et al. (2005b)
Monodehydroascorbate reductase (NADH)-like protein	*Arabidopsis thaliana*	Dixon et al. (2005b)
2-Cys peroxiredoxin	*Arabidopsis thaliana*	Dixon et al. (2005b)
Glutaredoxin	*Arabidopsis thaliana*	Dixon et al. (2005b)
Thioredoxin h2	*Populus trichocarpa*	Gelhaye et al. (2004)
Transport and transport ATPases		
H^+-ATPase	*Arabidopsis thaliana*	Dixon et al. (2005b)
ATP synthase	*Chlamydomonas reinhardtii*	Michelet et al. (2008)
Cytoskeleton		
Actin	*Arabidopsis thaliana*	Dixon et al. (2005b)
Tubulin α	*Arabidopsis thaliana*	Dixon et al. (2005b)
Tubulin β	*Arabidopsis thaliana*	Dixon et al. (2005b)
Other or unknown		
Oxygen-evolving enhancer protein 1 (OEE1)	*Arabidopsis thaliana*	Lee et al. (2004)
Guanylate kinase	*Arabidopsis thaliana*	Lee et al. (2004)
Pyruvate decarboxylase	*Arabidopsis thaliana*	Dixon et al. (2005b)
Sucrose synthase	*Arabidopsis thaliana*	Dixon et al. (2005b)
Methylenetetrahydrofolate dehydrogenase	*Arabidopsis thaliana*	Dixon et al. (2005b)
Cinnamoyl-CoA reductase-like protein	*Arabidopsis thaliana*	Dixon et al. (2005b)
γ-hydroxybutyrate dehydrogenase	*Arabidopsis thaliana*	Dixon et al. (2005b)
SAM:2-demethylmenaquinone methyltransferase	*Arabidopsis thaliana*	Dixon et al. (2005b)
Amylogenin	*Arabidopsis thaliana*	Dixon et al. (2005b)
ACC oxidase	*Arabidopsis thaliana*	Dixon et al. (2005b)
Mg-chelatase Chll-1	*Chlamydomonas reinhardtii*	Michelet et al. (2008)
Calreticulin	*Chlamydomonas reinhardtii*	Michelet et al. (2008)
Plastid lipid-associated protein 10	*Chlamydomonas reinhardtii*	Michelet et al. (2008)
Galactonolactone dehydrogenase	*Arabidopsis thaliana*	Leferink et al. (2009)
Inorganic pyrophosphatase	*Chlamydomonas reinhardtii*	Michelet et al. (2008)
Annexin 1	*Arabidopsis thaliana*	Konopka-Postupolska et al. (2009)

Table 3.4 Proteins undergoing S-glutathionylation in plant upon stress treatment

S-glutathionylated proteins in plants submitted to stress conditions			
Protein	Organism	Literature	Stress
Photosynthesis and photorespiration			
NAD(P)H-quinone oxidoreductase subunit H	*Triticum aestivum*	Gietler et al. (2016)	Dehydration
Protochlorophyllide reductase	*Triticum aestivum*	Gietler et al. (2016)	Dehydration
Chlorophyll a-b binding protein 1	*Triticum aestivum*	Gietler et al. (2016)	Dehydration
Protochlorophyllide reductase, chloroplastic	*Triticum aestivum*	Gietler et al. (2016)	Dehydration
Ribulose-1,5-bisphosphate carboxylase/oxygenase large subunit	*Triticum aestivum*	Gietler et al. (2016)	Dehydration
Replication and transcription			
Multiple organellar RNA editing factor 5	*Triticum aestivum*	Gietler et al. (2016)	Dehydration
Poly [ADP-ribose] polymerase 3	*Triticum aestivum*	Gietler et al. (2016)	Dehydration
Histone deacetylase HDT2	*Triticum aestivum*	Gietler et al. (2016)	Dehydration
Mediator of RNA polymerase II transcription subunit 26a	*Triticum aestivum*	Gietler et al. (2016)	Dehydration
AP2-like ethylene-responsive transcription factor SMZ	*Triticum aestivum*	Gietler et al. (2016)	Dehydration
Translation			
50S ribosomal protein L16	*Triticum aestivum*	Gietler et al. (2016)	Dehydration
40S ribosomal protein S10-1	*Triticum aestivum*	Gietler et al. (2016)	Dehydration
30S ribosomal protein S4	*Triticum aestivum*	Gietler et al. (2016)	Dehydration
40S ribosomal protein S2–1	*Triticum aestivum*	Gietler et al. (2016)	Dehydration
Proteolysis and posttranslational modifications of proteins			
ATP-dependent zinc metalloprotease FTSH 5	*Triticum aestivum*	Gietler et al. (2016)	Dehydration
Ubiquitin-like protein-NEDD8-like protein RUB3	*Triticum aestivum*	Gietler et al. (2016)	Dehydration
Membrane-anchored ubiquitin-fold protein 2	*Triticum aestivum*	Gietler et al. (2016)	Dehydration
Cysteine proteinase inhibitor 2	*Triticum aestivum*	Gietler et al. (2016)	Dehydration
Putative protein phosphatase 2C–like protein 45	*Triticum aestivum*	Gietler et al. (2016)	Dehydration
Calcium and calcium/calmodulin-dependent serine/threonine-protein kinase	*Triticum aestivum*	Gietler et al. (2016)	Dehydration
Amino acid metabolism			
Phenylalanine ammonia-lyase	*Triticum aestivum*	Gietler et al. (2016)	Dehydration

(continued)

Table 3.4 (continued)

S-glutathionylated proteins in plants submitted to stress conditions			
Protein	Organism	Literature	Stress
Lipid and fatty acid metabolism			
GDSL esterase/lipase EXL4	*Triticum aestivum*	Gietler et al. (2016)	Dehydration
Probable lipid phosphate phosphatase 4	*Triticum aestivum*	Gietler et al. (2016)	Dehydration
Phospholipase A1-II δ	*Triticum aestivum*	Gietler et al. (2016)	Dehydration
Chaperons			
Chloroplast envelope membrane 70 kDa heat shock-related protein	*Triticum aestivum*	Gietler et al. (2016)	Dehydration
Cpn 60	*Arabidopsis thaliana*	Michelet et al. (2005)	Light
Redox regulation			
Leaf-specific thionin BTH6	*Triticum aestivum*	Gietler et al. (2016)	Dehydration
Leaf-specific thionin DB4	*Triticum aestivum*	Gietler et al. (2016)	Dehydration
Thioredoxin M2	*Triticum aestivum*	Gietler et al. (2016)	Dehydration
Thioredoxin f	*Arabidopsis thaliana*	Michelet et al. (2005)	Light
	Spinacia oleracea	Michelet et al. (2005)	Light
Signal transduction			
F-box protein GID2	*Triticum aestivum*	Gietler et al. (2016)	Dehydration
Gibberellin-regulated protein 10	*Triticum aestivum*	Gietler et al. (2016)	Dehydration
Protein RALF-like 15	*Triticum aestivum*	Gietler et al. (2016)	Dehydration
Transport and transport ATPases			
Nonspecific lipid-transfer protein	*Triticum aestivum*	Gietler et al. (2016)	Dehydration
ATP synthase subunit α	*Triticum aestivum*	Gietler et al. (2016)	Dehydration
ATP synthase subunit β	*Triticum aestivum*	Gietler et al. (2016)	Dehydration
Cytoskeleton			
Actin-depolymerizing factor 10	*Triticum aestivum*	Gietler et al. (2016)	Dehydration
Other or unknown			
Oxygen-evolving enhancer protein 1 (OEE1)	*Triticum aestivum*	Gietler et al. (2016)	Dehydration
Root meristem growth factor 3	*Triticum aestivum*	Gietler et al. (2016)	Dehydration
Beta carbonic anhydrase 3	*Triticum aestivum*	Gietler et al. (2016)	Dehydration
Thiamine pyrophosphokinase 2	*Triticum aestivum*	Gietler et al. (2016)	Dehydration
Peroxisomal membrane protein 11-1	*Triticum aestivum*	Gietler et al. (2016)	Dehydration

7.1 Biotic Stresses

During biotic stresses, plants are able to activate systematic acquired resistance (SAR), which is a mechanism against wide spectrum of herbivorous insects and microbial pathogens. To effectively activate the system the cross talk between salicylic acid (SA) and jasmonic acid (JA) signaling pathways is needed. Some of the major players in this cross talk are the non-expressor of PR genes 1 (NPR1), WRKY4 factors, and MPK4. Oxidative stress, which accompanies plant infection, leads to S-nitrosylation of NPR1 transcription factor. During this process, a change of form from the oligomeric to monomeric, which is transported to the nucleus, is observed (Li 2014). In the nucleus, it interacts with TGA1, another redox-regulated transcription factor. TGA1 is part of basic leucine zipper (bZIP) transcription factors binding to a TGA-related motive family (Dietz 2014). In *A. thaliana*, 10 TGA transcription factor proteins were revealed, including TGA1–7, PERIANTHIA (PAN), TGA9, and TGA10. They are responsible for regulation of SAR and participation in basal resistance (Jakoby et al. 2002). After nitrosoglutathione (GSNO) treatment, it was showed that TGA1 residues Cys260, Cys266, and Cys287 can be both S-glutathionylated and S-nitrosylated and Cys172 only S-glutathionylated. Modification of Cys residues of TGA1 enhances its DNA-binding ability (Lindermayr et al. 2010; Dietz 2014). Therefore, S-glutathionylation of TGA1 is crucial for SAR regulation.

In *A. thaliana* leaves, glycine decarboxylase complex (GDC), a key enzyme of the photorespiratory C2 cycle in C3 plants, may be inhibited due to its S-nitrosylation and S-glutathionylation by GSNO, although it should be emphasized that S-nitrosylation was just short-time modification, which promoted S-glutathionylation. Moreover, it was proved that during pathogen attack, i.e., *Escherichia coli*, activity of GDC can be inhibited not only by its direct posttranslational modification but also by S-glutathionylation of elicitor harpin, a strong inducer of ROS and NO (Palmieri et al. 2010).

Studies on *Triticum aestivum* indicated that S-glutathionylation protects wheat against *Fusarium* evoked oxidative stress. Stress tolerance in co-stressed plants, like low Cd^{2+}, was accompanied with increased content of free protein thiols and increased ratio of free thiols to thiol disulfides. Moreover, crucial metabolic enzymes have been reported to be susceptible to redox regulation by glutathionylation, among which the most extensively studied were glycolytic enzymes, i.e., fructose-1,6-bisphosphate aldolase and 3-phosphoglycerate kinase, which play a central role in energy production and their modification may serve as a protection against oxidative damage (Mohapatra and Mittra 2016).

Recently, analysis of tomato proteome in response to *Pseudomonas syringae* infection revealed that S-thiolation concerned cellular carbohydrate metabolism, energy processes (plasma membrane adenosine triphosphate, ATP synthesis, photosynthesis, and pentose-phosphate shunt), gluconeogenesis, and cysteine biosynthetic process. Moreover, five redox-regulated proteins (ferredoxin, peroxidase 12-like, glutamine synthetase, cysteine synthase, and lactoylglutathione lyase-like) were involved in oxidation-reduction process (Balmant et al. 2015).

Increased glutathionylation under biotic stress implicates the role of this modification in the regulation of various processes and diverse metabolic and signaling pathways.

7.2 *Abiotic Stresses*

Majority of research on glutathionylation under the influence of oxidative stress relates to an artificial stress, i.e., induced by tert-butylhydroperoxide on culture of *A. thaliana* cells. In effect, 79 enzymes were either directly S-glutathionylated or formed a complex with S-glutathionylated polypeptides. Several of those enzymes were associated with sugar-metabolism-related proteins such as enolase, UDP-glucose pyrophosphorylase, UDP-glucose dehydrogenase and enzymes related to photosynthesis (RuBisCO binding protein subunit α), peptidases such as cytosolic aminopeptidase, leucine aminopeptidase, proline iminopeptidase, and many others including GST, 2-Cys PRX, and GRX (Dixon et al. 2005b).

More recently large-scale proteomic analyses of photosynthetic model organism *Chlamydomonas reinhardtii* permitted the identification of 225 proteins undergoing glutathionylation in response to H_2O_2 treatment. Majority of S-glutathionylated proteins were involved in Calvin-Benson cycle like phosphoribulokinase, glyceraldehyde-3-phosphate dehydrogenase, ribose-5-phosphate isomerase, and phosphoglycerate kinase (Zaffagnini et al. 2012b).

S-glutathionylation was examined in particular on the plants resistant to dehydration. In seeds of *Spinacia oleracea*, S-glutathionylation of an acyl carrier protein was observed through desiccation, but this process was rapidly reversed during imbibition. ACP is a small acidic cofactor which plays an essential role in plant lipid metabolism. Sulfhydryl groups of ACP subjected to oxidative reactions could be inactivated. Similar phenomenon was observed for the period of desiccation and rehydration of wheat grains, *Acer platanoides* seeds and in the resurrection plant *Boea hydroscopica* (Butt and Ohlrogge 1991; Navari-Izzo et al. 2000; Rhazi et al. 2003; Pukacka and Ratajczak 2007). In biological systems, these oxidation reactions are prevented or reversed through a variety of protective mechanisms including the glutathione/glutathione reductase system. (Colville and Kranner 2010).

Glutathionylation also plays an important role in the regulation of several signaling proteins including annexins which are cell membrane proteins dependent on Ca^{2+} (Mortimer et al. 2008). S-glutathionylation of annexin 1 (AnnAt1) in *A. thaliana* on both Cys111 and Cys239 alters its ability to bind Ca^{2+} which leads to the reduction or total inhibition of enzyme activity. AnnAt1 is S-glutathionylated in vivo in response to ABA treatment, which mimics hormonal reaction to abiotic stress, and consequently a decrease in the calcium affinity was observed (Clark et al. 2010). It is suggested that it may be ROS-based, calcium-dependent mechanism of alternation of the signal transduction pathway (Konopka-Postupolska et al. 2009).

Finally, S-glutathionylation has been implicated in the regulation of protein folding and stability. The activity of heat shock proteins (Hsp) increased into drought and high-temperature stress. Prokaryotic Hsp33 and eukaryotic Hsp70 were both activated by S-glutathionylation, which enhances the efficiency of their protective functions (Fedoroff 2006).

In response to abiotic stress, S-glutathionylation may also regulate the activity and function of nuclear proteins, including transcription factors, as well as affect chromatin structure and dynamics of the condensation process (Zagorchev et al. 2013). In addition, by perturbing their DNA-binding sites, glutathionylation inhibits the DNA-binding activity of several redox-sensitive transcription factors including c-Jun and c-Fos (Dietz 2014). It is a common mechanism in regulation of transcription factors in nonphotosynthetic organism like OxyR in *Escherichia coli*, Yap1 in yeast subunits p50 or p65 of NF-κB factor (Cooper et al. 2011), and nuclear proteins AP-1 and STAT3 in mammals (Xie et al. 2009; Zaffagnini et al. 2012a). Homologous thiol-based mechanism must be present in plants.

Studies performed by Gietler et al. (2016) on *Triticum aestivum* L. var. Zadra seedlings, tolerant and sensitive to dehydration, revealed that in response to water deficiency there was noticeable change in profiles of S-glutathionylated proteins; however, more prominent changes were observed in sensitive ones. In tolerant spring wheat seedlings, S-glutathionylation of 17 proteins was upregulated, whereas S-glutathionylation of 8 proteins was downregulated. The opposite direction of the changes was observed in sensitive seedlings, where S-glutathionylation of 23 proteins was downregulated and there was no observed upregulation (Fig. 3.5).

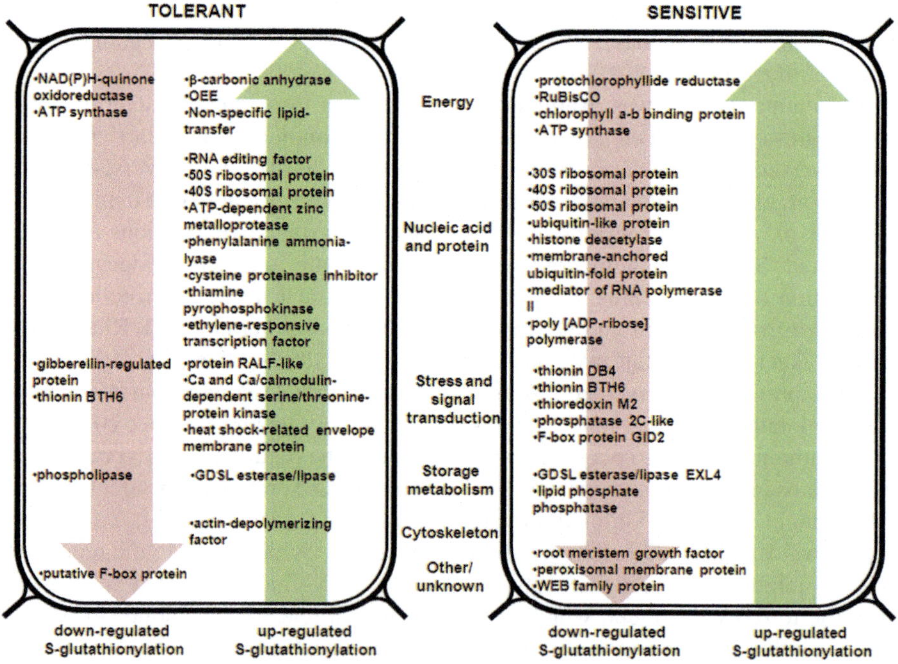

Fig. 3.5 S-glutathionylated up- and downregulated proteins in wheat seedlings (*Triticum aestivum* L.) tolerant and sensitive to dehydration

Downregulation of S-glutathionylation in sensitive seedlings may have a direct impact on loss of drought tolerance by lack of thiol group protection against irreversible oxidative modifications. Observed decrease in abundance of S-glutathionylated proteins can be caused by low GSH/GSSG ration in this type of seedlings. Changes in S-glutathionylation occasionally concerned the same proteins in tolerant and sensitive plants, i.e., S-glutathionylation of BTH6 thionin decreased in both types of wheat seedlings, but S-glutathionylation of 50S ribosomal protein L16, DB4 thionin, and GDSL esterase/lipase EXL4 increased in tolerant plants and decreased in sensitive plants (Gietler et al. 2016).

S-glutathionylation of specific proteins may be one of the mechanisms of wheat seedlings' tolerance to dehydration, especially the increased abundance of S-glutathionylation of cysteine proteinase inhibitor 2 and ATP-dependent zinc-metalloprotease (FTSH 5), which may regulate their proteolytic activity (Gietler et al. 2016). Similar codependence was observed in *Pisum sativum* where cysteine proteinase inhibitor 2 was crucial for desiccation tolerance (Wang et al. 2012).

S-glutathionylation may enhance energy saving in wheat seedlings by possible inhibition of activities of oxygen-evolving enhancer protein 1 (OEE1), a nonspecific lipid transfer protein, and beta carbonic anhydrase 3 protein, which were observed in drought-tolerant wheat seedlings (Gietler et al. 2016). It is part of mechanism of cellular metabolism adjustment at both, the metabolic and gene expression, levels (Bogdan and Zagdańska 2009; Gietler et al. 2016).

In response to dehydration, S-glutathionylation of calcium and calcium/calmodulin-dependent serine/threonine-protein kinase (CCaMK) and leaf-specific thionin DB4 increased. Modification of CCaMK, a protein involved in ABA-induced antioxidant defense, is particularly interesting, especially taking into consideration that it is activated by H_2O_2-dependent NO production (Ma et al. 2012; Gietler et al. 2016). CCaMK is a part of the Ca^{2+}/calmodulin-mediated signaling network, and thus it plays a role in the regulation not only of the plant response to environmental stress but also its growth and development. Furthermore, S-glutathionylation of mitochondrial multiple organelle RNA editing factor 5 and AP2-like ethylene-responsive transcription factor may trigger this signaling pathway during dehydration (Gietler et al. 2016).

It is worth mentioning that protein S-glutathionylation probably interplays with S-nitrosylation. Those modifications can be alternative in relation to each other or one can promote the other. In studies performed on dehydration-tolerant and dehydration-sensitive wheat seedlings during stress, it was found that five proteins underwent both types of modifications. The leaf-specific thionins BTH6 and DB4, chloroplastic 50S ribosomal protein L16, phospholipase A1-II delta, and chloroplastic thioredoxin M2 were both S-nitrosylated and S-glutathionylated upon water deficiency. It may be an example of fine mechanism of abiotic stress response, affecting plant metabolism during unfavorable environmental conditions (Gietler et al. 2016).

8 Other Thiol Modified Proteins and Their Involvement in Plant Growth and Development

Redox state of thiol groups is alternated during germination. ROS accumulation during this process may promote formation of disulfides bonds and of mixed disulfides between LMW thiols and cysteinyl residues (i.e., S-glutathionylation; Buchanan and Balmer 2005). S-glutathionylation of proteins during germination may on the one hand prevent them from irreversible oxidation and on the other hand regulate protein activity, folding, and susceptibility to degradation (El-Maarouf-Bouteau et al. 2013). Moreover, seed imbibition is initiated by many of ROS-associated mechanism including redox modifications of protein thiols (Bykova et al. 2011).

Germination is dependent on GA- and ABA-regulated pathways. Light signal promotes GA-mediated phytochrome interacting factor 3-like 5 degradation by proteasome. ABA mediates prevention of the proteasome-dependent degradation of protein ABA insensitive 5, which is a germination repressor. However, α and β subunits of 20S proteasome may undergo S-glutathionylation which proves that hormone signaling during germination may be regulated by redox homeostasis changes (Henmi et al. 2007).

Another similarly regulated protein related to plant development is blade-on-petiole proteins (BOP), which are involved in leaf and floral morphogenesis in *Arabidopsis*. Due to the fact that they are very similar to previously described NPR1, it is highly possible that their activity is controlled by Cys modification by its S-glutathionylation (Hepworth et al. 2005).

9 Conclusion and Future Perspectives

Protein S-thiolation is a candidate mechanism of protein regulation in response to changes in the intracellular redox potential. It is being revived as a mechanism of redox-mediated signal transduction as well as an adaptive cellular reaction protecting critical regulatory molecules from permanent loss of function.

The current state of knowledge about which proteins are undergoing S-glutathionylation in vivo in plants treated with environmental factors, as well as how particular proteins are affected by their S-thiolation, is not sufficient and requires further research toward understanding which among potentially S-glutathionylated proteins are involved in stress tolerance.

However, it is worth to emphasize that during the last decade a great advance in our knowledge of the relevance of S-thiolation reaction has been observed. It has been shown that this process causes an extremely extensive influence on different biochemical events and biological processes. Hence, it can be expected that the next decade will bring a better understanding of chemistry and biochemistry of protein thiolation.

Acknowledgement The authors would like to thank Professor Barbara Zagdańska for critical reading the manuscript and valuable discussion.

References

Abat JK, Deswal R (2009) Differential modulation of S-nitrosoproteome of *Brassica juncea* by low temperature: change in S-nitrosylation of Rubisco is responsible for the inactivation of its carboxylase activity. Proteomics 9:4368–4380

Abat JK, Mattoo AK, Deswal R (2008) S-nitrosylated proteins of a medicinal CAM plant *Kalanchoe pinnata*- ribulose-1,5-bisphosphate carboxylase/oxygenase activity targeted for inhibition. FEBS J 275:2862–2872

Asano T, Hayashi N, Kobayashi M, Aoki N, Miyao A, Mitsuhara I, Ichikawa H, Komatsu S, Hirochika H, Kikuchi S, Ohsugi R (2012) A rice calcium-dependent protein kinase OsCPK12 oppositely modulates salt-stress tolerance and blast disease resistance. Plant J 69:26–36

Bach RD, Dmitrenko O, Thorpe C (2008) Mechanism of thiolate-disulfide interchange reactions in biochemistry. J Org Chem 73(1):12–21

Bai XG, Yang LM, Tian MH, Chen JH, Shi JS, Yang YP, Hu X (2011) Nitric oxide enhances desiccation tolerance of recalcitrant *Antiaris toxicaria* seeds via protein S-nitrosylation and carbonylation. PLoS One 6:e20714

Balmant KM, Parker J, Yoo M-J, Zhu N, Dufresne C, Chen S (2015) Redox proteomics of tomato in response to *Pseudomonas syringae* infection. Hortic Res 2:15043

Barrett WC, De Gnore JP, Konig S, Fales HM, Keng YF, Zhang ZY, Yim MB, Chock PB (1999) Regulation of PTP1B via glutathionylation of the active site cysteine 215. Biochemistry 38:6699–6705

Bashir H, Ahmad J, Bagheri R, Nauman M, Qureshi MI (2013) Limited sulfur resource forces *Arabidopsis thaliana*to shift towards non-sulfur tolerance under cadmium stress. Environ Exp Bot 94:19–32

Baty JW, Hampton MB, Winterbourn CC (2002) Detection of oxidant sensitive thiol proteins by fluorescence labeling and two-dimensional electrophoresis. Proteomics 2:1261–1266

Bedhomme M, Adamo M, Marchand CH, Couturier J, Rouhier N, Lemaire SD, Zaffagnini M, Trost P (2012) Glutathionylation of cytosolic glyceraldehyde-3-phosphate dehydrogenase from the model plant *Arabidopsis thaliana* is reversed by both glutaredoxins and thioredoxins *in vitro*. Biochem J 445:337–347

Begara-Morales JC, Sánchez-Calvo B, Chaki M, Valderrama R, Mata-Pérez C, López-Jaramillo J, Padilla MN, Carreras A, Corpas FJ, Barroso JB (2014) Dual regulation of cytosolic ascorbate peroxidase (APX) by tyrosine nitration and S-nitrosylation. J Exp Bot 65:527–538

Belenghi B, Romero-Puertas MC, Vercammen D, Brackenier A, Inzé D, Delledonne M, Van Breusegem F (2007) Metacaspase activity of *Arabidopsis thaliana* is regulated by S-nitrosylation of a critical cysteine residue. J Biol Chem 282:1352–1358

Bhatt I, Tripathi BN (2011) Plant peroxiredoxins: catalytic mechanisms, functional significance and future perspectives. Biotechnol Adv 29(6):850–859

Bhattacharjee S, Deterding LJ, Jiang J, Bonini MG, Tomer KB, Ramirez DC, Mason RP (2007) Electron transfer between a tyrosyl radical and a cysteine residue in hemoproteins: spin trapping analysis. J Am Chem Soc 129:13493–13501

Birtić S, Ksas B, Genty B, Mueller MJ, Triantaphylides C, Havaux M (2011) Using spontaneous photon emission to image lipid oxidation patterns in plant tissues. Plant J 67:1103–1115

Biteau B, Labarre J, Toledano MB (2003) ATP-dependent reduction of cysteine-sulphinic acid by *S. cerevisiae* sulphiredoxin. Nature 425:980–984

Bogdan J, Zagdańska B (2009) Alterations in sugar metabolism coincide with a transition of wheat seedlings to dehydration intolerance. Environ Exp Bot 66:186–194

Broniowska KA, Diers AR, Hogg N (2013) S-nitrosoglutathione. Biochim Biophys Acta 1830:173–3181

Buchanan BB, Balmer Y (2005) Redox regulation: a broadening horizon. Annu Rev Plant Biol 56:187–220

Butt AD, Ohlrogge JB (1991) Acyl carrier protein is conjugated to glutathione in spinach seed. Plant Physiol 96(3):937–942

Bykova N, Rampitsch C (2013) Modulating protein function through reversible oxidation: redox-mediated processes in plants revealed through proteomics. Proteomics 13:579–596

Bykova NV, Hoehn B, Rampitsch C, Banks T, Stebbing JA, Fan T, Knox R (2011) Redox-sensitive proteome and antioxidant strategies in wheat seed dormancy control. Proteomics 11:865–882

Chae HZ, Oubrahim H, Park JW, Rhee SG, Chock PB (2012) Protein glutathionylation in the regulation of peroxiredoxins: a family of thiol-specific peroxidases that function as antioxidants, molecular chaperones, and signal modulators. Antioxid Redox Signal 16(6):506–523

Chen CM, Kristopeit SM (1981) Metabolism of cytokinins: dephosphorylation of cytokinin ribonucleotide by 5'-nucleotidase from wheat. Plant Physiol 67:494–498

Chuang MH, Wu MS, Lo WL, Lin JT, Wong CH, Chiou SH (2006) The antioxidant protein alkylhydroperoxide reductase of *Heliobacter pylori* switches from a peroxide reductase to a molecular chaperone function. Proc Natl Acad Sci U S A 103:2552–2557

Chung HS, Wang SB, Venkatraman V, Murray CI, Van Eyk JE (2013) Cysteine oxidative posttranslational modifications: emerging regulation in the cardiovascular system. Circ Res 112:382–392

Clark G, Konopka-Postupolska D, Hennig J, Roux S (2010) Is annexin 1 a multifunctional protein during stress responses? Plant Signal Behav 5:303–7

Colville L, Kranner I (2010) Desiccation tolerant plants as model systems to study redox regulation of protein thiols. Plant Growth Regul 62:241–255

Cooper AJ, Pinto JT, Callery PS (2011) Reversible and irreversible protein glutathionylation: biological and clinical aspects. Expert Opin Drug Metab Toxicol 7:891–910

Corpas FJ, Palma JM, del Río LA, Barroso JB (2013) Protein tyrosine nitration in higher plants grown under natural and stress conditions. Front Plant Sci 4:29

Couturier J, Jacquot JP, Rouhier N (2009) Evolution and diversity of glutaredoxins in photosynthetic organisms. Cell Mol Life Sci 66:2539–2557

Dalle-Donne I, Rossi R, Colombo G, Giustarini D, Milzani A (2009) Protein S-glutathionylation: a regulatory device from bacteria to humans. Trends Biochem Sci 34(2):85–96

de Pinto MC, Locato V, Sgobba A, Romero-Puertas MDC, Gadaleta C, Delledonne M, De Gara L (2013) S-nitrosylation of ascorbate peroxidase is part of programmed cell death signaling in tobacco bright yellow-2 cells. Plant Physiol 163:1766–1775

Díaz M, Achkor H, Titarenko E, Martínez MC (2003) The gene encoding glutathione-dependent formaldehyde dehydrogenase/GSNO reductase is responsive to wounding, jasmonic acid and salicylic acid. FEBS Lett 543:136–139

Diaz-Vivancos P, Dong Y, Ziegler K, Markovic J, Pallardó FV, Pellny TK, Verrier PJ, Foyer CH (2010) Recruitment of glutathione into the nucleus during cell proliferation adjusts whole-cell redox homeostasis in *Arabidopsis thaliana* and lowers the oxidative defence shield. Plant J 64(5):825–838

Diaz-Vivancos P, de Simone A, Kiddle G, Foyer CH (2015) Glutathione-linking cell proliferation to oxidative stress. Free Radic Biol Med 89:1154–1164

Dietz KJ (2011) Peroxiredoxins in plants and cyanobacteria. Antioxid Redox Signal 15(4):1129–1159

Dietz KJ (2014) Redox regulation of transcription factors in plant stress acclimation and development. Antioxid Redox Signal 21(9):1356–1372

Dixon DP, Lapthorn A, Edwards R (2002) Plant glutathione transferases. Genome Biol 3(3):REVIEWS3004

Dixon DP, Fordham-Skelton AP, Edwards R (2005a) Redox regulation of a soybean tyrosine-specific protein phosphatase. Biochemistry 44:7696–7703

Dixon DP, Skipsey M, Grundy NM, Edwards R (2005b) Stress-induced protein S-glutathionylation in *Arabidopsis*. Plant Physiol 138:2233–2244

El-Maarouf-Bouteau H, Meimoun P, Job C, Job D, Bailly C (2013) Role of protein and mRNA oxidation in seed dormancy and germination. Front Plant Sci 4:77

Elviri L, Speroni F, Careri M, Mangia A, di Toppi LS, Zottini M (2010) Identification of *in vivo* nitrosylated phytochelatins in *Arabidopsis thaliana* cells by liquid chromatography-direct electrospray-linear ion trap-mass spectrometry. J Chromatogr A 1217:4120–4126

Fares A, Rossignol M, Peltier JB (2011) Proteomics investigation of endogenous S-nitrosylation in *Arabidopsis*. Biochem Biophys Res Commun 416:331–336

Fedoroff N (2006) Redox regulatory mechanisms in cellular stress responses. Ann Bot 98(2):289–283

Feechan A, Kwon E, Yun BW, Wang Y, Pallas JA, Loake GJ (2005) A central role for S-nitrosothiols in plant disease resistance. Proc Natl Acad Sci U S A 102:8054–8059

Figueiredo-Pereira ME, Yakushin S, Cohen G (1998) Disruption of the intracellular sulfhydryl homeostasis by cadmium-induced oxidative stress leads to protein thiolation and ubiquitination in neuronal cells. J Biol Chem 273(12703):12709

Findlay VJ, Townsend DM, Morris TE, Fraser JP, He L, Tew KD (2006) A novel role for human sulfiredoxin in the reversal of glutathionylation. Cancer Res 66:6800–6806

Gallogly MM, Starke DW, Mieyal JJ (2009) Mechanistic and kinetic details of catalysis of thiol-disulfide exchange by glutaredoxins and potential mechanisms of regulation. Antioxid Redox Signal 11(5):1059–1081

Gao XH, Zaffagnini M, Bedhomme M, Michelet L, Cassier-Chauvat C, Decottignies P, Lemaire SD (2010) Biochemical characterization of glutaredoxins from *Chlamydomonas reinhardtii*: kinetics and specificity in deglutathionylation reactions. FEBS Lett 584:2242–2248

Gaupels F, Kuruthukulangarakoola GT, Durner J (2011) Upstream and downstream signals of nitric oxide in pathogen defence. Curr Opin Plant Biol 14:707–714

Gelhaye E, Rouhier N, Gérard J, Jolivet Y, Gualberto J, Navrot N, Ohlsson PI, Wingsle G, Hirasawa M, Knaff DB, Wang H, Dizengremel P, Meyer Y, Jacquot JP (2004) A specific form of thioredoxin h occurs in plant mitochondria and regulates the alternative oxidase. Proc Natl Acad Sci U S A 101(40):14545–14550

Gelhaye E, Rouhier N, Navrot N, Jacquot JP (2005) The plant thioredoxin system. Cell Mol Life Sci 62:24–35

Gietler M, Nykiel M, Orzechowski S, Fettke J, Zagdańska B (2016) Proteomic analysis of S-nitrosylated and S-glutathionylated proteins in wheat seedlings with different dehydration tolerances. Plant Physiol Biochem 108:507–518

Giles GI, Tasker KM, Jacob C (2001) Hypothesis: the role of reactive sulphur species in oxidative stress. Free Radic Biol Med 31:1279–1283

Giles GI, Tasker KM, Collins C, Giles NM, O'rourke E, Jacob C (2002) Reactive sulphur species: an *in vitro* investigation of the oxidation properties of disulphide S-oxides. Biochem J 364:579–585

Giles NM, Watts AB, Giles GI, Fry FH, Littlechild JA, Jacob C (2003) Metal and redox modulation of cysteine protein function. Chem Biol 10:677–693

Grant CM, Quinn KA, Dawes IW (1999) Differential protein S-thiolation of glyceraldehyde-3-phosphate dehydrogenase isoenzymes influences sensitivity to oxidative stress. Mol Cell Biol 19(2650):2656

Hell R, Wirtz M (2011) Molecular biology, biochemistry and cellular physiology of cysteine metabolism in *Arabidopsis thaliana*. Arabidopsis Book 9:e0154

Henmi K, Yanagida M, Ogawa K (2007) Roles of reactive oxygen species and glutathione in plant development. Int J Plant Dev Biol 1(2):185–193

Hepworth SR, Zhang Y, McKim S, Li X, Haughn GW (2005) BLADE-ON-PETIOLE-dependent signaling controls leaf and floral patterning in *Arabidopsis*. Plant Cell 17:1434–1448

Herrero E, de la Torre-Ruiz MA (2007) Monothiol glutaredoxins: a common domain for multiple functions. Cell Mol Life Sci 64(12):1518–1530

Hess DT, Matsumoto A, Nudelman R, Stamler JS (2001) S-Nitrosylation: spectrum and specificity. Nat Cell Biol 3:E46–E49

Hess DT, Matsumoto A, Kim SO, Marshall HE, Stamler JS (2005) Protein S-nitrosylation: purview and parameters. Nat Rev Mol Cell Biol 6:150–166

Hill BG, Bhatnagar A (2012) Protein S-glutathiolation: redox-sensitive regulation of protein function. J Mol Cell Cardiol 52(3):559–567

Hou Z, Wang L, Liu J, Hou L, Liu X (2013) Hydrogen sulfide regulates ethylene-induced stomatal closure in *Arabidopsis thaliana*. J Integr Plant Biol 55:277–289

Iglesias-Baena I, Barranco-Medina S, Lázaro-Payo A, López-Jaramillo FJ, Sevilla F, Lázaro JJ (2010) Characterization of plant sulfiredoxin and role of sulphinic form of 2-Cys peroxiredoxin. J Exp Bot 61:1509–1521

Iglesias-Baena I, Barranco-Medina S, Sevilla F, Lázaro J-J (2011) The dual-targeted plant sulfiredoxin retroreduces the sulfinic form of atypical mitochondrial peroxiredoxin. Plant Physiol 155:944–955

Ito H, Iwabuchi M, Ogawa K (2003) The sugar-metabolic enzymes aldolase and triose-phosphate isomerase are targets of glutathionylation in *Arabidopsis thaliana*: detection using biotinylated glutathione. Plant Cell Physiol 44:655–660

Jacob C, Anwar A (2008) The chemistry behind redox regulation with a focus on sulphur redox systems. Physiol Plant 133:469–480

Jacquot JP, Rouhier N, Gelhaye E (2002) Redox control by dithiol-disulfide exchange in plants. Part I. Chloroplastic systems. Ann N Y Acad Sci 973:508–519

Jakoby M, Weisshaar B, Dröge-Laser W, Vicente-Carbajosa J, Tiedemann J, Kroj T, Parcy F (2002) bZIP research group.bZIP transcription factors in *Arabidopsis*. Trends Plant Sci 7:106–111

Janssen-Heininger YM, Mossman BT, Heintz NH, Forman HJ, Kalyanaraman B, Finkel T, Stamler JS, Rhee SG, van der Vliet A (2008) Redox-based regulation of signal transduction: principles, pitfalls, and promises. Free Radic Biol Med 45:1–17

Jung CH, Thomas JA (1996) S-Glutahiolated hepatocyte proteins and insulin disulfides as substrates for reduction by glutaredoxin, thioredoxin, protein disulfide isomerase and glutathione. Arch Biochem Biophys 335:61–72

Kato H, Takemoto D, Kawakita K (2013) Proteomic analysis of S-nitrosylated proteins in potato plant. Physiol Plant 148:371–386

Kim Y, Jang HH, Lee JR, Sung NR, Lee HB, Lee DH, Lee DH, Park DJ, Kang CH, Chung WS, Lim CO, Yun DJ, Kim WY, Lee KO, Lee SY (2009) Oligomerization and chaperone activity of a plant 2-Cys peroxiredoxin in response to oxidative stress. Plant Sci 177:227–232

Klatt P, Lamas S (2000) Regulation of protein function by S-glutathiolation in response to oxidative and nitrosative stress. Eur J Biochem 267:4928–4944

Konopka-Postupolska D, Clark G, Goch G, Debski J, Floras K, Cantero A, Fijolek B, Roux S, Hennig J (2009) The role of annexin 1 in drought stress in *Arabidopsis*. Plant Physiol 150(3):1394–1410

Kranner I, Birtić S, Anderson K, Pritchard HW (2006) Glutathione half-cell reduction potential: a universal stress marker and modulator of programmed cell death? Free Radic Biol Med 40:2155–2165

Kranner I, Minibayeva FV, Beckett RP, Seal CE (2010) What is stress? Concepts, definitions and applications in seed science. New Phytol 188:655–673

Kubienová L, Kopečný D, Tylichová M, Briozzo P, Skopalová J, Šebela M, Navrátil M, Tâche R, Luhová L, Barroso JB, Petřivalský M (2013) Structural and functional characterization of a plant S-nitrosoglutathione reductase from *Solanum lycopersicum*. Biochimie 95:889–902

Kuruthukulangarakoola GT, Lindermayr C (2013) Regulation and function of protein S-nitrosylation in plant stress. In: Sarwat M, Ahmad A, Abdin MZ (eds) Stress signaling in plants: genomics and proteomics perspective, vol 1. Springer, New York, pp 123–148

Kuźniak E, Kazmierczak A, Wielanek M, Glowacki R, Kornas A (2013) Involvement of salicylic acid, glutathione and protein S-thiolation in plant cell death-mediated defence response of *Mesembryanthemum crystallinum* against *Botrytis cinerea*. Plant Physiol Biochem 63:30–38

Lai TS, Hausladen A, Slaughter TF, Eu JP, Stamler JS, Greenberg CS (2001) Calcium regulates S-nitrosylation, denitrosylation, and activity of tissue transglutaminase. Biochemistry 40:4904–4910

Lamotte O, Bertoldo JB, Besson-Bard A, Rosnoblet C, Aimé S, Hichami S, Terenzi H, Wendehenne D (2014) Protein S-nitrosylation: specificity and identification strategies in plants. Front Chem 2:114

Lee K, Lee J, Kim Y, Bae D, Kang KY, Yoon SC, Lim D (2004) Defining the plant disulfide proteome. Electrophoresis 25:532–541

Lee U, Wie C, Fernandez BO, Feelisch M, Vierling E (2008) Modulation of nitrosative stress by S-nitrosoglutathione reductase is critical for thermotolerance and plant growth in *Arabidopsis*. Plant Cell 20:786–802

Leferink NG, Van Duijn E, Barendregt A, Heck AJ, Van Berkel WJ (2009) Galactonolactone dehydrogenase requires a redox-sensitive thiol for optimal production of vitamin C. Plant Physiol 150:596–605

Leichert LI, Jakob U (2004) Protein thiol modifications visualized *in vivo*. PLoS Biol 2:1723–1737

Leterrier C, Vacher H, Fache MP, d'Ortoli SA, Castets F, Autillo-Touati A, Dargent B (2011) End-binding proteins EB3 and EB1 link microtubules to ankyrin G in the axon initial segment. Proc Natl Acad Sci U S A 108(21):8826–8831

Leterrier M, Airaki M, Palma JM, Chaki M, Barroso JB, Corpas FJ (2012) Arsenic triggers the nitric oxide (NO) and S-nitrosoglutathione (GSNO) metabolism in *Arabidopsis*. Environ Pollut 166:136–143

Li S (2014) Redox modulation matters: emerging functions for glutaredoxins in plant development and stress responses. Plant (Basel) 3(4):559–582

Li J, Jia H, Wang J, Cao Q, Wen Z (2014) Hydrogen sulfide is involved in maintaining ion homeostasis via regulating plasma membrane Na^+/H^+ antiporter system in the hydrogen peroxide-dependent manner in salt-stress *Arabidopsis thaliana* root. Protoplasma 251:899–912

Lin AH, Wang YQ, Tang JY, Xue P, Li CL, Liu LC, Hu B, Yang F, Loake GJ, Chu C (2012) Nitric oxide and protein S-nitrosylation are integral to hydrogen peroxide-induced leaf cell death in rice. Plant Physiol 158:451–464

Lindermayr C, Saalbach G, Durner J (2005) Proteomic identification of S-nitrosylated proteins in *Arabidopsis*. Plant Physiol 137:921–930

Lindermayr C, Saalbach G, Bahnweg G, Durner J (2006) Differential inhibition of *Arabidopsis* methionine adenosyltransferases by protein S-nitrosylation. J Biol Chem 281:4285–4291

Lindermayr C, Sell S, Muller B, Leister D, Durner J (2010) Redox regulation of the NPR1-TGA1 system of *Arabidopsis thaliana* by nitric oxide. Plant Cell 22:2894–2907

Lowther WT, Haynes AC (2011) Reduction of cysteine sulfinic acid in eukaryotic, typical 2-Cys peroxiredoxins by sulfiredoxin. Antioxid Redox Signal 15:99–109

Ma FF, Lu R, Liu HY, Shi B, Zhang JH, Tan MP, Zhang A, Jiang M (2012) Nitric oxide activated calcium/calmodulin-dependent protein kinase regulates the abscisic acid induced antioxidant defence in maize. J Exp Bot 63:4835–4847

Marino SM, Gladyshev VN (2010) Structural analysis of cysteine S-nitrosylation: a modified acid-based motif and the emerging role of trans-nitrosylation. J Mol Biol 395:844–859

Martinez-Ruiz A, Araujo IM, Izquierdo-Alvarez A, Hernansanz-Agustin P, Lamas S, Serrador JM (2013) Specificity in S-nitrosylation: a short-range mechanism for NO signaling? Antioxid Redox Signal 19:1220–1235

Matsumoto M, Ogawa K (2008) New insight into the Calvin cycle regulation – glutathionylation of fructose bisphosphate aldolase in response to illumination. In: Photosynthesis. Energy from the sun. Springer, Dordrecht, pp 871–874

Meyer Y, Belin C, Delorme-Hinoux V, Reichheld JP, Riondet C (2012) Thioredoxin and glutaredoxin systems in plants: molecular mechanisms, crosstalks, and functional significance. Antioxid Redox Signal 17:1124–1160

Michelet L, Zaffagnini M, Marchand C, Collin V, Decottignies P, Tsan P, Lancelin JM, Trost P, Miginiac-Maslow M, Noctor G, Lemaire SD (2005) Glutathionylation of chloroplast thioredoxin f is a redox signaling mechanism in plants. Proc Natl Acad Sci U S A 102:16478–16483

Michelet L, Zaffagnini M, Vanacker H, Le Marechal P, Marchand C, Schroda M, Lemaire SD, Decottignies P (2008) *In vivo* targets of S-thiolation in *Chlamydomonas reinhardtii*. J Biol Chem 283:21571–21578

Michelet L, Zaffagnini M, Morisse S, Sparla F, Pérez-Pérez ME, Francia F, Danon D, Marchand CH, Fermani S, Trost P, Lemaire SD (2013) Redox regulation of the Calvin–Benson cycle: something old, something new. Front Plant Sci 4:470

Mieyal JJ, Chock PB (2012) Posttranslational modification of cysteine in redox signaling and oxidative stress: focus on S-glutathionylation. Antioxid Redox Signal 16(6):471–475

Mishra M, Jiang H, Wu L, Chawsheen HA, Wei Q (2015) The sulfiredoxin-peroxiredoxin (Srx-Prx) axis in cell signal transduction and cancer development. Cancer Lett 366:150–159

Mohapatra S, Mittra B (2016) Protein glutathionylation protects wheat (*Triticum aestivum* Var. Sonalika) against *Fusarium* induced oxidative stress. Plant Physiol Biochem 109:319–325

Møller IM, Rogowska-Wrzesinska A, Rao RS (2011) Protein carbonylation and metal-catalyzed protein oxidation in a cellular perspective. J Proteome 74:2228–2242

Morisse S, Zaffagnini M, Gao XH, Lemaire SD, Marchand CH (2014) Insight into protein S-nitrosylation in *Chlamydomonas reinhardtii*. Antioxid Redox Signal 21:1271–1284

Mortimer JC, Laohavisit A, Macpherson N, Webb A, Brownlee C, Battey NH, Davies JM (2008) Annexins: multifunctional components of growth and adaptation. J Exp Bot 59:533–544

Nagy P (2013) Kinetics and mechanisms of thiol–disulfide exchange covering direct substitution and thiol oxidation-mediated pathways. Antioxid Redox Signal 18(13):1623–1641

Nagy P, Winterbourn CC (2010) Rapid reaction of hydrogen sulfide with the neutrophil oxidant hypochlorus acid to generate polusulfides. Chem Res Toxicol 23(10):1541–1543

Navari-Izzo F, Quartacci MF, Pinzino C, Rascio N, Vazzana C, Sgherri CLM (2000) Protein dynamics in thylakoids of the desiccation-tolerant plant *Boea hygroscopica* during dehydration and rehydration. Plant Physiol 124(3):1427–1436

Netto LES, de Oliveira MA, Monteiro G, Demasi APD, Cussiol JRR, Discola KF, Demasi M, Silva GM, Alves SV, Faria VG, Horta BB (2007) Reactive cysteine in proteins: protein folding, antioxidant defense, redox signaling and more. Comp Biochem Physiol C 146:180–193

Noguera-Mazon V, Lemoine J, Walker O, Rouhier N, Salvador A, Jacquot JP, Milzani A (2006) Glutathionylation induces the dissociation of 1-Cys D-peroxiredoxin non-covalent homodimer. J Biol Chem 281:31736–31742

Nordstrand K, Åslund F, Holmgren A, Otting G, Berndt KD (1999) NMR structure of *Escherichia coli* glutaredoxin 3-glutathione mixed disulfide complex: implications for the enzymatic mechanism. J Mol Biol 286(541):552

Obinger C, Regelsberger G, Strasser G, Burner U, Peschek GA (1997) Purification and characterization of a homodimeric catalase-peroxidase from the cyanobacterium *Anacystis nidulans*. Biochem Biophys Res Commun 235:545–552

Ortega-Galisteo AP, Rodriguez-Serrano M, Pazmino DM, Gupta DK, Sandalio LM, Romero-Puertas MC (2012) S-Nitrosylated proteins in pea (*Pisum sativum* L.) leaf peroxisomes: changes under abiotic stress. J Exp Bot 63:2089–2103

Palmieri L, Papaleo V, Porcelli V, Scarcia P, Gaita L, Sacco R (2010) Altered calcium homeostasis in autism-spectrum disorders: evidence from biochemical and genetic studies of the mitochondrial aspartate/glutamate carrier AGC1. Mol Psychiatry 15:38–52

Peltoniemi MJ, Karala AR, Jurvansuu JK, Kinnula VL, Ruddock LW (2006) Insights into deglutathionylation reactions. Different intermediates in the glutaredoxin and protein disulfide isomerase catalyzed reactions are defined by the gamma-linkage present in glutathione. J Biol Chem 281:33107–33114

Perazzolli M, Dominici P, Romero-Puertas MC, Zago ED, Zeier J, Sonoda M, Lamb C, Delledonne M (2004) *Arabidopsis* non-symbiotic haemoglobin AHb1 modulates nitric oxide bioactivity. Plant Cell 16:1–10

Perez-Bueno ML, Rahoutei J, Sajnani C, García-Luque I, Barón M (2004) Proteomic analysis of the oxygen evolving complex of photosystem II under biotic stress: studies on *Nicotiana benthamiana* infected with tobamoviruses. Proteomics 4:418–425

Pivato M, Fabrega-Prats M, Masi A (2014) Low-molecular-weight thiols in plants: functional and analytical implications. Arch Biochem Biophys 560:83–99

Pukacka S, Ratajczak E (2007) Ascorbate and glutathione metabolism during development and desiccation of orthodox and recalcitrant seeds of the genus *Acer*. Funct Plant Biol 34:601–613

Quijano C, Alvarez B, Gatti RM, Augusto O, Radi R (1997) Pathways of peroxynitrite oxidation of thiol groups. Biochemist J322:167–173

Reddie KG, Carroll KS (2008) Expanding the functional diversity of proteins through cysteine oxidation. Curr Opin Chem Biol 12(6):746–754

Reddy AS (2001) Molecular motors and their functions in plants. Int Rev Cytol 204:97–178

Rey P, Bécuwe N, Barrault M-B, Rumeau D, Havaux M, Biteau B, Toledano MB (2007) The *Arabidopsis thaliana* sulfiredoxin is a plastidic cysteine-sulfinic acid reductase involved in the photooxidative stress response. Plant J 49:505–514

Rhazi L, Cazalis R, Lemelin E, Aussenac T (2003) Changes in the glutathione thiol-disulfide status during wheat grain development. Plant Physiol Biochem 41:895–902

Rodríguez-Serrano M, Pazmiño DM, Sparkes I, Rochetti A, Hawes C, Romero-Puertas MC, Sandalio LM (2014) 2,4-Dichlorophenoxyacetic acid promotes S-nitrosylation and oxidation of actin affecting cytoskeleton and peroxisomal dynamics. J Exp 65(17):4783–4793

Romero-Puertas MC, Campostrini N, Matte A, Righetti PG, Perazzolli M, Zolla L, Roepstorff P, Delledonne M (2008) Proteomic analysis of S-nitrosylated proteins in *Arabidopsis thaliana* undergoing hypersensitive response. Proteomics 8:1459–1469

Roos G, Messens J (2011) Protein sulfenic acid formation: from cellular damage to redox regulation. Free Radic Biol Med 51:314–326

Rouhier N (2010) Plant glutaredoxins: pivotal players in redox biology and iron-sulphur centre assembly. New Phytol 186:365–372

Rouhier N, Lemaire SD, Jacquot JP (2008) The role of glutathione in photosynthetic organisms: emerging functions for glutaredoxins and glutathionylation. Annu Rev Plant Biol 59:143–166

Schnaubelt D, Queval G, Dong Y, Diaz-Vivancos P, Makgopa ME, Howell G, De Simone A, Bai J, Hannah MA, Foyer CH (2015) Low glutathione regulates gene expression and the redox potentials of the nucleus and cytosol in *Arabidopsis thaliana*. Plant Cell Environ 38(2):266–279

Serpa V, Vernal J, Lamattina L, Grotewold E, Cassia R, Terenzi H (2007) Inhibition of AtMYB2 DNA-binding by nitric oxide involves cysteine S-nitrosylation. Biochem Biophys Res Commun 361:1048–1053

Serrato AJ, Fernández-Trijueque J, Barajas-López J-D, Chueca A, Sahrawy M (2013) Plastid thioredoxins: a "one-for-all" redox-signaling system in plants. Front Plant Sci 4:463

Skaff O, Pattison DI, Davies MJ (2009) Hypothiocyanous acid reactivity with low-molecular-mass and protein thiols: absolute rate constants and assessment of biological relevance. Biochemist J422(1):111–117

Spadaro D, Yun B-W, Spoel SH, Chu C, Wang Y-Q, Loake GJ (2010) The redox switch: dynamic regulation of protein function by cysteine modifications. Physiol Plant 138:360–371

Staab CA, Ålander J, Brandt M, Lengqvist J, Morgenstern R, Grafström RC, Höög JO (2008) Reduction of S-nitrosoglutathione by alcohol dehydrogenase 3 is facilitated by substrate alcohols via direct cofactor recycling and leads to GSH-controlled formation of glutathione transferase inhibitors. Biochem J 413:493–504

Stamler JS, Lamas S, Fang FC (2001) Nitrosylation. The prototypic redox-based signaling mechanism. Cell 106(6):675–683

Suzuki N, Koussevitzky S, Mittler R, Miller G (2012) ROS and redox signalling in the response of plants to abiotic stress. Plant Cell Environ 35:259–270

Svensson BE (1993) Thiols as peroxidase substrates. Free Radic Biol Med 14(2):167–175

Tada Y, Spoel SH, Pajerowska-Mukhtar K, Mou Z, Song J, Wang C, Zuo J, Dong X (2008) Plant immunity requires conformational charges of NPR1 via S-nitrosylation and thioredoxins. Science 321:952–956

Tanou G, Job C, Rajjou L, Arc E, Belghazi M, Diamantidis G, Molassiotis A, Job D (2009) Proteomics reveals the overlapping roles of hydrogen peroxide and nitric oxide in the acclimation of citrus plants to salinity. Plant J 60:795–804

Tanou G, Filippou P, Belghazi M, Job D, Diamantidis G, Fotopoulos V, Molassiotis A (2012) Oxidative and nitrosative-based signaling and associated post-translational modifications orchestrate the acclimation of citrus plants to salinity stress. Plant J 72:585–599

Terrile MC, París R, Calderón-Villalobos LIA, Iglesias MJ, Lamattina L, Estelle M, Casalongué CA (2012) Nitric oxide influences auxin signaling through S-nitrosylation of the *Arabidopsis* transport inhibitor response 1 auxin receptor. Plant J 70(3):492–500

Trivedi MV, Laurence JS, Siahaan TJ (2009) The role of thiols and disulfides in protein chemical and physical stability. Curr Protein Pept Sci 10(6):614–625

Trujillo M, Alvarez B, Radi R (2016) One- and two-electron oxidation of thiols: mechanisms, kinetics and biological fates. Free Radic Res 50(2):150–171

Wang YQ, Feechan A, Yun BW, Shafiei R, Hofmann A, Taylor P, Xue P, Yang FQ, Xie ZS, Pallas JA, Chu CC, Loake GJ (2009a) S-nitrosylation of AtSABP3 antagonizes the expression of plant immunity. J Biol Chem 284(4):2131–2137

Wang L, Tsuda K, Sato M, Cohen JD, Katagiri F, Glazebrook J (2009b) *Arabidopsis* CaM binding protein CBP60g contributes to MAMP-induced SA accumulation and is involved in disease resistance against *Pseudomonas syringae*. PLoS Pathog 5:e1000301

Wang W-Q, Møller IM, Song S-Q (2012) Proteomic analysis of embryonic axis of *Pisum sativum* seeds during germination and identification of proteins associated with loss of desiccation tolerance. J Proteome 17:68–86

Winterbourn CC, Metodiewa D (1999) Reactivity of biologically important thiol compounds with superoxide and hydrogen peroxide. Free Radic Biol Med 27(3):322–328

Woo HA, Jeong W, Chang TS, Park KJ, Park SJ, Yang JS, Rhee SG (2005) Reduction of cysteine sulfinic acid by sulfiredoxin is specific to 2-cys peroxiredoxins. J Biol Chem 280:3125–3128

Xie Y, Kole S, Precht P, Pazin MJ, Bernier M (2009) S-glutathionylation impairs signal transducer and activator of transcription 3 activation and signaling. Endocrinology 150:1122–1131

Yun BW, Feechan A, Yin M, Saidi NB, Le Bihan T, Yu M, Moore JW, Kang JG, Kwon E, Spoel SH, Pallas JA, Loake GJ (2011) S-nitrosylation of NADPH oxidase regulates cell death in plant immunity. Nature 478(7368):264–268

Zaffagnini M, Bedhomme M, Groni H, Marchand CH, Puppo C, Gontero B, Cassier-Chauvat C, Decottingnies P, Lamaire SD (2012a) Glutathinonylation in the photosynthetic model organism *Chlamydomonas reinhardtii*: a proteomic survey. Mol Cell Proteomics 11(2):M111.014142

Zaffagnini M, Bedhomme M, Marchand CH, Couturier JR, Gao XH, Rouhier N, Trost P, Lemaire SD (2012b) Glutaredoxin S12: unique properties for redox signaling. Antioxid Redox Signal 16:17–32

Zaffagnini M, Bedhomme M, Marchard AH, Morisse S, Trost P, Lemaire S (2012c) Redox regulation in photosynthetic organisms: focus on glutathionylation. Antioxid Redox Signal 16(6):567–586

Zaffagnini M, Fermani S, Costa A, Lemaire SD, Trost P (2013) Plant cytoplasmic GAPDH: redox post-translational modifications and moonlighting properties. Front Plant Sci 4:450

Zagorchev L, Seal CE, Kranner I, Odjakova M (2012) Redox state of low-molecular-weight thiols and disulphides during somatic embryogenesis of salt-treated suspension cultures of *Dactylis glomerata* L. Free Radic Res 46:656–664

Zagorchev L, Seal CE, Kranner I, Odjakova M (2013) A central role for thiols in plant tolerance to abiotic stress. Int J Mol Sci 14(4):7405–7432

Zhang H, Xu Y, Joseph J, Kalyanaraman B (2008) Influence of intramolecular electron transfer mechanism in biological nitration, nitrosation, and oxidation of redox-sensitive amino acids. Methods Enzymol 440:65–94

Chapter 4
Plant Glutathione Peroxidases: Structural and Functional Characterization, Their Roles in Plant Development

Krisztina Bela, Sajid Ali Khan Bangash, and Jolán Csiszár

Abstract Glutathione peroxidase (GPX) enzymes (EC 1.11.1.9, EC 1.11.1.12 and EC 1.11.1.15) are widespread among eukaryotic organisms and have pivotal role in peroxide detoxification. Detailed phylogenetic analysis of thiol peroxidases, including eukaryotic peroxiredoxins (PRXs) and eukaryotic GPXs, has revealed that plant glutathione peroxidases are more closely related to mammalian hydroperoxide peroxidases than to fungal GPXs, and a more precise name for them should be glutathione peroxidase-like (GPXL) enzymes. Plant GPXLs are mostly monomeric proteins that use the thioredoxin (TRX) system more effectively than the glutathione system during the reduction of H_2O_2 and lipid peroxides. GPXLs were suggested to be a putative link between the glutathione-based and the thioredoxin-based detoxifying systems. They possess some functional overlaps with the PRXs and glutathione transferases (GSTs), with respect to the maintenance of H_2O_2 homeostasis by the elimination of peroxides, and are also involved in the regulation of the redox homeostasis by maintaining the thiol-disulfide balance. In addition GPXs have been shown to perform other functions, including acting as redox transducers in abscisic acid (ABA) regulated responses, drought stress signaling, and redox-associated modification of nuclear proteins. Plant GPXLs not only protect cells from stress-induced oxidative damage, but they have also implicated in the regulation of processes associated with plant growth and development. Among these processes, their involvement in plant regeneration and shoot organogenesis is discussed.

Keywords Glutathione • Glutathione peroxidases • Plant development • Thioredoxin

K. Bela (✉) • J. Csiszár
Department of Plant Biology, University of Szeged, Közép fasor 52, Szeged H-6726, Hungary
e-mail: belakriszti88@gmail.com

S.A.K. Bangash
Chemical Signalling, INRES, University of Bonn,
Friedrich-Ebert-Allee 144, Bonn D-53113, Germany

M.A. Hossain et al. (eds.), *Glutathione in Plant Growth, Development, and Stress Tolerance*, https://doi.org/10.1007/978-3-319-66682-2_4

Contents

1 Introduction

The glutathione peroxidases (GPXs; EC 1.11.1.9, EC 1.11.1.12, and EC 1.11.1.15) were discovered 60 years ago in erythrocytes and characterized as antioxidant enzymes (Mills 1957). Over the years, GPXs have been extensively studied and described as seleno- or nonseleno-proteins that reduce H_2O_2 or organic hydroperoxides to water or corresponding alcohols, using glutathione as the reducing substrate in mammalian cells (Herbette et al. 2007). In plants, homologues to one member of the animal GPX gene family (GPX4 or phospholipid hydroperoxide glutathione peroxidase (PHGPX)) have also been identified and isolated from several plant species. In 1985, Drotar et al. (1985) proved that GPX activity occurred in cultured plant cells, and more recently different isoforms of plant GPXs have been found in various subcellular compartments (Navrot et al. 2006; Margis et al. 2008; Attacha et al. 2017). The original ancestor of the GPX gene family is still unclear and with gene evolution has been shown to be nonlinear (Margis et al. 2008). According to their protein structures, GPXs can be divided into three main polyphyletic clusters: vertebrates and invertebrates, bacteria and fungi, and plant GPXs. Due to their structural similarity to animal GPXs, but different activities and substrate specificities, plant GPXs are described as glutathione peroxidase-like enzymes (GPXLs) (Attacha et al. 2017). Eight GPXLs have been identified in *Arabidopsis thaliana*, which have been reported to be expressed differentially in the cytosol, chloroplasts, Golgi, mitochondria, nucleus, and plasma membrane (Attacha et al. 2017). In contrast, *Oryza sativa* has only five GPXLs, predicted to be localized in the cytosol, chloroplasts, and mitochondria (Islam et al. 2015). Based upon phylogenetic analysis of plant GPXL protein sequences, it has been proposed that plant GPXLs can be divided into five major groups, with GPXLs with similar predicted subcellular localization clustered together (Islam et al. 2015; Ozyigit et al. 2016). Between GPXLs several segmental duplications have been found, indicating the possibility of gene duplication events (Ozyigit et al. 2016).

Plant glutathione peroxidase-like enzymes show altered substrate usage and lower peroxidase activity compared to true glutathione peroxidases. In addition, some plant glutathione transferases (GSTs) play important roles in abiotic and biotic stress responses and have greater GSH-dependent peroxidase activities, against H_2O_2 and organic peroxides, than GPXLs (Dixon and Edwards 2010).

2 Structure of Plant GPXLs

The protein structure of GPXs consists of central β-sheets surrounded by α-helices (Koh et al. 2007), and the structure is conserved among the GPX family members, with only a few exceptions arising due to differences in oligomerization. Some mammalian GPXs form tetramers, facilitated by their oligomerization loop between the α3-helix and β6-strand (Toppo et al. 2008), while in contrast the monomer PHGPXs and plant GPXLs do not contain oligomerization loops (Maiorino et al. 1995). However, *Populus trichocarpa* GPXL5 does not possess an oligomerization loop, but can form a non-covalent dimer with the help of hydrophobic and aromatic residues (Koh et al. 2007).

Even plant GPXs, which are closely related to animal PHGPXs (Margis et al. 2008), differ in the cysteine type and number of cysteines in their catalytic sites. PHGPXs are selenoenzymes and have a selenocysteine that is responsible for the peroxidase activity (Toppo et al. 2008; Tosatto et al. 2008; Toppo et al. 2009), while plant GPXLs are seleno-independent proteins and contain cysteines in their active site (Eshdat et al. 1997). In addition, animal GPXs possess two cysteines (one selenocysteine and one cysteine), while plant GPXLs contain three cysteines, only two of which, the peroxidatic and the resolving cysteines, take part in the catalytic cycle (Navrot et al. 2006; Koh et al. 2007). During the reduction of peroxide, a sulfenic acid is formed on the peroxidatic cysteine; after that the resolving cysteine can form an intramolecular disulfide bridge with the peroxidatic cysteine. This organization is not universal in all plant species, for example, the third cysteine is the resolving type in some plants, e.g., poplar (Navrot et al. 2006), but in Chinese cabbage it has been established that both the second and the third cysteine can be responsible for disulfide bridge formation (Jung et al. 2002). The intramolecular rearrangement of plant GPXLs during the catalytic cycle and regeneration is similar to that of the peroxiredoxins (PRXs), because of their substrate specificity, and these plant GPXLs are considered as a fifth group of peroxiredoxins (Navrot et al. 2006; Rouhier and Jacquot 2005). For regeneration, GPXs normally use reduced GSH, but may react also with reduced thioredoxin (TRX) (Koh et al. 2007). Plant GPXLs prefer TRX for the regeneration system (Jung et al. 2002; Iqbal et al. 2006) (Fig. 4.1).

3 Biochemical Characterization of Plant GPXL Enzymes

GPXs in animals are the major antioxidant enzymes responsible for protecting cells against oxidative stress via the reduction of H_2O_2 and organic hydroperoxides using reduced glutathione (Maiorino et al. 1995) or thioredoxin as substrates (Bjornstedt et al. 1994). In plants, H_2O_2 elimination is mainly carried out by ascorbate peroxidase (APX) and catalase (CAT) isoenzymes (Ozyigit et al. 2016), since plant GPXLs have lower peroxidase activities. A possible explanation for this was proposed by Maiorino et al. (1995) who demonstrated that replacing a selenocysteine for

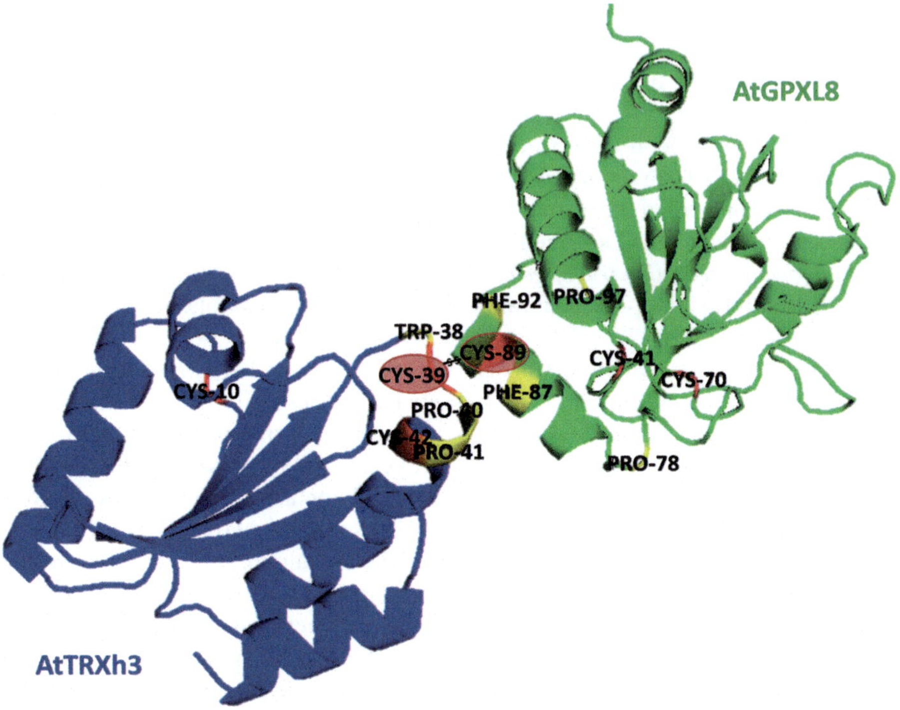

Fig. 4.1 Model of the *Arabidopsis* GPXL8 and TRXh3 complex interaction, 3D protein structures created by PyMOL software (PyMOL Molecular Graphics System, Version 1.8 Schrödinger, LLC). The model based on poplar GPXL5 by Koh et al. (2007). During the GPX regeneration, the CYS-39 of TRXh3 forms a disulfide bond with the CYS-89 of GPXL8, while aromatic residues could involve in the protein-protein interaction (Koh et al. 2007)

cysteine in an animal PHGPX, as found in plant GPXLs, leads to a drastic decrease in enzyme activity.

Herbette et al. (2002) investigated two plant GPXL isoenzymes from sunflower and tomato. They have found using GSH as a regeneration substrate that these enzymes showed low activities against organic hydroperoxides (v_{max} = 15.8–57.5 nmol*min^{-1}*mg^{-1}; K_m = 12.1–128 μM) and undetectable activities toward H_2O_2. Nonetheless, using purified *E. coli* TRX as an electron donor, these enzymes showed increased activities against organic hydroperoxides (v_{max} = 108.7–169.5 nmol*min^{-1}*mg^{-1}; K_m = 8.6–16.6 μM) and H_2O_2 (v_{max} = 147.1–153.8 nmol*min^{-1}*mg^{-1}; K_m = 13.7–13.9 μM) as well (Herbette et al. 2002). Similar substrate preference has been found for Chinese cabbage (Jung et al. 2002), poplar (Navrot et al. 2006), and *Arabidopsis* (Iqbal et al. 2006) GPXLs.

Most plant GPXLs prefer TRX as a reduction substrate, compared to GSH (Jung et al. 2002; Iqbal et al. 2006). Their connection to both GSH and TRX systems was also indicated by characterization of *Arabidopsis root meristemless1 (rml1)* mutant, which is severely limited in GSH synthesis capacity (Vernoux et al. 2000).

Transcriptome profiling of the *rml1-1* revealed altered expression of different GSH- and TRX-dependent genes, such as GSTs, TRXs, and GPXLs compared to wild type (Schnaubelt et al. 2015).

4 Involvement of GPXs in Signaling

Reactive oxygen species (ROS), such as superoxide radicals ($O_2^{\bullet -}$), hydrogen peroxide (H_2O_2), hydroxyl radicals ($OH^{\bullet}$), and singlet oxygen ($^{1}O_2$), are Janus-faced elements and are toxic to cells, causing oxidative damage to macromolecules, but can also function as signaling molecules. Among the ROS, H_2O_2 is a relatively stable molecule (with a longer half-life) with a selective reactivity that favors Cys-containing proteins and peptides. Oxidative modification of a Cys thiol group to a sulfenic acid (Cys-SOH) can act as a regulatory switch (Ma et al. 2007). However, while Cys-SOH formation is reversible, Cys-SOHs can react rapidly with other thiols (GSH) to form intra- or intermolecular disulfides (S-glutathionylation), protecting Cys-SOH against overoxidation (Waszczak et al. 2014). GPXs act as intermediates that can transfer the redox signals, allowing the oxidation of regulatory proteins that are otherwise not directly able to react with H_2O_2. The oxidation of cysteine-containing proteins involved in signaling, such as phosphatases, kinases, and transcription factors, may cause a change in their biological activity; therefore, different pathways can be initialized (Luo et al. 2005; Marinho et al. 2014).

The *S. cerevisiae* GPX3 is one of the most important antioxidant enzymes that modulates the activities of redox-sensitive thiol proteins, particularly those involved in signal transduction pathways and protein translocation. This nonselenoenzyme is able to interact with transcription factor Yap1 and responds to variations of H_2O_2 levels. Yap1 contains two cysteine-rich regulatory domains that can be oxidized by ROS or thiol-active electrophiles resulting in intramolecular disulfide bond formation and nuclear localization (Delaunay et al. 2002; Herrero et al. 2008). If the nuclear export of Yap1 is decreased and Yap1 is retained in the nucleus, it can regulate its target genes (Delaunay et al. 2000). As a result the Yap1 transcription factor activates the expression of antioxidant genes and so regulates hydroperoxide homeostasis in yeast (Delaunay et al. 2002), indicating that GPX3 can act as H_2O_2 receptor and redox transducer. Kho et al. (2006) found that GPX3 interacts also with methionine sulfoxide reductase (MXR1), through the formation of an intermolecular disulfide bond. MXR1 reverses the inactivation of proteins caused by the oxidation of critical methionine residues by reducing methionine sulfoxide to methionine. Under physiological conditions, Cys82 of GPX3 binds to Cys176 of MXR1. Upon oxidative stress, this disulfide bond is broken, and Cys82 of GPX3 is able to bind to Cys36-SOH through a thiol-disulfide exchange reaction. Accordingly, MXR1 is released and can repair oxidized proteins. The interaction between GPX3 and MXR1 may serve as an important and efficient regulatory link between ROS detoxification enzymes and repairing proteins (Kho et al. 2006).

The role of GPXs in signaling occurs in a wide range of organisms. Selenocysteine-containing GPXs are reactive with H_2O_2 and could take part in H_2O_2-based redox regulation. In mammalian sperm, GPX4 after oxidization by H_2O_2 acts as a signal transducer and is able to react with sperm mitochondria-associated cysteine-rich proteins, causing a reshuffling of the target protein cysteine residues, leading to the assembly and stabilization of spermatozoa mitochondrial capsule, and enhancing sperm motility (Maiorino et al. 2005). This signaling pathway is possibly regulated by the glutathione redox state, as the reduced glutathione is an antagonist regeneration substrate of the GPX4. Lack of GPX4 also promotes apoptotic signaling as the accumulation of phospholipid-hydroperoxides stimulates the release of apoptosis-inducing factor to the nucleus (Seiler et al. 2008). It has been shown that mammalian GPX1 has many cellular functions within the cell. It protects cells from oxidative damages, regulates metabolism and mitochondrial function, and controls cellular processes, such as apoptosis, growth, and signaling by modulating intracellular levels of hydrogen peroxide and the overall intracellular redox balance. GPX1 is implicated in inflammation processes, some cancers, and cardiovascular diseases (Lubos et al. 2011a; Lubos et al. 2011b). GPX1 also regulates insulin signaling by affecting the level of H_2O_2, which is required for the oxidative inactivation of protein tyrosine phosphatase 1β. Overexpression of the GPX1 enzyme leads to obesity and diabetes in mice (McClung et al. 2004). The non-selenocysteine containing mammalian GPX7 interacts with numerous proteins including 78-kDa glucose-regulated protein (GRP78) and protein disulfide isomerase (PDI). Upon oxidative stress, GPX7 is activated and transmits the disulfide bonding to specific proteins and turns on their activities to eliminate the stress, e.g., by facilitating protein folding in ER (Chen et al. 2016).

Plant GPX-like enzymes also play important roles in ROS-based signaling. According to the results of Miao et al. (2006), *Arabidopsis* GPXL3 functions as both a general scavenger and as an oxidative signal transducer specifically relaying the H_2O_2 signal in abscisic acid (ABA) and drought stress signaling. AtGPXL3 physically interacted with the 2C-type protein phosphatase ABA-insensitive 1 and 2 (ABI1 and ABI2) proteins. The redox states of both AtGPXL3 and ABI2 were found to be regulated by H_2O_2. The phosphatase activity of ABI2, measured in vitro, was reduced approximately fivefold by the addition of oxidized AtGPXL3. The reduced form of ABI2 was converted to the oxidized form by the addition of oxidized AtGPXL3 in vitro, which might mediate ABA and oxidative signaling. The *atgpxl3* mutation disrupted the ABA activation of calcium channels and the expression of ABA- and stress-responsive genes (Miao et al. 2006). The connection between a Ca^{2+}-mediated signal transduction pathway and the H_2O_2 activation of *GPXLs* was reported also in tobacco BY-2 cells (Kadota et al. 2005). The working of the oxidative stress-responsive putative voltage-dependent Ca^{2+}-permeable channel and the expression of peroxidases showed cell cycle dependence (Kadota et al. 2005). In *Arabidopsis thaliana*, GPXL8 enzyme localizes both in the cytosol and in the nucleus. Gaber et al. (2012) suggested that GPXL8 not only protects the cellular components against oxidative damages, but also has a role in redox modification of proteins, therefore taking part in nucleus signal transduction. The investigations of the role of tomato GPXL enzymes provided another example for involvement of GPXLs in signaling. Overexpression of tomato *GPXL5* led to accumulation of the RanBP1 (Ran-binding protein 1) signaling protein and a Calvin cycle-related protein, indicating a possible regulation of GPXL5 over photosynthetic processes (Herbette et al. 2005). Moreover, another SlGPXL has a function in apoptotic signaling, similarly to what was reported in animal GPX4 (Seiler et al. 2008). It can prevent the cells from Bcl-2-associated X protein (Bax)-induced cell death and is able to delay senescence by the reduction of phospholipid-hydroperoxides (Chen et al. 2004).

Because plant GPXLs have a weak activity toward H_2O_2, compared to APX and other peroxidases, they are regarded as being less important as peroxide scavengers (Eshdat et al. 1997), so it has been suggested that they may have other functions in plant cells (Milla et al. 2003; Herbette et al. 2007; Passaia and Margis-Pinheiro 2015). The most noteworthy discovery was that lack of GPXL function affected plant growth and development (Passaia et al. 2013; Passaia et al. 2014; Lima-Melo et al. 2016; Wang et al. 2017).

5 Role of GPXLs in Plant Development

Recently, there is increasing evidence for interactions between the glutathione and thioredoxin systems and the maintenance of the cellular redox homeostasis, which is critical for normal development, and successful organogenesis and regeneration of cultured cells (Marty et al. 2009; Bashandy et al. 2010; Lu and Holmgren 2014). Metabolic oxidation and GSH entrance into the nucleus is a crucial regulator of the cell cycle and cell differentiation (Diaz Vivancos et al. 2010). For the G1-S-phase transition of the cell cycle to occur, increased GSH levels are necessary; however, increased GSSG levels lead to arrested cell proliferation. *Arabidopsis rml1* mutants with severe GSH deficiency are unable to maintain the root apical meristem; however, the shoot apical meristem is not greatly affected, probably because of thioredoxin-dependent control (Diaz Vivancos et al. 2010). Moreover, the levels of GSH cause changes in antioxidant gene expression in *rml1* mutant plants. For example, the *GSTF11*, *GSTF14*, *GSTU20* and *GPXL1*, and *GPXL7* are downregulated, while *GSTF6*, *GSTF16*, *GSTU1*, *GSTU4*, *GSTU24*, *GSTU25*, and *GPXL6* are upregulated in *rml1* shoots, compared with wild-type plants (Schnaubelt et al. 2015). The glutathione redox potential has been suggested to act as a key determinant of cell death and dormancy in plants (Kranner et al. 2006). Furthermore, plants contain a large number of thioredoxins, which are some of the central players of thiol-disulfide homeostasis, and have important roles in plant growth, development, and chloroplast development. For example, *Arabidopsis trxh9* mutant plants are dwarfed with small yellowish leaves (Meng et al. 2010), *trxz* mutant seedlings have yellowish leaves and show limited growth, and without sucrose supplementation the mutation is lethal. Tobacco plants with silenced *TRXZ* gene also showed a similar phenotype (Arsova et al. 2010).

Utilizing the publicly available microarray and massively parallel signature sequencing (MPSS), data (https://www.genevestigator.com/gv/plant.jsp, http://mpss.udel.edu/at) revealed that expression profiles of *Arabidopsis* and rice *GPXL*s exhibit tissue and/

or organ specificities and are responsive to developmental stages (Bela et al. 2015; Islam et al. 2015). For example, the relatively high transcription level of *AtGPXL1*, *AtGPXL2*, *AtGPXL3*, and *AtGPXL6* in shoot apical meristems, seedlings, and rosette leaves suggests the physiological importance of the encoded isoenzymes in shoot development. The transcription of *AtGPXL2*, *AtGPXL3*, and *AtGPXL8* is activated under germination, while that of *AtGPXL1*, *AtGPXL4*, *AtGPXL6*, and *AtGPXL7* is repressed. During growth of pollen tube, the expression level of *AtGPXL7* and *AtGPXL8* increased; furthermore, a very high level of transcript amounts of *AtGPXL4* and *AtGPXL5* was found both in pollen and stamen (Bela et al. 2015). While in *Arabidopsis* the *AtGPXL1* and *AtGPXL6* showed high level of constitutive expression in silique, root, and inflorescence stage, in rice the *OsGPXL1* and *OsGPXL2* transcripts were the most abundant forms of GPXLs in all the analyzed tissues (Islam et al. 2015). The mitochondrial OsGPXL3 proved to be essential for normal *Oryza sativa* shoot development and seed production (Passaia et al. 2013) and for root development and photosynthesis (Lima-Melo et al. 2016). Mutation in *OsGPXL5* caused lower germination rate, reduced growth, and less filled grains compared to wild-type plants (Wang et al. 2017). Interestingly, 4-week-old *Arabidopsis* knockout mutants of GPXL7 under short-day conditions have greater rosette and under long-day photoperiod have more leaves than wild-type plants, indicating the role of GPXL7 in shoot development (Passaia et al. 2014). The relevance of AtGPXL7 in hormone-mediated root development, especially in lateral root development, was also demonstrated by using 1-naphtaleneacetic acid and synthetic strigolactone treatments (Passaia et al. 2014).

In silico searching for the *cis*-acting elements involved in different hormone regulation resulted in identification of auxin and methyl jasmonate response elements (in the 5′ regulator region of *AtGPXL2*, *AtGPXL6*, *AtGPXL7*, and *AtGPXL8*), gibberellin-responsive elements (in the upstream regulatory region of *AtGPXL6* and *AtGPXL8*), abscisic acid-responsive elements (in *AtGPXL1*, *AtGPXL2*, and *AtGPXL6*), and ethylene responsive element (in *AtGPXL1*, *AtGPXL4*, and *AtGPXL6)*. Additionally, meristem-specific *cis*-regulatory elements and seed development specific sequences were also identified in most of *AtGPXL* genes except for *AtGPXL4* (Bela et al. 2015).

Overexpression of a GPXL from *Citrus sinensis* displayed difficulties in the plant regeneration (Faltin et al. 2010). Because ROS homeostasis has essential role in the early stage of shoot organogenesis (Gupta and Datta 2003; Tian et al. 2003), and uncontrolled reduction of ROS leads to disruption of organ differentiation, this observation verifies the role of this enzyme in shoot organogenesis (Vernoux et al. 2000; Faltin et al. 2010).

6 Concluding Remarks

Biochemical properties of plant GPXs have several specific features: (1) they contain cysteine in their active site instead of the selenocysteine widespread in the living organisms, (2) they have lower peroxidase activity than Se-Cys-containing GPXs,

(3) several isoenzymes prefer to use TRX rather than GSH as a reductant compound, and (4) they may interact with regulatory proteins. Because of the substantial differences, and to avoid confusion resulting from nomenclature that is based on sequential homology with animal GPXs, plant GPXs are recently renamed as glutathione peroxidase-like (GPXL) enzymes (Attacha et al. 2017). Here we outlined that non-seleno GPXs could act as activated oxygen species sensors, hence being able to take part in ROS-mediated signaling pathways by oxidative modification of Cys thiol groups. In animals and yeast, increased numbers of GPXs have been proven to have important role in transcription factor activation or enzyme modification.

According to present knowledge, the thiol-dependent activities of plant GPXL enzymes propose their involvement in cellular redox homeostasis by modulating the disulfide state and maintaining the NADPH/NADP$^+$ balance. In the light of the recent findings, GPX and GPXL enzymes are not simply enzymatic compounds of the antioxidative defense. Plant GPXLs have crucial role in the correct plant development and in the maintenance of the redox state of cellular compounds either using GSH or TRX as substrate. They have a function in: (1) hormone-mediated growth of roots, (2) shoot organogenesis and plant regeneration, (3) flower and seed devel-

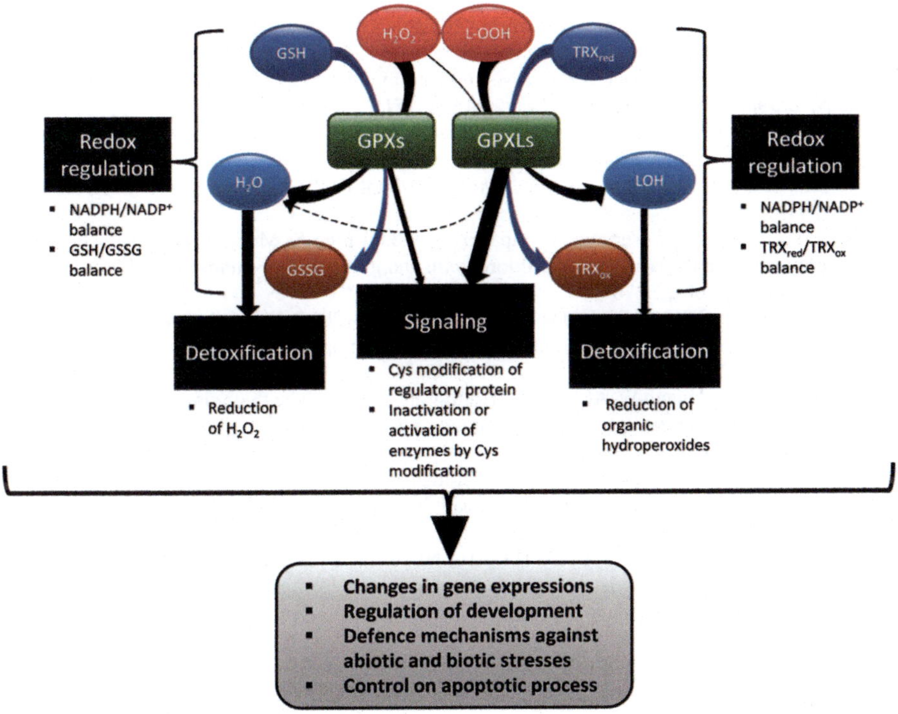

Fig. 4.2 Functions of glutathione peroxidases and glutathione peroxidase-like enzymes in cellular process. *GPXs* glutathione peroxidases, *GPXLs* glutathione peroxidase-like enzymes, *GSH* reduced glutathione, *GSSG* oxidized glutathione, *L-OOH* lipid-hydroperoxide, *LOH* lipid-alcohol, TRX_{red} reduced thioredoxin, TRX_{ox} oxidized thioredoxin

opment, and (4) suppression of programmed cell death. Although GPXLs are important elements of the normal development, until now only a few evidence have been found in plants about their functions as a signal transducer (Fig. 4.2). It is clear that there is still much more to discover about the full relevance of GPXs/GPXLs in plants. In the future, further studies will be required to characterize and identify the additional interaction partners of GPXs/GPXLs and thus get closer to a model for the whole signaling pathway network.

Acknowledgments We are grateful for the support of the Hungarian National Scientific Research Foundation [grant number OTKA K 105956]. S.A.K.B. thanks the Higher Education Commission (HEC) of Pakistan for a scholarship. K.B. received visiting scholarships from the Campus Mundi Program that is co-financed by the European Union and the Hungarian government. We thank José Manuel Ugalde for comments and suggestions that improved the manuscript. Last, but not least, we would like to express our gratitude to Dr. David J. Burritt for corrections and useful comments.

References

Arsova B, Hoja U, Wimmelbacher M, Greiner E, Ustun S, Melzer M, Petersen K, Lein W, Bornke F (2010) Plastidial thioredoxin z interacts with two fructokinase-like proteins in a thiol-dependent manner: evidence for an essential role in chloroplast development in *Arabidopsis* and *Nicotiana benthamiana*. Plant Cell 22:1498–1515

Attacha S, Solbach D, Bela K, Moseler A, Wagner S, Schwarzlander M, Aller I, Muller SJ, Meyer AJ (2017) Glutathione peroxidase-like enzymes cover five distinct cell compartments and membrane-surfaces in *Arabidopsis thaliana*. Plant Cell Environ. https://doi.org/10.1111/pce.12919

Bashandy T, Guilleminot J, Vernoux T, Caparros-Ruiz D, Ljung K, Meyer Y, Reichheld JP (2010) Interplay between the NADP-linked thioredoxin and glutathione systems in *Arabidopsis* auxin signaling. Plant Cell 22:376–391

Bela K, Horvath E, Galle A, Szabados L, Tari I, Csiszar J (2015) Plant glutathione peroxidases: emerging role of the antioxidant enzymes in plant development and stress responses. J Plant Physiol 176:192–201

Bjornstedt M, Xue J, Huang W, Akesson B, Holmgren A (1994) The thioredoxin and glutaredoxin systems are efficient electron donors to human plasma glutathione peroxidase. J Biol Chem 269:29382–29384

Chen S, Vaghchhipawala Z, Li W, Asard H, Dickman MB (2004) Tomato phospholipid hydroperoxide glutathione peroxidase inhibits cell death induced by Bax and oxidative stresses in yeast and plants. Plant Physiol 135:1630–1641

Chen YI, Wei PC, Hsu JL, Su FY, Lee WH (2016) NPGPx (GPx7): a novel oxidative stress sensor/transmitter with multiple roles in redox homeostasis. Am J Transl Res 8:1626–1640

Delaunay A, Isnard AD, Toledano MB (2000) H_2O_2 sensing through oxidation of the Yap1 transcription factor. EMBO J 19:5157–5166

Delaunay A, Pflieger D, Barrault MB, Vinh J, Toledano MB (2002) A thiol peroxidase is an H_2O_2 receptor and redox-transducer in gene activation. Cell 111:471–481

Diaz Vivancos P, Wolff T, Markovic J, Pallardo FV, Foyer CH (2010) A nuclear glutathione cycle within the cell cycle. Biochem J 431:169–178

Dixon DP, Edwards R (2010) Glutathione transferases. The Arabidopsis Book: e0131

Drotar A, Phelps P, Fall R (1985) Evidence for glutathione peroxidase activities in cultured plant cells. Plant Sci 42:35–40

Eshdat Y, Holland D, Faltin Z, Ben-Hayyim G (1997) Plant glutathione peroxidases. Physiol Plant 100:234–240

Faltin Z, Holland D, Velcheva M, Tsapovetsky M, Roeckel-Drevet P, Handa AK, Abu-Abied M, Friedman-Einat M, Eshdat Y, Perl A (2010) Glutathione peroxidase regulation of reactive oxygen species level is crucial for in vitro plant differentiation. Plant Cell Physiol 51:1151–1162

Gaber A, Ogata T, Maruta T, Yoshimura K, Tamoi M, Shigeoka S (2012) The involvement of Arabidopsis glutathione peroxidase 8 in the suppression of oxidative damage in the nucleus and cytosol. Plant Cell Physiol 53:1596–1606

Gupta SD, Datta S (2003) Antioxidant enzyme activities during in vitro morphogenesis of gladiolus and the effect of application of antioxidants on plant regeneration. Biol Plant 47:179–183

Herbette S, Lenne C, Leblanc N, Julien JL, Drevet JR, Roeckel-Drevet P (2002) Two GPX-like proteins from *Lycopersicon esculentum* and *Helianthus annuus* are antioxidant enzymes with phospholipid hydroperoxide glutathione peroxidase and thioredoxin peroxidase activities. Eur J Biochem 269:2414–2420

Herbette S, Menn AL, Rousselle P, Ameglio T, Faltin Z, Branlard G, Eshdat Y, Julien JL, Drevet JR, Roeckel-Drevet P (2005) Modification of photosynthetic regulation in tomato overexpressing glutathione peroxidase. Biochim Biophys Acta 1724:108–118

Herbette S, Roeckel-Drevet P, Drevet JR (2007) Seleno-independent glutathione peroxidases. FEBS J 274:2163–2180

Herrero E, Ros J, Belli G, Cabiscol E (2008) Redox control and oxidative stress in yeast cells. Biochim Biophys Acta 1780:1217–1235

Iqbal A, Yabuta Y, Takeda T, Nakano Y, Shigeoka S (2006) Hydroperoxide reduction by thioredoxin-specific glutathione peroxidase isoenzymes of *Arabidopsis thaliana*. FEBS J 273:5589–5597

Islam T, Manna M, Kaul T, Pandey S, Reddy CS, Reddy M (2015) Genome-wide dissection of Arabidopsis and rice for the identification and expression analysis of glutathione peroxidases reveals their stress-specific and overlapping response patterns. Plant Mol Biol Report 33:1413–1427

Jung BG, Lee KO, Lee SS, Chi YH, Jang HH, Kang SS, Lee K, Lim D, Yoon SC, Yun DJ, Inoue Y, Cho MJ, Lee SY (2002) A Chinese cabbage cDNA with high sequence identity to phospholipid hydroperoxide glutathione peroxidases encodes a novel isoform of thioredoxin-dependent peroxidase. J Biol Chem 277:12572–12578

Kadota Y, Furuichi T, Sano T, Kaya H, Gunji W, Murakami Y, Muto S, Hasezawa S, Kuchitsu K (2005) Cell-cycle-dependent regulation of oxidative stress responses and Ca^{2+} permeable channels NtTPC1A/B in tobacco BY-2 cells. Biochem Biophys Res Commun 336:1259–1267

Kho CW, Lee PY, Bae KH, Cho S, Lee ZW, Park BC, Kang S, Lee DH, Park SG (2006) Glutathione peroxidase 3 of *Saccharomyces cerevisiae* regulates the activity of methionine sulfoxide reductase in a redox state-dependent way. Biochem Biophys Res Commun 348:25–35

Koh CS, Didierjean C, Navrot N, Panjikar S, Mulliert G, Rouhier N, Jacquot JP, Aubry A, Shawkataly O, Corbier C (2007) Crystal structures of a poplar thioredoxin peroxidase that exhibits the structure of glutathione peroxidases: insights into redox-driven conformational changes. J Mol Biol 370:512–529

Kranner I, Birtic S, Anderson KM, Pritchard HW (2006) Glutathione half-cell reduction potential: a universal stress marker and modulator of programmed cell death? Free Radical Biol Med 40:2155–2165

Lima-Melo Y, Carvalho FE, Martins MO, Passaia G, Sousa RH, Neto MC, Margis-Pinheiro M, Silveira JA (2016) Mitochondrial GPX1 silencing triggers differential photosynthesis impairment in response to salinity in rice plants. J Integr Plant Biol 58:737–748

Lu J, Holmgren A (2014) The thioredoxin antioxidant system. Free Radic Biol Med 66:75–87

Lubos E, Kelly NJ, Oldebeken SR, Leopold JA, Zhang YY, Loscalzo J, Handy DE (2011a) Glutathione peroxidase-1 deficiency augments proinflammatory cytokine-induced redox signaling and human endothelial cell activation. J Biol Chem 286:35407–35417

Lubos E, Loscalzo J, Handy DE (2011b) Glutathione peroxidase-1 in health and disease: from molecular mechanisms to therapeutic opportunities. Antioxid Redox Signal 15:1957–1997

Luo D, Smith SW, Anderson BD (2005) Kinetics and mechanism of the reaction of cysteine and hydrogen peroxide in aqueous solution. J Pharm Sci 94:304–316

Ma LH, Takanishi CL, Wood MJ (2007) Molecular mechanism of oxidative stress perception by the Orp1 protein. J Biol Chem 282:31429–31436

Maiorino M, Aumann K-D, Brigelius-Flohé R, Doria D, van den Heuvel J, McCarthy J, Roveri A, Ursini F, Flohé L (1995) Probing the presumed catalytic triad of selenium-containing peroxidases by mutational analysis of phospholipid hydroperoxide glutathione peroxidase (PHGPx). Biol Chem 376:651–660

Maiorino M, Roveri A, Benazzi L, Bosello V, Mauri P, Toppo S, Tosatto SC, Ursini F (2005) Functional interaction of phospholipid hydroperoxide glutathione peroxidase with sperm mitochondrion-associated cysteine-rich protein discloses the adjacent cysteine motif as a new substrate of the selenoperoxidase. J Biol Chem 280:38395–38402

Margis R, Dunand C, Teixeira FK, Margis-Pinheiro M (2008) Glutathione peroxidase family – an evolutionary overview. FEBS J 275:3959–3970

Marinho HS, Real C, Cyrne L, Soares H, Antunes F (2014) Hydrogen peroxide sensing, signaling and regulation of transcription factors. Redox Biol 2:535–562

Marty L, Siala W, Schwarzlander M, Fricker MD, Wirtz M, Sweetlove LJ, Meyer Y, Meyer AJ, Reichheld JP, Hell R (2009) The NADPH-dependent thioredoxin system constitutes a functional backup for cytosolic glutathione reductase in *Arabidopsis*. Proc Natl Acad Sci U S A 106:9109–9114

McClung JP, Roneker CA, Mu W, Lisk DJ, Langlais P, Liu F, Lei XG (2004) Development of insulin resistance and obesity in mice overexpressing cellular glutathione peroxidase. Proc Natl Acad Sci U S A 101:8852–8857

Meng L, Wong JH, Feldman LJ, Lemaux PG, Buchanan BB (2010) A membrane-associated thioredoxin required for plant growth moves from cell to cell, suggestive of a role in intercellular communication. Proc Natl Acad Sci U S A 107:3900–3905

Miao Y, Lv D, Wang P, Wang XC, Chen J, Miao C, Song CP (2006) An Arabidopsis glutathione peroxidase functions as both a redox transducer and a scavenger in abscisic acid and drought stress responses. Plant Cell 18:2749–2766

Milla MAR, Maurer A, Huete AR, Gustafson JP (2003) Glutathione peroxidase genes in Arabidopsis are ubiquitous and regulated by abiotic stresses through diverse signaling pathways. Plant J 36:602–615

Mills GC (1957) Hemoglobin catabolism I. Glutathione peroxidase, an erythrocyte enzyme which protects hemoglobin from oxidative breakdown. J Biol Chem 229:189–197

Navrot N, Collin V, Gualberto J, Gelhaye E, Hirasawa M, Rey P, Knaff DB, Issakidis E, Jacquot JP, Rouhier N (2006) Plant glutathione peroxidases are functional peroxiredoxins distributed in several subcellular compartments and regulated during biotic and abiotic stresses. Plant Physiol 142:1364–1379

Ozyigit II, Filiz E, Vatansever R, Kurtoglu KY, Koc I, Ozturk MX, Anjum NA (2016) Identification and comparative analysis of H_2O_2-scavenging enzymes (ascorbate peroxidase and glutathione peroxidase) in selected plants employing bioinformatics approaches. Front Plant Sci 7:301

Passaia G, Margis-Pinheiro M (2015) Glutathione peroxidases as redox sensor proteins in plant cells. Plant Sci 234:22–26

Passaia G, Fonini LS, Caverzan A, Jardim-Messeder D, Christoff AP, Gaeta ML, de Araujo Mariath JE, Margis R, Margis-Pinheiro M (2013) The mitochondrial glutathione peroxidase GPX3 is essential for H_2O_2 homeostasis and root and shoot development in rice. Plant Sci 208:93–101

Passaia G, Queval G, Bai J, Margis-Pinheiro M, Foyer CH (2014) The effects of redox controls mediated by glutathione peroxidases on root architecture in *Arabidopsis thaliana*. J Exp Bot 65:1403–1413

Rouhier N, Jacquot J-P (2005) The plant multigenic family of thiol peroxidases. Free Rad Biol Med 38:1413–1421 https://doi.org/10.1016/j.freeradbiomed.2004.07.037

Schnaubelt D, Queval G, Dong Y, Diaz-Vivancos P, Makgopa ME, Howell G, De Simone A, Bai J, Hannah MA, Foyer CH (2015) Low glutathione regulates gene expression and the redox potentials of the nucleus and cytosol in *Arabidopsis thaliana*. Plant Cell Environ 38:266–279

Seiler A, Schneider M, Forster H, Roth S, Wirth EK, Culmsee C, Plesnila N, Kremmer E, Radmark O, Wurst W, Bornkamm GW, Schweizer U, Conrad M (2008) Glutathione peroxidase 4 senses and translates oxidative stress into 12/15-lipoxygenase dependent- and AIF-mediated cell death. Cell Metab 8:237–248

Tian M, Gu Q, Zhu M (2003) The involvement of hydrogen peroxide and antioxidant enzymes in the process of shoot organogenesis of strawberry callus. Plant Sci 165:701–707

Toppo S, Vanin S, Bosello V, Tosatto SC (2008) Evolutionary and structural insights into the multi-faceted glutathione peroxidase (GPx) superfamily. Antioxid Redox Signal 10:1501–1514

Toppo S, Flohe L, Ursini F, Vanin S, Maiorino M (2009) Catalytic mechanisms and specificities of glutathione peroxidases: variations of a basic scheme. Biochim Biophys Acta 1790:1486–1500

Tosatto SC, Bosello V, Fogolari F, Mauri P, Roveri A, Toppo S, Flohe L, Ursini F, Maiorino M (2008) The catalytic site of glutathione peroxidases. Antioxid Redox Signal 10:1515–1526

Vernoux T, Wilson RC, Seeley KA, Reichheld JP, Muroy S, Brown S, Maughan SC, Cobbett CS, Van Montagu M, Inze D, May MJ, Sung ZR (2000) The ROOT MERISTEMLESS1/CADMIUM SENSITIVE2 gene defines a glutathione-dependent pathway involved in initiation and maintenance of cell division during postembryonic root development. Plant Cell 12:97–110

Wang X, Fang G, Yang J, Li Y (2017) A thioredoxin-dependent glutathione peroxidase (OsGPX5) is required for rice normal development and salt stress tolerance. Plant Mol Biol Report 35:333–342

Waszczak C, Akter S, Eeckhout D, Persiau G, Wahni K, Bodra N, Van Molle I, De Smet B, Vertommen D, Gevaert K, De Jaeger G, Van Montagu M, Messens J, Van Breusegem F (2014) Sulfenome mining in *Arabidopsis thaliana*. Proc Natl Acad Sci U S A 111:11545–11550

Chapter 5
Plant Glutathione Peroxidases: Antioxidant Enzymes in Plant Stress Responses and Tolerance

Krisztina Bela, Sajid Ali Khan Bangash, Riyazuddin, and Jolán Csiszár

Abstract In contrast to other eukaryotic organisms, plants are unable to run away from unfavourable conditions; they must cope with different abiotic and biotic stress factors. Under abiotic and biotic stresses, the production of reactive oxygen and reactive nitrogen species (ROS and RNS) can damage the biological membranes, proteins and nucleic acids. However, plants have developed complex defence systems including different non-enzymatic and enzymatic antioxidants as shields to prevent the toxic effects of an increased amount of ROS and RNS. Glutathione peroxidases (GPXs) are important antioxidant enzymes in animals, but plants contain GPX-like (GPXLs) enzymes. In contrast to animal GPXs, plant GPXLs contain cysteine in their active site instead of selenocysteine, and most of them prefer thioredoxin as the electron donor rather than glutathione. In the last 25 years, many researches proved that plant GPXLs also are essential elements of plant stress responses and are important ROS scavengers. Overexpression of GPXLs in different plant species led to increased tolerance against drought, salt, osmotic, heavy metal and particularly oxidative stresses; however, in some cases, it caused decreased tolerance against biotic stresses. In this chapter, we focus on the importance of plant GPXLs in stress responses, highlighting the significance of distinct genes as possible candidates for genetic engineering to improve the yield of agricultural plants under unfavourable environment.

Keywords Abiotic stress • Antioxidant enzymes • Biotic stress • Glutathione peroxidases

K. Bela (✉) • Riyazuddin • J. Csiszár
Department of Plant Biology, University of Szeged, Közép fasor 52, Szeged H-6726, Hungary
e-mail: belakriszti88@gmail.com

S.A.K. Bangash
Chemical Signalling, INRES, University of Bonn,
Friedrich-Ebert-Allee 144, Bonn D-53113, Germany

M.A. Hossain et al. (eds.), *Glutathione in Plant Growth, Development, and Stress Tolerance*, https://doi.org/10.1007/978-3-319-66682-2_5

Contents

1 Introduction

Eukaryotic organisms live under constantly changing environmental conditions that could negatively affect their development and reproduction. In contrast to other eukaryotes, plants are sessile organisms, unable to run away from unfavourable conditions; thus they must cope with different abiotic and biotic stress factors. Environmental stresses represent the most limiting factors to agricultural productivity worldwide. Their influence is not only restricted on currently cultivated fields, but they also hamper the introduction of crop plants in non-cultivated areas. A global problem in the improvement of crop productivity is the large variation of annual crop yields due to unpredictable environmental stresses.

2 Environmental Stresses Impair the Development and Yield of Plants

There are two main categories of environmental factors: abiotic and biotic. Biotic stress occurs as a result of damage by other living organisms, for example, by bacteria, viruses, fungi, parasites, insects or weeds. These factors destroy more than 40% of all potential food production each year, despite the huge amount of pesticide or other non-chemical controls used (Pimentel and Greiner 1997).

On the other hand, abiotic stress occurs as the negative impacts of non-living factors on the organisms, for example, water deficiency or flooding, extreme temperature, salinity, insufficient nutrition, radiation and light intensity, mechanical effects, metals, chemicals and pollutants. Among the abiotic stress factors, one of the most limiting for crop production is water. The two ends of water supply are too much (flooding) or too little water (drought). The Food and Agriculture Organization (FAO) analysis in 2015 revealed that 37% of the damage and loss to crops and livestock is because of flood. Flood causes on plants the death of leaves, wilting or epinasty and finally the loss of production. From another side, according to the analysis of the World Resources Institute in 2013, 28% of the cultivated areas are exposed to high or extremely high drought stress, but in the case of some cultivated plants, this number is more extreme: 35% of maize fields, 43% of wheat fields and

57% of cotton fields suffer from high water deficit (Reig et al. 2013). The effects of drought stress on plants lead to reduced germination, development, photosynthesis and production. For example, in the case of sunflower in 2015, the production fell down by 5.5% compared with the 5-year average, mainly driven by strong drought-related yield decreases in the main producing European Member States – Bulgaria, Romania, France and Spain (FAO 2016). In order to maintain the agricultural production in these areas, farmers decide on irrigation. However, the extreme groundwater extraction is leading to the salinization of soils. Most of the crop plants are sensitive to salinity caused by high concentrations of salts in the soil. About 33% of the cultivated areas are affected by high salt stress, and this number is increasing by 10% of the rate annually (Jamil et al. 2011). Salt as an osmotic stress causes the same symptoms like drought stress, hence reducing the growth and development; moreover, salt imposes ion toxicity, too (Ashraf and Harris 2004). And besides water and salt stresses, 23% of the overall production losses are caused by extreme temperature, radiation, pollution and other factors (FAO 2017).

Every environmental stress may end up in oxidative stress, at least to some degree. The elevated level of reactive oxygen and nitrogen species (ROS and RNS), such as superoxide radicals ($O_2^{\bullet-}$), hydrogen peroxide (H_2O_2), hydroxyl radicals ($OH^{\bullet}$), singlet oxygen ($^{1}O_2$), nitric oxide ($NO^{\bullet}$) and peroxynitrite ($ONOO^-$), can cause damage to lipids, proteins and DNA (Mittler 2002; Luis et al. 2006).

3 Glutathione Peroxidases Are Versatile ROS Scavengers

Plants have developed complex antioxidant defence systems to counteract the deleterious effect of increased amount of ROS and RNS. This defence system comprises non-enzymatic and enzymatic components in different cellular compartments. Non-enzymatic components include the major redox buffers glutathione and ascorbate, as well as carotenoids, tocopherols and phenolic compounds. They are important cofactors of the enzymatic antioxidants and elements of redox homeostasis (Sharma et al. 2012). The enzymatic components of the defence system include several antioxidant enzymes such as superoxide dismutase (SOD), catalase (CAT), guaiacol peroxidase (POD), ascorbate peroxidase (APX), monodehydroascorbate reductase (MDHAR), dehydroascorbate reductase (DHAR), glutathione reductase (GR), glutathione transferase (GST) and glutathione peroxidase-like enzyme (GPXL). These enzymes and their isoenzymes are located in different subcellular compartments (Noctor and Foyer 1998). The activities of these enzymes are generally increased under stress conditions, and in several cases, their activities correlate well with enhanced tolerance (Foyer et al. 1997). Their role and mechanism in stress responses have been investigated intensively for several decades; however, relatively little is known about plant GPXLs.

The glutathione peroxidase (GPX) enzymes are non-heme thiol peroxidases that catalyse the reduction of H_2O_2 or organic hydroperoxides to water or the corresponding alcohols using reduced glutathione. Numerous GPXs characterized from various organisms revealed their role in ROS scavenging (Noctor et al. 2012; Yang et al. 2015). The mammalian GPXs are central components of the antioxidant defence system and contribute in the repair of biomembranes (Imai and Nakagawa 2003; Margis et al. 2008; Brigelius-Flohe and Maiorino 2013). The plant GPXLs are closely related to animal phospholipid hydroperoxide glutathione peroxidases; however, they contain cysteine instead of selenocysteine in their active site and prefer the thioredoxin (TRX) regenerating system rather than the glutathione system (Iqbal et al. 2006; Navrot et al. 2006; Herbette et al. 2007; Margis et al. 2008). Nevertheless, GPXLs have also an important role in the elimination of organic hydroperoxides and lipid peroxides (Milla et al. 2003; Bela et al. 2015). For example, *Arabidopsis* plants lacking GPXL8 contain an elevated level of the lipid peroxidation marker malondialdehyde (MDA) compared to wild-type plants after exposure to salt or osmotic stresses (Gaber 2011). They also participate in the H_2O_2 homeostasis. As another example, *Atgpx1* mutant *Arabidopsis* plants had higher basal foliar H_2O_2 levels than the wild-type plants under low-light condition, which were elevated even further under high-light stress (Chang et al. 2009).

4 Role of Plant Glutathione Peroxidases in Plant Stress Tolerance

Several reports provided direct or indirect evidence for the importance of GPXLs in different stress responses. Their very important feature is that they may protect proteins and DNA against oxidative stress. Gaber et al. (2012) proved that *Arabidopsis* plants overexpressing *AtGPXL8* accumulated less oxidized proteins and 8-oxo-2′-deoxyguanosine under oxidative stress. GPXLs also help to protect biological membranes by the reduction of lipid peroxides (Herbette et al. 2002; Jung et al. 2002; Chen et al. 2004; Iqbal et al. 2006; Navrot et al. 2006); thus, the plant GPXLs are thought to be part of the enzymatic antioxidant systems. Although GPXLs mainly take part in the elimination of organic hydroperoxides, in some cases, they also react with H_2O_2.

4.1 GPXLs in Oxidative Stress Responses

H_2O_2 is an important compound of the oxidative stress and is a component of signalling processes. About the role of GPXLs in signalling, see more details in Chap. 4 in this book ("Plant Glutathione Peroxidases: Structural and Functional Characterization and Their Roles in Plant Development").

It was reported that external H_2O_2 treatment resulted in elevated transcript levels of many *GPXL* genes in *Panax ginseng* (Kim et al. 2014) and in *Oryza sativa* (Li et al. 2000; Passaia et al. 2013). The dramatic increase of *OsGPXL* mRNA levels after the H_2O_2 treatment was reported originally by Li et al. (2000). Passaia et al. (2013) showed that all the five *OsGPXLs* were induced 2–8 h after the 10 μM H_2O_2 treatment. Islam et al. (2015) analysed separately the shoot and the root of rice during oxidative stress, and their results also proved the induction of *GPXL* genes by H_2O_2 treatment; however, *OsGPXL5* was induced only in root tissues. *Osgpxl3* knockdown mutant plants displayed short root and shoot phenotypes and increased H_2O_2 production in root tissues compared to wild-type plants (Passaia et al. 2013). These results indicate that in rice the OsGPXLs are important in H_2O_2 elimination. *AtGPXL3* was also induced by H_2O_2; moreover, *Atgpxl3* knockout mutants are more sensitive to H_2O_2 treatment, and the H_2O_2 level was elevated in these plants compared to wild type. These mutants showed delayed leaf development compared to wild-type plants on H_2O_2-containing media (Miao et al. 2006). However, *Panax ginseng PgGPXL1* expression increased only in the first 24 h of H_2O_2 exposure, and the *PgGPXL2* was parallely downregulated (Kim et al. 2014). In this sense, other authors have suggested that GPXLs may have a role not only in elimination but also in H_2O_2 perception and signalling (see more details in Chap. 4).

One important environmental factor that generates oxidative stress is the ground-level ozone, which enters the leaves through stomata during normal gas exchange. It is a strong oxidant which may cause several types of symptoms including chlorosis and necrosis. Ozone has significant effect on crop yield, and dicot species, like soybean, cotton and peanut, are more sensitive to yield loss caused by ozone than monocot species, such as sorghum, corn and wheat (Heagle 1989). Furthermore, *Nicotiana plumbaginifolia NpGPXL*, together with *CAT2* and *CAT3* genes, showed induction to ozone treatment, indicating the significance of GPXLs also in ozone stress (Willekens et al. 1994).

A series of experiments using paraquat also connect the GPXLs to oxidative stress responses, because it is one of the most widely used quick-acting and non-selective herbicides which generate ROS due to interaction with the free electrons originated from chloroplast photosystem I (PSI) (Upham and Hatzios 1987) or mitochondrial NADH:ubiquinone oxidoreductase (Complex I) (Tawara et al. 1996). The evolved superoxide then may attack biological membranes. Paraquat treatment also caused early transcriptional activation of *GPXLs* in different plants. Treatment of barley leaves dramatically increased the transcript level for cytosolic *HvGPXL1* and chloroplastidic *HvGPXL2* (Churin et al. 1999). Similar induction of *GPXL* was observed due to paraquat in *Raphanus sativus* in the light; however, the gene was downregulated in the dark (Yang et al. 2005). In a *Conyza bonariensis* paraquat-resistant biotype, an elevated transcript level of genes coding antioxidant enzymes, as well as the increase of SOD, APX, DHAR, MDAR, GR and GPXL enzyme activities, was observed after paraquat treatment (Ye and Gressel 2000). Expression of *Citrus sinensis GPXL* gene in *Escherichia coli* enhanced the tolerance against paraquat, but this tolerance depended on the growth stage (Holland et al. 1994). GPXL8 knockout and overexpressing *Arabidopsis* mutants showed decreased and increased tolerance against paraquat, respectively, that was correlated with enlarged and reduced root growth inhibition.

4.2 Involvement of GPXLs in Biotic Stress Responses

Plant diseases cause major economic losses for farmers worldwide, even though the plants have a complex defence system with constitutive and inducible components against pathogen attacks. The endogenous H_2O_2 accumulation is an important feature of the incompatible plant–pathogen interaction. The oxidative burst is a rapid production of large amount of ROS in response to external stimuli that overwhelms the cellular antioxidative defences (Wojtaszek 1997). However, a reasonable regulation of antioxidant systems is part of the signalling pathways, activating defence responses during pathogen attack (De Gara et al. 2003). Inoculation of *Nicotiana sylvestris* with GTAMV (green tomato atypical mosaic virus) resulted in induction of *GPXL* (Criqui et al. 1992), and similar induction was detectable in *Helianthus annuus* during *Plasmopara halstedii* infection (Roeckel-Drevet et al. 1998). Plant glutathione peroxidases are important even in response to insect attack. Colonization of *Zea mays* seedlings by aphids *Sitobion avenae* and *Rhopalosiphum padi* upregulated *ZmGPXL1* and *ZmGPXL3* genes and increased the GPXL enzyme activity (Sytykiewicz 2016). In *Panax ginseng*, *GPXLs* respond to biotic stress differently: *PgGPXL1* expression increased compared to control; conversely, *PgGPXL2* expression gradually decreased during *Colletotrichum gloeosporioides* pathogen attack (Kim et al. 2014). On the other hand, GPXLs do not always have a supportive role in biotic stress responses. Depletion of AtGPXL1 resulted in expanded lesions by *Pseudomonas syringae* on *Arabidopsis* leaves, and bacterial titres were 10 times lower compared to the wild type, where hypersensitive cell death was restricted to the area around the infection. These results showed that the depletion of AtGPXL1 activity improves resistance against virulent bacteria (Chang et al. 2009). *GPXL5* overexpression in tomato plants increased the size of necrotic areas during *Botrytis cinerea* infection. Thus, *GPXL* overexpression counteracted the plant defence response (Herbette et al. 2011).

4.3 GPXLs in Salt Stress Responses

Salinity is one of the most serious factors limiting the yield of agricultural crops. Salt stress causes water deficiency, ion toxicity, nutritional disorders, metabolism alterations, membrane damage and oxidative stress. Salt stress responses of plants include production of different osmolytes and chaperones, ion channel activation and induction of the antioxidant defence system (Carillo et al. 2011). The first salt stress associated *GPXL* was isolated from *Citrus sinensis*: a fast induction of *CsGPXL* was detectable after salt treatment (Avsian-Kretchmer et al. 1999), followed by an increase in the level of CsGPXL protein in cultured cells originating from different organs (Ben-Hayyim et al. 1993; Holland et al. 1993; Beeor-Tzahar et al. 1995). In *A. thaliana*, salt stress increased the transcript level of *AtGPXL1*, *AtGPXL2*, *AtGPXL4*, *AtGPXL5*, *AtGPXL6*, *AtGPXL7* and *AtGPXL8*, but *AtGPXL3* was not affected (Sugimoto and Sakamoto 1997; Milla et al. 2003; Gao et al. 2014).

Atgpxl8 knockout mutants were more sensitive to the salt treatment (Gaber 2011). In *Thellungiella salsuginea*, exposure to salt stress induced or repressed the *TsGPXLs* in organ-specific and tissue-specific manner (Gao et al. 2014). For example, *TsGPXL5*, *TsGPXL7* and *TsGPXL8* were induced in shoots, whereas in roots almost all glutathione peroxidase genes (*TsGPXL1*, *TsGPXL2*, *TsGPXL3*, *TsGPXL5*, *TsGPXL7* and *TsGPXL8*) showed induction by 300 mM NaCl (Gao et al. 2014). *T. salsuginea* is a close relative of *Arabidopsis* which represents a halophytic model for salt stress tolerance studies, but the role of GPXLs in salt stress tolerance has been also reported in crop plants.

Among three barley *GPXLs*, two were activated after salt treatment, the expression of *HvGPXL1* being much higher than *HvGPXL2* (Churin et al. 1999). Similarly, *OsGPXL1* transcript level increased rapidly (Kang et al. 2004), followed by elevated GPXL enzyme activity due to high salinity (Lima-Melo et al. 2016). Similar induction was observed in the case of *OsGPXL3*, but *OsGPXL2* and *OsGPXL4* were activated only in the roots (Islam et al. 2015). *OsGPXL5* was induced in root tissues, but its expression was reduced in shoot tissues; however, *Osgpx5* knockout lines showed increased sensitivity towards high concentration of salt (Wang et al. 2017).

In *Arabidopsis*, the expression of two different *Triticum aestivum TaGPXL* genes led to increased tolerance against salt stress (Zhai et al. 2013). Transgenic plants remained green, and the root inhibition by salinity was reduced; furthermore, germination rate also increased on salt-containing media compared to the wild-type plants. Evaluation of the background of this process revealed that *TaGPXL* overexpression caused an elevated transcript level of *SOS1* (Na^+–H^+ antiporter) and *RbohD* (NADPH oxidase) genes, but downregulated the *ABI1* and *ABI2* (2C protein phosphatases), suggesting a role for TaGPXLs in salt stress signalling (Zhai et al. 2013). Li et al. (2013) also proved the importance of TaGPXLs during high-salinity treatment, because *TaGPXL* transcript levels greatly increased after treatment, together with *MDHAR*, *DHAR* and *glutathione synthetase 3 (GS3)*, and parallely salt stress markedly raised the contents of both glutathione and ascorbate in the leaves of wheat seedlings. Also in *Panax ginseng*, in *Nelumbo nucifera* and in tea plants, *PgGPXL1*, *PgGPXL2*, *NnGPXL* and *CsGPXL2*, respectively, were upregulated by salt treatment (Diao et al. 2014; Fu 2014; Kim et al. 2014). However, in other plants, a role for GPXLs in salt stress response has not been reported; for example, none of the six *Lotus japonicus GPXL* genes were affected by short-term salt treatment, but only after 7 days (Ramos et al. 2009).

4.4 Involvement of GPXLs in Osmotic and Drought Stress Tolerance

One of the components of salt stress is osmotic stress, caused by the change in solute concentrations. So, it is not a surprise that most of the genes which were induced by salt stress are induced also by osmotic stress. For example, the *CsGPXL* in citrus (Ben-Hayyim et al. 1993), *HvGPXL1* and *HvGPXL2* in barley (Churin et al. 1999),

PgGPXL1 in ginseng (Kim et al. 2014), *CsGPXL2* in tea plants (Fu 2014) and *Arabidopsis GPXLs* have been described as osmotic stress-inducible (Milla et al. 2003; Gaber 2011). However, *AtGPXL2* transcript level interestingly decreased under osmotic stress caused by mannitol (Milla et al. 2003), and in contrast with the results during salt stress, *AtGPXL3* was activated (Miao et al. 2007). In *Thellungiella salsuginea*, *TsGPXL1*, *TsGPXL3*, *TsGPXL4* and *TsGPXL7* were significantly upregulated in shoots due to osmotic stress and in roots almost all *TsGPXL* genes, except for *TsGPXL1* (Gao et al. 2014).

The other major limiting factor in crop productivity is the drought. The physiological responses of plants to drought stress generally included the production of antioxidants, osmotic protective compounds and growth regulators (Farooq et al. 2009). In *Euphorbia esula*, among other antioxidant enzymes, GPXLs, GSTs and GR play important roles in plant defence mechanisms against drought (Anderson and Davis 2004). In rice, drought stress induced all of *OsGPXLs* to some degree. After 12 h from drying the seedlings on Whatman sheet, mRNA level is increased for *OsGPXL1*, *OsGPXL2*, *OsGPXL3* and *OsGPXL4* in shoots; however, for *OsGPXL5*, it was reduced. In the roots of rice seedling all the five *OsGPXLs* were activated after 12 h (Islam et al. 2015). Similarly, *OsGPXL1* was activated after removing the source of water from seedlings for 2 days (Kang et al. 2004). In contrast, expression of *OsGPXL4* and *OsGPXL5* was reduced when rice plants were grown without water for 15 days (Passaia et al. 2013). The role of poplar PtGPXLs during water deficit has been also described as not uniform. After 6 days of water withdrawal, the protein level of some PtGPXLs increased, whereas some decreased (Navrot et al. 2006). In case of *Arabidopsis* plants, the important role of the GPXL3 in drought stress responses was reported, because defects of *AtGPX3* reduced drought stress tolerance, whereas *AtGPX3* overexpression in transgenic plants enhanced drought stress resistance (Miao et al. 2006).

4.5 Role of GPXLs Under Low and High Temperatures

Extreme temperatures also cause serious damages in agricultural production. Temperature stresses in plants are classified into three types depending on the stressor, which may be high, chilling or freezing temperature. The three types induce different stress responses, the activation of antioxidant enzymes being part of all of them (Wang et al. 2017). The involvement of GPXLs in chilling stress (4 –10 °C) response of rice is controversial. After 16 h on 4 °C, the mRNA levels of *OsGPXL1* and *OsGPXL3* increased both in shoots and roots; however, *OsGPXL2*, *OsGPXL4* and *OsGPXL5* were downregulated (Islam et al. 2015). When plants were exposed to 10 °C for 24 h, the transcription of either the *OsGPXL1*, *OsGPXL3* or *OsGPXL5* was induced (Passaia et al. 2013), whereas, in other experimental systems, when plants were subjected to 4 °C for 3 days, *OsGPXL1* was not induced by 24 h, but only after 48 h, and the activation disappeared by 72 h (Kang et al. 2004). A similar response was observed in ginseng plants: after 8 h of chilling stress, both *PgGPXL1* and

PgGPXL2 were induced, but later the expression gradually fell down (Kim et al. 2014). Diao et al. (2014) investigated *Nelumbo nucifera GPXL* expression only in short-term chilling, and the expression of this gene increased within an hour, and this activation was maintained until 6 h. In *Arabidopsis* plants, only *AtGPXL6* was activated among the eight genes on 4 °C (Milla et al. 2003). In eggplant, this treatment activated *SmGPXL1* and *SmGPXL2* together with the expression of other genes coding antioxidant enzymes (*GSTs*, *GR*, *MDAR* and *DHAR*) (Chen et al. 2011). On the other hand, overexpression of tomato *SlGPXL5* protected the photosynthetic machinery from chilling treatment under moderate light (Herbette et al. 2005).

Heat stress caused a somewhat different response compared to chilling. For example, in rice plants, all *OsGPXLs* were activated in the shoots and/or roots by heat treatment, contrary to the downregulation observed by chilling (Islam et al. 2015; Wang et al. 2017), and in *Arabidopsis*, heat stress upregulated *AtGPXL1*, instead of *AtGPXL6*, which was induced by chilling stress (Milla et al. 2003). However, a similar *GPXL* gene expression pattern was found after short-term heat stress and chilling in *Nelumbo nucifera* (Diao et al. 2014) and in tea plants (Fu 2014).

4.6 Other Stresses

Mechanical stimulation can also happen during pathogen attack; however, it is considered to be an abiotic stress factor. *NsGPXL* in *Nicotiana sylvestris* and *HaGPXL* in *Helianthus annuus* were induced after wounding in the same way as in biotic stress (Criqui et al. 1992; Roeckel-Drevet et al. 1998). The mRNAs of tomato *GPXLs* were also accumulated after mechanical stimulation; however, the dynamics of the transcription of the two investigated genes were different: *SlGPXL1* was induced within an hour, whereas *SlGPXL2* activation was a bit slower, about 6 h after the treatment (Depège et al. 2000). Moreover, overexpression of *SlGPXL5* led to increased tolerance against wounding-induced growth inhibition (Herbette et al. 2011). Similar rapid gene activation was detectable after injuries in the case of *Nelumbo nucifera GPXL* (Diao et al. 2014). Interestingly, in rice cut-induced *OsGPXL* expression profile showed light dependency: the gene was induced after 12 h from wounding in dark condition, while under light condition this induction delayed to 24 h (Agrawal et al. 2002).

Unfortunately, we cannot turn a blind eye over anthropogenic factors like metal and other chemical pollutions. These factors are also harmful for plants and are continuously increasing factors affecting crop yield (Dukhovskis et al. 2003). As under other environmental stresses, antioxidants are really important elements in metal or chemical stress responses. As an example, the herbicide norflurazon inhibits the synthesis of carotenoids in plant leaves and in this way destructs pigment–protein complexes by photo-oxidation and blocks chloroplast development. Norflurazon treatment on barley leaves caused a dramatic increase in the mRNA level of *HvGPXL1* and *HvGPXL2* but decreased the level of *HvGPXL3* (Churin et al. 1999).

Different metals also affect differently depending on the plant studied. While iron caused the activation of *AtGPXL2*, *AtGPXL5* and *AtGPXL6* in *Arabidopsis* (Sugimoto and Sakamoto 1997; Milla et al. 2003) and *CsGPXL2* in tea plants (Fu 2014), aluminium induced only *AtGPXL6* (Sugimoto and Sakamoto 1997), but did not cause any changes in rice *OsGPXLs* (Passaia et al. 2013), and even downregulated all the *LjGPXLs* in *Lotus japonicus* (Ramos et al. 2009). Copper treatment also induced *AtGPXL2*, *AtGPXL5* and *AtGPXL6* in *Arabidopsis* (Sugimoto and Sakamoto 1997; Milla et al. 2003) and *CsGPXL2* in tea plants (Fu 2014), but decreased the protein level of some poplar PtGPXLs. Cadmium increased the level of particular poplar PtGPXLs (Navrot et al. 2006) and activated the lotus *LjGPXLs*, but later the degree of the induction decreased (Matamoros et al. 2015).

5 GPXL Mutants and Overexpressing Plants Harbour Altered Stress Tolerance

The function of plant GPXs in stress responses was extensively studied by employing transgenic plants engineered to enhance or reduce GPXL pools. Loss-of-function mutations of GPXL in many cases negatively affect the tolerance against environmental stresses in different plants. For example, in *A. thaliana*, depletion of *AtGPXL1* and *GPXL7* gene expression led to decreased tolerance against photo-oxidative stress; however, it increased the resistance against virulent *Pseudomonas syringae* (Chang et al. 2009). Defects of AtGPX3 reduced the drought stress tolerance. The mutants displayed impaired stomatal closure, faster water loss, and lower temperatures of leaves (Miao et al. 2006). Knockout mutation of *AtGPXL8* led to increased sensitivity to salt and osmotic stresses compared to wild type (Gaber 2011); furthermore, paraquat treatment affected the mutant plants more and caused suppressed root growth and higher level of oxidized proteins (Gaber et al. 2012). According to the results of the experiments performed using knockout mutants, the *Oryza sativa* OsGPXL1 mitochondrial enzyme is important for both phases of photosynthesis, root growth, water use efficiency and photorespiration under salinity (Lima-Melo et al. 2016), and depletion of OsGPXL5 also negatively affected the salt stress tolerance (Wang et al. 2017).

In accordance with the above results, overexpression of wheat *GPXL* genes in *Arabidopsis* enhanced early tolerance to high salt stress, and the transgenic plants showed higher germination rate and decreased growth inhibition by NaCl treatment (Zhai et al. 2013). In *Solanum lycopersicum*, overexpression of *SlGPXL5* seemed to protect the photosynthetic activities from chilling treatment under moderate light (Herbette et al. 2005); however, the transformed plants had significantly larger necrotic areas after *Botrytis cinerea* infection than wild-type plants. Thus, GPXL overexpression alleviated the abiotic stress and counteracted the plant defence response against biotic stress factors (Herbette et al. 2011).

6 Concluding Remarks

As global climate becomes more extreme, the abiotic stresses, the rapidly evolving pathogens and weeds cause more adverse environment for plants which can affect the productivity of crops. In the last two decades, many researches proved that plant GPXLs are essential elements of plant stress responses and are important ROS scavengers. Like animal GPXs, GPXLs are also able to reduce H_2O_2 and organic hydroperoxides to water or the corresponding alcohols. Not surprisingly, the involvement of plant glutathione peroxidases in stress responses has been reported in different plants. Overexpression of GPXLs in different plant species led to increased tolerance against abiotic stresses; however, in some cases, it caused decreased tolerance against biotic stress. It is clear that GPXLs could be promising candidates in the genetic engineering or traditional breeding to develop stress-resistant crop plants; however, further intensive research is needed to explore their connection to other elements of the antioxidant system and signalling.

Acknowledgements The authors are grateful for the support of the Hungarian National Scientific Research Foundation [grant number OTKA K 105956]. Riyazuddin is funded through the Stipendium Hungaricum Scholarship Programme in Hungary. S.A.K.B. thanks the Higher Education Commission (HEC) of Pakistan and the University of Agriculture Peshawar for the scholarship (grant number 360/SIBGE). K.B. received visiting scholarships from the Campus Mundi Programme that is co-financed by the European Union and the Hungarian government.

References

Agrawal GK, Rakwal R, Jwa N-S (2002) Cloning and characterization of a jasmonate inducible rice (*Oryza sativa L.*) peroxidase gene, *OsPOX*, against global signaling molecules and certain inhibitors of kinase-signaling cascade(s). Plant Sci 162:49–58

Anderson JV, Davis DG (2004) Abiotic stress alters transcript profiles and activity of glutathione S-transferase, glutathione peroxidase, and glutathione reductase in *Euphorbia esula*. Physiol Plant 120:421–433

Ashraf M, Harris P (2004) Potential biochemical indicators of salinity tolerance in plants. Plant Sci 166:3–16

Avsian-Kretchmer O, Eshdat Y, Gueta-Dahan Y, Ben-Hayyim G (1999) Regulation of stress-induced phospholipid hydroperoxide glutathione peroxidase expression in citrus. Planta 209:469–477

Beeor-Tzahar T, Ben-Hayyim G, Holland D, Faltin Z, Eshdat Y (1995) A stress-associated citrus protein is a distinct plant phospholipid hydroperoxide glutathione peroxidase. FEBS Lett 366:151–155

Bela K, Horvath E, Galle A, Szabados L, Tari I, Csiszar J (2015) Plant glutathione peroxidases: emerging role of the antioxidant enzymes in plant development and stress responses. J Plant Physiol 176:192–201

Ben-Hayyim G, Faltin Z, Gepstein S, Camoin L, Strosberg AD, Eshdat Y (1993) Isolation and characterization of salt-associated protein in citrus. Plant Sci 88:129–140

Brigelius-Flohe R, Maiorino M (2013) Glutathione peroxidases. Biochim Biophys Acta 1830:3289–3303

Carillo P, Annunziata MG, Pontecorvo G, Fuggi A, Woodrow P (2011) Salinity stress and salt tolerance. In: Abiotic stress in plants – mechanisms and adaptations. InTech, Rijeka, pp 21–38
Chang CC, Ślesak I, Jordá L, Sotnikov A, Melzer M, Miszalski Z, Mullineaux PM, Parker JE, Karpińska B, Karpiński S (2009) Arabidopsis chloroplastic glutathione peroxidases play a role in cross talk between photooxidative stress and immune responses. Plant Physiol 150:670–683
Chen S, Vaghchhipawala Z, Li W, Asard H, Dickman MB (2004) Tomato phospholipid hydroperoxide glutathione peroxidase inhibits cell death induced by Bax and oxidative stresses in yeast and plants. Plant Physiol 135:1630–1641
Chen S, Zimei L, Cui J, Jiangang D, Xia X, Liu D, Yu J (2011) Alleviation of chilling-induced oxidative damage by salicylic acid pretreatment and related gene expression in eggplant seedlings. Plant Growth Regul 65:101–108
Churin Y, Schilling S, Börner T (1999) A gene family encoding glutathione peroxidase homologues in *Hordeum vulgare* (barley). FEBS Lett 459:33–38
Criqui MC, Jamet E, Parmentier Y, Marbach J, Durr A, Fleck J (1992) Isolation and characterization of a plant cDNA showing homology to animal glutathione peroxidases. Plant Mol Biol 18:623–627
De Gara L, de Pinto MC, Tommasi F (2003) The antioxidant systems vis-à-vis reactive oxygen species during plant–pathogen interaction. Plant Physiol Biochem 41:863–870
Depège N, Varenne M, Boyer N (2000) Induction of oxidative stress and GPX-like protein activation in tomato plants after mechanical stimulation. Physiol Plant 110:209–214
Diao Y, Xu H, Li G, Yu A, Yu X, Hu W, Zheng X, Li S, Wang Y, Hu Z (2014) Cloning a glutathione peroxidase gene from *Nelumbo nucifera* and enhanced salt tolerance by overexpressing in rice. Mol Biol Rep 41:4919–4927
Dukhovskis P, Juknys R, Brazaityte A, Zukauskaite I (2003) Plant response to integrated impact of natural and anthropogenic stress factors. Russ J Plant Physiol 50:147–154
FAO (2015) The impact of disasters on agriculture and food security. Food and Agriculture Organization of the United Nations, Rome
FAO (2016) Agriculture, forestry and fishery statistics, Statistical books. Food and Agriculture Organization of the United Nations, Luxembourg: Belgium
FAO (2017) The impact of disasters on agriculture: addressing the information gap. Food and Agriculture Organization of the United Nations, Rome
Farooq M, Wahid A, Kobayashi N, Fujita D, Basra S (2009) Plant drought stress: effects, mechanisms and management. In: Sustainable agriculture. Springer, Dordrecht, pp 153–188
Foyer CH, Lopez-Delgado H, Dat JF, Scott IM (1997) Hydrogen peroxide-and glutathione-associated mechanisms of acclimatory stress tolerance and signalling. Physiol Plant 100:241–254
Fu J-Y (2014) Cloning of a new glutathione peroxidase gene from tea plant (*Camellia sinensis*) and expression analysis under biotic and abiotic stresses. Bot Stud 55:1–6
Gaber A (2011) Arabidopsis glutathione peroxidase 8 is a key enzyme in response to environmental stresses. Arab J Biotechnol 14:213–224
Gaber A, Ogata T, Maruta T, Yoshimura K, Tamoi M, Shigeoka S (2012) The involvement of Arabidopsis glutathione peroxidase 8 in the suppression of oxidative damage in the nucleus and cytosol. Plant Cell Physiol 53:1596–1606
Gao F, Chen J, Ma T, Li H, Wang N, Li Z, Zhang Z, Zhou Y (2014) The glutathione peroxidase gene family in *Thellungiella salsuginea*: genome-wide identification, classification, and gene and protein expression analysis under stress conditions. Int J Mol Sci 15:3319–3335
Heagle AS (1989) Ozone and crop yield. Annu Rev Phytopathol 27:397–423
Herbette S, Lenne C, Leblanc N, Julien JL, Drevet JR, Roeckel-Drevet P (2002) Two GPX-like proteins from *Lycopersicon esculentum* and *Helianthus annuus* are antioxidant enzymes with phospholipid hydroperoxide glutathione peroxidase and thioredoxin peroxidase activities. Eur J Biochem 269:2414–2420
Herbette S, Menn AL, Rousselle P, Ameglio T, Faltin Z, Branlard G, Eshdat Y, Julien JL, Drevet JR, Roeckel-Drevet P (2005) Modification of photosynthetic regulation in tomato overexpressing glutathione peroxidase. Biochim Biophys Acta 1724:108–118

Herbette S, Roeckel-Drevet P, Drevet JR (2007) Seleno-independent glutathione peroxidases. FEBS J 274:2163–2180

Herbette S, Labrouhe DTD, Drevet JR, Roeckel-Drevet P (2011) Transgenic tomatoes showing higher glutathione peroxidase antioxidant activity are more resistant to an abiotic stress but more susceptible to biotic stresses. Plant Sci 180:548–553

Holland D, Ben-Hayyim G, Faltin Z, Camoin L, Strosberg AD, Eshdat Y (1993) Molecular characterization of salt-stress-associated protein in citrus: protein and cDNA sequence homology to mammalian glutathione peroxidases. Plant Mol Biol 21:923–927

Holland D, Faltin Z, Perl A, Ben-Hayyim G, Eshdat Y (1994) A novel plant glutathione peroxidase-like protein provides tolerance to oxygen radicals generated by paraquat in *Escherichia coli*. FEBS Lett 337:52–55

Imai H, Nakagawa Y (2003) Biological significance of phospholipid hydroperoxide glutathione peroxidase (PHGPx, GPx4) in mammalian cells. Free Radic Biol Med 34:145–169

Iqbal A, Yabuta Y, Takeda T, Nakano Y, Shigeoka S (2006) Hydroperoxide reduction by thioredoxin-specific glutathione peroxidase isoenzymes of *Arabidopsis thaliana*. FEBS J 273:5589–5597

Islam T, Manna M, Kaul T, Pandey S, Reddy CS, Reddy M (2015) Genome-wide dissection of *Arabidopsis* and rice for the identification and expression analysis of glutathione peroxidases reveals their stress-specific and overlapping response patterns. Plant Mol Biol Report 33:1413–1427

Jamil A, Riaz S, Ashraf M, Foolad M (2011) Gene expression profiling of plants under salt stress. Crit Rev Plant Sci 30:435–458

Jung BG, Lee KO, Lee SS, Chi YH, Jang HH, Kang SS, Lee K, Lim D, Yoon SC, Yun DJ, Inoue Y, Cho MJ, Lee SY (2002) A Chinese cabbage cDNA with high sequence identity to phospholipid hydroperoxide glutathione peroxidases encodes a novel isoform of thioredoxin-dependent peroxidase. J Biol Chem 277:12572–12578

Kang S-G, Jeong HK, Suh HS (2004) Characterization of a new member of the glutathione peroxidase gene family in *Oryza sativa*. Mol Cells 17:23–28

Kim YJ, Jang MG, Noh HY, Lee HJ, Sukweenadhi J, Kim JH, Kim SY, Kwon WS, Yang DC (2014) Molecular characterization of two glutathione peroxidase genes of *Panax ginseng* and their expression analysis against environmental stresses. Gene 535:33–41

Li WJ, Feng H, Fan JH, Zhang RQ, Zhao NM, Liu JY (2000) Molecular cloning and expression of a phospholipid hydroperoxide glutathione peroxidase homolog in *Oryza sativa*. Biochim Biophys Acta 1493:225–230

Li G, Peng X, Wei L, Kang G (2013) Salicylic acid increases the contents of glutathione and ascorbate and temporally regulates the related gene expression in salt-stressed wheat seedlings. Gene 529:321–325

Lima-Melo Y, Carvalho FE, Martins MO, Passaia G, Sousa RH, Neto MC, Margis-Pinheiro M, Silveira JA (2016) Mitochondrial GPX1 silencing triggers differential photosynthesis impairment in response to salinity in rice plants. J Integr Plant Biol 58:737–748

Luis A, Sandalio LM, Corpas FJ, Palma JM, Barroso JB (2006) Reactive oxygen species and reactive nitrogen species in peroxisomes. Production, scavenging, and role in cell signaling. Plant Physiol 141:330–335

Margis R, Dunand C, Teixeira FK, Margis-Pinheiro M (2008) Glutathione peroxidase family – an evolutionary overview. FEBS J 275:3959–3970

Matamoros MA, Saiz A, Penuelas M, Bustos-Sanmamed P, Mulet JM, Barja MV, Rouhier N, Moore M, James EK, Dietz KJ, Becana M (2015) Function of glutathione peroxidases in legume root nodules. J Exp Bot 66:2979–2990

Miao Y, Lv D, Wang P, Wang XC, Chen J, Miao C, Song CP (2006) An *Arabidopsis* glutathione peroxidase functions as both a redox transducer and a scavenger in abscisic acid and drought stress responses. Plant Cell 18:2749–2766

Miao Y, Guo J, Liu E, Li K, Dai J, Wang P, Chen J, Song C (2007) Osmotically stress-regulated the expression of glutathione peroxidase 3 in *Arabidopsis*. Chin Sci Bull 52:127–130

Milla MAR, Maurer A, Huete AR, Gustafson JP (2003) Glutathione peroxidase genes in *Arabidopsis* are ubiquitous and regulated by abiotic stresses through diverse signaling pathways. Plant J 36:602–615

Mittler R (2002) Oxidative stress, antioxidants and stress tolerance. Trends Plant Sci 7:405–410
Navrot N, Collin V, Gualberto J, Gelhaye E, Hirasawa M, Rey P, Knaff DB, Issakidis E, Jacquot JP, Rouhier N (2006) Plant glutathione peroxidases are functional peroxiredoxins distributed in several subcellular compartments and regulated during biotic and abiotic stresses. Plant Physiol 142:1364–1379
Noctor G, Foyer CH (1998) Ascorbate and glutathione: keeping active oxygen under control. Annu Rev Plant Biol 49:249–279
Noctor G, Mhamdi A, Chaouch S, Han Y, Neukermans J, Marquez-Garcia B, Queval G, Foyer CH (2012) Glutathione in plants: an integrated overview. Plant Cell Environ 35:454–484
Passaia G, Fonini LS, Caverzan A, Jardim-Messeder D, Christoff AP, Gaeta ML, de Araujo Mariath JE, Margis R, Margis-Pinheiro M (2013) The mitochondrial glutathione peroxidase GPX3 is essential for H_2O_2 homeostasis and root and shoot development in rice. Plant Sci 208:93–101
Pimentel D, Greiner A (1997) Environmental and socio-economic costs of pesticide use. In: Techniques for reducing pesticide use: economic and environmental benefits. Wiley, New York
Ramos J, Matamoros MA, Naya L, James EK, Rouhier N, Sato S, Tabata S, Becana M (2009) The glutathione peroxidase gene family of *Lotus japonicus*: characterization of genomic clones, expression analyses and immunolocalization in legumes. New Phytol 181:103–114
Reig P, Shiao T, Gassert F (2013) Aqueduct water risk framework. Working paper. Washington, DC: World Resources Institute. Available online at http://www.wri.org/publication/aqueduct-water-risk-framework
Roeckel-Drevet P, Gagne G, Labrouhe D, Tourvieille D, Dufaure JP, Nicolas P, Drevet J (1998) Molecular characterization, organ distribution and stress-mediated induction of two glutathione peroxidase-encoding mRNAs in sunflower (*Helianthus annuus*). Physiol Plant 103:385–394
Sharma P, Jha AB, Dubey RS, Pessarakli M (2012) Reactive oxygen species, oxidative damage, and antioxidative defense mechanism in plants under stressful conditions. J Bot 2012:1–26
Sugimoto M, Sakamoto W (1997) Putative phospholipid hydroperoxide glutathione peroxidase gene from *Arabidopsis thaliana* induced by oxidative stress. Genes Genet Syst 72:311–316
Sytykiewicz H (2016) Transcriptional reprogramming of genes related to ascorbate and glutathione biosynthesis, turnover and translocation in aphid-challenged maize seedlings. Biochem Syst Ecol 69:236–251
Tawara T, Fukushima T, Hojo N, Isobe A, Shiwaku K, Setogawa T, Yamane Y (1996) Effects of paraquat on mitochondrial electron transport system and catecholamine contents in rat brain. Arch Toxicol 70:585–589
Upham BL, Hatzios KK (1987) Counteraction of paraquat toxicity at the chloroplast level. Z Naturforsch C 42:824–828
Wang X, Fang G, Yang J, Li Y (2017) A thioredoxin-dependent glutathione peroxidase (OsGPX5) is required for rice normal development and salt stress tolerance. Plant Mol Biol Report 35:333–342
Willekens H, Van Camp W, Van Montagu M, Inze D, Langebartels C, Sandermann H Jr (1994) Ozone, sulfur dioxide, and ultraviolet B have similar effects on mRNA accumulation of antioxidant genes in *Nicotiana plumbaginifolia* L. Plant Physiol 106:1007–1014
Wojtaszek P (1997) Oxidative burst: an early plant response to pathogen infection. Biochem J 322:681–692
Yang XD, Li WJ, Liu JY (2005) Isolation and characterization of a novel PHGPx gene in *Raphanus sativus*. Biochim Biophys Acta 1728:199–205
Yang SL, Yu PL, Chung KR (2015) The glutathione peroxidase-mediated reactive oxygen species resistance, fungicide sensitivity and cell wall construction in the citrus fungal pathogen *Alternaria alternata*. Environ Microbiol 18:923–935
Ye B, Gressel J (2000) Transient, oxidant-induced antioxidant transcript and enzyme levels correlate with greater oxidant-resistance in paraquat-resistant *Conyza bonariensis*. Planta 211:50–61
Zhai CZ, Zhao L, Yin LJ, Chen M, Wang QY, Li LC, Xu ZS, Ma YZ (2013) Two wheat glutathione peroxidase genes whose products are located in chloroplasts improve salt and H_2O_2 tolerances in *Arabidopsis*. PLoS One 8:e73989

Chapter 6
Glutathione as a Key Player in Plant Abiotic Stress Responses and Tolerance

Vittoria Locato, Sara Cimini, and Laura De Gara

Abstract Adverse environmental conditions, such as drought, salinity, high temperature, and toxic metal accumulation, affect plant growth and fitness. Plants have evolved a number of interconnected molecular pathways to defend themselves against different abiotic stresses. In these metabolic networks, redox signaling plays a pivotal role in determining plant tolerance to stress and survival. Glutathione/glutathione disulfide is one of the most versatile redox couples in metabolism. It directly or indirectly buffers the cellular redox state, by acting as enzyme cofactor, controlling the oxido-reduction of other thiols and participating in post-translational protein modifications under both physiological and stress conditions. Glutathione also plays a key role as a conjugating agent in detoxification against xenobiotics or metabolites which need to be sequestered within the vacuole. Glutathione also acts as a signal controlling gene expression and cell cycle progression. These features highlight the importance of glutathione in regulating plant growth and development as well as in conferring tolerance to plants subjected to stress. This chapter describes the involvement of this multifaceted molecule in plant abiotic stress responses.

Keywords Abiotic stress • Ascorbate–glutathione cycle • Reactive oxygen species • Redox homeostasis • Signaling • Thiol

Contents

V. Locato (✉) • S. Cimini • L. De Gara
Unit of Food Science and Human Nutrition, Department of Medicine, Campus Bio-Medico University of Rome, via Alvaro del Portillo 21, 00128 Rome, Italy
e-mail: v.locato@unicampus.it

M.A. Hossain et al. (eds.), *Glutathione in Plant Growth, Development, and Stress Tolerance*, https://doi.org/10.1007/978-3-319-66682-2_6

1 Introduction

Plants, like all living organisms, are exposed to predictable and unpredictable environmental changes. Due to their sessile habits, plant growth and survival depend on the plant's ability to modulate the metabolism in order to counteract such environmental changes. Almost all adverse environmental conditions lead to cell oxidative damage, which is caused by the overproduction of reactive oxygen species (ROS) in the stress-exposed tissues. Interestingly, within a strict range of concentrations, ROS, such as H_2O_2, also play important signaling roles, as they are involved in the activation of defense responses against biotic and abiotic stresses. Indeed, how a species/variety copes with environmental stress is often due to its capacity to sense alterations in the ROS concentration and to trigger the adequate metabolic adjustments.

In plant defense responses, it is also crucial that stress-exposed cells maintain ROS under threshold levels, as well as recovering from the oxidative damage induced by the overwhelming ROS. Plants thus have enzymatic and nonenzymatic networks that modulate the ROS levels. The efficiency of these networks varies in relation to stress intensity, target tissue or cell compartment, plant developmental stage, as well as plant genotype (De Gara et al. 2010; de Pinto et al. 2015). It also depends on the epigenetic signatures that modify the plant's capacity to respond to certain stress conditions (Centomani et al. 2015).

Glutathione is probably one of the most important metabolites involved in the defense responses against environmental stresses. In fact, glutathione and its related enzymes evolved very early in aerobic organisms, probably together with dioxygenic photosynthesis (Deponte 2013). Owing to its cysteine (Cys) moiety, the tripeptide GSH[1] (γ-L-glutamyl-L-cysteinylglycine) is a versatile redox molecule. However, its role goes beyond ROS scavenging and redox homeostasis. Glutathione is the major form of organic sulfur transported in phloem (Mendoza-Cózatl et al. 2008). It plays a pivotal role in the interaction between plants and symbiotic nitrogen-fixing bacteria, in the compartmentalization and neutralization of xenobiotics and heavy metals, and in the vacuolar transport of secondary metabolites (Cheng et al. 2017; Noctor et al. 2012). GSH has also been suggested as the main donor of the reduced sulfur group for glucosinolate biosynthesis in *Arabidopsis* (Parisy et al. 2006). This nonexhaustive list of processes involving glutathione highlights its role in different plant defense strategies.

In Sect. 2, the importance of the glutathione metabolic network in plant defense response is briefly outlined. The protective role of glutathione and its dependent enzymes in specific abiotic stress is then discussed in this chapter. The focus is on

[1] In this chapter, the acronym GSH indicates the thiol (reduced) form of glutathione, while GSSG indicates the disulfide (oxidized) form. When the term "glutathione" is used, no distinction is made between the two forms or it refers to the whole GSH/GSSG pool.

three major abiotic stresses – drought, salinity, and heavy metal contamination – since their importance has been increasing under climate change and as consequence of anthropic activities.

2 Glutathione Metabolic Network in Plant Defense Responses

Under physiological conditions, glutathione cellular concentration is in the millimolar range; however, several parameters affect its concentration. When Cys (the main limiting factor in glutathione biosynthesis) is available, glutathione levels are increased by several stressing conditions. Heavy metals, drought, light, and jasmonic acid affect the expression of glutathione biosynthetic genes (Noctor et al. 2012). Infection of *Arabidopsis thaliana* Col-0 with avirulent bacteria *Pseudomonas syringae PstAvrB* induces an increase in the glutathione pool (Hussain et al. 2016). On the other hand, the increase in H_2O_2 observed in cells exposed to stress also controls the glutathione catabolism and protein turnover/activity (Noctor et al. 2012).

Glutathione is synthetized in its reduced form (GSH).When GSH acts as an electron donor, glutathione disulfide (GSSG) is produced. The basal GSH/GSSG ratio is about 20:1 (Mhamdi et al. 2010). This ratio can decrease significantly when plants are exposed to stress. Several papers have underlined the significance of different glutathione levels and the redox state in specific subcellular compartments. Cytosol is probably the cellular compartment with the highest level of glutathione in reduced state. On the other hand, vacuole accumulates mainly GSSG and is probably the cellular compartment where the catabolism of this metabolite occurs (Queval et al. 2011).

A dynamic flux of GSH between nucleus and cytoplasm is required for the correct progression of the cell cycle (Fig. 6.1; Diaz-Vivancos et al. 2010). Stress alters the glutathione concentration in different cellular compartments and disrupts its physiological fluxes between organelles. Wounding and drought decrease the cytosolic ratio GSH/GSSG (Meyer et al. 2007; Jubany-Mari et al. 2010). The exposure of tobacco BY2 cells to ophiobolin A, a sesterpenoid produced by phytopathogenic fungi, arrests the cell cycle probably by altering the intracellular partitioning of glutathione between the nuclei and cytoplasm (Locato et al. 2015).

Glutathione metabolism involves several different proteins and enzymes. In *Arabidopsis*, according to the TAIR website (www.arabidopsis.org), about 150 genes are involved in glutathione metabolism. Different classes of proteins are involved in glutathione use, with glutathione S-transferases (GSTs) representing the most numerous group. Functionally, GSTs are classified as RX:glutathione R-transferase, where R stands for an electrophilic group, including heterocyclic, aliphatic, or aromatic molecules, and X stands for the leaving group (sulfate, nitrile, or halide ions). In the reaction, the glutathionyl moiety is conjugated with the R group. GSTs play a significant role in anabolic and catabolic pathways (Noctor et al. 2011).

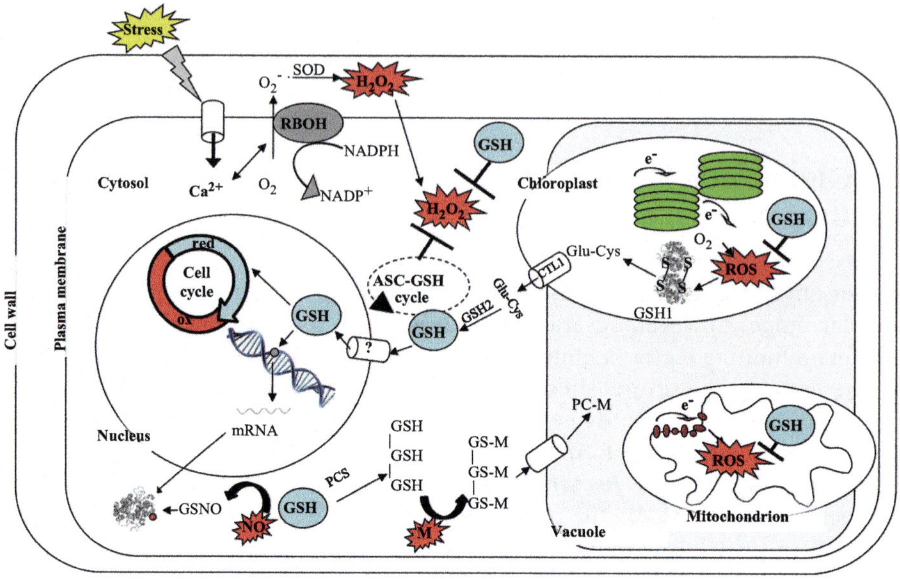

Fig. 6.1 The role of GSH in major protective defense mechanisms activated by plant cell against environmental stress. The figure reports the main roles of GSH in defense responses activated by plants against environmental stresses. GSH can take part in defensive strategies by modulating cell cycle progression (e.g., in xenobiotic detoxification), gene expression (directly or activating redox-sensitive transcriptional factors), cell compartment redox state (by itself or by regulating ASC redox state in the ASC–GSH cycle), detoxification (e.g., in metal chelation directly or through PC synthesis), and enzyme activity (by post-translational modifications and regulating redox state). *ASC* ascorbate, *GSH* glutathione, *ROS* reactive oxygen species, *M* metals, *GSNO* *S*-nitrosoglutathione, *SOD* superoxide dismutase, *RBOH* respiratory burst oxidase homolog protein (NADPH oxidase), *GSH1* γ-glutamylcysteine synthetase, *GSH2* glutathione synthetase, *PCS* phytochelatin synthase, *CTL1 Plasmodium falciparum* chloroquine resistance transporter-like 1

GSTs are regulated by metabolites mostly correlated to the onset of environmental stress. In plants, they thus have a very high structural and functional diversity, with different forms in several cellular compartments. Glyoxalases are another class of enzymes involved in plant defense to stress that catalyze the conjugation of toxic oxo-aldehydes to GSH (Noctor et al. 2011; see Sect. 3.2).

Glutaredoxins (GRXs) are also a large family of glutathione-dependent plant enzymes. GRXs are divided into three classes (I, II, and III) on the basis of a four-amino acid motif in the active site sequence. These enzymes catalyze the thiol–disulfide exchange and are thus involved in the regeneration of thiol-containing enzymes. Due to their functions and substrate specificity, several GRX isoenzymes can overlap thioredoxins (Grant 2001). In mammals, GRXs and thioredoxins are both involved in DNA synthesis (Zahedi Avval and Holmgren 2009). GRXs also catalyze deglutathionylation, modulate ion channels, and participate in iron homeostasis and in the assembly of iron–sulfur clusters (Noctor et al. 2011; Deponte 2013).

Studies on transformed plants that alter the expression of GRXs clearly indicate the involvement of many of these enzymes in resistance or tolerance to a plethora of different stresses, although their roles in normal plant development have also been reported (Noctor et al. 2011). GSH peroxidase is probably the first GSH-dependent enzyme detected to be involved in the defense processes.

GSH peroxidase (GPX) activity was first detected in mammal erythrocytes in 1957 (Mills 1957). In plants the most important H_2O_2-scavenging peroxidase is ascorbate peroxidase (APX) rather than GPX. In fact, the GPX Km for H_2O_2 is high, particularly in comparison with APX. However, in 1997, GPXs were also characterized in plants (Eshdat et al. 1997). These enzymes use both thioredoxin and GSH as the electron donor (thioredoxin often being a more efficient reductant). Plant GPXs efficiently scavenge organic peroxides, supporting the hypothesis that their main physiological role in plants is to regulate the level of lipid peroxides rather than to scavenge H_2O_2 (Noctor et al. 2011).

The ascorbate–glutathione cycle (also known as the Foyer–Halliwell–Asada cycle from the name of the researchers who discovered this network of reactions) is another pathway through which GSH is involved in defense processes. In this cycle, glutathione is the physiological electron donor designed to recycle ascorbate (ASC) from its fully oxidized form dehydroascorbate (DHA). GSSG reductase (GR) uses pyridine nucleotide as the electron donor to regenerate GSH. Interestingly, the ascorbate–glutathione cycle has been detected in almost all the cell compartments, and its involvement in cell protection against oxidative stress has been well established in plants (Fig.6.1; Locato et al. 2009, 2016; De Gara et al. 2010).

Apart from being a substrate of specific enzymes, glutathione also modulates the cellular metabolism by being involved in reversible post-transcriptional modifications, such as S-glutathionylation and S-nitrosylation. These redox-sensitive modifications of protein cysteinyl residues regulate the activity of specific enzymes involved in plant defense responses (Fig.6.1; de Pinto et al. 2013; Dixon et al. 2005; Locato et al. 2015; see Sect. 3.1).

3 Glutathione Role in Abiotic Stress

3.1 *Drought*

Water scarcity is one of the ongoing impacts of climate change and is expected to lead to a drastic reduction in crop productivity. Plants respond to drought mainly by reducing endogenous water loss by stomatal closure. Drought-induced stomatal closure is mainly regulated by the hormone, abscisic acid (ABA). When plants experience a water deficit, ABA accumulates in the xylem sap and moves to the guard cells where a signaling network is activated, causing the guard cells to shrink and thus the stomata to close. ABA directly activates a signal transduction pathway involving the activation of outward anion channels (slow anion channel-associated 1) which lead to membrane hyperpolarization, thus inhibiting potassium inward channels (KAT1). Osmotic changes occurring in guard cells as a consequence of the ionic channel regulation by ABA cause stomatal closure (reviewed by Mittler and Blumwald 2015).

The ABA signaling network involves ROS production in the apoplast and an increase in Ca^{2+} in the cytosol of guard cells (Pei et al. 2000; Kwak et al. 2003). ROS production occurs by the activation of a plasma-membrane NADPH oxidase (RBOHF; respiratory burst oxidase homolog F protein; Drerup et al. 2013) and an apoplastic superoxide dismutase (SOD; Sirichandra et al. 2009). It then leads to H_2O_2 accumulation in the apoplastic and intracellular compartments (Fig. 6.1). H_2O_2 in turn seems to activate Ca^{2+}channels, and the consequent Ca^{2+}cytosolic peak further activates RBOHF. This then triggers an amplification loop in ROS–Ca^{2+} signaling, promoting ABA-induced stomatal closure (Drerup et al. 2013). In fact, ROS accumulation seems to promote ABA sensitivity in guard cells.

Several papers have highlighted the role of GSH in ABA-induced stomatal closure (Jahan et al. 2008; Okuma et al. 2011; Akter et al. 2012; Akter et al. 2013). Okuma et al. (2011) reported that ABA reduced GSH content in cell guards as part of its signaling pathway, leading to stomatal closure. Indeed, chemical treatment that reduces GSH content in guard cells seems to enhance guard cell sensitivity to ABA (Okuma et al. 2011; Akter et al. 2012). *Arabidopsis* GSH-deficient mutant *cad2-1*, lacking γ-glutamylcysteine synthetase, the first enzyme in GSH biosynthesis, also shows enhanced ABA-dependent stomata closure (Okuma et al. 2011). In another GSH-defective mutant *chl-1* (chlorinal-1; defective in light-harvesting complexes), treatments with GSH monoethyl ester restored the GSH content of guard cells and the wild-type phenotype, thus confirming the role of GSH in modulating guard cell sensitivity to ABA (Okuma et al. 2011; Jahan et al. 2008). In *cad2-1* mutants, GSH depletion promotes apoplastic H_2O_2 accumulation and in turn plasma-membrane Ca^{2+} channel sensitivity to ROS, thereby increasing guard cell sensitivity to ABA signaling (Munemasa et al. 2013). Thus, impairment in GSH biosynthesis in guard cells seems to inhibit light-induced stomatal opening (Jahan et al. 2016).

Nitric oxide (NO) is another reactive species involved in the ABA-induced signaling pathway that promotes stomatal closure (Bright et al. 2006). In fact, ABA-induced ROS production triggers an increase in NO in guard cells. NO may also be responsible for the negative feedback of ABA-induced stomata closure (Wang et al. 2015). The ABA signaling pathway requires the activation of a Ser/Thr kinase named OPEN STOMATA 1 (OST1), which is involved in the activation of slow anion channel-associated 1 and RBOHF (Mittler and Blumwald 2015). OST1 can be inhibited by S-nitrosylation, as a consequence of a NO increase mediated by ABA signaling (Wang et al. 2015). Glutathione may play a role in this process since S-nitrosoglutathione (GSNO) represents a storage NO compound in the cells which can act as a trans-nitrosylating agent (Fig.6.1; de Pinto et al. 2013; Locato et al. 2016). GSNO levels in cells are regulated by the enzyme GSNO reductase (GSNOR), which in turn controls cell thiol levels (Locato et al. 2016). *Arabidopsis thaliana* GSNOR-defective mutants (*gsnor1-3*) overaccumulate S-nitroso-thiols (SNO) in guard cells and are insensitive to ABA-induced stomatal closure (Wang et al. 2015).

It has been demonstrated that *gsnor1-3* plants present constitutive S-nitrosylation of OST1 with consequent inhibition of the ABA signaling pathway (Wang et al. 2015). This mechanism is important in preventing mild drought-triggered hyperactivation of ABA-dependent stomatal closure. In fact, prolonged stomatal closure reduces plant gas exchanges and restrains CO_2 availability, thus affecting photosynthesis and in turn plant growth and survival. Stomatal closure also reduces the CO_2/O_2 rate, forcing RUBISCO to shift from carboxylase to oxygenase activity and consequently increasing photorespiration, which is another source of ROS (Noctor et al. 2002).

Proteomic studies have highlighted the role of different enzymes that use GSH as a substrate in defense responses to drought. It has been shown that in *Zea mays* and *Brassica napus* plants, the GST level increases under drought (Zhao et al. 2016; Wang et al. 2016). GSTs are enzymes involved in xenobiotic detoxification and oxidative protection (reviewed by Dixon et al. 2002). They catalyze GSH conjugation to xenobiotics and peroxides, potentially regulating cell GSH availably and thus its redox-buffering capability (Noctor et al. 2012). *Arabidopsis thaliana atgst17* mutants show increased tolerance to drought (Chen et al. 2012). GST17 is one of the most active GSTs in *Arabidopsis*, and *atgst17*-defective mutants show a higher GSH content compared to WT. Curiously, these plants are more tolerant to drought, which, Chen et al. (2012) suggest, depends on the fact that the increased GSH level activates ABA synthesis and in turn the ABA-protective effect against drought.

3.2 *Salinity*

The high concentration of salts in the soil is another adverse environmental condition that limits the productivity of arable land. Soil salinity is mainly related to high levels of NaCl. Studying plant defense responses to salt stress is complicated because an excess of salts in the soil subjects plants to both osmotic stress (short-term effect) and ionic toxicity (long-term effect).

Osmotic stress hinders the roots from absorbing water due to the high solute concentration in the soil. Plants rapidly respond to osmotic stress by reducing water loss through stomatal closure. However, plant tolerance to salt stress may be mainly due to the ability of plants to maintain the K^+/Na^+ ratio within a physiological range (reviewed by Munns and Tester 2008). Indeed, ionic toxicity seems above all to be due to the hyperaccumulation of Na^+ in plant tissues. Consequently, cellular K^+ content decreases since Na^+ competes with K^+ for intracellular transport. Reduced cellular K^+ levels have a deleterious effect on different metabolic pathways (Gupta and Huang 2014).

Plant tolerance to salinity is also related to a plant's capacity to counteract salinity-induced oxidative stress, since ROS are triggered by salt stress conditions. Demidchik et al. (2010) showed that the cell K^+ levels decrease as a consequence of salt stress conditions, also because ROS activate guard cell outward rectifying potassium (GORK) channels. In fact when high ROS levels are produced under salinity, the consequent ion leakage derived from the activation of GORK channels leads to programmed cell death (Demidchik et al. 2010). Thus, *Brassica napus*, a *Brassica* species that is tolerant to salt, shows a higher ability to retain K^+ by reducing root K^+ permeable channel sensitivity to ROS (Chakraborty et al. 2016). The role of ROS in the signaling of plant defense responses activated by salt has also been reported. *Arabidopsis thaliana* double mutants *atrbohd1-atrbohf1* and *atrbohd2-atrbohf2*, lacking RBOHD1-RBOHF1 and RBOHD2-RBOHF2, show reduced ROS production and in turn a decreased tolerance to salt stress compared to WT plants. It seems that ROS promote salt tolerance in WT plants by activating molecular mechanisms, thus increasing K^+/Na^+ ratio under salinity (Ma et al. 2012). A salt-responsive ethylene-responsive factor (SERF1) activated by the ROS produced under salt stress has also been identified. SERF1 seems to trigger a molecular cascade involving MAPK gene expression regulation which promotes salt tolerance (Schmidt et al. 2013).

All these data underline the importance of the ROS/antioxidant balance under salt stress, since low ROS production activates defense responses and ROS hyperaccumulation triggers cell death (Demidchik et al. 2010,2014). However, in this context, the role of GSH as a key redox-buffering compound is not clear. To the best of our knowledge, only a few papers have dealt with the involvement of the GSH network in the defense response to salt. High GSH levels have been correlated to a higher salt tolerance (Zagorchev et al. 2013 and references therein). In line with this, sulfur supplementation has been found to increase salt stress tolerance by increasing GSH levels and reducing salt-dependent oxidative stress in mustard (Fatma et al. 2014,2016). Sulfur assimilation under salt stress has also been found to improve by supplementation with sodium nitroprusside (SNP), which is a NO donor (Fatma et al. 2016). In fact, NO apparently stimulates GSH biosynthesis by the activation of γ-glutamylcysteine synthetase (Innocenti et al. 2007). NO also acts as a signal in defense responses triggered by salt activating the proton pump as well as the Na^+/H^+ antiport in tonoplasts responsible for Na^+ sequestration into the vacuole (Zhang et al. 2006). Thus, a cross talk between NO and GSH in salt stress responses has been proposed. GSNO, as the cellular and systemic reservoir of NO, seems to play a role in this process (Fatma et al. 2016).

Dinler et al. (2014) suggested that different GSTs play a role in plant defense responses activated by NO signaling triggered under salt stress. However, the role of GST in salt stress acclimation has not been clarified (Moons 2003; Chen et al. 2012; Dinler et al. 2014). The expressions of different GST genes are stimulated by salt stress in soybean leaves. On the other hand, the regulation of GST enzymatic activity under stress conditions appears more complex as it fluctuates throughout the treatment time (Dinler et al. 2014). GST has been indicated as a putative target of S-nitrosylation in both the roots and leaves of citrus plants subjected to salt stress. This thus suggests that this redox-dependent post-translational modification influences GST activity in response to salinity (Tanou et al. 2012).

The ectopic expression of rice and tomato GSTs in *Arabidopsis thaliana* appears to increase the tolerance to salt stress in transgenic plants (Sharma et al. 2014; Xu et al. 2015). The expression of tomato LeGSTU2 in *Arabidopsis* plants subjected to

NaCl reduced oxidative and osmotic stress by increasing the activity of antioxidant enzymes, such as APX and monodehydroascorbate reductase (MDHAR), as well as the proline level (Xu et al. 2015). On the other hand, the suppression of GST17 in *Arabidopsis* defective mutants increases tolerance to both drought and salt stresses, suggesting the different involvement of specific GSTs in plant response to stress (Chen et al. 2012). The ectopic expression of dehydroascorbate reductase (DHAR) also increases salt tolerance in plants (Eltayeb et al. 2006; Ushimaru et al. 2006). DHAR is a GST that uses transient GSH conjugation as part of its catalytic mechanism to regenerate the reduced form of ASC. Thus, ASC recycling is part of the GSH metabolic network that enables plants to cope with the oxidative stress caused by salinity (reviewed by Gallie 2013).

GPX is another actor in the GSH network and plays an important role in plant response to salt stress (Islam et al. 2015; Pilarska et al. 2016). GPXs are well-known enzymes that protect cells from ROS-dependent damage. Plant GPXs use GSH and thioredoxin as a substrate to reduce hydroperoxides (Passaia and Margis-Pinheiro 2015 and references therein). In addition, GPXs are possible redox sensors, which are involved in ROS-triggered signaling pathways (Passaia and Margis-Pinheiro 2015). In the salt-resistant *Eutrema salsugineum* (*Thellungiella salsuginea*), GPX was the only antioxidant enzyme analyzed that showed a higher activity compared to the salt-sensitive *Arabidopsis thaliana* under control and salt conditions (Pilarska et al. 2016). The ectopic expression of a *Pennisetum glaucum* GPX has also been shown to increase rice tolerance to salinity (Islam et al. 2015).

GSH is involved in the defense mechanisms activated against salt stress, also participating in the methylglyoxal (MG) detoxification route. MG is a cytotoxic compound, mainly derived from carbohydrate and amino acid catabolism, which also increases within cells as a consequence of salt stress (Yadav et al. 2005). Methylglyoxal is detoxified by the glyoxalase system which classically consists in two enzymes: glyoxalase I, which converts GSH and MG to *S*-D-lactoylglutathione, and glyoxalase II, which converts *S*-D-lactoylglutathione into D-lactate and GSH (Singla-Pareek et al. 2003).

A novel route involving a unique enzyme, glyoxalase III, has also been discovered in plants (Ghosh et al. 2016). Interestingly, a glyoxalase II, which is likely to be responsive to the GSH level, has been identified in rice, and its activity appears to be correlated to salt tolerance in rice (Singla-Pareek et al. 2008). The overexpression of this enzyme in tobacco improved the salinity tolerance of transgenic plants by increasing their photosynthesis efficiency and antioxidant defenses, including the maintenance of high GSH levels under stress (Ghosh et al. 2014).

3.3 Heavy Metals

In the last two centuries, the increasing anthropogenic activities, mainly related to metallurgic industry and agricultural practices such as mining and the use of fertilizers and sewage sludge, have all released abundant heavy metals in the soil. Metals persist in the environment since they are nonbiodegradable and, when they are absorbed by crops, they affect the crop growth and viability as well as threaten food security (Alloway 2012).

The reduction in growth caused by metal uptake seems to be mainly related to metal ion competition with essential cations, normally being absorbed by plants. Although some metals such as Fe, Mn, Cu, Zn, and Ni are plant micronutrients, as they are cofactors of many enzymes, they become toxic when their levels go above plant needs (Anjum et al. 2015a). On the other hand, nonessential metals such as Cd, Pb, Hg, As, and Ag are also phytotoxic at low concentrations and can poison humans since in the diet they have been correlated to various pathologies (Jarüp 2003; Järup and Akesson 2009).

Metal phytotoxicity seems to be mainly correlated to the overproduction of metal-induced ROS. The mechanism by which metals induce oxidative stress in plants depends on their particular nature. Metals that have more than one oxidation state are redox-active and can participate in the Fenton reaction, leading to the formation of hydroxyl radicals (OH·). In turn, as OH· are the major reactive ROS radicals, they promote cell oxidative damage, leading to metabolic impairment and possibly cell death. On the other hand, nonredox active metals can indirectly promote ROS production. They compete with essential cations in the catalytic site of many enzymes, causing their inactivation and consequently metabolic impairment, also leading to the failure of the antioxidant defense mechanisms. The cation displacement in the proteins involved in the electron transfer of photosynthesis and the respiratory chain increases ROS production in chloroplasts and mitochondria (Fig. 6.1; Cuypers et al. 2016 and references therein). Metals also compete with the cofactors of many enzymes involved in ROS scavenging, such as SOD and APX, thus affecting ROS detoxification (Kliebenstein et al. 1998; Jespersen et al. 1997).

Plants have evolved various strategies to counteract metal toxicity. The ability of plants to prevent metal translocation to the shoots may be positively correlated with metal tolerance. Indeed the first defense strategy against metal accumulation consists in restraining metal absorption by the roots. Metals can thus be sequestered into the rhizosphere by cell wall components, such as pectins and extensins, or root exudates which block them outside the roots (reviewed by Hernández et al. 2015). When these mechanisms fail, metal ions enter cells through essential ion channels where they can then be excluded by active plasma-membrane transporters.

In order to reduce metal concentration in the cytosol, plants have also evolved a number of ligands that chelate metals within cells. The resulting metal complexes are translocated into the vacuole where they are detoxified (Anjum et al. 2015b and references therein). The metal-chelating agents consist of organic acids, amino acids and their derivates, as well as thiol compounds, such as GSH, metallothioneins (MTs), and phytochelatins (PCs). MTs are proteins rich in Cys residues, whereas PCs are GSH oligomers (with GSH residues ranging from 2 to 11) whose formation is catalyzed by PC synthases (PCS), a class of enzymes that seems to be directly activated by metals (Fig. 6.1; Vatamaniuk et al. 2000).

Free Cys is not a good candidate to chelate metals since it is oxidized by them and then released, with a consequent increased reactivity of reduced metals. On the other hand, Cys incorporation into polymers, such as GSH, MT, and PC, leads to a more efficient metal chelation by the formation of stabilized metal–thiol interactions (Jozefczak et al. 2012). GSH has been shown to play a major role in metal tolerance on its own or as a substrate for PC synthesis. Several studies have reported that when GSH biosynthesis is inhibited by L-buthionine sulfoximine (BSO), plant sensitivity to metal increases (Jozefczak et al. 2012; Flores-Cáceres et al. 2015), whereas plant treatments that increase GSH endogenous levels enhance metal tolerance (Wang et al. 2011; Guan et al. 2015).

Plant exposure to Cd induces GSH and PC synthesis by promoting the expression of the genes coding for GSH biosynthetic enzymes. GSH synthesis occurs in two steps: the first reaction is catalyzed by γ-glutamylcysteine synthetase (GSH1), an enzyme found in plastids. The second reaction is catalyzed by GSH synthetase (GSH2) and occurs in the cytosol (Fig. 6.1). The first reaction seems to be the limiting step in GSH biosynthesis, and GSH1 activity is also negatively regulated by the GSH level. This may depend on the fact that GSH1 is a redox-sensitive enzyme that works in a homodimeric form whose assembly is due to the formation of two disulfide bonds (Fig. 6.1). Thus, under metal-induced oxidative conditions *Arabidopsis* GSH1 is activated (Hicks et al. 2007). Sugar beet transgenic plants expressing *Streptococcus thermophilus* γ-glutamylcysteine synthetase-glutathione synthetase (StGCS-GS), with limited negative feedback control, show increased tolerance to Cu, Cd, and Zn with increased GSH and PC levels (Liu et al. 2015). Thus, a rice mutant sensitive to As and Cd shows lower GSH and PC synthetic capability (Yang et al. 2016). This mutant is defective in CRT (*Plasmodium falciparum* chloroquine resistance transporter)-like 1 (OsCTL1), which is a transporter located in the plastidial envelope of WT plants. OsCTL1 is a homolog of *Arabidopsis thaliana* CTL transporters which are known to be responsible for the efflux of γ-glutamylcysteine from plastids to cytosol that is required for GSH synthesis (Fig. 6.1; Maughan et al. 2010).

Some research groups suggested that increased PC synthesis reduces the concentration of free GSH in the cell, removing the negative GSH feedback on its synthesis, thus in turn promoting GSH production in the plants exposed to metals (Jozefczak et al. 2012 and references therein). This mainly occurs when plants are also well supplemented with sulfur or exogenous GSH, since limiting sulfur conditions may constrain metal tolerance by reducing GSH availability (Noctor et al. 2012). *Nicotiana tabacum* expressing *Arabidopsis thaliana* AtPCS1 shows enhanced Cd tolerance when plants were treated with exogenous GSH; on the contrary, transgenic plants appear to be hypersensitive to Cd in the absence of GSH supplementation (Brunetti et al. 2011). Chen et al. (2016) identified and characterized an *Arabidopsis thaliana* zinc-finger transcription factor (ZAT6), whose expression was activated by Cd. Chen et al. demonstrated that ZAT6 directly activates GSH1 gene expression by directly binding the GSH1 promoter and also increases the expression of GSH2, PCS1, and PCS2 under Cd exposure. *Arabidopsis* plants overexpressing ZAT6 show increased tolerance to Cd stress, since ZAT6 coordinates the stimulation of GSH and PC syntheses.

Cross talk between ethylene, S assimilation, and GSH is also a key element in metal-activated plant defense. Ethylene production depends on the availability of the S-containing methionine (its precursor), and its release is increased in many plant species exposed to metals (Keunen et al. 2016). This hormone seems to promote S assimilation under metal stress by increasing the activity of ATP sulfurylase, the first enzyme involved in sulfur assimilation, and thus promoting an increase in plant GSH levels and consequent tolerance to metals (reviewed by Keunen et al. 2016). Many studies report that plants increase S assimilation under metal stress. Cd appears to induce the expression of genes coding for sulfur transporters (SULTR) in *Arabidopsis*, promoting S uptake and assimilation at the transcriptional level (Yamaguchi et al. 2016). *Arabidopsis sultr1;1-sultr1;2* double mutants lack highly efficient root sulfate transporters. These mutants show reduced GSH biosynthesis under Cd and limiting S conditions. In order to promote Cd chelation, the produced GSH is used for PC synthesis in the mutant plants. This reduces free GSH availability and in turn cell antioxidant capability against Cd-induced oxidative stress, leading to enhanced Cd sensitivity in the double mutants (Liu et al. 2015).

High GSH levels have been found in metal-hyperaccumulating plants, which are a metal-tolerant species possibly used in phytoremediation due to their ability to absorb and withstand high levels of metals and thus reduce metal pollution in the soil. It has been suggested that the hyperaccumulator species preferentially translocate metal from roots to shoots, where PC–metal complexes are detoxified in the vacuole (Mendoza-Cózatl et al. 2011). He et al. (2015) transformed poplar plants in order to obtain transgenic plants with increased GSH biosynthetic capability, as a possible phytoremediation strategy. In these plants, GSH seems to control Cd transport and detoxification genes at transcriptional levels. GSH exogenous application also promotes the expression of transcription factors implicated in the stress response in tomato plants exposed to Cd stress (Hasan et al. 2016). GSH application to oilseed rape roots also increases Cd tolerance by promoting metal exclusion from the roots and Cd translocation to the shoots (Nakamura et al. 2013).

Increased PC synthesis does not seem to be adopted by metal-tolerant plants. As previously mentioned, PC hyperaccumulation under metal stress may decrease cell GSH availability, with a huge consequence on the cell antioxidant shield and a negative correlation with stress tolerance (Seth et al. 2012). In fact, it is widely accepted that the cell redox state is mainly related to the GSH/GSSG balance, as GSH is a major soluble redox-buffering compound (Noctor et al. 2012). Increased metal sensitivity in *pad2-1*, GSH-defective mutants, and *vitc2-1*, ASC-defective ones, is likely mainly related to intracellular GSH depletion rather than reduced ASC levels (Koffler et al. 2014). However, in *Triticum durum*, plants supplemented with GL, the last precursor of ASC synthesis, Cd tolerance was increased. Interestingly, apart from increasing plant ASC levels, GL treatment slightly increases GSH levels and reduces PC accumulation. This thus suggests that, in wheat GL-supplemented plants, GSH is further recruited to manage oxidative conditions under Cd stress by controlling the ASC redox state (Paradiso et al. 2008). Jozefczak et al. (2015) highlighted that compensatory mechanisms activated by *cad2-1*, GSH-deficient mutants, are less effective against metal stress than those activated by *vitc1-1*, ASC-deficient plants. The higher Cd tolerance observed in *vitc1-1* depended on their increased Cd chelation capability triggered by enhanced GSH and PC syntheses.

4 Conclusions and Future Perspectives

The evidence reported in this chapter only partially describes the importance of glutathione metabolism in plant life. An increasing body of data highlights its involvement in the plant–environment interaction, as well as in plant development, similarly to what occurs in all aerobic organisms. It is not a coincidence that glutathione metabolism represents one of the most ancient defense lines, the evolution of which probably promotes di-oxygenic photosynthesis. Due to its cysteine moiety, its metabolism is based on sulfur biochemistry. It has been argued that the ability of glutathione to interact with several nucleophile molecules makes it a very versatile adapter molecule (Deponte 2013). In the last decade, glutathione has been considered as a real hub controlling and regulating several different metabolic pathways, from hormone biosynthesis to xenobiotic detoxification (Noctor et al. 2012).

Several studies have also underlined the importance of regulating GSH fluxes among specific cell compartments. The different patterns of GSH portioning within cells can modulate the redox state of specific compartments and may be part of the signaling pathways involved in defense responses, also including cell cycle regulation (Diaz-Vivancos et al. 2010; Locato et al. 2015). However, this issue has been poorly investigated in plants.

Modulating specific steps in the glutathione metabolism by metabolic engineering could be a strategy used to increase plants' capacity to withstand a large array of environmental stresses. However, the multifunctionality and versatility of this metabolite could complicate the identification of a sustainable biotechnological approach, as has also been suggested for ASC metabolism bioengineering (Locato et al. 2013). The recent identification of a quantitative trait locus (QTL) governing a complex phenotype is a possible alternative strategy for improving the crop productivity and nutritional value of plant-derived food matrices (Locato et al. 2013). This may also be true of glutathione metabolic signatures, in terms of increasing plant defense responses and consequently crop availability for human nutrition.

Acknowledgments The authors acknowledge the funding from the Ministero dell'Istruzione dell'Università e della Ricerca PRIN n. 20153NM8RM

References

Akter N, Sobahan MA, Uraji M, Ye W, Hossain MA, Mori IC et al (2012) Effects of depletion of glutathione on abscisic acid and methyl jasmonate induced stomata closure in *Arabidopsis thaliana*. Biosci Biotechnol Biochem 76:2032–2037

Akter N, Okuma E, Sobahan MA, Uraji M, Munemasa S, Nakamura Y et al (2013) Negative regulation of methyl jasmonate-induced stomatal closure by glutathione in *Arabidopsis*. J Plant Growth Regul 32:208–215

Alloway BJ (2012) Sources of heavy metals and metalloids in soils. In: Alloway BJ (ed) Heavy metals in soils. Springer, Heidelberg, pp 15–50

Anjum NA, Singh HP, Khan MIR, Masood A, Per TS, Negi A et al (2015a) Too much is bad–an appraisal of phytotoxicity of elevated plant-beneficial heavy metal ions. Environ Sci Pollut Res 22:3361–3382

Anjum NA, Hasanuzzaman M, Hossain MA, Thangavel P, Roychoudhury A, Gill SS, MAM R, Adam V, Fujita M, Kizek R, Duarte AC, Pereira E, Ahmad I (2015b) Jacks of metal/metalloid chelation trade on plants-an overview. Front Plant Sci 6:192

Bright J, Desikan R, Hancock JT, Weir IS, Neill SJ (2006) ABA induced NO generation and stomatal closure in Arabidopsis are dependent on H_2O_2 synthesis. Plant J 45:113–122

Brunetti P, Zanella L, Proia A, De Paolis A, Falasca G, Altamura MM, Sanità di Toppi L, Costantino P, Cardarelli M (2011) Cadmium tolerance and phytochelatin content of *Arabidopsis* seedlings overexpressing the phytochelatin synthase gene AtPCS1. J Exp Bot 62:5509–5519

Centomani I, Sgobba A, D'Addabbo P, Dipierro N, Paradiso A, De Gara L, Dipierro S, Viggiano L, de Pinto MC (2015) Involvement of DNA methylation in the control of cell growth during heat stress in tobacco BY-2 cells. Protoplasma 252:1451–1459

Chakraborty K, Bose J, Shabala L, Shabala S (2016) Difference in root K^+ retention ability and reduced sensitivity of K^+-permeable channels to reactive oxygen species confer differential salt tolerance in three brassica species. J Exp Bot 67(15):4611–4625

Chen JH, Jiang HW, Hsieh EJ, Chen HY, Chien CT, Hsieh HL, Lin TP (2012) Drought and salt stress tolerance of an Arabidopsis glutathione S-transferase U17 knockout mutant are attributed to the combined effect of glutathione and abscisic acid. Plant Physiol 158:340–351

Chen J, Yang L, Yan X, Liu Y, Wang R, Fan T, Ren Y, Tang X, Xiao F, Liu Y, Cao S (2016) Zinc-finger transcription factor ZAT6positively regulates cadmium tolerance through the glutathione-dependent pathway in Arabidopsis. Plant Physiol 171(1):707–719

Cheng G, Karunakaran R, East AK, Munoz-Azcarate O, Poole PS (2017) Glutathione affects the transport activity of Rhizobium leguminosarum 3841 and is essential for efficient nodulation. FEMS Microbiol Lett. https://doi.org/10.1093/femsle/fnx045

Cuypers A, Hendrix S, Amaral dos Reis R, De Smet S, Deckers J, Gielen H, Jozefczak M, Loix C, Vercampt H, Vangronsveld J, Keunen E (2016) Hydrogen peroxide, signaling in disguise during metal Phytotoxicity. Front Plant Sci 7:470

De Gara L, Locato V, Dipierro S, de Pinto MC (2010) Redox homeostasis in plants. The challenge of living with endogenous oxygen production. Resp Physiol Neurobiol 173:S13–S19

de Pinto MC, Locato V, Sgobba A, Romero-Puertas Mdel C, Gadaleta C, Delledonne M, De Gara L (2013) S-nitrosylation of ascorbate peroxidase is part of programmed cell death signaling in tobacco bright yellow-2 cells. Plant Physiol 163:1766–1775

de Pinto MC, Locato V, Paradiso A, De Gara L (2015) Role of redox homeostasis in thermotolerance under a climate change scenario. Ann Bot 116:487–496

Demidchik V, Cuin TA, Svistunenko D, Smith SJ, Miller AJ, Shabala S, Sokolik A, Yurin V (2010) Arabidopsis root K^+-efflux conductance activated by hydroxyl radicals: single channel properties, genetic basis and involvement in stress-induced cell death. J CellSci 123:1468–1479

Demidchik V, Straltsova D, Medvedev SS, Pozhvanov GA, Sokolik A, Yurin VJ (2014) Stress-induced electrolyte leakage: the role of K^+-permeable channels and involvement in programmed cell death and metabolic adjustment. J Exp Bot 65(5):1259–1270

Deponte M (2013) Glutathione catalysis and the reaction mechanisms of glutathione-dependent enzymes. Biochim Biophys Acta 1830:3217–3266

Diaz-Vivancos P, Dong Y-P, Ziegler K, Markovic J, Pallardó F, Pellny TK, Verrier P, Foyer CH (2010) Recruitment of glutathione into the nucleus during cell proliferation adjusts whole cell redox homeostasis in Arabidopsis thaliana and lowers the oxidative defence shield. Plant J 64:825–838

Dinler BS, Antoniou C, Fotopoulos V (2014) Interplay between GST and nitric oxide in the early response of soybean (*Glycine max* L.) plants to salinity stress. J Plant Physiol 171(18):1740–1747
Dixon DP, Davies BG, Edwards E (2002) Functional divergence in the glutathione transferase superfamily in plants. J Biol Chem 277:30859–30869
Dixon DP, Skipsey M, Grundy NM, Edwards R (2005) Stress induced protein S-glutathionylation in Arabidopsis. Plant Physiol 138:2233–2244
Drerup MM, Schlücking K, Hashimoto K, Manishankar P, Steinhorst L, Kuchitsu K, Kudla J (2013) The calcineurin B-like calcium sensors CBL1 and CBL9 together with their interacting protein kinase CIPK26 regulate the Arabidopsis NADPH oxidase RBOHF. Mol Plant 6:559–569
Eltayeb AE, Kawano N, Badawi GH, Kaminaka H, Sanekata T, Morishima I, Shibahara T, Inanaga S, Tanaka K (2006) Enhanced tolerance to ozone and drought stresses in transgenic tobacco overexpressing dehydroascorbate reductase in cytosol. Physiol Plant 127:57–65
Eshdat Y, Holland D, Faltin Z, Ben-Hayyim G (1997) Plant glutathione peroxidases. Physiol Plant 100:234–240
Fatma M, Asgher M, Masood A, Khan NA (2014) Excess sulfur supplementation improves photosynthesis and growth in mustard under salt stress through increased production of glutathione. Environ Exp Bot 107:55–63
Fatma M, Masood A, Per TS, Khan NA (2016) Nitric oxide alleviates salt stress inhibited photosynthetic performance by interacting with sulfur assimilation in mustard. Front Plant Sci 7:521
Flores-Cáceres ML, Hattab S, Hattab S, Boussetta H, Banni M, Hernández LE (2015) Specific mechanisms of tolerance to copper and cadmium are compromised by a limited concentration of glutathione in alfalfa plants. Plant Sci 233:165–173
Gallie DR (2013) The role of l-ascorbic acid recycling in responding to environmental stress and in promoting plant growth. J Exp Bot 64:433–443
Ghosh A, Pareek A, Sopory SK, Singla-Pareek SL (2014) A glutathione responsive rice glyoxalase II,OsGLYII-2, functions in salinity adaptation by maintaining better photosynthesis efficiency and antioxidant pool. Plant J 80:93–105
Ghosh A, Kushwaha HR, Hasan MR, Pareek A, Sopory SK, Singla-Pareek SL (2016) Presence of unique glyoxalase III proteins in plants indicates the existence of shorter route for methylglyoxal detoxification. Sci Rep 6:18358
Grant CM (2001) Role of the glutathione/glutaredoxin and thioredoxin systems in yeast growth and response to stress conditions. Mol Microbiol 39:533–541
Guan C, Ji J, Jia C, Guan W, Li X, Jin C, Wang G (2015) A GSHS-like gene from Lycium chinense may be regulated by cadmium-induced endogenous salicylic acid and overexpression of this gene enhances tolerance to cadmium stress in Arabidopsis. Plant Cell Rep 34(5):871–884
Gupta B, Huang B (2014) Mechanism of salinity tolerance in plants: physiological, biochemical, and molecular characterization. Int J Genomics:701596. 18 pages
Hasan MK, Liu C, Wang F, Ahammed GJ, Zhou J, Xu MX, Yu JQ, Xia XJ (2016) Glutathione-mediated regulation of nitric oxide, S-nitrosothiol and redox homeostasis confers cadmium tolerance by inducing transcription factors and stress response genes in tomato. Chemosphere 161:536–545
He J, Li H, Ma C, Zhang Y, Polle A, Rennenberg H, Cheng X, Luo ZB (2015) Overexpression of bacterial γ-glutamylcysteine synthetase mediates changes in cadmium influx, allocation and detoxification in poplar. New Phytol 205(1):240–254
Hernández LE, Sobrino-Plata J, Montero-Palmero MB, Carrasco-Gil S, Flores-Cáceres ML, Ortega-Villasante C, Escobar C (2015) Contribution of glutathione to the control of cellular redox homeostasis under toxic metal and metalloid stress. J Exp Bot 66(10):2901–2911
Hicks LM, Cahoon RE, Bonner ER, Rivard RS, Sheffield J, Jez JM (2007) Thiol-based regulation of redox-active glutamate-cysteine ligase from *Arabidopsis thaliana*. Plant Cell 19:2653–2661
Hussain J, Chen J, Locato V, Sabetta W, Behera S, Cimini S, Griggio F, Martínez-Jaime S, Graf A, Bouneb M, Pachaiappan R, Fincato P, Blanco E, Costa A, De Gara L, Bellin D, de Pinto

MC, Vandelle E (2016) Constitutive cyclic GMP accumulation in Arabidopsis Thaliana compromises systemic acquired resistance induced by an avirulent pathogen by modulating local signals. Sci Rep 6:36423

Innocenti G, Pucciariello C, LeGleuher M, Hopkins J, de Stefano M, Delledonne M et al (2007) Glutathione synthesis is regulated by nitric oxide in *Medicago truncatula* roots. Planta 225:1597–1602

Islam T, Manna M, Reddy MK (2015) Glutathione peroxidase of Pennisetum Glaucum (PgGPx) is a functional Cd^{2+}dependent peroxiredoxin that enhances tolerance against salinity and drought stress. PLoS One 10(11):e0143344

Jahan MS, Ogawa K, Nakamura Y, Shimoishi Y, Mori IC, Murata Y (2008) Deficient glutathione in guard cells facilitates abscisic acid-induced stomatal closure but does not affect light-induced stomatal opening. Biosci Biotechnol Biochem 72:2795–2798

Jahan MS, Nozulaidi M, Khairi M, Mat N (2016) Light-harvesting complexes in photosystem II regulate glutathione-induced sensitivity of Arabidopsis guard cells to abscisic acid. J Plant Physiol 195:1–8

Järup L (2003) Hazards of heavy metal contamination. Brit Med Bull 68:167–182

Järup L, Akesson A (2009) Current status of cadmium as an environmental health problem. Toxicol Appl Pharmacol 238:201–208

Jespersen HM, Kjaersgard IVH, Ostergaard L, Welinder KG (1997) From sequence analysis of three novel ascorbate peroxidases from Arabidopsis thaliana to structure, function and evolution of seven types of ascorbate peroxidase. Biochem J 326:305–310

Jozefczak M, Remans T, Vangronsveld J, Cuypers A (2012) Glutathione is a key player in metal-induced oxidative stress defenses. Int J Mol Sci 13:3145–3175

Jozefczak M, Bohler S, Schat H, Horemans N, Guisez Y, Remans T, Vangronsveld J, Cuypers A (2015) Both the concentration and redox state of glutathione and ascorbate influence the sensitivity of Arabidopsis to cadmium. Ann Bot 116(4):601–612

Jubany-Mari T, Alegre-Batlle L, Jiang K, Feldman LJ (2010) Use of a redox-sensing GFP (c-roGFP1) for real-time monitoring of cytosol redox status in *Arabidopsis thaliana* water-stressed plants. FEBS Lett 584:889–897

Keunen E, Schellingen K, Vangronsveld J, Cuypers A (2016) Ethylene and metal stress: small molecule, big impact. Front Plant Sci 7:23

Kliebenstein DL, Monde R-A, Last RL (1998) Superoxide dismutase in Arabidopsis: an eclectic enzyme family with disparate regulation and protein localization. Plant Physiol 118:637–650

Koffler BE, Polanschütz L, Zechmann B (2014) Higher sensitivity of pad2-1 and vtc2-1 mutants to cadmium is related to lower subcellular glutathione rather than ascorbate contents. Protoplasma 251(4):755–769

Kwak JM, Mori IC, Pei ZM, Leonhardt N, Torres MA, Dangl JL, Bloom RE, Bodde S, Jones JD, Schroeder JI (2003) NADPH oxidase *AtrbohD* and *AtrbohF* genes function in ROS-dependent ABA signaling in Arabidopsis. EMBO J 22:2623–2633

Liu D, An Z, Mao Z, Ma L, Lu Z (2015) Enhanced heavy metal tolerance and accumulation by transgenic sugar beets expressing *Streptococcus thermophilus* StGCS-GS in the presence of cd, Zn and cu alone or in combination. PLoS One 10(6):e0128824

Locato V, de Pinto MC, De Gara L (2009) Different involvement of the mitochondrial, plastidial and cytosolic ascorbate-glutathione redox enzymes in heat shock responses. Physiol Plant 135(3):296–306

Locato V, Cimini S, De Gara L (2013) Strategies to increase vitamin C in plants: from plant defense perspective to food biofortification. Front Plant Sci 4:152

Locato V, Novo Uzal E, Cimini S, Zonno MC, Evidente A, Micera A, Foyer CH, De Gara L (2015) Low concentrations of the toxin ophiobolin A lead to an arrest of the cell cycle and alter the intracellular partitioning of glutathione between the nuclei and cytoplasm. J Exp Bot 66:2991–3000

Locato V, Paradiso A, Sabetta W, De Gara L, de Pinto MC (2016) Nitric oxide and reactive oxygen species in PCD signaling. Adv Bot Res 77:165–192

Ma L, Zhang H, Sun L, Jiao Y, Zhang G, Miao C, Hao F (2012) NADPH oxidase AtrbohD and AtrbohF function in ROS-dependent regulation of Na^+/K^+ homeostasis in *Arabidopsis* under salt stress. J Exp Bot 63:305–317

Maughan SC, Pasternak M, Cairns N, Kiddle G, Brach T, Jarvis R, Haas F, Nieuwland J, Lim B, Müller C (2010) Plant homologs of the *Plasmodium falciparum* chloroquine-resistance transporter, PfCRT, are required for glutathione homeostasis and stress responses. Proc Natl Acad Sci USA 107:2331–2336

Mendoza-Cózatl DG, Butko E, Springer F, Torpey JW, Komives EA, Kehr J, Schroeder JI (2008) Identification of high levels of phytochelatins, glutathione and cadmium in the phloem sap of Brassica napus. A role for thiol-peptides in the long-distance transport of cadmium and the effect of cadmium on iron translocation. Plant J 54:249–259

Mendoza-Cózatl DG, Jobe TO, Hauser F, Schroeder JI (2011) Long distance transport, vacuolar sequestration, tolerance, and transcriptional responses induced by cadmium and arsenic. Curr Opin Plant Biol 14:554–562

Meyer AJ, Brach T, Marty L, Kreye S, Rouhier N, Jacquot JP, Hell R (2007) Redox-sensitive GFP in *Arabidopsis thaliana* is a quantitative biosensor for the redox potential of the cellular glutathione redox buffer. Plant J 52:973–986

Mhamdi A, Hager J, Chaouch S, Queval G, Han Y, Taconnat L, Saindrenan P, Gouia H, Issakidis-Bourguet E, Renou JP, Noctor G (2010) Arabidopsis GLUTATHIONE REDUCTASE 1 plays a crucial role in leaf responses to intracellular H_2O_2 and in ensuring appropriate gene expression through both salicylic acid and jasmonic acid signalling pathways. Plant Physiol 153:1144–1160

Mills GC (1957) Hemoglobin catabolism. I. Glutathione peroxidase, an erythrocyte enzyme which protects hemoglobin from oxidative breakdown. J Biol Chem 229:189–197

Mittler R, Blumwald E (2015) The roles of ROS and ABA in systemic acquired acclimation. Plant Cell 27:64–70

Moons A (2003) OsGSTU3 and OsGSTU4, encoding tau class glutathione S-transferases, are heavy metal and hypoxic stress-induced and differentially salt stress-responsive in rice roots. FEBS Lett 553:427–432

Munemasa S, Muroyama D, Nagahashi H, Nakamura Y, Mori IC, Murata Y (2013) Regulation of reactive oxygen species-mediated abscisic acid signaling in guard cells and drought tolerance by glutathione. Front Plant Sci 4:472

Munns R, Tester M (2008) Mechanisms of salinity tolerance. Annu Rev Plant Biol 59:651–681

Nakamura S, Suzui N, Nagasaka T, Komatsu F, Ishioka NS, Ito-Tanabata S, Kawachi N, Rai H, Hattori H, Chino M, Fujimaki S (2013) Application of glutathione to roots selectively inhibits cadmium transport from roots to shoots in oilseed rape. J Exp Bot 64(4):1073–1081

Noctor G, Veljovic-Jovanovic S, Driscoll S, Novitskaya L, Foyer CH (2002) Drought and oxidative load in the leaves of C3plants: a predominant role for photorespiration? Ann Bot 89(7):841–850

Noctor G, Queval G, Mhamdi A, Chaouch S, Foyer CH (2011) Glutathione. The Arabidopsis Book 9:e0142. https://doi.org/10.1043/tab.0142. Elocation-id: e0142

Noctor G, Mhamdi A, Chaouch S, Han Y, Neukermans J, Márquez-García B, Queval G, Foyer CH (2012) Glutathione in plants: an integrated overview. Plant Cell Environ 35:454–484

Okuma E, Jahan MS, Munemasa S, Hossain MA, Muroyama D, Islam MM, Ogawa K, Watanabe-Sugimoto M, Nakamura Y, Shimoishi Y, Mori IC, Murata Y (2011) Negative regulation of abscisic acid-induced stomatal closure by glutathione in *Arabidopsis*. J Plant Physiol 168:2048–2055

Paradiso A, Berardino R, de Pinto MC, Sanità di Toppi L, Storelli MM, Tommasi F, De Gara L (2008) Increase in ascorbate-glutathione metabolism as local and precocious systemic responses induced by cadmium in durum wheat plants. Plant Cell Physiol 49(3):362–374

Passaia G, Margis-Pinheiro M (2015) Glutathione peroxidases as redox sensor proteins in plant cells. Plant Sci 234:22–26

Parisy V, Poinssot B, Owsianowski L, Buchala A, Glazebrook J, Mauch F (2006) Identification of PAD2 as a γ-glutamylcysteine synthetase highlights the importance of glutathione in disease resistance in Arabidopsis. Plant J 49:159–172

Pei ZM, Murata Y, Benning G, Thomine S, Klüsener B, Allen GJ, Grill E, Schroeder JI (2000) Calcium channels activated by hydrogen peroxide mediate abscisic acid signalling in guard cells. Nature 406:731–734

Pilarska M, Wiciarz M, Jaji'c I, Kozieradzka-Kiszkurno M, Dobrev P, Vanková R, Niewiadomska E (2016) A different pattern of production and scavenging of reactive oxygen species in halophytic Eutrema salsugineum (*Thellungiella salsuginea*) plants in comparison to *Arabidopsis thaliana* and its relation to salt stress signaling. Front Plant Sci 7:1179

Queval G, Jaillard D, Zechmann B, Noctor G (2011) Increased intracellular H_2O_2 availability preferentially drives glutathione accumulation in vacuoles and chloroplasts. Plant Cell Environ 34:21–32

Schmidt R, Mieulet D, Hubberten HM, Obata T, Hoefgen R, Fernie AR et al (2013) Salt-responsive ERF1 regulates reactive oxygen species-dependent signaling during the initial response to salt stress in rice. Plant Cell 25:2115–2131

Seth C, Remans T, Keunen E, Jozefczak M, Gielen H, Opdenakker K et al (2012) Phytoextraction of toxic metals: a central role for glutathione. Plant Cell Environ 35:334–346

Sharma R, Sahoo A, Devendran R, Jain M (2014) Over-expression of a rice tau class glutathione S-transferase gene improves tolerance to salinity and oxidative stresses in Arabidopsis. PLoS One 9:e92900

Singla-Pareek SL, Reddy M, Sopory SK (2003) Genetic engineering of the glyoxalase pathway in tobacco leads to enhanced salinity tolerance. Proc Natl Acad Sci U S A 100:14672–14677

Singla-Pareek SL, Yadav SK, Pareek A, Reddy M, Sopory SK (2008) Enhancing salt tolerance in a crop plant by overexpression of glyoxalase II. Transgenic Res 17:171–180

Sirichandra C, Gu D, Hu HC et al (2009) Phosphorylation of the Arabidopsis AtrbohF NADPH oxidase by OST1 protein kinase. FEBS Lett 583:2982–2986

Tanou G, Filippou P, Belghazi M, Job D, Diamantidis G, Fotopoulos V, Molassiotis A (2012) Oxidative and nitrosative-based signaling and associated post-translational modifications orchestrate the acclimation of citrus plants to salinity stress. Plant J 72(4):585–599

Ushimaru T, Nakagawa T, Fujioka Y, Daicho K, Naito M, Yamauchi Y, Nonaka H, Amako K, Yamawaki K, Murata N (2006) Transgenic *Arabidopsis* plants expressing the rice dehydroascorbate reductase gene are resistant to salt stress. J Plant Physiol 163:1179–1184

Vatamaniuk OK, Mari S, Lu YP, Rea PA (2000) Mechanism of heavy metal ion activation of phytochelatin (PC) synthase: blocked thiols are sufficient for PC synthase-catalyzed transpeptidation of glutathione and related thiol peptides. J Biol Chem 275:31451–31459

Wang F, Chen F, Cai Y, Zhang G, Wu F (2011) Modulation of exogenous glutathione in ultrastructure and photosynthetic performance against Cd stress in the two barley genotypes differing in Cd tolerance. Biol Trace Elem Res 144(1-3):1275–1288

Wang P, Du Y, Hou YJ, Zhao Y, Hsu CC, Yuan F, Zhu X, Tao WA, Song CP, Zhu JK (2015) Nitric oxide negatively regulates abscisic acid signaling in guard cells by S-nitrosylation of OST1. Proc Natl Acad Sci U S A 112(2):613–618

Wang L, Jin X, Li Q, Wang X, Li Z, Wu X (2016) Comparative proteomics reveals that phosphorylation of β carbonic anhydrase 1might be important for adaptation to drought stress in Brassica napus. Sci Rep 6:39024

Xu J, Xing X-J, Tian Y-S, Peng R-H, Xue Y, Zhao W et al (2015) Transgenic *Arabidopsis* plants expressing tomato glutathione S-transferase showed enhanced resistance to salt and drought stress. PLoS One 10(9):e0136960

Yadav SK, Singla-Pareek SL, Ray M, Reddy MK, Sopory SK (2005) Methylglyoxal levels in plants under salinity stress are dependent on glyoxalase I and glutathione. Biochem Biophys Res Commun 337:61–67

Yamaguchi C, Takimoto Y, Ohkama-Ohtsu N, Hokura A, Shinano T, Nakamura T, Suyama A, Maruyama-Nakashita A (2016) Effects of cadmium treatment on the uptake and translocation of sulfate in *Arabidopsis thaliana*. Plant Cell Physiol 57(11):2353–2366

Yang J, Gao MX, Hu H, Ding XM, Lin HW, Wang L, Xu JM, Mao CZ, Zhao FJ, Wu ZC (2016) OsCLT1, a CRT-like transporter 1, is required for glutathione homeostasis and arsenic tolerance in rice. New Phytol 211(2):658–670

Zagorchev L, Seal CE, Kranner I, Odjakova M (2013) A central role for thiols in plant tolerance to abiotic stress. Int J Mol Sci 14(4):7405–7432

Zahedi Avval F, Holmgren A (2009) A molecular mechanisms of thioredoxin and glutaredoxin as hydrogen donors for mammalian S phase ribonucleotide reductase. J Biol Chem 284:8233–8240

Zhang Y, Wang L, Liu Y, Zhan Q, Wei Q, Zhang W (2006) Nitric oxide enhances salt tolerance in maize seedlings through increasing activities of proton-pump and Na^+/H^+ antiport in the tonoplast. Planta 224:545–555

Zhao Y, Wang Y, Yang H, Wang W, Wu J, Hu X (2016) Quantitative proteomic analyses identify ABA-related proteins and signal pathways in maize leaves under drought conditions. Front Plant Sci 7:1827

Chapter 7
The Regulatory and Signaling Roles of Glutathione in Modulating Abiotic Stress Responses and Tolerance

Wang Chen, Leng Xiangpeng, Zhang Wenying, and Fang Jinggui

Abstract Glutathione (GSH) plays a key role in the maintenance of tissue antioxidant defence and in the regulation of redox-sensitive signal transduction. The size of the GSH pool and its redox status are strongly correlated with the tolerance of plants. The effect of GSH on plant stress resistance is achieved mainly through the cycle of ascorbate (AsA)-GSH producing dehydroascorbate (DHA) re-reduction, in which GSH is intermediate in the recycling of H_2O_2 reduction and the recycling of GSH is regulated by the glutathione reductase (GR) enzyme. Under oxidative stress, H_2O_2 and reactive oxygen species (ROS) can be reduced by GSH, while GSH is oxidized to oxidized glutathione (GSSG). Under normal physiological conditions, GSSG can be reduced to GSH by GR in the presence and with the involvement of NADPH, thus creating the redox cycle. AsA/DHA, GSH/GSSG, NADPH/NADP are the three most important interconnected and coordinated redox pairs in plant cells, and plants can scavenge ROS and strengthen their resistance to abiotic stresses by modulating the status of redox pairs; thus the redox metabolism pathways are initiated by diverse environmental stresses. In addition, GSH can regulate the ROS signal transduction pathway, and ROS levels depend on the redox status of redox pairs in plant cells. GSH can activate the genes involved in phytoalexin biosynthesis to repress the accumulation of plant toxin. H_2O_2 can strengthen this effect through the up-regulation of GSSG. When H_2O_2 reaches a higher level, the biosynthesis of GSH is stimulated, while GSH, on the other hand, can mediate the response of the plant to the H_2O_2 signal.

Keywords Stress response • Glutathione • Regulatory role • ROS • Signal

---

W. Chen • L. Xiangpeng • Z. Wenying • F. Jinggui (✉)
College of Horticulture, Nanjng Agricultural University, Jiangsu Nanjing City 210095, China
e-mail: Fanggg@njau.edu.cn

M.A. Hossain et al. (eds.), *Glutathione in Plant Growth, Development, and Stress Tolerance*, https://doi.org/10.1007/978-3-319-66682-2_7

Contents

1 Introduction

Plants respond to environmental stresses by regulating metabolic pathways that act to counteract resulting cellular damage. Regulation of the intracellular redox environment is critical in cellular physiology for influencing signaling pathways and cell fate in response to stress. As part of their response to these environmental stresses, plants can produce glutathione (GSH), which acts as an antioxidant by inhibiting reactive oxygen species (ROS) and is involved in the ascorbate-glutathione (AsA-GSH) cycle that eliminates peroxides (Noctor and Foyer 1998; Rouhier et al. 2008). GSH is made up of glutamate (Glu), cysteine (Cys), and glycine (Gly), and is found in the vast majority of prokaryotic and eukaryotic cells. GSH and GSH-associated metabolism play critical roles in protecting cells from oxidative and other forms of stress (Noctor et al. 1998; Tausz et al. 2004, Akram et al. 2017), which are summarized in a schematic overview in Fig. 7.1 (Zagorchev et al. 2013). From Fig. 7.1, it can be seen that the roles of GSH mainly involve in plant disease resistance, cell proliferation, root development, salt tolerance, protection against chilling damage, and the metabolic detoxification of a range of xenobiotics, herbicides, air pollutants (sulfur dioxide and ozone), and heavy metals (Galant et al. 2011). GSH realizes its physiological functions through its metabolism and signal transduction pathways, and thus it is essential to understand these pathways to gain insight into the regulatory and signal roles of GSH in modulation of abiotic stress responses and tolerance.

The diverse abiotic stresses of plants usually exhibit non-specific stress responses of cross-tolerance to various stresses of resistant plant species (Verslues et al. 2006). It has been reported that temperature stress (Ohama et al. 2017), heavy metals (Keunen et al. 2011), salt stress, and a water deficit (Miller et al. 2010) can all lead to increased production of ROS, with downstream alterations of oxidative signaling. GSH is a key water-soluble antioxidant and plays a central part in ROS scavenging through the AsA-GSH cycle and as an electron donor to GSH peroxidase (GPX) (Zagorchev et al. 2013). Furthermore, GSH is a key ROS scavenger and major cellular redox buffer, and thus it is a pivotal part of the stress signaling pathways and has important roles in the regulation of plant stress responses (Zagorchev et al. 2013). In this chapter we discuss the roles of GSH in the maintenance of tissue antioxidant defense and in the regulation of the redox-sensitive signal transduction, and the correlation between the tolerance of plants and the size of the GSH pool and

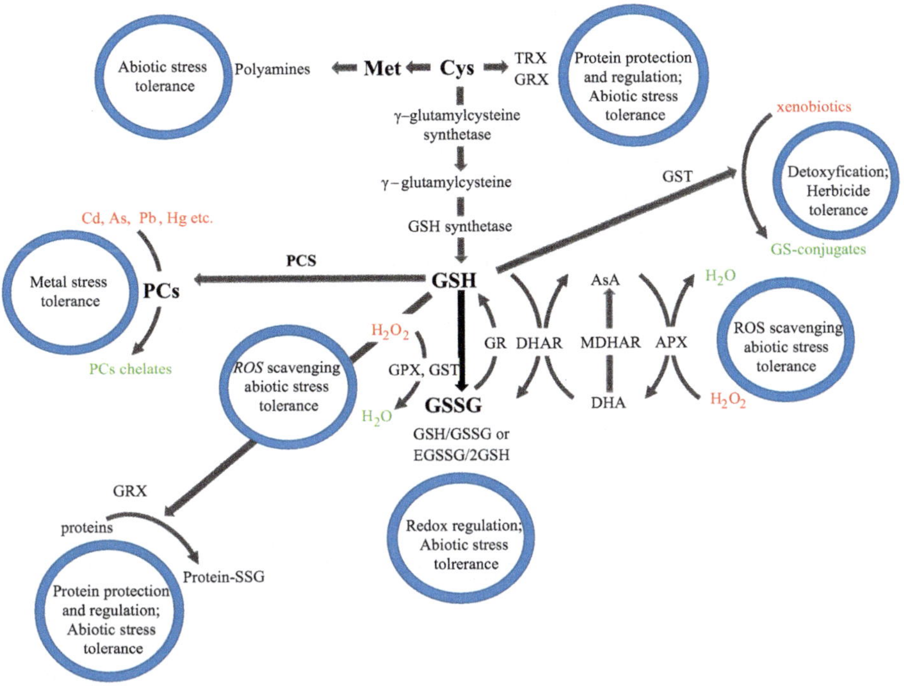

Fig. 7.1 Overview of the roles of thiols in plant tolerance to abiotic stress (Zagorchev et al. 2013). Potential roles of and the significance for abiotic stress tolerance are depicted by blue circles next to SH-adducts. Potential deleterious compounds are shown in red and their adducts in green font. *Met* methionine, *Cys* cysteine, *TRX* thioredoxin, *GRX* glutaredoxin, *PC* phytochelatins, *PCS* phytochelatin synthetase, *GSH* glutathione, *GSSG* oxidized glutathione, *GST* glutathione-S-transferase, *GPX* glutathione peroxidase, *GR* glutathione reductase, *DHAR* dehydroascorbate reductase, *AsA* ascorbate, *MDHAR* monodehydroascorbate reductase, *DHA* dehydroascorbate, *APX* ascorbate peroxidase

its redox status, and describe plant resistance to abiotic stress through the ROS signal transduction pathway mediated by GSH.

2 Manipulating Glutathione (GSH) Levels in GSH Metabolism

2.1 Plant Glutamate-Cysteine Ligase

GSH is synthetized by two (Fig. 7.2) enzymatic reaction steps. First, γ-glutamylcysteine (γ-EC) is formed in an ATP-dependent reaction, catalyzed by glutamate-cysteine ligase (GCL), also known as γ-glutamylcysteine synthetase (γ-ECS; EC 6.3.2.2; GSH1), which is the rate-limiting reaction. Glutathione synthetase (GSH-S or GS; EC 6.3.2.3; GSH2) then catalyzes the addition of Gly to γ-EC

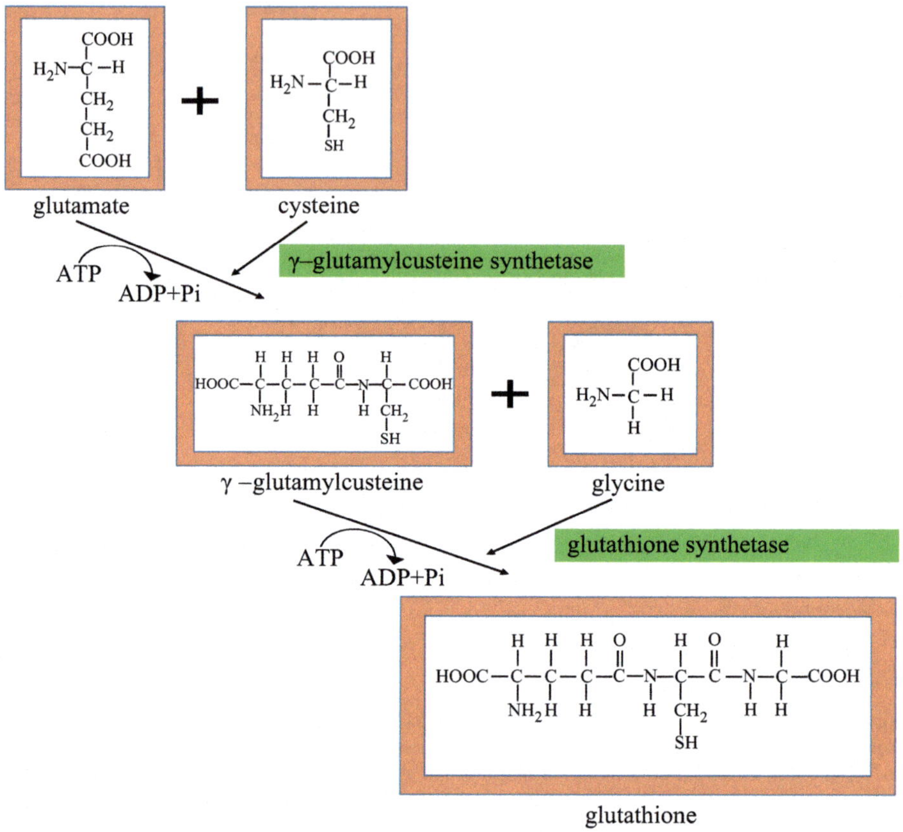

Fig. 7.2 Schematic representations depicting the pathway of glutathione biosynthesis from constituent amino acids. Two sequential ATP-dependent reactions allow the synthesis of γ-glutamylcysteine (γ-EC) from L-glutamate and L-cysteine, followed by the formation of GSH by addition of glycine (Gly) to the C-terminal end of γ-EC (Modified from Noctor et al. 1998)

(Noctor et al. 2012). In the first step, GSH1 catalyzes the linking of Cys with Glu to form γ-EC, whereas in the second step, Gly is linked to γ-EC by GSH2 to form the final product GSH. GSH synthesis can occur in the cytosol, chloroplasts, and mitochondria (Zechmann and Müller 2010), and both enzymes are encoded by single genes with alternate transcription start sites that are associated with their subcellular location (Wachter et al. 2005). Recently published results confirmed that the concentrations of GSH increased at least transiently in plants exposed to copper (Cu) (Leng et al. 2015), cadmium (Cd) (Gill et al. 2012), lead (Pb) (Estrella-Gómez et al. 2012), salt (Zagorchev et al. 2012), nutrients (Goiris et al. 2015), drought (Pyngrope et al. 2013), cold (Jiang et al. 2013), waterlogging (Alhdad et al. 2013), and heat stress (Kumar et al. 2013).

Of the two enzymes, GCL appears to be rate-limiting. Exposure to heavy metals increases the levels of GCL mRNA in *Brassica juncea* and activates transcription of

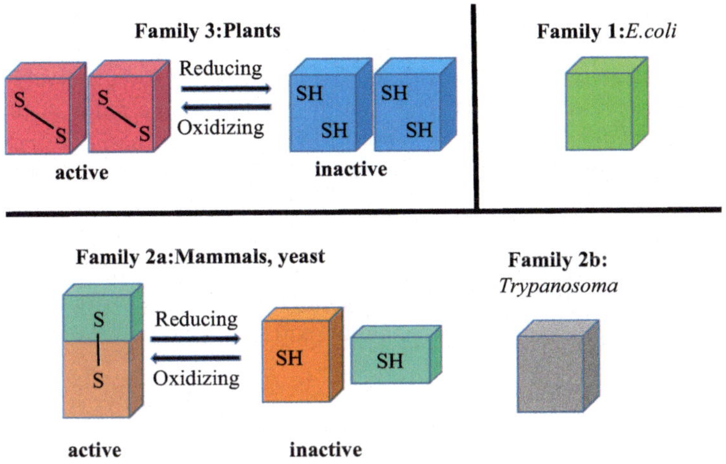

Fig. 7.3 Overview of glutamate-cysteine ligase. Oligomeric organization and redox regulation of the three types of GCL are shown. The three distinct families are non-plant eukaryotes (mammals, yeast, and trypanosomes), γ-proteobacteria (*Escherichia coli: E. coli*), and plants (*Arabidopsis*) and α-proteobacteria (Rhizobium)

both GCL and GS in Arabidopsis (*Arabidopsis thaliana*) (Schafer et al. 1997; Xiang and Oliver 1998). Overexpression of *Escherichia coli* (*E. coli*) GCL in plants improves tolerance to Cd and arsenic (As), demonstrating the importance of this enzyme in heavy metal protection (Dhankher et al. 2002). Bioinformatic analysis of the GCL genes from multiple species suggests that these sequences group into three families (Fig. 7.3), non-plant eukaryotes (mammals, yeast, and trypanosomes), γ-proteobacteria (*Escherichia coli*), and plants (*Arabidopsis*) and α-proteobacteria (*Rhizobium*) (Copley and Dhillon 2002; Galant et al. 2011). Sequence comparisons within each family show similarities, but pairwise comparisons between groups display no statistically significant relationships (Copley and Dhillon 2002).

Of the three GCL families, the enzymes from the non-plant eukaryotes have been the most studied. For instance, the mammalian and Drosophila GCL consist of a 70-kDa catalytic or heavy subunit and a 30-kDa regulatory or light subunit (Fraser et al. 2002; Maher 2005; Hothorn et al. 2006). The catalytic subunit catalyzes the formation of γ-EC and is inhibited by GSH, whereas the regulatory subunit increases the affinity of the enzyme for Glu and decreases the inhibitory effect of GSH. The GCL from *Trypanosoma brucei* and the mammalian catalytic subunit share a 45 % amino acid identity, but the *T. brucei* GCL functions as a monomer (Lueder and Phillips 1996). Similarly, the *E. coli* GCL also functions as a monomeric protein (Hibi et al. 2004).

Biochemical studies of the GCL from plants revealed that these proteins are regulated by the redox environment through a mechanism that differs from the heterodimeric enzymes (Jez et al. 2004; Hothorn et al. 2006; Gromes et al. 2008). Kinetic analysis of *Arabidopsis* GCL showed that the enzyme was inactivated by buthionine sulfoximine, a potent inhibitor of mammalian GCL, and used a random ter-reactant kinetic mechanism with a preferred order of binding for catalysis (Jez et al. 2004). Mechanistic studies of the plant GCL demonstrated that a reversible disulfide bond formation alters catalytic activity with the oxidized protein more actively than the reduced protein (Jez et al. 2004). The control of GCL by the redox state offers a simple post-translational control mechanism of GSH biosynthesis in plants. GSH maintains an intracellular redox balance, and the redox regulation of GCL also provides a control switch for GSH production. Under oxidizing conditions, the demand for GSH increases and GCL is activated. As the concentration of GSH increases, the cellular environment shifts to a more reduced potential and GCL activity decreases. Direct control of the rate-limiting enzyme in GSH production provides a post-translational switch for responding to intracellular oxidative signals.

2.2 *Plant GSH Synthetase*

GS catalyzes the ATP-dependent addition of Gly to γ-EC in the second step of GSH synthesis. Structural and functional characterization of GS from bacteria demonstrates that this enzyme functions as a tetramer (Yamaguchi et al. 1993), whereas the mammalian, yeast, and plant GS act as dimers (Polekhina et al. 1999; Gogos and Shapiro 2002; Jez and Cahoon 2004). GS is localized both to chloroplasts and cytosol in plant cells (Hell and Bergmann 1988), but preferentially GS is localized in cytosol from the abundance of mRNA splice variants (Wachter et al. 2005). Additionally, Hell and Bergmann (1988) reported the allocation of 24 % of the gsh2 activity to photoheterotrophic cell cultures of tobacco (*Nicotiana tabacum*) plastids, whereas pea (*Pisum sativum*) plastids were reported to exhibit 47–69 % gsh2 activity (Klapheck et al. 1987). It has also been reported that GS in *Arabidopsis* is encoded by a single copy nuclear gene with alternate transcription start sites leading to either plastid-targeted or cytosolic protein (Wachter et al. 2005). The GS from *Arabidopsis*, wheat (*Triticum aestivum*), maize (*Zea mays*), and various legumes share about a 40 % amino acid sequence identity with the human and yeast homologs (Galant et al. 2011).

Unlike other organisms, some plants synthesize GSH analogs in which β-alanine, serine, or Glu replaces Gly in the tripeptide. Multiple legumes use β-alanine instead of Gly to synthesize homoglutathione (hGSH) (Klapheck et al. 1995; Matamoros et al. 1999; Moran et al. 2000). Hydroxymethyl-glutathione (R = serine) has been isolated from rice (*Oryza sativa*) and Agrostis (*Agrostis stolonifera*, a grass species; Klapheck et al. 1994). In maize, the synthesis of GSH-like peptides with Glu replacing Gly is exclusively induced by exposure to Cd (Meuwly et al. 1995).

Of these alternate GSH analogs, hGSH is the most understood. To date, hGSH is found in 14 different legumes (pea; sweet pea; alfalfa, *Medicago sativa*; soybean, *Glycine max* (L.) Merr; bean; mungbean, *Vigna radiata* (L.) Wilczek; lentil, *Lablab purpureus* (L.) Sweet; chickpea, *Cicer arietinum;* cowpea, *Vigna unguiculata* (L.) Walp; red clover, *Trifolium pratense;* sweet clover; Italian clover; blue fenugreek, *Trigonella foenum-graecum;* and runner bean, *Phaseolus coccineus*); however, two additional legumes (broad bean, *Vicia faba* and lupine, *Lupinus micranthus*) appear to lack this GSH analog in all assayed tissue types (Matamoros et al. 1999). Additionally, cowpea and pea have hGSH in roots and nodules, but not in leaves (Moran et al. 2000). In soybean, leaves and seeds contain 50- to 200-fold and 135-fold more hGSH than GSH, respectively (Matamoros et al. 1999). Soybean nodules contain fourfold more hGSH than GSH, while the roots contain nearly 80-fold more hGSH (Matamoros et al. 1999). For biosynthesis, GSH and hGSH share a common first reaction, i.e., the synthesis of γ-EC, but the chemical diversity results from altered specificity for the nucleophilic amino acid in the second reaction of the pathway. The soybean hGSH synthetase (hGS) is related by about 70 % sequence identity to the soybean GS. Because the soybean genome appears to have undergone two rounds of genome duplication, hGS likely arose from GS by divergent evolution after the first duplication event (Frendo et al. 2001). In *Arabidopsis*, there is a single gene encoding for GS, whereas the soybean genome contains two GS and two hGS genes, with each pair sharing 87 % and 93 % identity, respectively.

2.3 Plant GSH Reductase

GSH reductase (GR) belongs to a group of flavoenzymes and contains an essential disulfide group, where one mole of NADPH is required to reduce GSSG to GSH for every mole of GSSG (Fig. 7.4) (Gill et al. 2013). Regarding GR catalytic

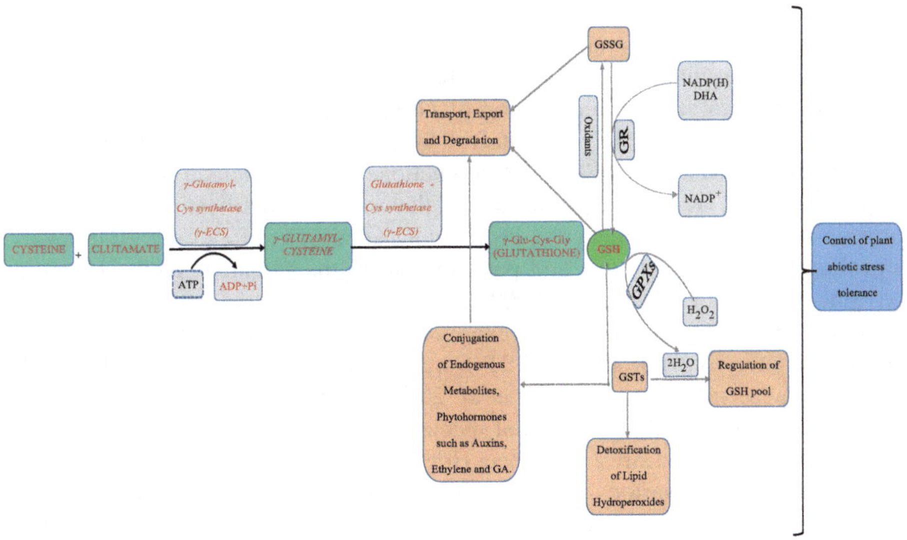

Fig. 7.4 Simplified scheme of biosynthesis and multifunctional role-based integration of the reduced and oxidized glutathione (viz., GSH and GSSG, respectively) with GSH-associated major enzymatic-antioxidant defense system components including glutathione reductase (GR), glutathione peroxidases (GPXs), and glutathione S-sulfotransferase (GSTs), and their cumulative significance for the control of plant abiotic stress responses (Modified from Gill et al. 2013)

mechanisms, it is to be emphasized here that the availability of substrate availability largely controls GR redox inter-conversions when compared to the reduced GR; more stability is exhibited by the oxidized GR form, which can tolerate divalent metal ions including Zn^{2+}, Cu^{2+}, and Fe^{2+}. The catalytic mechanism involves two steps: first, the flavin moiety is reduced by NADPH, the flavin is oxidized, and a redox-active disulfide bridge is reduced to produce a thiolate anion and a Cys. The reduction of GSSG via thioledisulfide interchange reactions is mainly involved in the second step (Ghisla and Massey 1989).

As a component of the AsA-GSH pathway, GR plays an important role in ROS detoxification and GSH regeneration, and confers abiotic stress tolerance in plants (Hasanuzzaman et al. 2010). Increased GR activity confers stress tolerance and has the ability to alter the redox state of important components of the electron transport chain. The major involvement of GR in conferring stress tolerance is the recycling of GSH and the maintenance of the GSH/GSSG ratio in the plant cell (Gill et al. 2013). To this end, an exhaustive literature search has revealed differential modulation of GR activity in different plant species exposed to metal-metalloids, salinity, drought, and various abiotic stresses.

2.4 Plant GSH-S-Transferases

GSTs (EC 2.5.1.18) comprise an extensive family of proteins with a great variety of functions. Up to 90 genes encoding GSTs are transcribed in different plant species, most of which are differentially induced by stress, and they are involved in important roles in enzymatic thiol-dependent ROS scavenging mechanisms (Chi et al. 2011). GSTs catalyze the conjugation of GSH to an electrophilic substrate (Chi et al. 2011); for example, they can catalyze the conversion of H_2O_2 at the expense of GSH, thereby producing GSSG. Known functions in plants include conjugation and sequestration of xenobiotics, transport of flavonoids, detoxification of ROS and organic radicals, programmed cell death under conditions of elevated ROS levels, signaling through flavonoids, and participation in the fumarate synthesis during tyrosine catabolism (Dixon et al. 2002).

Overexpression or heterologous expression of GSTs can contribute to abiotic stress tolerance. A chloroplastic GST from *Prosopis juliflora* improved drought stress tolerance in tobacco (George et al. 2010). GSTs are also good examples of genes from economically unimportant species that could be used for transformation of crops with enhanced stress tolerance. However, plant stress tolerance relies on an intricate network of various pathways, and changing only one part may produce unexpected results. For example, an *Arabidopsis* GST knockout mutant accumulated far more GSH than the wild type and showed much higher tolerance to salt and drought stress (Chen et al. 2012). In summary, GSTs comprise a large family of GSH-dependent enzymes that are involved in numerous stress-responsive mechanisms, mediating GSH functions in plant cells.

3 Redox State in the Diverse GSH Oxidation Reduction Cycles

Through the AsA-GSH cycle, also known as the Foyer-Halliwell-Asada cycle, superoxide can be converted to H_2O_2 by superoxide dismutase (SOD, EC 1.15.1.1). H_2O_2 is then further detoxified by ascorbate peroxidase (APX, EC 1.11.1.11), whereby monodehydroascorbate (MDHA) is produced, which can spontaneously dismutate to AsA and DHA (Foyer and Noctor 2011). MDHA can also be reduced to AsA by NAD(P)H-dependent monodehydroascorbate reductase (MDHAR, EC 1.6.5.4) and DHA reduced by GSH-dependent dehydroascorbate reductase (DHAR, EC 1.8.5.1). Abiotic stresses can simultaneously increase both the mRNA transcripts and activity of both enzymes (Eltelib et al. 2011), and over-expression of DHAR rather than MDHAR has been associated with stress tolerance (Yin et al. 2010). However, a salt-inducible MDHAR from the halophyte, *Avicennia marina*, may suffice to induce salt tolerance in transgenic plants (Kavitha et al. 2010). Evidence for the cooperation of various enzymes in the AsA-GSH cycle was demonstrated for transgenic tobacco plants, in which only those transformants that had DHAR and GR up-regulated showed increased tolerance to salt and cold stress (Le Martret et al. 2011). Therefore, the DHAR pathway may be less effective during sulfur starvation or in GSH-deficient mutants, but a shift from the GSH-dependent DHAR to GSH-independent MDHAR ascorbate regeneration under sulfur deficiency was not supported by experimental data. On the contrary, MDHAR activity was shown to decrease in sulfur-starved plants.

Enhanced GSH-dependent DHA reduction also improved tolerance to oxidative stress in DHAR transformants of *Arabidopsis* (Wang et al. 2010). Interestingly, these transformants showed constitutively high DHAR activity, but also increased concentrations of both AsA and GSH compared to the wild type. *E. coli*, expressing a tomato (*Lycopersicon esculentum* Mill) DHAR gene, also showed enhanced tolerance to H_2O_2, although mRNA levels in tomato plants slightly increased during cold treatment and decreased during treatment with abscisic acid (ABA) (Kabir et al. 2011), and only mechanical wounding significantly increased the DHAR mRNA levels. An increase in the activity of two differentially expressed tomato DHARs was reported in response to salt and drought stress (Kabir and Wang 2010). A decrease in GR activity can affect the GSH/GSSG ratio, but also decrease the AsA pool and impact on the AsA redox state (Ding et al. 2009), with an overall decrease in stress tolerance (Yousuf et al. 2012). Increased GR activity was reported in response to various stresses, such as salt and Cd, both in leaves and in roots and in all subcellular fractions. As mentioned earlier (Le Martret et al. 2011), an increase in GR activity alone is not sufficient to confer stress tolerance. More likely, a coordinated and finely regulated action of all enzymes of the AsA-GSH cycle in conjunction with that of other ROS-processing enzymes in all cell compartments is required for plant stress tolerance.

In an earlier study on grapevines (*Vitis vinifera*), the AsA-GSH cycle, the GPX pathway, and the peroxiredoxin (PRX)/ thioredoxin (TRX) pathway were identified,

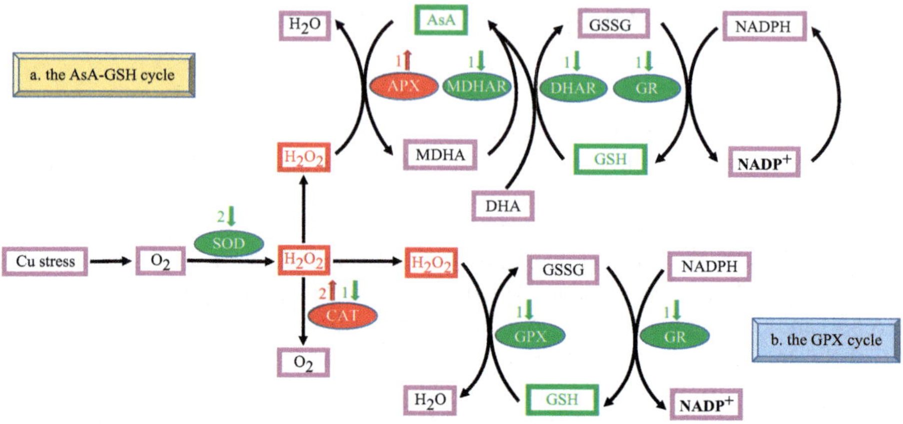

Fig. 7.5 Pathways for reactive oxygen species (ROS) scavenging under Cu stress in plants. Red regions indicate up-regulation and green regions denote down-regulation. (**a**) The ascorbate-glutathione (AsA-GSH) cycle. (**b**) The glutathione peroxidase (GPX) cycle. Superoxide dismutase (SOD) acts as the first line of defense converting $O_2^{\bullet-}$ into H_2O_2. Catalase (CAT), ascorbate peroxidases (APX) and glutathione peroxidase (GPX) then detoxify H_2O_2. AsA and GSH are antioxidants in green. *DHA* dehydroascorbate, *GSH* glutathione, *GSSG* oxidized glutathione, *GR* glutathione reductase, *MDHAR* monodehydroascorbate reductase, *DHAR* dehydroascorbate reductase

all of which also played essential roles in the defense system against ROS and scavenge H_2O_2 by Cu stress (Leng et al. 2015). As shown in Fig. 7.5, in total four genes, including one up-regulated gene and three down-regulated genes, are related to AsA-GSH cycle; in addition, seven GPX cycle genes (SOD, two down-regulated genes; CAT, two up-regulated genes, and one down-regulated gene; GPX, one down-regulated gene; GR, one down-regulated gene) were identified under Cu stress. In addition, from the transcriptome data, many genes encoding these enzymes involved in ROS generation and the scavenging system were significantly changed by waterlogging and drought stresses, and the number of changed genes is more than those in salt stress. The components of the antioxidant defense system included enzymatic and non-enzymatic antioxidants. Enzymatic antioxidants include APX, MDHAR, DHAR, and GR, and non-enzymatic antioxidants are GSH, AsA (both water soluble), carotenoids, and tocopherols (lipid soluble). In waterlogging stress, 119 transcripts with differentially-expressed profiles were identified as genes encoding enzymes in the ROS scavenging system. They were categorized into AsA-GSH cycle (48), GPX (one), GST pathway (63), and the PRX pathway (eight) (Fig. 7.6).

Previous studies have confirmed that the AsA-GSH cycle and the GPX pathway also play an essential role in the defense system against ROS and scavenge H_2O_2 through waterlogging stress. In total, five AsA-GSH cycle genes (including APX, GR, GRX, MDHAR, and DHAR) were identified (Fig. 7.6), including one up-regulated gene and three down-regulated genes. In the GPX pathway, 63 GSTs (13 up-regulated, 11 down-regulated) were identified. Furthermore, there were one up-regulated and three down-regulated PRX in the PRX pathway. In addition, studies also indicated that all three alternative oxidases (AOs) were significantly up-regulated

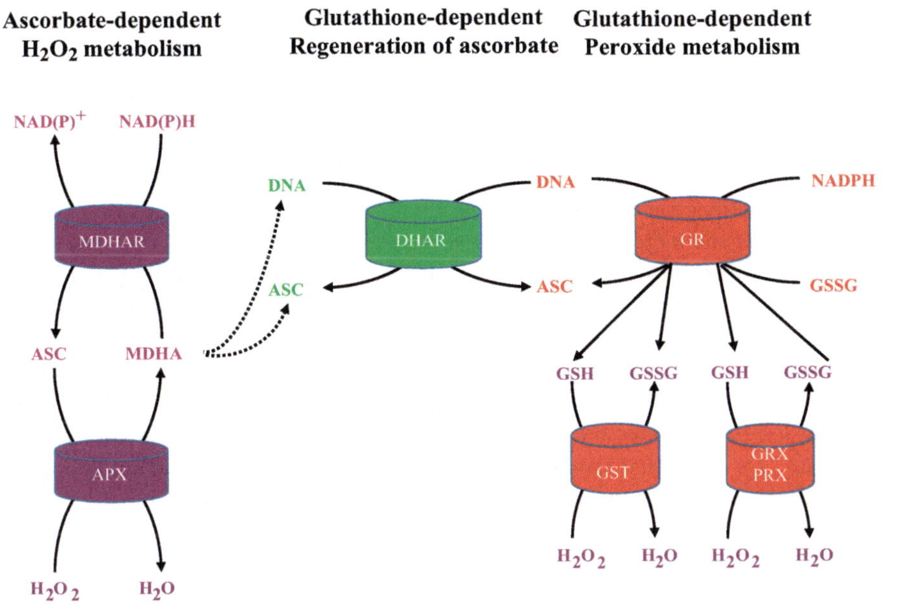

Trait	Description	No. of up-regulate	No. of down-regulate	No. of no change	No. of total
AsA-GSH cycle	APX	0	2	3	5
	MDHAR	0	0	7	7
	DHAR	0	2	0	2
	GR	1	0	1	2
	GRX	2	8	22	32
GPX pathway	GST	13	11	39	63
PRX pathway	PRX	1	3	4	8

Fig. 7.6 Glutathione-dependent reactive oxygen species (ROS) generation/scavenging system in grapevine suffering from waterlogging stress. Below the figure are each numbers of genes encoding enzymes in the ROS scavenging system. Red indicates up-regulation and green denotes down-regulation. *APX* ascorbate peroxidases, *ASC* ascorbate, *DHAR* dehydroascorbate reductase, *MDHA(R)* monodehydroascorbate (reductase), *GR* glutathione reductase, *GSH* glutathione, *GSSG* oxidized glutathione, *GST* glutathione S-transferase, *GRX* glutaredoxins, *PRX* peroxiredoxin

in cyanide-resistant respiration. Two polyphenol oxidases (PPOs) were ubiquitous copper-containing enzymes and also were significantly up-regulated compared with a control (Leng et al. 2015).

4 Regulatory Roles and Signals of GSH in Plant Stress Response

4.1 Redox Status of GSH Involved in Plant Stress Response

The GSH/GSSG redox couple is involved in several physiologic processes in plants under both optimal and stress conditions. Several studies of plants that overexpressed γ-ECS or transgenic plants expressing bacterial γ-ECS evaluated its effect on metal tolerance based on the assumption that higher levels of GSH and phytochelatins (PC) will lead to more efficient metal sequestration. Interestingly, several authors reported that increased expression of this enzyme may not be related to stress tolerance, especially Cd tolerance. Overexpression of γ-ECS in *Arabidopsis* (Xiang et al. 2001) did not enhance resistance to Cd stress, despite increased levels of GSH and PC. Similarly, the expression of bacterial γ-ECS in *Arabidopsis* did not enhance Cd tolerance and even caused Cd sensitivity (Li et al. 2005), although some resistance to As and mercury (Hg) was observed. Transgenic cotton wood (*Anemone vitifolia Buch*) also showed enhanced As tolerance (LeBlanc et al. 2011). The overall conclusion was that Cd tolerance may rely on other factors than tolerance to other metals.

Although increased levels of GSH may be required for metal tolerance, it seems that tolerant genotypes tend to keep their γ-ECS levels lower than sensitive ones. Similarly, long-term water deficit in *Vigna radiata* caused a decrease in both γ-ECS activity and its transcript levels in roots and even lower activity, combined with higher mRNA levels during the recovery period. This is not in agreement with the assumption that abiotic stress tolerance involves an increase in γ-ECS abundance and activity along with increases in Cys and GSH concentrations, as shown for salt stress (Nazar et al. 2011). Overall, in most, if not all, of the cited studies, GSH expression and activity increased simultaneously with that of γ-ECS. Loss of function of any of the two enzymes proved to be lethal to early developmental stages, and GSH deficiency resulted in increased sensitivity to Cd in *Arabidopsis* (Lim et al. 2011).

HGSH is a homologue of GSH, characteristic for legumes in which the C-terminal Gly is substituted by β-Ala and, overall, has the same functions as GSH. It is regarded as an important regulator of root nodule formation, symbiotic interactions, and nitrogen fixation (El Msehli et al. 2011). It serves functions in the transport of reduced sulfur and has antioxidant activity, as does GSH. Legumes are also capable of synthesizing hGSHs in response to heavy metal stress (Sobrino-Plata et al. 2009). A study of hGSH synthetase transcript levels and activity and hGSH concentration in root nodules exposed to different stresses (Loscos et al. 2008) did not reveal any significant increase in hGSH synthetase, except for mRNA levels that rose upon Cd and H_2O_2 treatments. In contrast, Cruz de Carvalho et al. (2010) found increased mRNA levels in the leaves of a drought-tolerant cowpea cultivar, encoding hGSH synthetase during drought stress and desiccation, revealing different stress response patterns than in roots. The distribution of GSH and hGSH in different plant organs is species-specific and determined by the differential expression of the corresponding synthetase genes (Becana et al. 2010). Other authors reported that a glyphosate-resistant soybean line showed an increase in hGSH concentrations upon glyphosate treatment, whereas in a glyphosate-sensitive line, it remained unchanged or decreased slightly; more importantly, the hGSH/hGSSG ratio remained higher in the resistant line (Diaz Vivancos et al. 2011). The latter is an example of a complete overlap of hGSH and GSH functions, which is not surprising, as hGSH replaces GSH in soybean. A comparative study between soybean and white lupin (*Lupinus albus*) subjected to Cd and As

(Vázquez et al. 2009) showed that hGSH levels (in soybean) increased significantly more than GSH levels (in white lupin). Clemente et al. (2012) also suggested that GSH and hGSH play distinct roles in plant development and stress response, based on the differential hormone-mediated expression and activity of GSH and hGSH synthetases. Hence, GSH and hGSH have largely the same functions, although in some cases the two homologues may be differentially affected by stress in plant species that are able to synthesize both hGSH and GSH, but the importance of such effects is still to be elucidated.

4.2 GSH Redox Signals in Plant Abiotic Stress

GSH could be involved in regulating the induction of plant adaptive or death processes by signals from metabolic pathways and the environment. The interaction between ROS and antioxidants may provide the action point between signals from metabolic pathways and the environment (Foyer and Noctor 2005). The GSH/GSSG redox couple may have evolved for the adjustment of the cellular redox state and redox signaling (Noctor and Foyer 1998). Several regulatory and structural genes controlled by the redox status and ROS signaling have been identified in mutant and transgenic *Arabidopsis* and in wild-type plants treated with dithiothreitol or ROS-generating agents using transcript profiling, which could clarify the function of the redox network (Gadjev et al. 2006; Kolbe et al. 2006). This network controls the level of ROS by integrating signals from different cell compartments during abiotic stress, and the GSH/GSSG couple participates in its fine tuning (Meyer 2008).

The redox state of the GSH/GSSG couple is changed under abiotic stress conditions, because GSH is oxidized by the removal of the accumulating H_2O_2 under abiotic stress conditions. Stress-induced changes in the H_2O_2 content, and subsequently in the GSH/GSSG ratio, have a central role in signaling (Foyer et al. 1997; Neill et al. 2002; Dietz 2008; Quan et al. 2008). It was reported that the interaction between H_2O_2 and GSH in stress signaling was found in mungbeans, in which exogenous H_2O_2 increased both GSH levels and chilling tolerance (Yu et al. 2003). The low GSH concentration in the pad2–1 mutant (mutation in the γ-ECS gene) did not affect the transcript abundance of γ-ECS and GSHS (Parisy et al. 2007). The involvement of GSH in redox signaling has been confirmed by the observation that inter- and intracellular GSH pools are linked by transport across the membranes, the rate of which could be similar to that of synthesis, as is the case for the chloroplast envelope (Noctor et al. 2002). The GSH/GSSG couple is able to modify the activity of various compounds (enzymes, regulatory proteins) directly through their reduction/oxidation state. The indirect regulation of proteins by the GSH/GSSG couple may occur due to cross-talk between GSH/GSSG and other redox systems through glutathionylation or thiol-disulfide transition, which may have a role in signaling and responses to abiotic stress (Rausch et al. 2007; Ying et al. 2007).

4.3 GSH in Modulation of Transmitting ROS Signal

GSH could be considered as a candidate transmitter of intracellular ROS signals, and possesses three relevant properties: (1) GSH is highly reduced under optimal conditions; (2) shifts toward a more GSSG status are well described in response to increased intracellular ROS availability; and (3) mechanisms exist that are theoretically able to link such shifts to an altered redox state, and therefore biological activity, of target proteins. There is a good correlation between the expected intracellular H_2O_2 availability and the status of the GSH pool. Enhanced ROS availability has less impact on the AsA-DHA ratio than on the redox status of the GSH pool (Queval et al. 2009; Mhamdi et al. 2010a, b). Plant cells might be able to maintain very high cytoplasmic AsA-DHA ratios simultaneously with low GSH/GSSG ratios, presumably because of efficient GSH-independent pathways of AsA regeneration and/or the difference in redox potential between the GSH/GSSG and AsA/DHA couples.

To date, direct evidence for a role for GSH in transmitting H_2O_2 signals is scarce. GSH-GSSG ratios were decreased by increased H_2O_2 availability or by lowered GSH recycling capacity. Mutants in which one of the two *Arabidopsis* GR genes (encoding a mainly cytosolic isoform) is knocked out (*gr1*) are aphenotypic and show no evidence of generalized oxidative stress, despite an increase in leaf GSSG, indicating that plants can tolerate mild constitutive perturbation of the GSH redox state without adverse consequences on growth and development (Marty et al. 2009; Mhamdi et al. 2010a). However, when compared with identical conditions, *gr1* shows gene expression patterns that partly recapitulate those driven by H_2O_2 in *cat2* (Mhamdi et al. 2010a). Moreover, introduction of the *gr1* mutation into the *cat2* background causes marked modulation of H_2O_2-associated transcript profiles. This observation points to a significant role for GSH status in transmitting a subset of the signals derived from intracellular H_2O_2.

Maintenance of low GSSG concentrations under optimal conditions could confer high sensitivity in signal transduction. It would allow relatively small ROS-triggered departures from this highly reduced state to be perceived as significant changes in redox potential by sensitive proteins. A second important factor is GSH concentration, which in itself affects GSH redox potential. Even if the GSH/GSSG ratio does not change, decreased concentration causes the redox potential to increase (i.e., become more positive). The decreased reduction state of GRXs caused by a more positive GSH redox potential could explain observations in plants that both lack NADPH-dependent TRX reductase and are partly deficient in GSH (Reichheld et al. 2007; Bashandy et al. 2010). However, quantification of changes in GSH redox potential caused by stress or mutations has produced relatively modest values (about 20 mV), and it is still unclear whether such adjustments are a major part of the mechanism of ROS-dependent signaling through the GSH pool.

4.4 GSH in Cross-Talking with Other Signaling Pathways

The GSH/GSSG couple and H_2O_2 may interact with other signaling pathways during the stress response (Fig. 7.7). Nitric oxide (NO), an important regulatory molecule, affects H_2O_2 concentration due to the inhibition of Cat and ascorbate peroxidase (APX) (Clark et al. 2000), whereas exogenous H_2O_2 activated nitric oxide (NO) synthesis in tobacco (De Pinto et al. 2006), suggesting a bidirectional interaction between the two compounds. NO also influenced GSH synthesis, as demonstrated in *Medicago trunculata* roots in which the GSH level and γ-ECS and GSHS gene expressions were increased by NO (Innocenti et al. 2007). During the interaction of GSH with NO, S-nitrosoglutathione (GSNO) is formed in a reaction that may interconnect the ROS- and reactive nitrogen-based signaling pathways (Neill et al. 2002).

Another possibility for the activation of protective mechanisms through H_2O_2 and the GSH/GSSG couple is based on their interaction with Ca^{2+} (Fig. 7.7). H_2O_2 treatment alone or combined with a low temperature increased the Ca^{2+} concentration in tobacco (Price et al. 1994), which could have a role in the Ca^{2+}-dependent regulation of the enzymes. In maize, the interaction of Ca^{2+} and ROS was observed during the induction of the antioxidant system by ABA, and it was concluded that Ca^{2+} can be found both before and after ROS in the signaling pathway related to oxidative stress (Jiang and Zhang 2003). It was also reported that Ca^{2+} enhanced both the GSH concentration and the stress tolerance in rice (Lu et al. 1999). However, in tobacco, GSH and GSSG treatment resulted in a rapid, transient increase in the Ca^{2+} level, suggesting that GSH may be involved in the activation of Ca^{2+}-dependent protein kinases in the early part of stress-induced signaling pathways (Go'mez et al. 2004).

The effect of abiotic stresses on H_2O_2, GSH, and GSSG concentrations may be transmitted by various plant hormones (Fig. 7.7). As has also been observed in chilled maize (Janda et al. 1999), salicylic acid (SA) stimulated the formation of ROS in *Arabidopsis* subjected to salt or osmotic stress (Borsani et al. 2001). GSH

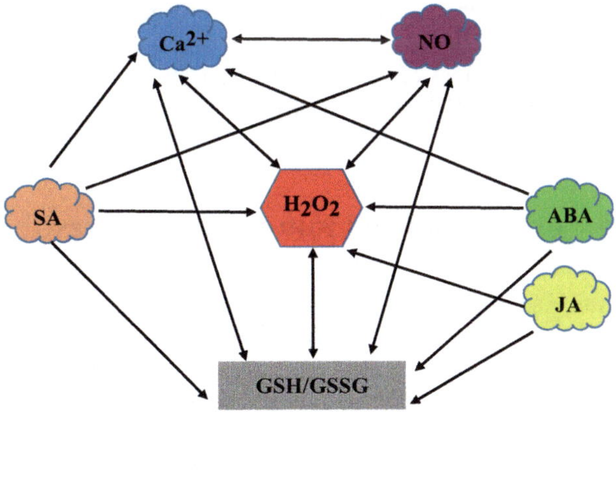

Fig. 7.7 A model for possible cross-talking between redox and other signaling pathways. The various signal transducers (Ca^{2+}, NO) and plant hormones may affect the GSH level and/or GSH/GSSG (oxidized glutathione) ratio directly or through H_2O_2. *ABA* abscisic acid, *JA* jasmonic acid, *SA* salicylic acid, *GSH* glutathione

and GR were affected by SA in a soybean cell suspension (Knörzer et al. 1999) and SA increased the GR activity in rice leaves (Ganesan and Thomas 2001). SA induced various alterations in γ-EC and GSH contents, GR activity, and γ-ECS and GR transcript levels in two maize genotypes with different levels of stress tolerance (Kellos et al. 2008); the γ-EC and GSH concentrations were increased in both genotypes by exogenous application of SA. Consistent with this observation, the overexpression of a gene coding for an enzyme involved in SA degradation caused a decrease in both the GSH concentration and the resistance to oxidative stress in rice (Kusumi et al. 2006). SA-induced NO production was found in *Arabidopsis*, and Ca^{2+} accumulation was a component of the signaling cascade (Zottini et al. 2007).

Another stress hormone, ABA, induced changes in ROS concentration in *Arabidopsis*, activating the Ca^{2+} channels of the cell membranes and increasing the Ca^{2+} concentration in cytosol (Murata et al. 2001). The connection between the redox state of the cells and H_2O_2 and ABA was shown in the *Arabidopsis* mutant GSH peroxidase 3 (*ATGPX3*), in which the addition of oxidized *ATGPX3* protein in vitro converted the protein phosphatase. The connection between the redox state of the cells and H_2O_2 and ABA was shown in the *Arabidopsis* mutant *ATGPX3* in which the addition of oxidized *ATGPX3* protein in vitro converted the protein phosphatase described in ABA insensitive2 (*ABI2*) mutants to its oxidized form. *ABI2*, in turn, influences Ca^{2+} channels and stomatal closure (Miao et al. 2006). In two maize genotypes differing in their stress tolerance, ABA differentially affected the GSH content, GSH/GSSG ratio, GR activity, and γ-ECS transcript level (Kellos et al. 2008). ABA may also affect the GSH/GSSG ratio and redox signaling (Pastori and Foyer 2002).

Like SA and ABA, jasmonic acid (JA) also regulated gene expression through H_2O_2, as found in tobacco (Mur et al. 2006). In addition, JA influenced GSH concentration and the genes involved in GSH metabolism in *Arabidopsis* (Sasaki-Sekimoto et al. 2005). As with SA, ethylene, and NO, JA also increased the transcript level of GST, suggesting that the various plant growth regulators interact (Moons 2005). The order of the components in the above signaling pathway described may vary, and some components may be absent or additional ones may be present depending on environmental effects, plant species, organs, and cell types. Multidirectional forward and backward interactions responsible for the regulation of metabolic pathways may exist between the compounds displayed in the figure to ensure the most effective protection against environmental stress (Noctor 2006; Dietz 2008; Miller et al. 2008).

5 Conclusion and Future Perspectives

Being sessile and sensitive organisms, plants inevitably encounter a variety of abiotic stresses in nature. Abiotic stress largely leads to accelerated generation of ROS, while GSH is the significant scavenger of ROS. The effect of GSH on plant stress resistance is achieved mainly through the cycle of AsA-GSH, which includes the three important interconnected and coordinated redox pairs of AsA/DHA, GSH/GSSG, NADPH/NADP, and by whose redox status modulated through the GSH antioxidant system, plants could scavenge ROS and strengthen their resistance to abiotic stresses. GSH can regulate the ROS signal transduction pathway, as well as cross-talk with other signaling pathways, and ROS levels depend on the redox status of redox pairs in plant cells.

Over the past few decades, although recent progress of research into GSH has been achieved, there are still many open questions: its regulatory function in abiotic stress still remains to be elucidated, the molecular mechanisms of GSH stress tolerance are not fully understood, and the data available are incomplete. Currently, many studies are expected to be carried out in the future, and they could involve several aspects: (i) how influential changes in the abundance or redox states of these compounds are in determining plant function or responses to the environment could be investigated, even though the changes in ascorbate and GSH status can exert a powerful influence on plant function; (2) the function of these enzymes related to the GSH metabolic process needs to be further recognized; (3) the role of subcellular changes in the redox state of the GSH/GSSG couple in stressed plants should be ascertained; (4) the effect of changes in the redox state of the GSH/GSSG couple on the transcript, protein, and metabolite profiles and on post-translational modification of proteins should be further investigated; (5) how GSH evolved from being an antioxidant to being a key intermediate in multiple-signaling networks and the interaction of GSH/GSSG with other signaling molecules during the stress response should be investigated; (6) a well integrative multi-approach (the molecular, physiological, and molecular-genetic including the application of transgene technology) should be adopted to identify underlying vital mechanisms that are related to the regulatory and signaling roles of GSH in modulating abiotic stress responses and tolerance.

References

Akram S, Siddiqui MN, Hussain BMN, Bari MAA, Mostofa MG, Hossain MA, Tran LSP (2017) Exogenous glutathione modulates salinity tolerance of soybean [*Glycine max* (L.) Merrill] at reproductive stage. J Plant Growth Regul. https://doi.org/10.1007/s00344-017-9691-9

Alhdad GM, Seal CE, Al-Azzawi MJ, Flowers TJ (2013) The effect of combined salinity and waterlogging on the halophyte *Suaeda maritima*: the role of antioxidants. Environ Exp Bot 87:120–125

Bashandy T, Guilleminot J, Vernoux T, Caparros-Ruiz D, Ljung K, Meyer Y, Reichheld JP (2010) Interplay between the NADP-linked thioredoxin and glutathione systems in *Arabidopsis* auxin signaling. Plant Cell 22:376–391

Becana M, Matamoros MA, Udvardi M, Dalton DA (2010) Recent insights into antioxidant defenses of legume root nodules. New Phytol 188:960–976

Borsani O, Valpuesta V, Botella MA (2001) Evidence for a role of salicylic acid in the oxidative damage generated by NaCl and osmotic stress in *Arabidopsis* seedlings. Plant Physiol 126:1024–1030

Chen JH, Jiang HW, Hsieh EJ, Chen HY, Chien CT, Hsieh HL, Lin TP (2012) Drought and salt stress tolerance of an *Arabidopsis* glutathione S-transferase U17 knockout mutant are attributed to the combined effect of glutathione and abscisic acid. Plant Physiol 158:340–351
Chi Y, Cheng Y, Vanitha J, Kumar N, Ramamoorthy R, Ramachandran S, Jiang SY (2011) Expansion mechanisms and functional divergence of the glutathione S-transferase family in sorghum and other higher plants. DNA Res 18:1–16
Clark D, Durner J, Navarre DA, Klessig DF (2000) Nitric oxide inhibition of tobacco catalase and ascorbate peroxidase. Mol Plant-Microbe Interact 13:1350–1384
Clemente MR, Bustos-Sanmamed P, Loscos J, James EK, Pérez-Rontomé C, Navascués J, Gay M, Becana M (2012) Thiol synthetases of legumes: Immunogold localization and differential gene regulation by phytohormones. J Exp Bot 63:3923–3934
Copley SD, Dhillon JK (2002) Lateral gene transfer and parallel evolution in the history of glutathione biosynthesis genes. Genome Biol 3:Research 0025. 1
Cruz de Carvalho MH, Brunet J, Bazin J, Kranner I, d' Arcy-Lameta A, Zuily-Fodil Y, Contour-Ansel D (2010) Homoglutathione synthetase and glutathione synthetase in drought-stressed cowpea leaves: expression patterns and accumulation of low-molecular-weight thiols. J Plant Physiol 167:480–487
De Pinto MC, Paradiso A, Leonetti P, De Gara L (2006) Hydrogen peroxide, nitric oxide and cytosolic ascorbate peroxidase at the crossroad between defence and cell death. Plant J 48:784–795
Dhankher OP, Li Y, Rosen BP, Shi J, Salt D, Senecoff JF, Sashti NA, Meagher RB (2002) Engineering tolerance and hyperaccumulation of arsenic in plants by combining arsenate reductase and gamma-glutamylcysteine synthetase expression. Nat Biotechnol 20:1140–1145
Diaz Vivancos P, Driscoll SP, Bulman CA, Ying L, Emami K, Treumann A, Mauve C, Noctor G, Foyer CF (2011) Perturbations of amino acid metabolism associated with glyphosate-dependent inhibition of shikimic acid metabolism affect cellular redox homeostasis and alter the abundance of proteins involved in photosynthesis and photorespiration. Plant Physiol 157:256–268
Dietz KJ (2008) Redox signal integration: from stimulus to networks and genes. Physiol Plant 133:459–468
Ding S, Lu Q, Zhang Y, Wen X, Zhang L, Lu C (2009) Enhanced sensitivity to oxidative stress in transgenic tobacco plants with decreased glutathione reductase activity leads to a decrease in ascorbate pool and ascorbate redox state. Plant Molr Biol 69:577–592
Dixon DP, Lapthorn A, Edwards R (2002) Plant glutathione transferases. Genome Biol 3:Reviews3004. 1
El Msehli S, Lambert A, Baldacci-Cresp F, Hopkins J, Boncompagni E, Smiti SA, Hérouart D, Frendo P (2011) Crucial role of (homo)glutathione in nitrogen fixation in *Medicago truncatula* nodules. New Phytol 192:496–506
Eltelib HA, Badejo AA, Fujikawa Y, Esaka M (2011) Gene expression of monodehydroascorbate reductase and dehydroascorbate reductase during fruit ripening and in response to environmental stresses in acerola (*Malpighia glabra*). J Plant Physiol 168:619–627
Estrella-Gómez NE, Sauri-Duch E, Zapata-Pérez O, Santamaría JM (2012) Glutathione plays a role in protecting leaves of *Salvinia minima* from Pb^{2+} damage associated with changes in the expression of SmGS genes and increased activity of GS. Environ Exp Bot 75:188–194
Foyer CH, Noctor G (2005) Redox homeostasis and antioxidant signaling: a metabolic link between stress perception and physiological responses. Plant Cell 17:1866–1875
Foyer CH, Noctor G (2011) Ascorbate and glutathione: the heart of the redox hub. Plant Physiol 155:2–18
Foyer CH, Lopez-Delgado H, Dat JF, Scott IM (1997) Hydrogen peroxide and glutathione-associated mechanisms of acclamatory stress tolerance and signaling. Physiol Plant 100:241–254
Fraser JA, Saunders RDC, McLellan LI (2002) *Drosophila melanogaster* glutamate-cysteine ligase activity is regulated by a modifier subunit with a mechanism of action similar to that of the mammalian form. J Biol Chem 277:1158–1165
Frendo P, Jiménez MJ, Mathieu C, Duret L, Gallesi D, Van de Sype G, Hérouart D, Puppo A (2001) A *Medicago truncatula* homoglutathione synthetase is derived from glutathione synthetase by gene duplication. Plant Physiol 126:706–715

Gadjev I, Vanderauwera S, Gechev TS, Laloi C, Minkov IN, Shulaev V, Apel K, Inze D, Mittler R, van Breusegem F (2006) Transcriptomic footprints disclose specificity of reactive oxygen species signaling in *Arabidopsis*. Plant Physiol 141:436–445

Galant A, Preuss ML, Cameron JC, Jez JM (2011) Plant glutathione biosynthesis: diversity in biochemical regulation and reaction products. Front Plant Sci 2:45

Ganesan V, Thomas G (2001) Salicylic acid response in rice: influence of salicylic acid on H_2O_2 accumulation and oxidative stress. Plant Sci 160:1095–1106

George S, Venkataraman G, Parida A (2010) A chloroplast-localized and auxin-induced glutathione S-transferase from phreatophyte *Prosopis juliflora* confer drought tolerance on tobacco. J Plant Physiol 167:311–318

Ghisla S, Massey V (1989) Mechanisms of flavoprotein-catalyzed reactions. Eur J Biochem 181:1–17

Gill SS, Khan NA, Tuteja N (2012) Cadmium at high dose perturbs growth, photosynthesis and nitrogen metabolism while at low dose it up regulates sulfur assimilation and antioxidant machinery in garden cress (*Lepidium sativum* L.) Plant Sci 182:112–120

Gill SS, Anjum NA, Hasanuzzaman M, Gill R, Trivedi DK, Ahmad I, Pereira E, Tuteja N (2013) Glutathione and glutathione reductase: a boon in disguise for plant abiotic stress defense operations. Plant Physiol Bioch 70:204–212

Gogos A, Shapiro L (2002) Large conformational changes in the catalytic cycle of glutathione synthase. Structure 10:1669–1676

Goiris K, Van Colen W, Wilches I, León-Tamariz F, Cooman LD, Muylaert K (2015) Impact of nutrient stress on antioxidant production in three species of microalgae. Algal Res 7:51–57

Gomez LD, Noctor G, Knight MR, Foyer CH (2004) Regulation of calcium signaling and gene expression by glutathione. J Exp Bot 55:1851–1859

Gromes R, Hothorn M, Lenherr ED, Rybin V, Scheffzek K, Rausch T (2008) The redox switch of gamma-glutamylcysteine ligase via a reversible monomer-dimer transition is a mechanism unique to plants. Plant J 54:1063–1075

Hasanuzzaman M, Hossain MA, Fujita M (2010) Physiological and biochemical mechanisms of nitric oxide induced abiotic stress tolerance in plants. Am J Plant Physiol 5:295–324

Hell R, Bergmann L (1988) Glutathione synthetase in tobacco suspension cultures: catalytic properties and localization. Physiol Plant 72:70–76

Hibi T, Nii H, Nakatsu T, Kimura A, Kato H, Hiratake J, Oda J (2004) Crystal structure of g-glutamylcysteine synthetase: insights into the mechanism of catalysis by a key enzyme for glutathione homeosta-sis. Proc Natl Acad Sci U S A 101:15052–15057

Hothorn M, Wachter A, Gromes R, Stuwe T, Rausch T, Scheffzek K (2006) Structural basis for the redox control of plant glutamate cysteine ligase. J Biol Chem 281:27557–27565

Innocenti G, Pucciariello C, Le Gleuher M, Hopkins J, Stefano M, Delledonne M, Puppo A, Baudouin E, Frendo P (2007) Glutathione synthesis is regulated by nitric oxide in *Medicago truncatula* roots. Planta 225:1597–1602

Janda T, Szalai G, Tari I, Paldi E (1999) Hydroponic treatment with salicylic acid decreases the effects of chilling injury in maize (*Zea mays* L.) plants. Planta 208:175–180

Jez JM, Cahoon RE (2004) Kinetic mechanism of glutathione synthetase from *Arabidopsis thaliana*. J Biol Chem 279:42726–42731

Jez JM, Cahoon RE, Chen SX (2004) *Arabidopsis thaliana* glutamate-cysteine ligase: functional properties, kinetic mechanism, and regulation of activity. J Biol Chem 279:33463–33470

Jiang M, Zhang J (2003) Cross-talk between calcium and reactive oxygen species originated from NADPH oxidase in abscisic acid-induced antioxidant defence in leaves of maize seedlings. Plant Cell Environ 26:929–939

Jiang YP, Huang LF, Cheng F, Zhou YH, Xia XJ, Mao WH, Shi K, Yu JQ (2013) Brassinosteroids accelerate recovery of photosynthetic apparatus from cold stress by balancing the electron partitioning, carboxylation and redox homeostasis in cucumber. Physiol Plant 148:133–145

Kabir MH, Wang MH (2010) Expression pattern of two dehydroascorbate reductase genes from tomato (*Solanum lycopersicum L.*) in response to stress. J Korean Soc Appl Biol Chem 53:668–676

Kabir MH, Han W, Wang MH (2011) Environmental stress response of a dehydroascorbate reductase gene from tomato, and its protective role in *Escherichia coli*. Hort Environ Biotech 52:621–628

Kavitha K, George S, Venkataraman G, Parida A (2010) A salt-inducible chloroplastic monodehydroascorbate reductase from halophyte *Avicennia marina* confers salt stress tolerance on transgenic plants. Biochimie 92:1321–1329

Kellos T, Tımar I, Szilagyi V, Szalai G, Galiba G, Kocsy G (2008) Stress hormones and abiotic stresses have different effects on antioxidants in maize lines with different sensitivity. Plant Biol 10:563–572

Keunen E, Remans T, Bohler S, Vangronsveld J, Cuypers A (2011) Metal-induced oxidative stress and plant mitochondria. Int J Mol Sci 12:6894–6918

Klapheck S, Latus C, Bergmann L (1987) Localization of glutathione synthetase and distribution of glutathione in leaf cells of *Pisum sativum* L. J Plant Physiol 131:123–131

Klapheck S, Fliegner W, Zimmer I (1994) Hydroxymethyl- phytochelatins [(γ-glutamylcysteine) n-serine] are metal-induced peptides of the Poaceae. Plant Physiol 104:1325–1332

Klapheck S, Schlunz S, Bergmann L (1995) Synthesis of phytochelatins and homo-phytochelatins in *Pisum Sativum* L. Plant Physiol 107:515–521

Knörzer OC, Lederer B, Durner J, Boger P (1999) Antioxidative defense activation in soybean cells. Physiol Plant 107:294–302

Kolbe A, Oliver SN, Fernie AR, Stitt M, van Dongen JT, Geigenberger P (2006) Combined transcript and metabolite profiling of *Arabidopsis* leaves reveals fundamental effects of the thioldisulphide status on plant metabolism. Plant Physiol 141:412–422

Kumar S, Asif MH, Chakrabarty D, Tripathi RD, Dubey RS, Trivedi PK (2013) Differential expression of rice lambda class GST gene family members during plant growth, development, and in response to stress conditions. Plant Mol Biol Rep 31:569–580

Kusumi K, Yaeno T, Kojo K, Hirayama M, Hirokawa D, Yara A, Iba K (2006) The role of salicylic acid in the glutathione-mediated protection against photooxidative stress in rice. Physiol Plant 128:651–666

Le Martret B, Poage M, Shiel K, Nugent GD, Dix PJ (2011) Tobacco chloroplast transformants expressing genes encoding dehydroascorbate reductase, glutathione reductase, and glutathione-S-transferase, exhibit altered anti-oxidant metabolism and improved abiotic stress tolerance. Plant Biotechnol J 9:661–673

LeBlanc MS, Lima A, Montello P, Kim T, Meagher RB, Merkle S (2011) Enhanced arsenic tolerance of transgenic eastern cottonwood plants expressing gamma-glutamylcysteine synthetase. Int J Phytoremediation 13:657–673

Leng XP, Jia HF, Sun X, Shangguan LF, Mu Q, Wang BJ, Fang JQ (2015) Comparative transcriptome analysis of grapevine in response to copper stress. Sci Rep 5:17749

Li Y, Dhankher OP, Carreira L, Balish RS, Meagher RB (2005) Arsenic and mercury tolerance and cadmium sensitivity in *Arabidopsis* plants expressing bacterial γ-glutamylcysteine synthetase. Environ Toxicol Chem 24:1376–1386

Lim B, Meyer AJ, Cobbett CS (2011) Development of glutathione-deficient embryos in *Arabidopsis* is influenced by the maternal level of glutathione. Plant Biol 13:693–697

Loscos J, Matamoros MA, Becana M (2008) Ascorbate and homoglutathione metabolism in common bean nodules under stress conditions and during natural senescence. Plant Physiol 146:1282–1292

Lu SY, Li YC, Guo ZF, Li BS, Li MQ (1999) Enhancement of drought resistance of rice seedlings by calcium treatment. Chin J Rice Sci 13:161–164

Lueder DV, Phillips MA (1996) Characterization of Trypanosoma bru-cei γ-glutamylcysteine synthetase, an essential enzyme in the biosynthesis of trypanothione. J Biol Chem 271:17485–17490

Maher P (2005) The effects of stress and aging on glutathione metabolism. Ageing Res Rev 4:288–314

Marty L, Siala W, Schwarzla¨nder M, Fricker MD, Wirtz M, Sweetlove LJ, Meyer Y, Meyer AJ, Reichheld JP, Hell R (2009) The NADPH-dependent thioredoxin system constitutes a functional backup for cytosolic glutathione reductase in *Arabidopsis*. Proc Natl Acad Sci U S A 106:9109–9114

Matamoros MA, Moran JF, Iturbe-Ormaetxe I, Rubio MC, Becana M (1999) Glutathione and homoglutathione synthesis in legume root nodules. Plant Physiol 121:879–888

Meuwly P, Thibault P, Schwan AL, Rauser WE (1995) Three families of thiol peptides are induced by cadmium in maize. Plant J 7:391–400

Meyer AJ (2008) The integration of glutathione homeostasis and redox signaling. J Plant Physiol 165:1390–1403

Mhamdi A, Hager J, Chaouch S, Queval G, Han Y, Taconnat L, Saindrenan P, Gouia H, Issakidis-Bourguet E, Renou JP, Noctor G (2010a) *Arabidopsis* GLUTATHIONE REDUCTASE 1 plays a crucial role in leaf responses to intracellular hydrogen peroxide and in ensuring appropriate gene expression through both salicylic acid and jasmonic acid signaling pathways. Plant Physiol 153:1144–1160

Mhamdi A, Queval G, Chaouch S, Vanderauwera S, Van Breusegem F, Noctor G (2010b) Catalase function in plants: a focus on *Arabidopsis* mutants as stress-mimic models. J Exp Bot 61:41974220

Miao Y, Dong LV, Wang PC, Wang XC, Chen J, Miao C, Song CP (2006) An *Arabidopsis* glutathione peroxidase functions as both a redox transducer and a scavenger in abscisic acid and drought stress responses. Plant Cell 18:2749–2766

Miller G, Shulaev V, Mittler R (2008) Reactive oxygen signaling and abiotic stress. Physiol Plant 133:481–489

Miller G, Suzuki N, Ciftci-Yilmaz S, Mittler R (2010) Reactive oxygen species homeostasis and signaling during drought and salinity stresses. Plant Cell Environ 33:453–467

Moons A (2005) Regulatory and functional interactions of plant growth regulators and plant glutathione S-transferases (GSTs). Plant Hormones 72:155–202

Moran JF, Iturbe-Ormaetxe I, Matamoros MA, Rubio MC, Clemente MR, Brewin NJ, Becana M (2000) Glutathione and homoglutathione synthetases of legume nodules. Cloning, expression, and subcellular localization. Plant Physiol 124:1381–1392

Mur LAJ, Kenton P, Atzorn R, Miersch O, Wasternack C (2006) The outcomes of concentration-specific interactions between salicylate and jasmonate signaling include synergy, antagonism, and oxidative stress leading to cell death. Plant Physiol 140:249–262

Murata Y, Pei Z-M, Mori IC, Schroeder J (2001) Abscisic acid activation of plasma membrane Ca^{2+} channels in guard cells requires cytosolic NAD(P)H and is differentially disrupted upstream and downstream of reactive oxygen species production in abi1-1 and abi2-1 protein phosphatase 2C mutants. Plant Cell 13:2513–2523

Nazar R, Iqbal N, Masood A, Syeed S, Khan NA (2011) Understanding the significance of sulfur in improving salinity tolerance in plants. Environ Exp Bot 70:80–87

Neill SJ, Desikan R, Clarke A, Hurat RD, Hancock JT (2002) Hydrogen peroxide and nitric oxide as signaling molecules in plants. J Exp Bot 53:1237–1247

Noctor G (2006) Metabolic signaling in defence and stress: the central role of soluble redox couples. Plant Cell Environ 29:409–425

Noctor G, Foyer CH (1998) Ascorbate and glutathione: keeping active oxygen under control. Annu Rev Plant Physiol Plant Mol Biol 49:249–279

Noctor G, Arisi ACM, Jouanin L, Kunert KJ, Rennenberg H, Foyer CH (1998) Glutathione: biosynthesis, metabolism and relationship to stress tolerance explored in transformed plants. J Exp Bot 49:623–647

Noctor G, Gomez L, Vanacker H, Foyer CH (2002) Interactions between biosynthesis, compartmentation and transport in the control of glutathione homeostasis and signaling. J Exp Bot 53:1283–1304

Noctor G, Mhamdi A, Chaouch S, Han Y, Neukermans J, Marquez-Garcia B, Queval G, Foyer CH (2012) Glutathione in plants: an integrated overview. Plant Cell Environ 35:454–484

Ohama N, Sato H, Shinozaki K, Yamaguchi-Shinozaki K (2017) Transcriptional regulatory network of plant heat stress response. Trends Plant Sci 22:53–65

Parisy V, Poinssot B, Owsianowski L, Buchala A, Glazebrook J, Mauch F (2007) Identification of PAD2 as a c-glutamylcysteine synthetase highlights the importance of glutathione in disease resistance of Arabidopsis. Plant J 49:159–172

Pastori GM, Foyer CH (2002) Common components, networks, and pathways of cross-tolerance to stress. The central role of "redox" and abscisic acid-mediated controls. Plant Physiol 129:460–468

Polekhina G, Board PG, Gali RR, Rossjohn J, Parker MW (1999) Molecular basis of glutathione synthetase deficiency and a rare gene permutation event. EMBO J 18:3204–3213

Price AH, Taylor A, Ripley SJ, Griffiths A, Trewavas AJ, Knight MR (1994) Oxidative signals in tobacco increase cytosolic calcium. Plant Cell 6:1301–1310

Pyngrope S, Bhoomika K, Dubey RS (2013) Reactive oxygen species, ascorbate-glutathione pool, and enzymes of their metabolism in drought-sensitive and tolerant indica rice (*Oryza sativa L.*) seedlings subjected to progressing levels of water deficit. Protoplasma 250:585–600

Quan LJ, Zhang B, Shi WW, Li HY (2008) Hydrogen peroxide in plants: a versatile molecule of the reactive oxygen species network. J Integr Plant Biol 50:2–16

Queval G, Thominet D, Vanacker H, Miginiac-Maslow M, Gakiere B, Noctor G (2009) H_2O_2-activated up-regulation of glutathione in *Arabidopsis* involves induction of genes encoding enzymes involved in cysteine synthesis in the chloroplast. Mol Plant 2:344–356

Rausch T, Gromes R, Liedschulle V, Mu¨ller I, Bogs J, Galovic V, Wachter A (2007) Novel insight into the regulation of GSH biosynthesis in higher plants. Plant Biol 9:565–572

Reichheld JP, Khafif M, Riondet C, Droux M, Bonnard G, Meyer Y (2007) Inactivation of thioredoxin reductases reveals a complex interplay between thioredoxin and glutathione pathways in development. Plant Cell 19(6):1851–1865

Rouhier N, Lemaire SD, Jacquot JP (2008) The role of glutathione in photosynthetic organisms: emerging functions for glutaredoxins and glutathionylation. Annu Rev Plant Biol 59:143–166

Sasaki-Sekimoto Y, Taki N, Obayashi T, Aono M, Matsumoto F, Sakurai N, Suzuki H, Hirai MY, Noji M, Saito K, Masuda T, Takamiya K, Shibata D, Ohta H (2005) Coordinated activation of metabolic pathways for antioxidants and defence compounds by jasmonates and their roles in stress tolerance in *Arabidopsis*. Plant J 44:653–668

Schafer HJ, Greiner S, Rausch T, Haag-Kerwer A (1997) In seedlings of the heavy metal accumulator *Brassica Juncea* Cu^{2+} differentially affects transcript amounts for gamma-glutamylcysteine synthetase (gamma-ECS) and metallothionein (MT2). FEBS Lett 404:216–220

Sobrino-Plata J, Ortega-Villasante C, Flores-Cáceres ML, Escobar C, Del Campo FF, Hernández LE (2009) Differential alterations of antioxidant defenses as bioindicators of mercury and cadmium toxicity in alfalfa. Chemosphere 77:946–954

Tausz M, Sircelj H, Grill D (2004) The glutathione system as a stress marker in plant ecophysiology: is a stress-response concept valid? J Exp Bot 55:1955–1962

Vázquez S, Goldsbrough P, Carpena RO (2009) Comparative analysis of the contribution of phytochelatins to cadmium and arsenic tolerance in soybean and white lupin. Plant Physiol Bioch 47:63–67

Verslues PE, Agarwal M, Katiyar-Agarwal S, Zhu JH, Zhu JK (2006) Methods and concepts in quantifying resistance to drought, salt and freezing, abiotic stresses that affect plant water status. Plant J 45:523–539

Wachter A, Wolf S, Steininger H, Bogs J, Rausch T (2005) Differential targeting of GSH1 and GSH2 is achieved by multiple transcription initiation: implications for the compartmentation of glutathione biosynthesis in the *Brassicaceae*. Plant J 41:15–30

Wang Z, Xiao Y, Chen W, Tang K, Zhang L (2010) Increased vitamin C content accompanied by an enhanced recycling pathway confers oxidative stress tolerance in *Arabidopsis*. J Int Plant Biol 52:400–409

Xiang C, Oliver DJ (1998) Glutathione metabolic genes coordinately respond to heavy metals and jasmonic acid in *Arabidopsis*. Plant Cell 10:1539–1550

Xiang C, Werner BL, Christensen EM, Oliver DJ (2001) The biological functions of glutathione revisited in *Arabidopsis* transgenic plants with altered glutathione levels. Plant Physiol 126:564–574

Yamaguchi H, Kato H, Hata Y, Nishioka T, Kimura A, Oda J, Katsube Y (1993) Three-dimensional structure of the glutathione synthetase from *Escherichia coli* B at 2.0 Å resolution. J Mol Biol 229:1083–1100

Yin L, Wang S, Eltayeb AE, Uddin MI, Yamamoto Y, Tsuji W, Takeuchi Y, Tanaka K (2010) Overexpression of dehydroascorbate reductase, but not monodehydroascorbate reductase, confers tolerance to aluminum stress in transgenic tobacco. Planta 231:609–621
Ying J, Clavreul N, Sethuraman M, Adachi T, Cohen RA (2007) Thiol oxidation in signaling and response to stress: detection and quantification of physiological and pathophysiological thiol modifications. Free Radic Bio Med 43:1099–1108
Yousuf PY, Hakeem KUR, Chandna R, Ahmad P (2012) In: Ahmad P, Prasad MNV (eds) Role of glutathione reductase in plant abiotic stress. In abiotic stress responses in plants. Springer, New York, pp 149–158
Yu CW, Murphy TM, Lin CH (2003) Hydrogen-peroxide induced chilling tolerance in mung beans mediated through ABA independent glutathione accumulation. Funct Plant Biol 30:955–963
Zagorchev L, Seal CE, Kranner I, Odjakova M (2012) Redox state of low-molecular-weight thiols and disulphides during somatic embryogenesis of salt-treated suspension cultures of *Dactylis glomerata* L. Free Radic Res 46:656–664
Zagorchev L, Seal CE, Kranner I, Odjakova M (2013) A central role for thiols in plant tolerance to abiotic stress. Int J Mol Sci 14:7405–7432
Zechmann B, Müller M (2010) Subcellular compartmentation of glutathione in dicotyledonous plants. Protoplasma 246:15–24
Zottini M, Costa A, de Michele R, Ruzzene M, Carimi F, Lo Schiavo F (2007) Salicylic acid activates nitric oxide synthesis in *Arabidopsis*. J Exp Bot 58:1397–1405

Chapter 8
Exogenous Glutathione-Mediated Abiotic Stress Tolerance in Plants

Fangbin Cao, Manman Fu, Runfeng Wang, Pedro Diaz-Vivancos, and Mohammad Anwar Hossain

Abstract Glutathione (GSH), a major non-protein low-molecular-weight thiol tripeptide in plant cells, is involved in a variety of life processes, including cell differentiation, removal of free radicals and hydroperoxides, thiol-disulfide exchange, and the synthesis of phytochelatin. Along with its oxidized form (GSSG), GSH plays key roles in maintaining cellular redox homeostasiss and signaling, as well as in defense reactions. As a component of ascorbate-glutathione (AsA-GSH) and glyoxalase pathways, GSH is involved in the regulation of hydrogen peroxide and methylglyoxal levels, ensuring their signaling functions, which are necessary for normal growth, development, and stress tolerance. In plants, GSH metabolism also plays important functions in determining the degree of expression of defense-related genes during abiotic and biotic stresses. Plants easily uptake exogenously applied GSH, which is transported into cellular compartments inducing a series of physiological and biochemical processes, including the modulation of abiotic stress tolerance. Recent studies have shown the multiple roles of exogenous GSH in improving abiotic stress tolerance through the regulation of multiple stress responsive pathways; however, the precise molecular mechanisms of exogenous GSH-induced abiotic stress tolerance are largely unknown. This chapter provides an overview to highlight the involvement of exogenous GSH in modulating abiotic stress tolerance. We also highlight the possible mechanisms of uptake and transport of the exogenously applied GSH under stressful conditions.

F. Cao (✉) • M. Fu • R. Wang
Department of Agronomy, College of Agriculture and Biotechnology, Zijingang Campus, Zhejiang University, Hangzhou 310058, People's Republic of China
e-mail: caofangbin@zju.edu.cn

P. Diaz-Vivancos
Department of Plant Breeding, CEBAS-CSIC, 30100 Murcia, Spain

M.A. Hossain (✉)
Department of Genetics and Plant Breeding, Bangladesh Agricultural University, Mymensingh 2202, Bangladesh

Laboratory of Plant Nutrition and Fertilizers, Graduate School of Agricultural and Life Sciences, University of Tokyo, Tokyo, Japan
e-mail: anwargpb@bau.edu.bd

M.A. Hossain et al. (eds.), *Glutathione in Plant Growth, Development, and Stress Tolerance*, https://doi.org/10.1007/978-3-319-66682-2_8

Keywords Exogenous glutathione • Reactive oxygen species • Methylglyoxal • Abiotic stress tolerance • Tolerance mechanisms

Contents

1 Introduction

Abiotic stresses seriously restrict plant growth and development through the unrestrained accumulation of reactive oxygen species (ROS) and reactive carbonyl species (RCS), which can cause oxidation of lipids, proteins, inactivation of enzymes, and DNA damage, and finally cell death (Miller et al. 2010; Avery 2011; Hoque et al. 2012a; Biswas and Mano 2015; Hossain et al. 2015). Plants have developed efficient enzymatic and non-enzymatic defense systems to counter the deleterious effects of ROS and RCS as well as to maintain its optimum level in order to trigger specific protective responses needed to ensure normal growth and development (Hossain et al. 2011, 2015; Baxter et al. 2014; del Río 2015; Li et al. 2017). Recently, the role of glutathione (GSH; γ-L-glutamyl-L-cysteinylglycine) has attracted considerable interest from the scientific community due to its broad range of functions in plant growth, development, and stress tolerance (Chen et al. 2012; Cheng et al. 2015; Noctor et al. 2012; Munné-Bosch et al. 2013). GSH refers only to the reduced glutathione, whereas the term glutathione refers to the total pool (GSH plus glutathione disulphide; GSSG). Glutathione is present in various plant tissues in concentrations up to 2–3 mM; it plays an important role in many life processes, such as cell differentiation, enzymatic regulation, cell signaling, and cell death, and acts as an antioxidant (Srivalli and Khanna-Chopra 2008; Diaz-Vivancos et al. 2010, 2015; Cai et al. 2011a; Chen et al. 2012; Schnaubelt et al. 2013). Furthermore, glutathione is used as a marker of oxidative stress, acts as a major reservoir of reduced sulfur, and plays crucial roles in biotic and abiotic stress responses and tolerance in plants (Tausz et al. 2004; Zechman et al. 2014; Cheng et al. 2015).

GSH is synthesized by the sequential addition of cysteine and glutamate followed by the addition of glycine via two ATP-dependent steps catalyzed by γ-glutamylcysteine synthetase (γ-ECS) and GSH synthetase (GSHS), respectively (Noctor et al. 2002). γ-ECS is located in plastids whereas GSHS is located in cytosol and plastids, and both are encoded by single-copy genes in *Arabidopsis* (Cairns et al. 2006). In *Arabidopsis thaliana*, knockout mutations of γ-ECS and GSHS induce embryo and seedling lethal phenotypes, respectively (Cairns et al. 2006; Pasternak et al. 2008), whereas overexpression of *γ-ECS* and *GSHS* significantly elevate GSH content and abiotic stress tolerance (Zhu et al. 1999; Liedschulte et al. 2010; Choe et al. 2013). Müller et al. (2004)

used electron microscopic immunogold cytochemistry to investigate the distribution of GSH in plant cells and reported that the highest level of GSH was found in mitochondria in different plant tissues. As a component of the ascorbate-glutathione (AsA-GSH) cycle and glyoxalase cycle, GSH is involved in removing excess hydrogen peroxide (H_2O_2) and methylglyoxal (MG) levels as well as in the regulation of their signaling functions (Szalai et al. 2009; Hossain and Fujita 2009; Hossain et al. 2010, 2011; Baxter et al. 2014; Mostofa et al. 2015a, b; Hoque et al. 2016; Li et al. 2017). Along with its oxidized form (GSSG), the GSH system plays a key role in maintaining cellular redox homoeostasis and is also considered as a redox sensor of environmental stimuli (Cairns et al. 2006; Szalai et al. 2009). In addition, GSH can also modulate gene expression, cell division, reproductive growth and development, and protein activity (Foyer et al. 2001; Zechmann et al. 2011; Noctor et al. 2012; Marquez-Garcia et al. 2014). Cai et al. (2011b) found that application of exogenous GSH affects the accumulation pattern of many proteins under cadmium (Cd) stress in rice (*Oryza sativa* L.), and showed a genotypic- dependent effect. Besides its antioxidant functions, GSH is also the direct precursor of phytochelatins (PCs), which play key roles in heavy metal sequestration, chelation, and tolerance (Zhu et al. 1999; Hossain et al. 2012; Clemens and Ma 2016). Although significant progress has been made in learning about the multiple roles of GSH in abiotic stress tolerance, many aspects of GSH-mediated abiotic stress responses remain elusive. This chapter concentrates on the functions of exogenous GSH in defense against different abiotic stresses, and also briefly describes how exogenous GSH is absorbed and transported in regulating abiotic stress tolerance.

2 Glutathione Metabolism-Related Enzymes Conferring Abiotic Stress Tolerance

As an important non-protein sink of reduced sulfur, glutathione content is significantly affected by abiotic stresses in plants. Glutathione utilizing and regenerating enzymes such as glutathione reductase (GR), glutathione peroxidases (GPXs), glutathione *S*-transferases (GSTs), dehydroascorbate reductase (DHAR), glyoxalase I (Gly I), glyoxalase II (Gly II), and phytochelatin synthase (PCS) play central roles in scavenging abiotic stress-induced accumulation of ROS and MG as well as in the sequestration of toxic heavy metals into the vacuoles. An overview of the multiple functions of the glutathione and its related enzymes during abiotic stress conditions are shown in Fig. 8.1.

2.1 Glutathione Reductase

Glutathione reductase (GR; EC 1.8.1.7) belongs to the NADPH-dependent oxidoreductase family and plays key roles in plant cell defense against ROS by reducing GSSG to GSH (Gill et al. 2013). Edwards et al. (1990) purified and isolated different subcellular isoforms of GR and detected GR activity in mitochondrial, cytosolic, and chloroplastic

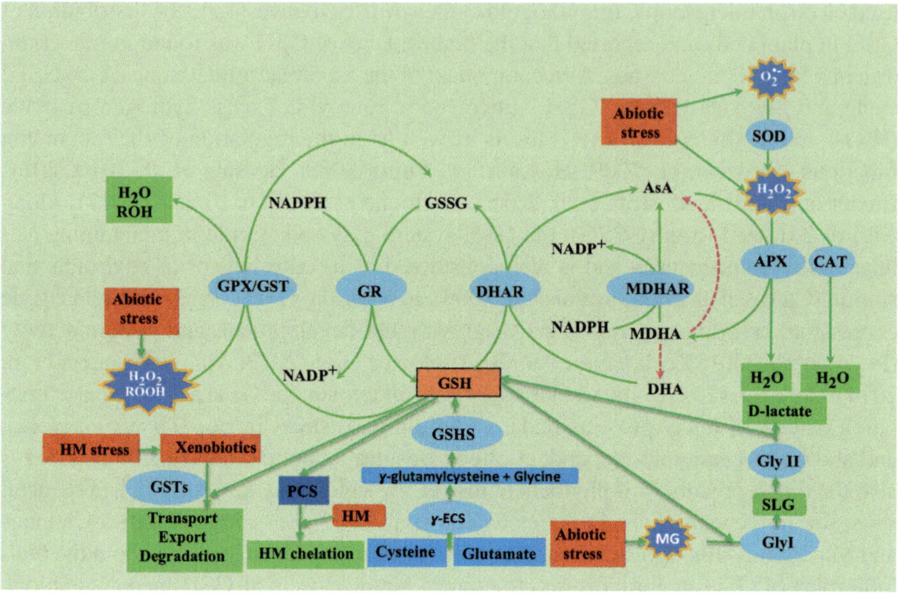

Fig. 8.1 Comprehensive scheme of GSH synthesis, interaction of GSH with its utilizing and regenerating enzymes in improving abiotic stress tolerance through stringent regulation of ROS and MG levels and heavy metal detoxification and chelation (modified from Hossain et al. 2012). GSH is synthesized from its constituent amino acids through two ATP-dependent reactions catalyzed by γ-ECS and GSHS. Superoxide ($O_2\bullet^-$) produced in cells is converted to H_2O_2 by SOD. H_2O_2 is then directly converted to H_2O by CAT or converted to H_2O by APX at the expense of AsA, depending on the cell compartment. The oxidized forms of ascorbic acid (MDHA and DHA) produced during the process are then converted to AsA by MDHAR and DHAR. GSSG is converted to GSH by GR at the expense of NADPH. GPX and GST catalyze the reduction of ROOH and H_2O_2, including lipid peroxides, to H_2O or alcohols. GSTs also catalyze the conjugation of metal-induced xenobioics and its transport into vacuoles. PCs sequester the metal to form a complex that is then transported into the vacuole. MG is degraded to form D-lactate through the glyoxalase pathway by the action of the enzymes Gly I and Gly II, which are GSH-dependent. R may be an aliphatic, aromatic, or heterocyclic group. *AsA* ascorbate, *DHA* dehydroascorbate, *γ-ECS* γ-glutamylcysteine synthetase, *HM* heavy metal, *NADPH* nicotinamide adenine dinucleotide phosphate, *APX* ascorbate peroxidase, *SOD* superoxide dismutase, *CAT* catalase, *GPX* glutathione peroxidase, *GR* glutathione reductase, *GSHS* GSH synthetase, *GSTs* glutathione *S*-transferases, *MG* methylglyoxal, *MDHA* monodehydroascorbate, *GSSG* oxidized glutathione, *GSH* reduced glutathione, *Gly II* glyoxalase II, *Gly I* glyoxalase I, *NADPH* nicotinamide adenine dinucleotide phosphate, *PCS* Phytochelatin synthase, *PCs* phytochelatins, *SLG S*- D-lactoylglutatione. For further discussion see the text

fractions of pea (*Pisum sativum* L.). GR is encoded by two genes, *GR1* and *GR2*. *GR1* encodes the protein that is detected in peroxisome and cytosol, while *GR2* encodes mitochondrial and chloroplastic GR (Kataya and Reumann 2010; Noctor et al. 2012). The positive function of GR in plant cells against abiotic stress has been widely reported. The major function of GR in conferring abiotic stress tolerance is the recycling of GSH and the maintaining of GSH/GSSG homeostasis (Noctor et al. 2012; Gietler et al. 2016). For instance, GR activity was increased in many plant species under abiotic stresses, such as heavy metal stress (Dazy et al. 2009), chilling (Turan and Eknekci 2011), salinity (Yazici et al. 2007), drought (Rapala-Kozik et al. 2008), and dehydration tolerance

(Gietler et al. 2016). However, some studies also reported that GR activity was decreased or not changed under abiotic stresses (Almeselmani et al. 2006; Hossain et al. 2010). It has been reported that the *GR1* deletion mutant showed normal growth and development whereas the *GR2* deletion mutant produced a lethal phenotype and showed growth arrest (Diaz-Vivancos et al. 2015). By using chloroplastic GR RNAi plants, Ding et al. (2009) showed that the activity of GR is very important for maintaining glutathione and the ascorbate pool under oxidative stress conditions. Additionally, it has been reported that the knockdown of *GR2* leads to early leaf senescence in *Arabidopsis* due to elevated levels of H_2O_2 and altered glutathione status (Ding et al. 2016a). Recently, Yin et al. (2017) reported that transgenic plants over-expressing the GR gene showed higher aluminum toxicity tolerance by increasing the ROS and RCS detoxification.

2.2 Glutathione peroxidases

Glutathione peroxidases (GPXs; EC 1.11.1.9) are a family of enzymes that catalyze the reduction of H_2O_2 and organic hydroperoxides using GSH as a reducing reagent (Diao et al. 2014). Unlike animals, most GPXs in plants are non-selenium dependent (Diao et al. 2014). Plant GPXs have been recognized as the fifth class of peroxiredoxins and are expressed in various subcellular compartments, such as cytosol, mitochondria, endoplasmic reticulum, and chloroplasts (Milla et al. 2003; Navrot et al. 2006). In *Arabidopsis*, eight related protein GPX isoenzymes, termed AtGPX1–AtGPX8, have been identified (Gaber et al. 2012). Gaber et al. (2012) found that the transcript and protein levels of AtGPX8, localized at the nucleus and cytosol, were up-regulated under oxidative damage induced by high-light stress and paraquat. Expression of *CsGPX2* was significantly up-regulated in *Camellia sinensis* plants under many abiotic stresses, including heavy metal, drought, and salinity (Fu 2014). Chen et al. (2004) reported that a tomato phospholipid hydroperoxide GPX (LePHGPX) not only inhibited cell death induced by oxidative stress in yeast, but also inhibited heat, salt, and Bax (a pro-apoptotic member of the Bcl-2 family of proteins) induced programmed cell death in *Nicotiana tabacum*. Transgenic tomato plants over-expressing GPX gene showed improved abiotic stress tolerance (Herbette et al. 2011) and transgenic *Arabidopsis* over-expressing *AtGPX8* showed higher oxidative stress tolerance and maintained cellular redox homeostasis (Gaber et al. 2012).

2.3 Glutathione S-transferases

Glutathione *S*-transferases (GSTs; EC 2.5.1.18) are a ubiquitous superfamily of enzymes that play important roles in many detoxification reactions (Dixon and Edwards 2010; Kumar et al. 2013). GSTs are also GSH-dependent detoxifying enzymes and constitute more than 1% of soluble protein in the leaves of maize (Marrs 1996). GSTs have multifunctional roles in plant cells: they catalyze GSH-dependent biotransformation processes, serve as binding and carrier protein for intracellular transport, and catalyze conjugation reactions (Edwards et al. 2000). The GSH-based transferase activity is involved in the conjugation of GSH with cytotoxic, electrophilic, and hydrophobic substrates (Soranzo et al. 2004). Functioning as GPX, plant GSTs can catalyze the reduction of hydroperoxides to less harmful alcohols and safeguard the protein function from oxidative damage, whereas its DHAR activity is involved in the maintenance of the redox homeostasis by regenerating AsA from DHA (Dixon and Edwards 2010). In plants, GSTs can be induced by abiotic stresses and elevated GST levels, contributing to maintaining the cell redox homeostasis (Kumar et al. 2013; Cao et al. 2014). Cao et al. (2014) found that GST activity was increased >50% in Cd-tolerant barley genotype, while no change in a sensitive genotype was observed under Cd toxicity. Kumar et al. (2013) reported that the expression of *OsGSTL2* in *Arabidpsis* provides tolerance for salt, osmotic, cold, and heavy metal stress. Similarly, transgenic tobacco over-expressing the sweet orange (*Citrus sinensis*) tau type glutathione transferases (*CsGSTUs*) showed higher salt, drought, and herbicide tolerance (Lo Cicero et al. 2015).

2.4 Dehydroascorbate Reductase

Dehydroascorbate reductase (DHAR; EC 1.8.5.1), the GSH-dependent enzyme in the AsA-GSH pathway, maintains the redox pool of ascorbate by recycling dehydroascorbate (DHA) to AsA and limits ROS-induced damage in plant cells (Gallie 2013; Noshi et al. 2016). Apart from recycling of DHA, this enzyme also plays diverse roles in plant growth and different plant physiological processes such as photosynthesis (Chen and Gallie 2008). In guard cells, the signaling function of H_2O_2 is regulated by both its AsA content and DHAR (Gallie 2013). Under abiotic stress conditions, susceptible plants showed lower DHAR activity and AsA/DHA ratio than tolerant plants (Mittova et al. 2003a, b; Ren et al. 2016). Transgenic plants over-expressing *DHAR* also showed higher abiotic stress tolerance that is accompanied by higher DHAR activity, AsA levels, as well as ascorbate redox state (Kim et al. 2014). Importantly, *DHAR* knock-down mutant showed higher sensitivity to high light stress due to a lower AsA level as well as DHAR activity; however, the redox state of GSH was markedly affected. These results suggest that both AsA and GSH redox states are altered by DHAR (Noshi et al. 2016). Additionally, under stressful conditions, when MDHAR activity is inhibited, the DHAR activity increases and acts as a functional back-up to maintain the cellular redox balance (Locato et al. 2009; Noshi et al. 2016). Hence, higher AsA content by AsA recycling through DHAR optimizes the AsA utilization and modulates abiotic oxidative stress tolerance.

2.5 Glyoxalase System Enzymes

Thy glyoxalase system is a ubiquitous GSH-dependent detoxification system in plants. In this system the glyoxalase I (Gly I; EC 4.4.1.5) and glyoxalase II (Gly II; EC 3.1.2.6) enzymes transform MG, a cytotoxic compound produced in ample amounts under stressful conditions, to D-lactate in different cellular organelles through two steps of irreversible reactions (Hossain et al. 2011; Hoque et al. 2016). In the first step, MG reacts with GSH forming hemithioacetal that is then converted to *S*-D-lactoylglutathione (SLG) in a reaction catalyzed by Gly I. In the second step, SLG is converted to D-lactate by the enzyme Gly II, being then the GSH regenerated in the system (Fig. 8.1). Recently, it has been reported that glyoxalase III can detoxify MG to D-lactate without of the participation of GSH (Ghosh et al. 2016). Although the glyoxalase system is involved in various plant physiological processes, its involvement in plant abiotic stress response and tolerance is considered crucial (Hossain et al. 2009, 2014a, b; Hossain and Fujita 2009; Kaur et al. 2014; Hoque et al. 2016). The glyoxalase system not only regulates MG levels in plants under stressful conditions but also regulates glutathione redox state through the recycling of GSH. A higher level of cellular GSH and GSH/GSSG ratio are required for stress defense against oxidative stress (Yadav et al. 2005a, b; Noctor et al. 2012). A large number of studies have shown a close link between the antioxidant and glyoxalase systems in plants (Hossain et al. 2010, 2011; Mostofa et al. 2015a, b; Nahar et al. 2015a, b, c). Recent studies in plants further demonstrated the diverse roles of this pathway in plant abiotic stress tolerance through the regulation of MG and ROS levels, allowing their signaling functions and improving stress tolerance through the expression of stress responsive genes (Hoque et al. 2012b, 2016; Li et al. 2017).

2.6 Phytochelatin Synthase

Phytochelatins [PCs; (γ-Glu-Cys)n-Gly (n = 2–11)] are widely accepted as the best-characterized heavy metal chelators and the major product for heavy metal detoxification and tolerance in plants, fungi, and other living organisms (Chia et al. 2013). PCs are cysteine-rich polypeptides that have high affinity for heavy metals (Lee and Hwang 2015). PCs are synthesized by the action of phytochelatin synthase (PCS) in cytosol with GSH as the precursor. Both GSH and PCs chelate heavy metals and metalloids such as Cd, copper (Cu), and arsenic (As), facilitating their sequestration into vacuoles (Cobbett and Goldsbrough 2002; Pilon-Smits 2005). In *Arabidopsis*, there are two genes encoding PCs, *AtPCS1* and *AtPCS2*. *AtPCS1* has been reported as the major player in PC synthesis, while the expression level of *AtPCS2* is much lower than *AtPCS1* in most tissues (Cobbett and Goldsbrough 2002; Blum et al. 2007). Meanwhile, AtPCS1 was ubiquitously present in *Arabidopsis* seedlings, while AtPCS2 was only found in the root tip (Blum et al. 2010). Blum et al. (2007) found that AtPCS1 had two cellular functions, mediating toxic heavy metal tolerance and GSH-conjugate degradation. Transgenic plants over-expressing *Arabidopsis* PCS gene (*AtPCS1*) in a non-accumulator plant *N. tabacum* improved Cd stress tolerance, and this response was further enhanced through the application of exogenous GSH (Pomponi et al. 2006). Besides heavy metals, PCs also play important roles in salinity, drought, heat, and UV-B tolerance (Chaurasia et al. 2016).

3 Uptake and Transport of Exogenously Applied GSH in Plant System

Uptake and transport of glutathione play central roles in many life processes, including sulfur assimilation, developmental processes, and tolerance against abiotic and biotic stresses. Glutathione-specific uptake systems have been found in plasma membranes of plant cells (Foyer et al. 2001). GSH uptake was observed in both protoplasts and cells (Noctor et al. 2012). Jamai et al. (1996) found that GSH was taken up by one saturable transporter with Km of 0.4 mM, while GSSG showed two systems with Km of 0.7 μM and 3.7 mM. In addition, it was also suggested that GSH and GSSG were taken up through proton symport. GSH uptake can be suppressed by GSSG and GS conjugates, while GSSG uptake can also be inhibited by GSH and GS conjugates (Zhang et al. 2004). Zhang et al. (2004) complemented a GSH-deficient yeast mutant with a GSH transporter cDNA from *O. sativa* and observed a strong increase in GSH uptake. Furthermore, the uptake activity showed a linear increase in the first 2–3 h. Noctor et al. (2000) incubated intact wheat chloroplasts with 100 and 1 μM ^{35}S-labelled GSH and found a time-dependent uptake within the initial 15 min. GSH concentration increased in all tissues of bean seedlings roots exposed to 1 mM GSH (Kumar et al. 2010). Moreover, GSH content in roots, leaves, and apex was increased 22-, 5-, and 3.5-fold after 4-h treatment, respectively. The results demonstrated that GSH is translocated to shoot and root systems through xylem.

GSH has been identified as a major form of long distance transport of reduced sulfur in xylem and phloem in plants, and can be readily exchanged between xylem and phloem in both directions (Schneider et al. 1994; Zhang et al. 2004). Different studies have suggested that GSH transport systems are present in membranes with fast exchange rates (Noctor et al. 2002; Tausz et al. 2004). The first high affinity GSH transporter (Hgt1p) was identified in *Scacharomyces cerevisiae* (Bourbouloux et al. 2000). In *Arabidopsis*, there are nine *Hgt1* homologues located in different chromosomes, and the homologues were also found in cotton (*Gossypium* sp.) and rice (*O. sativa)* (Foyer et al. 2001).

Intracellular transport between cytosol and organelles plays key roles in maintaining GSH homeostasis. Chloroplasts can synthesize GSH, and also uptake GSH from cytosol (Foyer et al. 2001). Noctor et al. (2012) suggested that γ-EC is produced exclusively in chloroplast, and then converted to GSH in chloroplast or transported to cytosol where the GSH can be transported to different organelles, including chloroplasts. Maughan et al. (2010) also reported that GSH biosynthesis was regulated by plastids and identified a plastid thiol transporter homologous to the *Plasmodium falciparum* chloroquine-resistance transporter (*Pf*CRT) in *Arabidopsis*. Arabidopsis mutants of the transporters were GSH-deficient and heavy metal-sensitive. In addition, knockout of the transporter family led to GSSG accumulation in cytosol, but not in plastids. In accordance with the literature, we suggest that exogenous GSH can be taken up through the root or leaf, then transported to different tissues via xylem and phloem, and finally transported to cytosol and different organelles via GSH transporters, which then play a positive role against abiotic stresses in plants.

4 Roles of Exogenous GSH in Modulating Abiotic Stress Tolerance

Abiotic stresses, in general, induce an overproduction of ROS and MG in plant cells and seriously limit different plant physiological process such as plant growth and development, leading to a reduced yield (El-Shabrawi et al. 2010; Saito et al. 2011; Hussain et al. 2016). Although GSH biosynthesis can be induced by abiotic stresses, this process can also be inhibited under serious stress conditions (Zhou et al. 2017). However, application of exogenous GSH can effectively compensate the decrease of endogenous GSH and improve abiotic stress tolerance in plants (Cai et al. 2010, 2011a; Zhou et al. 2017). A few recent studies focused on the effects of exogenous GSH in heavy metal, salinity, drought, heat, chilling, and low nutrient stresses through the assessment of different biochemical parameters related to stress tolerance (Chen et al. 2010; Cai et al. 2011a; Mostofa et al. 2014a; Nahar et al. 2015a, b, c; Hussain et al. 2016; Akram et al. 2017; Zhou et al. 2017). In the following section we will discuss the possible roles and mechanisms of exogenous GSH-mediated abiotic stress tolerance in plants.

4.1 Salinity Stress

Soil salinity is worldwide an increasing constraint in agricultural production, and nearly 20% of irrigated land has been affected by salinity in the world (Yamaguchi and Blumwald 2005). Oxidative stress and MG stress is also an important phenomenon of salinity (Mittova et al. 2003a, b; El-Shabrawi et al. 2010; Mostofa et al. 2015a; Akram et al. 2017). It has been suggested that the salt-tolerant genotypes displayed higher endogenous GSH concentrations than the susceptible genotypes in rice, tomato, and groundnut (Mittova et al. 2003a; El-Shabrawi et al. 2010; Kumar et al. 2010). The analysis of salt-tolerant and salt-sensitive cultivars also showed that the endogenous GSH levels and GSH-utilizing and regenerating enzymes are key factors in improving salt stress tolerance. Mittova et al. (2003a, b) showed that salt-tolerant *Lycopersicon pennellii* showed higher GSH biosynthesis, GSH content, GSH/GSSG ratio, and higher GST and GPX activities when compared to the salt-sensitive *L. esculentum* genotype. The tolerant genotype also showed lower H_2O_2 and malondialdehyde (MDA, lipid peroxidation marker) levels as compared to the sensitive one. GSH is of intrinsic importance in the prevention of salt-induced oxidative stress in *L. pennellii*, a mechanism that may also be employed by other salt-tolerant species. Subsequently, El-Shabrawi et al. (2010) showed that the salt-tolerant rice genotype (Pokkali) maintained a higher GSH and GSH/GSSG ratio, as well as Gly I, Gly II, SOD, CAT, peroxidase (POX), and GPX activities as compared to the salt-sensitive genotype (IR64). The tolerant genotype also showed lower ROS accumulation and ROS-induced DNA damage. These findings suggested the intrinsic function of GSH and proved that the coordinate induction of GSH biosynthesis and GSH-metabolizing enzymes is correlated with salt stress tolerance.

A large number of recent studies also elucidated the role of exogenous GSH in conferring salinity tolerance in tomato (*Solanum lycopersicum* L.), mung bean (*Vigna radiata* L.), rice (*Oryza sativa* L.), and cotton through enhancing antioxidant and glyoxalase pathway enzyme activities, GSH content, and photosynthetic capacity (Wang et al. 2014a; Nahar et al. 2015a; Hussain et al. 2016; Akram et al. 2017; Ibrahim et al. 2017). By using contrasting rice cultivars (Pokkal, salt tolerance, and Peta, salt sensitive), Wang et al. (2014a) showed that the application of either GSH or AsA modulates the salt-induced oxidative stress tolerance. Under salt stress, rice seedlings supplemented with GSH or AsA displayed lower ROS and MDA content, as well as higher endogenous levels of GSH and AsA and higher SOD, APX, and GR activities than non-treated salt-stressed seedlings. Nahar et al. (2015a) showed the importance of GSH in modulating salt stress tolerance in mung beans by analyzing ROS and MG metabolism. An abrupt increase in ROS, MG, and MDA levels was found in response to salt stress. The relative water content (RWC), chlorophyll (Chl), and AsA content, as well as the GSH/GSSG ratio was decreased by salt stress. The activities of CAT, DHAR, MDHAR, and Gly I decreased whereas the activities of APX, SOD, GST, GR and GPX increased. Seedlings treated with GSH + salt treatment resulted in better salt-induced (short-term) oxidative stress tolerance as indicated by lower ROS and MG levels; higher RWC, Chl, AsA, GSH, and GSH/GSSG ratio, and induced ROS and MG detoxification systems (Nahar et al. 2015a). Recently Zhou et al. (2017) confirmed the positive roles of exogenous GSH in improving salt-induced oxidative stress tolerance in tomato (*S. lycopersicum* L. cv. Zhongshu No. 4). Exogenous application of GSH increased the transcript level of GSH synthesis and metabolizing enzymes such as γ -ECS, GSHS, GST, GPX, and GR, the content of intracellular GSH and AsA, and the GSH/GSSG and AsA/DHA ratios in salt-stressed plants and in salt-stressed plants treated with buthionine sulfoximine (BSO, inhibitor of GSH synthesis key enzyme γ -ECS). Application of GSH also enhanced the activities of SOD, CAT, POD, and enzymes related to the AsA-GSH cycle including APX, DHAR, MDHAR, and GR, and decreased the content of H_2O_2 and $O_2{\bullet}^-$, and lipid peroxidation levels. Consequently, Ibrahim et al. (2017) showed the positive impact of exogenous GSH in modulating salt stress tolerance in cotton by using the contrasting salt-sensitive ‘Zhongmian 41’ and salt-tolerant ‘Zhong 9806’ cultivars. The application of salt stress (150 mM NaCl) produced a significant decrease in morphological (root and shoot characteristics), physiological (photosynthetic rate), and biochemical (MDA and chlorophyll levels) traits, and an altered leaf/root ultrastructure. Applications of exogenous GSH mitigated those deleterious effects, with a greater influence noticed in the salt-sensitive genotype.

Apart from improving salinity stress tolerance at the seedling stage, our recent study also showed that the application of exogenous GSH improves salinity stress tolerance in rice at the reproductive stage. Imposition of salt stress (200 mM NaCl) at the flowering stage resulted in a significant decrease in yield and yield-attributing traits, and a greater decrease was found in the salt-sensitive genotypes. Application of exogenous GSH improves salt stress tolerance as indicated by higher effective tillers per plant, number of filled grains per panicle, spikelet fertility, 100-seed weight, and seed yield per plants as compared to non-treated salt-stressed seedlings. The beneficial effects of exogenous GSH were higher in salt-susceptible genotypes as compared to the salt-tolerant genotypes (Hussain et al. 2016). Subsequently, we further proved that exogenous GSH improved salinity stress tolerance at seedling as well as at reproductive stage in soybean [*Glycine max* (L.) Merrill]. The imposition of salt stress at reproductive stage decreased the yield and yield-contributing traits. Application of exogenous GSH improved plant height, number of branches per plant, number of pods per plant, number of seeds per pod, number of seeds per plant, 100-seed weight, and yield per plant. Importantly, application of exogenous GSH at seedling stage also improved the oxidative stress tolerance as indicated by lower H_2O_2 and MDA levels (Akram et al. 2017). The above studies clearly demonstrated the diverse function of exogenous GSH in modulating salt stress tolerance through the regulation of multiple stress responsive pathways.

4.2 *Drought Stress*

Drought- or water stress-induced excessive accumulation of ROS due to impairment of photosynthesis has been well documented in plants (reviewed in Cruz de Carvalho 2008). Plenty of studies have shown that increased synthesis or recycling of GSH and high GSH/GSSG ratio might be essential for drought resistance in plants (Selote and Khanna-Chopra 2004; Gorantla et al. 2007; Garg et al. 2012; Cheng et al. 2015; Nahar et al. 2015b). Drought tolerant wheat cultivar showed a higher GSH redox pool due to higher GSH biosynthesis and AsA-GSH cycle enzyme activities as compared to sensitive cultivar (Garg et al. 2012). Expressed sequence tags (ESTs) analysis of drought-tolerant indica rice (Nagina 22) genotype also showed a high expression of GSH- and AsA-related stress defence genes such as *GSTs*, *GPX*, *Gly I*, and *APX* (Gorantla et al. 2007). Imposition of drought stress at the panicle development stage showed that the drought-tolerant genotype (N22) showed higher GSH and AsA levels and higher antioxidant enzyme (GR, SOD, APX) activities as compared to the sensitive genotype (Selote and Khanna-Chopra 2004). The function of GSH in modulating drought stress tolerance through the regulation of ROS and MG detoxification systems by using exogenous GSH has also been reported (Nahar et al. 2015b). Imposition of drought stress (−0.7 Mpa) in mung bean (*V. radiata* L.) seedlings resulted in a decrease in plant biomass, AsA content, GSH/GSSG ratio, DHAR, MDHAR, and CAT activities, but increased MDA, $O_2^{\bullet-}$, H_2O_2, proline, and MG content. The activities of Gly I and Gly II were also increased under drought stress. Application of exogenous GSH significantly alleviated drought-induced oxidative damage through enhancing the capacity of glyoxalase and antioxidant systems (Nahar et al. 2015b). Recently, Chen et al. (2012) reported that Arabidopsis *GST U17*-knockout mutant had higher drought and salinity stress tolerance due to higher accumulation of GSH and abscisic acid

(ABA). To explore how the mutant accumulated ABA, wild type plants were treated with exogenous GSH, and it was found that these plants accumulated higher ABA than those grown in the absence of GSH. Moreover, GSH-treated plants were more tolerant to salinity and drought, suggesting an interaction between GSH and ABA in increasing plant fitness under stressful conditions (Chen et al. 2012). More recently it has been reported that GSH modulates salt and drought stress tolerance by direct effects on global transcriptional changes as well as on ABA and JA biosynthesis and signaling (Cheng et al. 2015).

4.3 Heavy Metal Stress

Heavy metal or metalloid stress negatively affects plant growth and development and alters the physiological, biochemical, and molecular plant processes (reviewed in Hossain et al. 2012). The roles of GSH in modulating heavy metal or metalloid stress tolerance have been well documented in plants (Hossain et al. 2012; Anjum et al. 2014; Zhou et al. 2016). Studies with heavy metal tolerant or hyperaccumulator plants showed that the biosynthesis of GSH and the activities of GSH-regenerating and utilizing enzymes have significant effects on heavy metal tolerance. Iannelli et al. (2002) showed that high GSH and AsA content as well as APX, CAT, GR, GST, and GPX activities are key players in Cd tolerance in *Phragmites australis*. Recent transcriptomic analysis using low or high Cd-accumulating genotypes also showed the important roles of GSH in Cd stress tolerance (Zhou et al. 2016). Additionally, it has been reported that GSH-mediated ROS and MG metabolism are also involved in heavy metal tolerance in plants (Singla-Pareek et al. 2006; Hossain et al. 2010; Chen et al. 2010; Cai et al. 2011a; Mostofa et al. 2015b).

Numerous recent studies using exogenous application of GSH in barley, rice, cotton, and tobacco under different heavy metal toxicity conditions have shown the key role of GSH in heavy metal tolerance (Table 8.1). Our previous studies suggested that genotypic differences in Cd tolerance could be positively linked to the endogenous GSH content. Similarly, alleviation of Cd stress by exogenous GSH was significantly associated with increased endogenous GSH (Chen et al. 2010; Cai et al. 2011a). For instance, Cai et al. (2011a) investigated the effect of 50 μM GSH treatment on PCs, GSH, and cysteine content, and photosynthetic performance in different rice genotypes submitted to 5 and 50 μM Cd stresses. Exogenous GSH significantly increased GSH and PCs in the roots after 5 d exposure to 5 μM Cd, whereas GSH, cysteine, and PCs content decreased in plants submitted to 50 μM Cd. Nevertheless, external GSH markedly increased chlorophyll content, net photosynthetic rate, Fv/Fm, and effective PSII quantum yield, but decreased quantum yield of regulated energy dissipation and coefficient of non-photochemical quenching in both genotypes, compared with Cd treatments. Hasan et al. (2016) reported that foliar application of GSH significantly increased PCs content under Cd stress in tomato. GSH can also sequestrate heavy metal into cell walls (Hasan et al. 2016). Exogenous GSH also significantly alleviated Cr^{6+}-induced growth inhibition via

Table 8.1 Alleviating effects of exogenous GSH in modulating heavy metals stress response and tolerance

Plant species	Stress imposed	Factors responsible for exogenous GSH-mediated stress tolerance	Reference
Barley (*Hordeum vulgare)*	Cadmium	Exogenous GSH improved the capacity of antioxidant defense system and photosynthesis, ameliorated Cd-induced damage on ultrastructure, and reduced Cd-induced ROS accumulation and Cd concentration	Chen et al. (2010), Wang et al. (2011)
Tomato (*S. lycopersicum*)	Cadmium	Upregulated transcript level of several transcription factor and increased nitric oxide and S-nitrosothiol content, GSH:GSSG and AsA:DHA ratio, and sequestration of Cd into vacuoles and cell wall	Hasan et al. (2016)
Cotton (Gossypium spp.)	Cadmium	Alleviated Cd-induced growth inhibition, photosynthesis reduction, ROS accumulation, and microstructure damage of chloroplast	Daud et al. (2016)
Tobacco (Nicotiana tabacum)	Cadmium, Copper and Zinc	Elevated chlorophyll and rubisco content, but decreased rubisco activity except Cu	Son et al. (2014)
Rice (Oryza sativa)	Cadmium	Increased endogenous GSH, mineral element and chlorophyll content, induced up-regulation of PCs, regulated antioxidant defense enzyme activity and Cd-tolerant-related protein expression level and reduced Cd accumulation	Cai et al. (2010, 2011a, b), Cao et al. (2013a, 2015)
Rice (Oryza sativa)	Chromium	Increased secretion of organic acids and cell viability, changed the forms of Cr ions and distribution, alleviated Cr-induced damage on ultrastructure of root cell and chloroplast	Qiu et al. (2013)
Rice (Oryza sativa)	Copper	Decreased ROS and proline content, regulated antioxidant and MG detoxification system and reduced Cu uptake	Mostofa et al. (2014a)
Rice (Oryza sativa)	Cadmium+ Chromium	Reduced Cr uptake and translocation, improved H^+-ATPase activity and Fe, Zn and Mn uptake and translocation, regulated antioxidant enzyme activity and repressed MDA accumulation	Cao et al. (2013b)

increasing GSH concentration and secretion of organic acids (Qiu et al. 2013). GSH can alter forms of Cr ions in rhizosphere and their distribution among different sub-cellular components (Qiu et al. 2013).

Heavy metal injury is mainly attributed to the over-accumulation of ROS, including H_2O_2, superoxide radical ($O_2{\bullet}^-$), and hydroxyl radical (•OH). Several studies have showed that exogenous GSH reduced ROS accumulation through counteracting heavy metal-induced alterations of certain antioxidant enzymes and maintaining increased AsA/DHA and GSH/GSSG ratios (Chen et al. 2010; Cao et al. 2013a, b; Mostofa et al. 2014; Hasan et al. 2016). For instance, exogenous GSH significantly decreased $O_2^{\bullet -}$, H_2O_2 and MDA content in Cd-treated barley via: increasing extracellular GSH reduction, bringing root GPX, DHAR, and MDHAR activities down towards control levels, and increasing APX and CAT activities (Chen et al. 2010). External GSH also markedly increased MnSOD, sAPX, and tAPX activities, and up-regulated the expression level of certain APX and CAT isoenzymes, compared with Cd-treated plants. Similar results were also found in rice under copper (Cu) and combined Cd and chromium (Cr) stresses (Cao et al. 2013b; Mostofa et al. 2014a).

Exogenous GSH has the ability to decrease heavy metal uptake and transport, and ameliorate heavy metal-induced damage on root/leaf ultrastructure (Cai et al. 2010, 2011a; Wang et al. 2011; Cao et al. 2013a, b, 2015; Mostofa et al. 2014a). As mentioned above, GSH can be involved in regulating gene and protein expression. Cai et al. (2011b) investigated the effect of external GSH on 2-D based protein profiles under Cd stress in rice and found several proteins which levels were decreased by Cd treatment but increased in GSH + Cd-treated plants. These proteins included aminopeptidase N, clpA/clpB family protein, glycolipid transfer protein-like, and heat shock proteins. Hasan et al. (2016) found that foliar spray of exogenous GSH induced Cd tolerance; this response is related to the up-regulation of several transcription factors, including MYB transcription factors and ethylene-responsive transcription factors.

Based on the above discussion, it can be concluded that the mechanisms by which exogenous GSH alleviates heavy metal toxicity are mainly related to: the scavenging of the induced ROS production by regulating the antioxidant system; converting to PCs, which transport heavy metals into the vacuole; increasing photosynthesis performance; inducing organic acids secretion, which can chelate heavy metals; decreasing heavy metal uptake and transport and maintaining ion homeostasis; and up-regulating the expression of stress response genes.

4.4 Heat Stress

High temperature-induced oxidative and MG stress, and the involvement of GSH in improving heat stress tolerance have been well documented in plants (Mostofa et al. 2014b; Nahar et al. 2015c). Several studies using tolerant and susceptible genotypes have also shown the importance of GSH and its related enzymes in improving heat stress tolerance. Heat-tolerant wheat genotype (C 306) showed high SOD, CAT, APX, GR, and POD activities in response to heat stress at various stages (vegetative, anthesis, and 15 days after anthesis) of plant growth, whereas the activities of CAT, GR, and POX showed a significant decrease in the susceptible genotype (PBW 343). The level of H_2O_2 was also higher in the susceptible genotype due to the imposition of heat stress at various growth stages (Almeselmani et al. 2009). Heat acclimation-induced thermotolerance studies in tall fescue (*Festuca arundinacea*) and perennial rye grass (*Lolium perenee*) also showed the importance of GSH and AsA in improving heat-induced oxidative stress tolerance (Xu et al. 2006). Wang et al. (2014b) showed that heat stress tolerance of wheat (*Triticum aestivum* L.) was associated with high GR, SOD, and POD activities. Recent studies also showed the positive roles of exogenous GSH in modulating heat stress tolerance. Nahar et al. (2015c) reported the intrinsic functions of GSH in conferring short-term heat stress tolerance in mung beans (*Vigna radiata* L.). Heat stress (42 °C, 24–48 h) resulted in a severe oxidative stress and MG stress as indicated by high H_2O_2, MG, $O_2^{\bullet -}$, Pro, MDA content and lipoxygenase (LOX) activity as well as lower chlorophyll and relative water content. The activities of MDHAR, DHAR, GPX, CAT, and Gly I decreased whereas the activities of APX, GR, and GST increased. Importantly, pre-treatment with exogenous GSH led to improved stress tolerance as indicated by lower ROS, MG, and MDA levels and LOX activity. The endogenous level of GSH and GSH/GSSG ratio was higher in GSH-pretreated heat-stressed seedlings. Most of the enzymes of anti-oxidative and glyoxalase systems showed higher activities as compared to heat-stressed seedlings. These findings demonstrated the positive roles of exogenous GSH in improving short-term heat stress tolerance. Recently, Ding et al. (2016b) also showed that the application of exogenous GSH improved heat stress tolerance in cucumber (*Cucumis sativus* L.) seedlings by regulating morphological, physiological, and biochemical parameters. Heat stress resulted in a significant decrease in plant height, shoot growth characteristics, chlorophyll content, and lower photosynthetic rates, whereas increased plant growth, chlorophyll content, and photosynthetic rates were observed in the GSH-treated heat-stressed seedlings. Proline content increased in response to heat stress but a greater increase in Pro content was observed in GSH-treated seedlings. Importantly, heat stress led to severe oxidative stress as indicated by lower GSH content, GSH/GSSG ratio, and higher $O_2^{\bullet -}$ and MDA levels, whereas GSH-treated plants showed lower oxidative damage and higher GSH levels and GSH/GSSG ratio. GSH-treated heat-stressed plants also showed a significant increase in the activities of APX, POD, and GR as compared to heat-stressed seedlings. Heat stress significantly reduces the expression of most of the calvin cycle enzymes whereas a significant increase in these enzymes was observed in GSH-treated heat-stressed plants (Ding et al. 2016b). The above findings clearly demonstrated the multiple functions of GSH in plant growth and heat stress tolerance.

4.5 Cold Stress

Cold stress that includes chilling and/or freezing temperatures adversely affects plant growth and development, with GSH and its related enzymes playing an important role in regulating cold temperature-induced oxidative stress tolerance (Walker and McKersie 1993; Kocsy et al. 2001; Yu et al. 2002, 2003; Kaur et al. 2008; Li et al. 2013; Ao et al. 2013a, b). Chill-tolerant tomato (*L. hirsutum*) showed higher GSH content, GSH/GSSG ratio, and GR activity than the sensitive *L. esculentum* (Walker and McKersie 1993). A strong relationship between tissue

GSH content and chill tolerance has been reported in maize (*Zea mays*) and wheat (*Triticum aestivum* L.) (reviewed in Kocsy et al. 2001). Yu et al. (2002, 2003) reported that cold-acclimation or H_2O_2-induced chill tolerance is associated with higher GSH biosynthesis and antioxidant enzyme activities. Later, Hung et al. (2007) reported in a chill-sensitive mung bean (*Vigna radiata* L.) cultivar that H_2O_2 treatment induced a chill tolerance comparable to cold acclimation, and this response was correlated with increased GSH content. Opposite results were observed if the seedlings were pre-treated with a GSH biosynthetic inhibitor, buthionine sulfoximine (BSO). Cold-tolerant chickpea (*Cicer arietinum* L.) breeding lines showed higher activities of CAT, APX, and GR as compared to sensitive lines (Kaur et al. 2008). Li et al. (2013) showed that cold-priming induced cold tolerance in *Jatropha curcas* L. that was associated with higher activities of APX, GPX, GR, and SOD as well as higher GSH, AsA, GSH, Pro, and glycinebetaine (GB) levels. Consequently, Ao et al. (2013a, b) found that the cold-acclimation induced cold stress tolerance was due to increased AsA and GSH content, higher POD, CAT, SOD, APX, and GR activities, and higher expression levels of Pro and GB biosynthetic genes. The exogenous application of GSH also improved chill stress tolerance. In this context, Lukatkin and Anjum (2014) observed that the application of 100 μM exogenous GSH decreased $O_2^{\bullet-}$ generation, electrolyte leakage, and lipid peroxidation intensity, and improved chill stress tolerance in cucumber (*Cucumis sativus* L.).

4.6 Low or Excessive Nutrient Stress

Like other abiotic stresses, low or excessive amounts of essential nutrients also induce oxidative stress in plants, affecting sustainable agricultural production (Ruiz et al. 2003; Cervilla et al. 2007; Han et al. 2009). GSH and its associated enzymes were also found to play key roles in protecting plants from nutrient deficiency and toxicity stress (Ruiz et al. 2003; Cervilla et al. 2007; Han et al. 2009; Ramírez et al. 2013). The activities of SOD and AsA-GSH cycle enzymes and the levels of GSH and AsA increased in response to boron (B) deficiency stress in citrus (Han et al. 2009). However, the synthesis of GSH was inhibited due to excess B in sunflower (*Helianthus annuus* L.), whereas the application of exogenous GSH or cysteine significantly reduces B toxicity as indicated by a similar foliar biomass to that of control plants (Ruiz et al. 2003). Cervilla et al. (2007) also showed that the higher synthesis of GSH and AsA and the activities of AsA-GSH cycle enzymes play an important role in B-toxicity tolerance in tomato (*Solanum lycopersicum* L.). Iron (Fe) is required for many biological processes in plants, such as photosynthesis, electron transport, and nucleic acid synthesis (Ramírez et al. 2013). Fe deficiency can also induce oxidative stress as indicated by a decreased GSH level and higher accumulation of ROS in leaf and root tissues as well as higher chlorophyll degradation (Ramírez et al. 2013). Exogenous application of GSH alleviated Fe deficiency-induced chlorosis and restricted the over-accumulation of ROS. Additionally, exogenous GSH recovered the activity of APX to control level and preserved the level of ferredoxin2 (Ramírez et al. 2013). Later, Shanmugam et al. (2015) used a GSH-deficient mutant to investigate the mechanism of Fe-deficiency tolerance in *Arabidopsis*. The results showed that the mutant accumulated lower Fe than the wild type because of a lower expression level of Fe uptake-related genes under the Fe-deficiency condition. They also found that the nitric oxide-mediated induction of these genes was dependent of GSH addition in the mutant under the Fe-limited condition (Shanmugam et al. 2015). The results suggested that GSH supplementation can maintain cell redox homeostasis, activate the expression of Fe-uptake related genes, and increase internal Fe availability under Fe deficiency condition.

5 Conclusion and Perspectives

Oxidative and MG stress are a common characteristic of abiotic stresses, and pose a serious threat for normal plant growth and development, restricting full genetic potential to deal with the stress. Importantly, the pathways of GSH biosynthesis, transport, and metabolism have been well established in plants and plenty of research evidences suggest that the redox-state of glutathione is at the hub of the stress-signaling pathways modulating abiotic stress response and tolerance. Moreover, exogenous GSH plays an essential role during abiotic stress tolerance at various stages of plant growth. In summary, under abiotic stress conditions, GSH is mainly involved in: (1) antioxidant defense and ROS signaling, (2) MG detoxification and MG signaling, (3) direct or indirect metal chelation and sequestration, (4) increasing the expression of genes related to abiotic stress tolerance or nutrient transport, (5) xenobiotic detoxification, and (6) protecting cellular structures and reproductive development. However, key questions related to how exogenous GSH is absorbed and transported in different cell compartments, including cytoplasm, chloroplast, and mitochondria, need further investigation. In addition, GSH can induce ABA accumulation under stressful conditions. Nevertheless, the interaction of GSH with other plant hormones (such as jasmonic acid, salicylic acid, and ethylene) and signaling compounds like nitric oxide and Ca^{2+} needs to be elucidated. Additionally, more studies are needed to apply the current knowledge in practical agricultural production and breeding.

Acknowledgements This work was supported by the National Natural Science Foundation of China (31501233) and the China Postdoctoral Science Foundation funded project (2015M570513, 2016T90542). PDV thanks CSIC and the Spanish Ministry of Economy and Competitiveness for his 'Ramon & Cajal' research contract, co-financed by FEDER funds. The last author thankfully acknowledges the postdoctural fellowship from the Japan Society for the Promotion of Science (JSPS).

References

Akram S, Siddiqui MN, Hussain BMN, Bari MAA, Mostofa MG, Hossain MA, Tran LSP (2017) Exogenous glutathione modulates salinity tolerance of soybean [*Glycine max* (L.) Merrill] at reproductive stage. J Plant Growth Regul 36:877–888

Almeselmani M, Deshmukh PS, Sairam RK, Kushwaha SR, Singh TP (2006) Protective role of antioxidant enzymes under high temperature stress. Plant Sci 171:382–388

Almeselmani M, Deshmukh PS, Sairam RK (2009) High temperature stress tolerance in wheat genotypes: role of antioxidant defence enzymes. Acta Agron Hung 57:1–14

Anjum NA, Hasanuzzaman M, Hossain MA, Thangavel P, Roychoudhury A, Rodrigo M, Adam V, Fujita M, Kizek R, Duarte A, Pereira E, Ahmad I (2014) Jacks of metal/metalloid chelation trade in plants- an overview. Front Plant Sci 6:192

Ao PX, Li ZG, Fan DM, Gong M (2013a) Involvement of antioxidant defense system in chill hardening-induced chilling tolerance in *Jatropha curcas* seedlings. Acta Physiol Plant 35:153–160

Ao PX, Li ZG, Fan DM, Gong M (2013b) Involvement of compatible solutes in chill hardening-induced chilling tolerance in *Jatropha curcas* seedlings. Acta Physiol Plant 35:3457–3464

Avery SV (2011) Molecular targets of oxidative stress. Biochem J 434:201–210

Baxter A, Mittler R, Suzuki N (2014) ROS as key players in plant stress signaling. J Exp Bot 65:1229–1240

Biswas MS, Mano J (2015) Lipid peroxide-derived short-chain carbonyls mediate hydrogen peroxide-induced and salt-induced programmed cell death in plants. Plant Physiol 168:885–898

Blum R, Beck A, Korte A, Stengel A, Letzel T, Lendzian K, Grill E (2007) Function of phytochelatin synthase in catabolism of glutathione-conjugates. Plant J 49:740–749

Blum R, Meyer KC, Wunschmann J, Lendzian KJ, Grill E (2010) Cytosolic action of phytochelatin synthase. Plant Physiol 153:159–169

Bourbouloux A, Shahi P, Chakladar A, Delrot S, Bachhawat AK (2000) Hgt1p, a high affinity glutathione transporter from the yeast *Saccharomyces cerevisiae*. J Biol Chem 275:13259–13265

Cai Y, Lin L, Cheng W, Zhang G, Wu F (2010) Genotypic dependent effect of exogenous glutathione on Cd-induced changes in cadmium and mineral uptake and accumulation in rice seedlings (*Oryza sativa*). Plant Soil Environ 56:524–533

Cai Y, Cao F, Cheng W, Zhang G, Wu F (2011a) Modulation of exogenous glutathione in phytochelatins and photosynthetic performance against Cd stress in the two rice genotypes differing in Cd tolerance. Biol Trace Elem Res 143:1159–1173

Cai Y, Cao F, Wei K, Zhang G, Wu F (2011b) Genotypic dependent effect of exogenous glutathione on Cd-induced changes in proteins, ultrastructure and antioxidant defense enzymes in rice seedlings. J Hazard Mater 192:1056–1066

Cairns NG, Pasternak M, Wachter A, Cobbett CS, Meyer AJ (2006) Maturation of arabidopsis seeds is dependent on glutathione biosynthesis within the embryo. Plant Physiol 141:446–455

Cao F, Liu L, Ibrahim W, Cai Y, Wu FB (2013a) Alleviating effects of exogenous glutathione, glycinebetain, brassinosteroids and salicylic acid on cadmium toxicity in rice seedlings (*Oryza Sativa*). Agrotechnol 2:1

Cao F, Wang N, Zhang M, Dai H, Dawood M, Zhang G, Wu F (2013b) Comparative study of alleviating effects of GSH, Se and Zn under combined contamination of cadmium and chromium in rice (*Oryza sativa*). Biometals 26:297–308

Cao F, Chen F, Sun H, Zhang G, Chen ZH, Wu F (2014) Genome-wide transcriptome and functional analysis of two contrasting genotypes reveals key genes for cadmium tolerance in barley. BMC Genomics 15:611

Cao F, Cai Y, Liu L, Zhang M, He X, Zhang G, Wu F (2015) Differences in photosynthesis, yield and grain cadmium accumulation as affected by exogenous cadmium and glutathione in the two rice genotypes. Plant Growth Regul 75:715–723

Cervilla LM, Blasco B, Ríos JJ, Romero L, Ruiz JM (2007) Oxidative stress and antioxidants in tomato (*Solanum lycopersicum*) plants subjected to boron toxicity. Ann Bot 100:747–756

Chaurasia N, Mishra Y, Chatterjee A, Rai R, Yadav S, Rai LC (2016) Overexpression of phytochelatin synthase (PCS) enhances abiotic stress tolerance by altering the proteome of transformed *Anabaena* sp. PCC 7120. Protoplasma 254:1715. https://doi.org/10.1007/s00709-016-1059-7
Chen Z, Gallie DR (2008) Dehydroascorbate reductase affects nonphotochemical quenching and photosynthetic performance. J Biol Chem 283:21347–21361
Chen S, Vaghchhipawala Z, Li W, Asard H, Dickman MB (2004) Tomato phospholipid hydroperoxide glutathione peroxidase inhibits cell death induced by Bax and oxidative stresses in yeast and plants. Plant Physiol 135:1630–1641
Chen F, Wang F, Wu F, Mao W, Zhang G, Zhou M (2010) Modulation of exogenous glutathione in antioxidant defense system against Cd stress in the two barley genotypes differing in Cd tolerance. Plant Physiol Biochem 48:663–672
Chen JH, Jiang HW, Hsieh EJ, Chen HY, Chien CT, Hsieh HL, Lin TP (2012) Drought and salt stress tolerance of an Arabidopsis glutathione S-transferase U17 knockout mutant are attributed to the combined effect of glutathione and abscisic acid. Plant Physiol 158:340–351
Cheng MC, Ko K, Chang WL, Kuo WC, Chen GH, Lin TP (2015) Increased glutathione contributes to stress tolerance and global translational changes in Arabidopsis. Plant J 83:926–939
Chia JC, Yang CC, Sui YT, Lin SY, Juang RH (2013) Tentative identification of the second substrate binding site in Arabidopsis phytochelatin synthase. PLoS One 8:e82675
Choe YH, Kim YS, Kim IS, Bae MJ, Lee EJ, Kim YH, Park HM, Yoon HS (2013) Homologous expression of gamma-glutamylcysteine synthetase increases grain yield and tolerance of transgenic rice plants to environmental stresses. J Plant Physiol 170:610–618
Clemens S, Ma JF (2016) Toxic heavy metal and metalloid accumulation in crop plants and foods. Annu Rev Plant Biol 67:489–512
Cobbett C, Goldsbrough P (2002) Phytochelatins and metallothioneins: roles in heavy metal detoxification and homeostasis. Annu Rev Plant Biol 53:159–182
Cruz de Carvalho MH (2008) Drought stress and reactive oxygen species: production, scavenging and signaling. Plant Signal Behav 3:156–165
Daud MK, Mei L, Azizullah A, Dawood M, Ali I, Mahmood Q, Ullah W, Jamil M, Zhu SJ (2016) Leaf-based physiological, metabolic, and ultrastructural changes in cultivated cotton cultivars under cadmium stress mediated by glutathione. Environ Sci Pollut Res 23:15551–15564
Dazy M, Masfaraud JF, Ferard JF (2009) Induction of oxidative stress biomarkers associated with heavy metal stress in *Fontinalis antipyretica* Hedw. Chemosphere 75:297–302
Del Río LA (2015) ROS and RNS in plant physiology: an overview. J Exp Bot 66:2827–2837
Diao Y, Xu H, Li G, Yu A, Yu X, Hu W, Zheng X, Li S, Wang Y, Hu Z (2014) Cloning a glutathione peroxidase gene from *Nelumbo nucifera* and enhanced salt tolerance by overexpressing in rice. Mol Biol Rep 41:4919–4927
Diaz-Vivancos P, Wolff T, Markovic J, Pallardo FV, Foyer CH (2010) A nuclear glutathione cycle within the cell cycle. Biochem J 431:169–178
Diaz-Vivancos P, de Simone A, Kiddle G, Foyer CH (2015) Glutathione-linking cell proliferation to oxidative stress. Free Radic Biol Med 89:1154–1164
Ding SH, Lu QT, Zhang Y, Yang ZP, Wen XG, Zhang LX, Lu CM (2009) Enhanced sensitivity to oxidative stress in transgenic tobacco plants with decreased glutathione reductase activity leads to a decrease in ascorbate pool and ascorbate redox state. Plant Mol Biol 69:577–592
Ding S, Wang L, Yang Z, Lu Q, Wen X, Lu C (2016a) Decreased glutathione reductase2 leads to early leaf senescence in *Arabidopsis*. J Inter Plant Biol 58:29–47
Ding X, Jiang Y, He L, Zhou Q, Yu J, Hui D, Huang D (2016b) Exogenous glutathione improves high root-zone temperature tolerance by modulating photosynthesis, antioxidant and osmolytes systems in cucumber seedlings. Sci Rep 6:35424
Dixon DP, Edwards R (2010) Glutathione *S*-transferases. The Arabidopsis Book 8:e0131
Edwards EA, Rawsthorne S, Mullineaux PM (1990) Subcellular distribution of multiple forms of glutathione reductase in leaves of pea (*Pisum sativum* L.) Planta 180:278–284
Edwards R, Dixon DP, Walbot V (2000) Plant glutathione S-transferase: enzymes with multiple functions in sickness and in health. Trends Plant Sci 5:193–198

El-Shabrawi H, Kumar B, Kaul T, Reddy MK, Singla-Pareek SL, Sopory SK (2010) Redox homeostasis, antioxidant defense, and methylglyoxal detoxification as markers for salt tolerance in Pokkali rice. Protoplasma 245:85–96
Foyer CH, Theodoulou FL, Delrot S (2001) The functions of inter- and intracellular glutathione transport systems in plants. Trends Plant Sci 6:486–492
Fu JY (2014) Cloning of a new glutathione peroxidase gene from tea plant (*Camellia sinensis*) and expression analysis under biotic and abiotic stresses. Bot Stud 55:7
Gaber A, Ogata T, Maruta T, Yoshimura K, Tamoi M, Shigeoka S (2012) The involvement of Arabidopsis glutathione peroxidase 8 in the suppression of oxidative damage in the nucleus and cytosol. Plant Cell Physiol 53:1596–1606
Gallie DR (2013) The role of L-ascorbic acid recycling in responding to environmental stress and in promoting plant growth. J Exp Bot 64:433–443
Garg B, Jaiswal JP, Misra S, Tripathi BN, Prasad M (2012) A comprehensive study on dehydration-induced antioxidative responses during germination of Indian bread wheat (*Triticum aestivum* L. em Thell) cultivars collected from different agroclimatic zones. Physiol Mol Biol Plant 18:217–228
Ghosh A, Kushwaha HR, Hasan MR, Pareek A, Sopory SK, Singla-Pareek SL (2016) Presence of unique glyoxalase III proteins in plants indicates the existence of shorter route for methylglyoxal detoxification. Sci Rep 6:18358
Gietler M, Nykiel M, Zagdańska BM (2016) Changes in the reduction state of ascorbate and glutathione, protein oxidation and hydrolysis leading to the development of dehydration intolerance in *Triticum aestivum* L. seedlings. Plant Growth Regul 79:287
Gill SS, Anjum NA, Hasanuzzaman M, Gill R, Trivedi DK, Ahmad I, Pereira E, Tuteja N (2013) Glutathione and glutathione reductase: a boon in disguise for plant abiotic stress defense operations. Plant Physiol Biochem 70:204–212
Gorantla M, Babu P, Lachagari VBR, Reddy A, Wusirika R, Bennetzen JL, Reddy AR (2007) Identification of stress-responsive genes in an indica rice (*Oryza sativa* L.) using ESTs generated from drought-stressed seedlings. J Exp Bot 58:253–265
Han S, Tang N, Jiang HX, Yang LT, Li Y, Chen LS (2009) CO_2 assimilation, photosystem II photochemistry, carbohydrate metabolism and antioxidant system of citrus leaves in response to boron stress. Plant Sci 176:143–153
Hasan MK, Liu C, Wang F, Ahammed GJ, Zhou J, Xu MX, Yu JQ, Xia XJ (2016) Glutathione-mediated regulation of nitric oxide, S-nitrosothiol and redox homeostasis confers cadmium tolerance by inducing transcription factors and stress response genes in tomato. Chemosphere 161:536–545
Herbette S, de Labrouhe DT, Drevet JR, Roeckel-Drevet P (2011) Transgenic tomatoes showing higher glutathione peroxydase antioxidant activity are more resistant to an abiotic stress but more susceptible to biotic stresses. Plant Sci 180:548–553
Hoque TS, Uraji M, Tuya A, Nakamura Y, Murata Y (2012a) Methylglyoxal inhibits seed germination and root elongation and up-regulates transcription of stress-responsive genes in ABA-dependent pathway in *Arabidopsis*. Plant Biol 14:854–858
Hoque TS, Uraji M, Ye W, Hossain MA, Nakamura Y, Murata Y (2012b) Methylglyoxal-induced stomatal closure accompanied by peroxidase-mediated ROS production in *Arabidopsis*. J Plant Physiol 169:979–986
Hoque TS, Hossain MA, Mostofa MG, Burritt DJ, Fujita M, Tran LSP (2016) Methylglyoxal: an emerging signaling molecule in plant abiotic stress responses and tolerance. Front Plant Sci 7:1341
Hossain MA, Fujita M (2009) Purification of glyoxalase I from onion bulbs and molecular cloning of its cDNA. Biosci Biotech Biochem 73:2007–2013
Hossain MA, Hossain MZ, Fujita M (2009) Stress-induced changes of methylglyoxal level and glyoxalase I activity in pumpkin seedlings and cDNA cloning of glyoxalase I gene. Aust J Crop Sci 3(2):53–64
Hossain MA, Hasanuzzaman M, Fujita M (2010) Up-regulation of antioxidant and glyoxalase systems by exogenous glycinebetaine and proline in mung bean confer tolerance to cadmium stress. Physiol Mol Biol Plants 16:259–272
Hossain MA, Teixeira da Silva JA, Fujita M (2011) Glyoxalase system and reactive oxygen species detoxification system in plant abiotic stress response and tolerance: an intimate relationship.

In: Shanker A, Venkateswarlu B (eds) Abiotic stress in plants-mechanisms and adaptations. INTECH-Open Access Publisher, Rijeka, pp 235–266

Hossain MA, Piyatida P, Teixeira da Silva JA, Fujita M (2012) Molecular mechanism of heavy metal toxicity and tolerance in plants: central role of glutathione in detoxification of reactive oxygen species and methylglyoxal and in heavy metal chelation. J Bot 2012:872875

Hossain MA, Mostofa MG, Burritt DJ, Fujita M (2014a) Modulation of reactive oxygen species and methylglyoxal detoxification systems by exogenous glycinebetaine and proline improves drought tolerance in mustard (*Brassica juncea L.*). Int J Plant Biol Res 2(2):2014

Hossain MA, Hoque MA, Burritt DJ, Fujita M (2014b) Proline protects plants against abiotic oxidative stress: biochemical and molecular mechanisms. In: Ahmad P (ed) Oxidative damage to plants. Elsevier, Amsterdam, pp 477–522

Hossain MA, Bhattacharjee S, Armin SM, Qian P, Xin W, Li H-Y, Burritt DJ, Fujita M, Tran LSP (2015) Hydrogen peroxide-priming modulates abiotic oxidative stress tolerance: insights from ROS detoxification and scavenging. Front Plant Sci 6:420

Hung SH, Wang CC, Veselinov SV, Alexieva V, Yu CW (2007) Repetition of hydrogen peroxide treatment induced a chilling tolerance comparable to cold acclimation in mung bean. J Amer Soc Hort Sci 132:770–776

Hussain BMN, Akram S, Raffi SA, Burritt DJ, Hossain MA (2016) Exogenous glutathione improves salinity stress tolerance in rice (*Oryza sativa* L.). Plant Gene Trait 7:1–17

Iannelli MA, Pietrini F, Fiore L, Petrilli L, Massacci A (2002) Antioxidant response to cadmium in *Phragmites australis* plants. Plant Physiol Bioch 40:977–982

Ibrahim W, Ahmed IM, Chen X, Wu F (2017) Genotype-dependent alleviation effects of exogenous GSH on salinity stress in cotton is related to improvement in chlorophyll content, photosynthetic performance, and leaf/root ultrastructure. Environ Sci Pollut Res 24:9417–9427

Jamai A, Tommasini R, Martinoia E, Delrot S (1996) Characterization of glutathione uptake in broad bean leaf protoplast. Plant Physiol 111:1145–1152

Kataya ARA, Reumann S (2010) Arabidopsis glutathione reductase 1 is dually targeted to peroxisomes and the cytosol. Plant Signal Behav 5:171–175

Kaur G, Kumar S, Nayyar H, Upadhyaya HD (2008) Cold stress injury during the pod-filling phase in chickpea (*Cicer arietinum* L.). Effects on quantitative and qualitative components of seeds. J Agron Crop Sci 194:457–464

Kaur C, Ghosh A, Pareek A, Sopory SK, Singla-Pareek SL (2014) Glyoxalases and stress tolerance in plants. Biochem Soc Trans 42:485–490

Kim YS, Kim IS, Shin SY, Park TH, Park HM, Kim YH, Lee GS, Kang HG, Lee SH, Yoon HS (2014) Overexpression of dehydroascorbate reductase confers enhanced tolerance to salt stress in rice plants (*Oryza sativa* L. Japonica). J Agron Crop Sci 200:444–456

Kocsy G, Galiba G, Brunold C (2001) Role of glutathione in adaptation and signalling during chilling and cold acclimation in plants. Physiol Plant 113:158–164

Kumar B, Singla-Pareek SL, Sopory SK (2010) Glutathione homeostasis: crucial for abiotic stress tolerance in plants. In: Pareek A, Sopory SK, Bohnert HJ (eds) Abiotic stress adaptation in plants. Springer, Dordrecht, pp 263–282

Kumar S, Asif MH, Chakrabarty D, Tripathi RD, Dubey RS, Trivedi PK (2013) Expression of a rice Lambda class of glutathione S-transferase, *OsGSTL2*, in *Arabidopsis* provides tolerance to heavy metal and other abiotic stresses. J Hazard Mater 248–249:228–237

Lee BD, Hwang S (2015) Tobacco phytochelatin synthase (NtPCS1) plays important roles in cadmium and arsenic tolerance and in early plant development in tobacco. Plant Biotechnol Rep 9:107–114

Li ZG, Yuan LX, Wang QL, Ding ZL, Dong CY (2013) Combined action of antioxidant defense system and osmolytes in chilling shock-induced chilling tolerance in *Jatropha curcas* seedlings. Acta Physiol Plant 35:2127–2136

Li ZG, Duan XQ, Min X, Zhou ZH (2017) Methylglyoxal as a novel signal molecule induces the salt tolerance of wheat by regulating the glyoxalase system, the antioxidant system, and osmolytes. Protoplasma. https://doi.org/10.1007/s00709-017-1094-z

Liedschulte V, Wachter A, Zhigang A, Rausch T (2010) Exploiting plants for glutathione (GSH) production: uncoupling GSH synthesis from cellular controls results in unprecedented GSH accumulation. J Plant Biotechnol 8:807–820

Lo Cicero L, Madesis P, Tsaftaris A, Lo Piero AR (2015) Tobacco plants over-expressing the sweet orange tau glutathione transferases (*CsGSTUs*) acquire tolerance to the diphenyl ether herbicide fluorodifen and to salt and drought stresses. Phytochemistry 116:69–77

Locato V, de Pinto MC, De Gara L (2009) Different involvement of the mitochondiral, plastidal and cytosolic ascorbate-glutathione redox enzymes in heat shock responses. Phyiol Plant 135:296–306

Lukatkin AS, Anjum NA (2014) Control of cucumber (*Cucumis sativus* L.) tolerance to chilling stress-evaluating the role of ascorbic acid and glutathione. Front Environ Sci 2:62

Marquez-Garcia B, Njo M, Beeckman T, Goormachtig S, Foyer CH (2014) A new role for glutathione in the regulation of root architecture linked to strigolactones. Plant Cell Environ 37:488–498

Marrs KA (1996) The functions and regulation of glutathione S-transferases in plants. Annu Rev Plant Biol 47:127–158

Maughan SC, Pasternak M, Cairns N, Kiddle G, Brach T et al (2010) Plant homologs of the *Plasmodium falciparum* chloroquine-resistance transporter, PfCRT, are required for glutathione homeostasis and stress responses. Proc Natl Acad Sci U S A 107:2331–2336

Milla MA, Maurer A, Huete AR, Gustafson JP (2003) Glutathione peroxidase genes in *Arabidopsis* are ubiquitous and regulated by abiotic stresses through diverse signaling pathways. Plant J 36:602–615

Miller G, Suzuki N, Ciftci-Yilmaz S, Mittler R (2010) Reactive oxygen species homeostasis and signalling during drought and salinity stresses. Plant Cell Environ 33:453–467

Mittova V, Tal M, Volokita M, Guy M (2003a) Up-regulation of the leaf mitochondrial and peroxisomal antioxidative systems in response to salt-induced oxidative stress in the wild salt-tolerant tomato species *Lycopersicon pennellii*. Plant Cell Environ 26:845–856

Mittova V, Theodoulou FL, Kiddle G, Gomez L, Volokita M, Tal M, Foyer CH, Guy M (2003b) Coordinate induction of glutathione biosynthesis and glutathione-metabolizing enzymes is correlated with salt tolerance in tomato. FEBS Lett 554:417–421

Mostofa MG, Seraj ZI, Fujita M (2014a) Exogenous sodium nitroprusside and glutathione alleviate copper toxicity by reducing copper uptake and oxidative damage in rice (*Oryza sativa* L.) seedlings. Protoplasma 251:1373–1386

Mostofa MG, Yoshida N, Fujita M (2014b) Spermidine pretreatment enhances heat tolerance in rice seedlings through modulating antioxidative and glyoxalase systems. Plant Growth Regul 73:31–44

Mostofa MG, Hossain MA, Fujita M (2015a) Trehalose pretreatment induces salt tolerance in rice (*Oryza sativa* L.) seedlings: oxidative damage and co-induction of antioxidant defense and glyoxalase systems. Protoplasma 252:461–475

Mostofa MG, Hossain MA, Fujita M, Tran LS (2015b) Physiological and biochemical mechanisms associated with trehalose-induced copper-stress tolerance in rice. Sci Rep 5:11433

Müller M, Zechmann B, Zellnig G (2004) Ultrastructural localization of glutathione in *Cucurbita pepo* plants. Protoplasma 223:213–219

Munné-Bosch S, Queval G, Foyer CH (2013) The impact of global change factors on redox signaling underpinning stress tolerance. Plant Physiol 161:5–19

Nahar K, Hasanuzzaman M, Alam MM, Fujita M (2015a) Roles of exogenous glutathione in antioxidant defense system and methylglyoxal detoxification during salt stress in mung bean. Biol Plant 59:745–756

Nahar K, Hasanuzzaman M, Alam MM, Fujita M (2015b) Glutathione-induced drought stress tolerance in mung bean: coordinated roles of the antioxidant defence and methylglyoxal detoxification systems. AoB Plants 7:plv069

Nahar K, Hasanuzzaman M, Alam MM, Fujita M (2015c) Exogenous glutathione confers high temperature stress tolerance in mung bean (*Vigna radiata* L.) by modulating antioxidant defense and methylglyoxal detoxification system. Environ Exp Bot 112:44–54

Navrot N, Collin V, Gualberto J, Gelhaye E, Hirasawa M, Rey P, Knaff DB, Issakidis E, Jacquot JP, Rouhier N (2006) Plant glutathione peroxidases are functional peroxiredoxins distributed

in several subcellular compartments and regulated during biotic and abiotic stresses. Plant Physiol 142:1364–1379
Noctor G, Veljovic-Jovanovic S, Foyer CH (2000) Peroxide processing in photosynthesis: antioxidant coupling and redox signalling. Philos Trans R Soc Lond Ser B Biol Sci 355:1465–1475
Noctor G, Gomez L, Vanacker H, Foyer CH (2002) Interactions between biosynthesis, compartmentation and transport in the control of glutathione homeostasis and signalling. J Exp Bot 53:1283–1304
Noctor G, Mhamdi A, Chaouch S, Han Y, Neukermans J, Marquez-Garcia B, Queval G, Foyer CH (2012) Glutathione in plants: an integrated overview. Plant Cell Environ 35:454–484
Noshi M, Hatanaka R, Tanabe N, Terai Y, Maruta T, Shigeoka S (2016) Redox regulation of ascorbate and glutathione by a chloroplastic dehydroascorbate reductase is required for high-light stress tolerance in Arabidopsis. Biosci Biotechnol Biochem 80:870–877
Pasternak M, Lim B, Wirtz M, Hell R, Cobbett CS, Meyer AJ (2008) Restricting glutathione biosynthesis to the cytosol is sufficient for normal plant development. Plant J 53:999–1012
Pilon-Smits E (2005) Phytoremediation. Annu Rev Plant Biol 56:15–39
Pomponi M, Censi V, Di Girolamo V, De Paolis A, di Toppi LS, Aromolo R, Costantino P, Cardarelli M (2006) Overexpression of Arabidopsis phytochelatin synthase in tobacco plants enhances Cd^{2+} tolerance and accumulation but not translocation to the shoot. Planta 223:180–190
Qiu B, Zeng F, Cai S, Wu X, Haider SI, Wu F, Zhang G (2013) Alleviation of chromium toxicity in rice seedlings by applying exogenous glutathione. J Plant Physiol 170:772–779
Ramírez L, Bartoli CG, Lamattina L (2013) Glutathione and ascorbic acid protect *Arabidopsis* plants against detrimental effects of iron deficiency. J Exp Bot 64:3169–3178
Rapala-Kozik M, Kowalska E, Ostrowska K (2008) Modulation of thiamine metabolism in *Zea mays* seedlings under conditions of abiotic stress. J Exp Bot 59:4133–4143
Ren J, Sun LN, Zhang QY, Song XS (2016) Drought tolerance is correlated with the activity of antioxidant enzymes in *Cerasus humilis* seedlings. Biomed Res Int 2016:9851095
Ruiz JM, Rivero RM, Romero L (2003) Preliminary studies on the involvement of biosynthesis of cysteine and glutathione in the resistance to boron toxicity in sunflower plants. Plant Sci 165:811–817
Saito R, Yamamoto H, Makino A, Sugimoto T, Miyake C (2011) Methylglyoxal functions as Hill oxidant and stimulates the photoreduction of O_2 at photosystem I: a symptom of plant diabetes. Plant Cell Environ 34:1454–1464
Schnaubelt D, Schulz P, Hannah MA, Yocgo RE, Foyer CH (2013) A phenomics approach to the analysis of the influence of glutathione on leaf area and abiotic stress tolerance in *Arabidopsis thaliana*. Front Plant Sci 4:416
Schneider A, Schatten T, Rennenberg H (1994) Exchange between phloem and xylem during long distance transport of glutathione in spruce trees (*Picea abies*[Karst.] L). J Exp Bot 45:457–462
Selote DS, Khanna-Chopra R (2004) Drought-induced spikelet sterility is associated with an inefficient antioxidant defence in rice panicles. Physiol Plant 121:462–471
Shanmugam V, Wang YW, Tsednee M, Karunakaran K, Yeh KC (2015) Glutathione plays an essential ro in nitric oxide-mediated iron-deficiency signaling and iron-deficiency tolerance in *Arabidopsis*. Plant J 84:464–477
Singla-Pareek SL, Yadav SK, Pareek A, Reddy MK, Sopory SK (2006) Transgenic tobacco overexpressing glyoxalase pathway enzymes grow and set viable seeds in zinc-spiked soils. Plant Physiol 140:613–623
Son JA, Narayanankutty DP, Roh KS (2014) Influence of exogenous application of glutathione on rubisco and rubisco activase in heavy metal-stressed tobacco plant grown in vitro. Saudi J Biol Sci 21:89–97
Soranzo N, Sari Gorla M, Mizzi L, De Toma G, Frova C (2004) Organisation and structural evolution of the rice glutathione S-transferase gene family. Mol Gen Genomics 271:511–521
Srivalli S, Khanna-Chopra R (2008) Role of glutathione in abiotic stress tolerance. In: Khan NA, Singh S, Umar S (eds) Sulfur assimilation and abiotic stress in plants. Springer, Berlin/Heidelberg, pp 207–225
Szalai G, Kellős T, Galiba G, Kocsy G (2009) Glutathione as an antioxidant and regulatory molecule in plants under abiotic stress conditions. J Plant Growth Regul 28:66–80

Tausz M, Sircelj H, Grill D (2004) The glutathione system as a stress marker in plant ecophysiology: is a stress-response concept valid? J Exp Bot 55:1955–1962

Turan Ö, Ekmekçi Y (2011) Activities of photosystem II and antioxidant enzymes in chickpea (*Cicer arietinum* L.) cultivars exposed to chilling temperatures. Acta Physiol Plant 33:67–78

Walker MA, McKersie BD (1993) Role of ascorbate glutathione antioxidant system in chilling resistance of tomato. J Plant Physiol 141:234–239

Wang F, Chen F, Cai Y, Zhang G, Wu F (2011) Modulation of exogenous glutathione in ultrastructure and photosynthetic performance against Cd stress in the two barley genotypes differing in Cd tolerance. Biol Trace Elem Res 144:1275–1288

Wang R, Liu S, Zhou F, Hua C (2014a) Exogenous ascorbic acid and glutathione alleviate oxidative stress induced by salt stress in the chloroplasts of *Oryza sativa* L. Z Naturforsch C 69:226–236

Wang X, Cai J, Liu FL, Dai TB, Cao WX, Wollenweber B, Jiang D (2014b) Multiple heat priming enhances thermo-tolerance to a later high temperature stress via improving subcellular antioxidant activities in wheat seedlings. Plant Physiol Biochem 74:185–192

Xu S, Li J, Zhang X, Wei H, Cui L (2006) Effects of heat acclimation pretreatment on changes of membrane lipid peroxidation, antioxidant metabolites, and ultrastructure of chloroplasts in two cool-season turfgrass species under heat stress. Environ Exp Bot 56:274–285

Yadav SK, Singla-Pareek SL, Ray M, Reddy MK, Sopory SK (2005a) Methylglyoxal levels in plants under salinity stress are dependent on glyoxalase I and glutathione. Biochem Biophys Res Commun 337:61–67

Yadav SK, Singla-Pareek SL, Ray M, Reddy MK, Sopory SK (2005b) Transgenic tobacco plants overexpressing glyoxalase enzymes resist an increase in methylglyoxal and maintain higher reduced glutathione levels under salinity stress. FEBS Lett 579:6265–6271

Yamaguchi T, Blumwald E (2005) Developing salt-tolerant crop plants: challenges and opportunities. Trends Plant Sci 10:615–620

Yazici I, Türkan I, Sekmen AH, Demiral T (2007) Salinity tolerance of purslane (*Portulaca oleracea* L.) is achieved by enhanced antioxidative system, lower level of lipid peroxidation and proline accumulation. Environ Exp Bot 61:49–57

Yin L, Mano J, Tanaka K, Wang S, Zhang M, Deng X, Zhang S (2017) High level of reduced glutathione contributes to detoxification of lipid peroxide-derived reactive carbonyl species in transgenic Arabidopsis overexpressing glutathione reductase under aluminum stress. Physiol Plant DOI. https://doi.org/10.1111/ppl.12583

Yu CW, Murphy TM, Sung WW, Lin CH (2002) H_2O_2 treatment induces glutathione accumulation and chilling tolerance in mung bean. Funct Plant Biol 29:1081–1087

Yu CW, Murphy TM, Lin CH (2003) Hydrogen peroxide-induced chilling tolerance in mung beans mediated through ABA-independent glutathione accumulation. Funct Plant Biol 30:955–963

Zechmann B, Koffler BE, Russell SD (2011) Glutathione synthesis is essential for pollen germination *in vitro*. BMC Plant Biol 11:54

Zhang MY, Bourbouloux A, Cagnac O, Srikanth CV, Rentsch D, Bachhawat AK, Delrot S (2004) A novel family of transporters mediating the transport of glutathione derivatives in plants. Plant Physiol 134:482–491

Zhou Q, Guo JJ, He CT, Shen C, Huang YY, Chen JX, Guo JH, Yuan JG, Yang ZY (2016) Comparative transcriptome analysis between low- and high-cadmium-accumulating genotypes of pakchoi (*Brassica* chinensis L.) in response to cadmium stress. Environ Sci Technol 50:6485–6494

Zhou Y, Wen Z, Zhang J, Chen X, Cui J, Xub W, Liu HY (2017) Exogenous glutathione alleviates salt-induced oxidative stress intomato seedlings by regulating glutathione metabolism, redox status, and the antioxidant system. Sci Hort 220:90–101

Zhu YL, Pilon-Smits EAH, Tarun AS, Weber SU, Jouanin L, Terry N (1999) Cadmium tolerance and accumulation in Indian mustard is enhanced by overexpressing γ-glutamylcysteine synthetase. Plant Physiol 121:1169–1177

Chapter 9
Structure, Evolution and Functional Roles of Plant Glutathione Transferases

Evangelia Chronopoulou, Farid S. Ataya, Fotini Pouliou, Fereniki Perperopoulou, Nikolaos Georgakis, Irini Nianiou-Obeidat, Panagiotis Madesis, Elisavet Ioannou, and Nikolaos E. Labrou

Abstract Plant cytosolic glutathione transferases (EC 2.5.1.18, GSTs) are essential enzymes involved in multiple and diverse functions which are crucial to xenobiotic detoxification, hormone signalling, redox homeostasis, plant metabolism, growth regulation and adaptation to abiotic and biotic stresses. GSTs are capable of catalysing the conjugation of reduced glutathione (γ-L-Glu-L-Cys-Gly; GSH), via the sulphydryl group, to electrophilic centres on a vast number of molecules, both endogenous and xenobiotic, including herbicides, leading to their detoxification. Recent progress of plant proteomics, genomics and transcriptomics projects has allowed the identification, classification and evolutionary analysis of a large number of GST isoenzymes and has provided new knowledge and insights into their *in planta* function and catalytic role. This chapter focuses on plant GSTs and attempts to give an overview of their evolution, catalytic function and structural biology.

Keywords Plant glutathione transferase • Structure and phylogenetic relations • Catalytic functions

E. Chronopoulou • F. Pouliou • F. Perperopoulou • N. Georgakis • E. Ioannou • N.E. Labrou (✉)
Laboratory of Enzyme Technology, Department of Biotechnology, School of Food, Biotechnology and Development, Agricultural University of Athens, 75 Iera Odos Street, GR-11855 Athens, Greece
e-mail: lambrou@aua.gr

F.S. Ataya
Department of Biochemistry, College of Science, King Saud University, P.O. Box 2455, Riyadh 11451, Saudi Arabia

I. Nianiou-Obeidat
Laboratory of Genetics and Plant Breeding, School of Agriculture, Forestry and Natural Environment, Aristotle University of Thessaloniki, P.O. Box 261, Thessaloniki GR-54124, Greece

P. Madesis
Institute of Applied Biosciences, CERTH, 6th km Charilaou-Thermis Road, P.O. Box 361, Thermi GR-57001, Thessaloniki, Greece

M.A. Hossain et al. (eds.), *Glutathione in Plant Growth, Development, and Stress Tolerance*, https://doi.org/10.1007/978-3-319-66682-2_9

Contents

1 Introduction

Glutathione transferases (GSTs, E.C. 2.5.1.18) are phase II detoxifying enzymes, which catalyse the conjugation of tripeptide glutathione (GSH) to electrophilic sites on a wide range of mainly hydrophobic compounds (Chronopoulou et al. 2015; Labrou et al. 2015). The GSH conjugates that are formed by the catalytic action of GSTs are eliminated from the cytoplasm, through vacuolar sequestration and apoplastic deposition in plants (Cummins et al. 2011; Chronopoulou et al. 2015). The membrane transport of GSH conjugates is achieved by specific ATP-binding cassette (ABC) transporters that function as GS-X pumps.

GSTs are expressed in a tissue-specific manner and development stage, ranging from embryogenesis to senescence (Schröder 2001; Lan et al. 2009; Jain et al. 2010; Li et al. 2017). Specific biological roles that have been attributed to GSTs involve plant development and stress response adaptation, which are achieved through catalytic or non-catalytic functions (Labrou et al. 2015). For example, GSTs are expressed in a wide range of biotic and abiotic stresses, such as pathogen attack, xenobiotic compounds, heavy metals and ultraviolet (UV) radiation, salicylic acid, ethylene, nitric oxide, salt, osmosis and heat (Cummins et al. 1997; Roxas et al. 1997; Kampranis et al. 2000; Thom et al. 2002; Cho and Kong 2007; Basantani & Srivastava, 2007; Allocati et al., 2009; George et al. 2010; Rezaei et al., 2013; Csiszar et al., 2014; Chen et al. 2016; Skopelitou et al. 2015; Kao et al. 2016; Skopelitou et al. 2017). Safeners have been also found to induce GST activity (Cummins et al. 2009).

GSTs can be considered as moonlighting proteins (Jeffery 2009; Huberts and van der Klei 2010) because of their capability to get involved in different biological mechanisms (Chronopoulou et al. 2014), regulating a range of cell functions, such as cellular detoxification, stress response and adaptation, cell proliferation, oxidative factor scavenging, redox homeostasis, cell death regulation and secondary metabolite biosynthesis (Cummins et al. 1997; Roxas et al. 1997; Kampranis et al. 2000; Thom et al. 2002; Cho and Kong 2007; George et al. 2010; Chen et al. 2016).

GSTs' catalytic functions include GSH-dependent reactions such as conjugation to electrophilic substrates, hydroperoxidase activity that scavenge reactive organic hydroperoxides, isomerase activity, thioltransferase activity and dehydroascorbate reductase activity (Chronopoulou et al. 2014; Labrou et al. 2015; Axarli et al. 2016;

Skopelitou et al. 2015, 2017). GSTs also perform non-catalytic roles and function as transporters of a wide range of endogenous (i.e. auxins and cytokinins) and exogenous ligands, facilitating their delivery and distribution to specific receptors or compartments (Sheehan et al. 2001; Lu and Atkins 2004; Ginsberg et al. 2009; Dixon et al. 2010).

2 The Phylogeny of Plant GSTs

GSTs constitute an ancient enzyme superfamily that is believed to have evolved from a thioredoxin-like ancestor gene in response to the development of stress conditions. It has been proposed that exon shuffling, gene duplication, alternative splicing, swapping, mutagenesis and probably other unknown mechanisms have led to considerable sequence diversification and therefore functional heterogeneity and evolution of GSTs (Pickett and Lu, 1989; Dixon et al., 2002b; Frova 2003, 2006; Basantani & Srivastava, 2007; Allocati et al., 2009; Lan et al. 2009). The phylogeny of plant GSTs is quite complex, as there are three different superfamilies: cytosolic, mitochondrial and microsomal MAPEG (Pearson 2005). Recent plant genome analyses showed that the plant GSTs can be divided in 14 distinct classes, namely, tau (U), phi (F), theta (T), zeta (Z), lambda (L), γ-subunit of the eukaryotic translation elongation factor 1B (EF1Bγ), dehydroascorbate reductase (DHAR), metaxin, tetrachloro-hydroquinone dehalogenase (TCHQD), Ure2p and microsomal prostaglandin E synthase type 2 (mPGES-2) (Liu et al. 2013; Lallement et al. 2014a, b; Mashiyama et al., 2014; Jia et al. 2016). Recently, three new classes have been identified in plants: hemerythrin (GSTH), iota (GSTI) and glutathionyl-hydroquinone reductases (GHRs) (Yang et al. 2014). In Fig. 9.1, a phylogenetic tree of GST classes found in plants is shown. The sequences used for the construction of phylogenetic relationships were selected from model plant organisms including a gymnosperm (*Pinus tabulaeformis*), several angiosperms (*Arabidopsis thaliana, Glycine max, Phaseolus vulgaris, Triticum sativa, Oryza sativa, Setaria italica*), a lycophyte (*Selaginella moellendorffii*) and a moss (*Physcomitrella patens*).

Tau and phi GSTs are the largest group in plants. For instance, 42 of the 55 GSTs in *Arabidopsis thaliana* are classified as tau and phi. Phi class is often presented in the literature as plant-specific; however, basidiomycetes also possess GSTFs (Morel et al. 2013). GSTH, GSTI and Ure2p are only limited to non-vascular plants. GHRs, which have been found in bacteria, haloarchaea and fungi, constitute a newly described GST class that catalyse the reduction of glutathionyl-quinones. Among the plant GSTs, the sequences of the theta and the zeta classes have high degree of similarity (Dixon et al., 2002b, Edwards and Dixon, 2005; Mohsenzadeh et al. 2011). The lambda class of plant GSTs has no detectable GSH-conjugating activity towards the common xenobiotic substrates; thus, their functions are yet obscure (Chan and Lam 2014). Another class of plant GSTs that carry out different functions is the DHARs. DHARs function presumably as GSH-dependent reductases (Dixon and Edwards 2010a). The number of DHAR genes that can be found in plant species

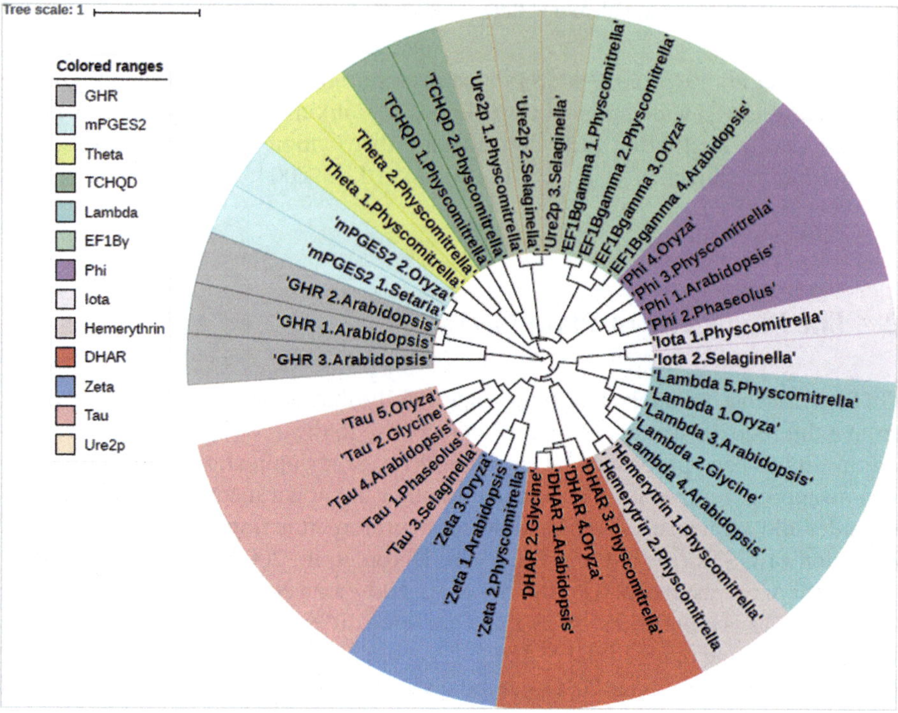

Fig. 9.1 Phylogenetic analysis of plant GSTs. The sequences used are those identified in *Arabidopsis thaliana*, *Oryza sativa*, *Phaseolus vulgaris*, *Glycine max*, *Physcomitrella patens* and *Selaginella moellendorffii* (Lan et al. 2009; Liu et al. 2013; Chronopoulou et al. 2014; Lallement et al. 2014a). Sequences were aligned with CLUSTAL Omega (Sievers et al. 2011), and phylogenetic tree was constructed with Geneious 9.1.2 (http://www.geneious.com, Kearse et al. 2012) with UPGMA tree-building method. Various classes can be distinguished: phi (GSTF), tau (GSTU), lambda (GSTL), theta (GSTT), dehydroascorbate reductase (DHAR), elongation factor 1Bγ (EF1Bγ), glutathionyl-hydroquinone reductase (GHR), hemerythrin (GSTH), iota (GSTI), zeta (GSTZ), microsomal prostaglandin E synthase type 2 (mPGES-2), tetrachloro-hydroquinone dehalogenase (TCHQD) and Ure2p. The accession numbers of proteins that were used for this phylogenetic tree are the following: Phi_1. *Arabidopsis thaliana* (CAA72973.1), Phi_2. *Phaseolus vulgaris* (AEX37999.1), Phi_3. *Physcomitrella patens* (AFZ39127.1), Phi_4. *Oryza sativa* (ABF93846.1). Tau_1. *Phaseolus vulgaris* (AEX38000.1), Tau_2. *Glycine max* (NP_001236312.1), Tau_3. *Selaginella moellendorffii* (EFJ09791.1), Tau_4. *Arabidopsis thaliana* (AAG52384.1), Tau_5. *Oryza sativa* (AAQ02687.1), Lambda_1.*Oryza sativa* (AAF70831.1), Lambda_2. *Glycine max* (AJE59635.1), Lambda_3. *Arabidopsis thaliana* (NP_191064.1), Lambda_4. *Arabidopsis thaliana* (NP_001119157.1), Lambda_5. *Physcomitrella patens* (AFZ39126.1), Theta_1. *Physcomitrella patens* (AFZ39142.1), Theta_2. *Physcomitrella patens* (AFZ39143.1), DHAR_1. *Arabidopsis thaliana* (AAF98403.1), DHAR_2. *Glycine max* (AJE59632.1), DHAR_3. *Physcomitrella patens* (AFZ39123.1), DHAR_4. *Oryza sativa* (AAL71856.1), EF1Bgamma_1. *Physcomitrella patens* (AFZ39147.1), EF1Bgamma_2. *Physcomitrella patens* (AFZ39148.1), EF1Bgamma_2. *Arabidopsis thaliana* (BAH56923.1), EF1Bgamma_1. *Oryza sativa* (LOC_Os02g12794), GHR_1. *Arabidopsis thaliana* (NP_199315), GHR_2. *Arabidopsis thaliana* (NP_001031671.1), GHR_3. *Arabidopsis thaliana* (NP_568632.1), Hemethrin_1. *Physcomitrella patens* (AFZ39150.1), Hemethrin_2. *Physcomitrella* patens (AFZ39151.1), Iota_1. *Physcomitrella patens* (AFZ39144.1), Iota_2. *Selaginella moellendorffii* (XP_002968645.1), Zeta_1. *Arabidopsis thaliana* (AAO60039.1), Zeta_2. *Physcomitrella patens* (AFZ39146.1), Zeta_3. *Oryza sativa* (ABA96700.2), mPGES2_1. *Setaria italica* (XP_004969028.1), mPGES2_2. *Oryza sativa* (CAH67930.1), TCHQD_1. *Physcomitrella patens* (AFZ39137.1), TCHQD_2. *Physcomitrella patens* (AFZ39138.1), Ure2p_1. *Physcomitrella patens* (AFZ39145.1), Ure2p_2. *Selaginella moellendorffii* (EFJ21054.1), Ure2p_3. *Selaginella moellendorffii* (EFJ21057.1)

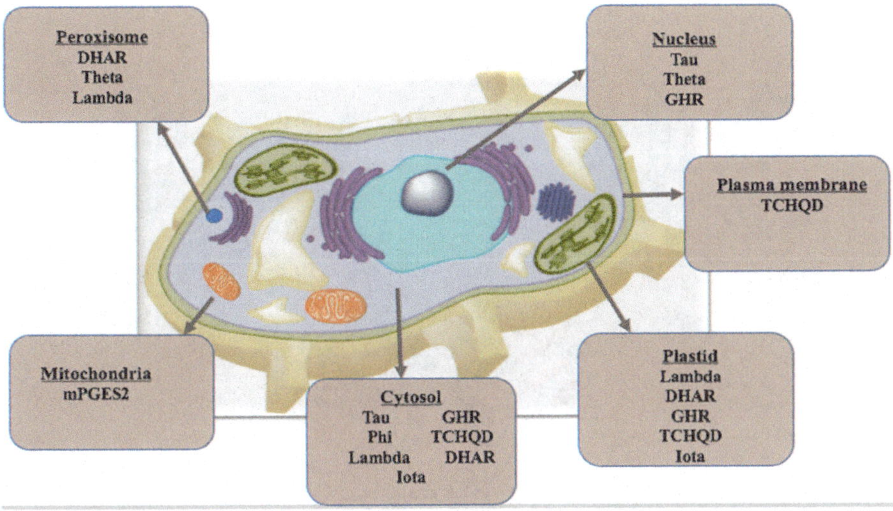

Fig. 9.2 Subcellular localization of plant GSTs that belong to different classes (Dixon et al. 2009; Liu et al. 2013; Lallement et al. 2014a)

is usually between two and three, while there are genes coding for chloroplastic enzymes (CPFC-active site motif) and for enzymes that are presumably cytosolic (CPFS-active site motif) (Lallement et al. 2014a). DHAR-GSTs have also essential role in GSH–ascorbic acid cycle. GSTs of the TCHQD class catalyse a reductive dehalogenation reaction. The sequence similarity of this class to the other GSTs ranges between 26% and 30%, with the greater similarity found in the N-terminal region (Arora and Bae 2014).

The knowledge of subcellular localization of GSTs is an important tool for establishing their biological function in plant cell. Most plant GSTs are cytosolic; however nuclear or apoplastic localization has been documented (Dixon et al. 1998; Labudda and Azam 2014). In addition, GSTL and DHAR genes have been reported to be expressed in peroxisomes and plastids (Dixon et al. 2009; Lallement et al. 2014a) (Fig. 9.2).

3 The Structure of Plant GSTs

The crystal structures of over 300 soluble GSTs from different prokaryotic and eukaryotic organisms have been resolved by X-ray crystallography. Despite the fact that there is a high diversity at sequence level, their three-dimensional structures are remarkably homologue. Among them, the structures of several members of plant GSTs (e.g. tau, phi, zeta, lambda and GHR classes) have been characterized in detail (Fig. 9.3). Examples include the phi class GST from *A. thaliana* (Reinemer et al. 1996; Thom et al. 2001); the tau class GSTs from wheat (Thom et al. 2002),

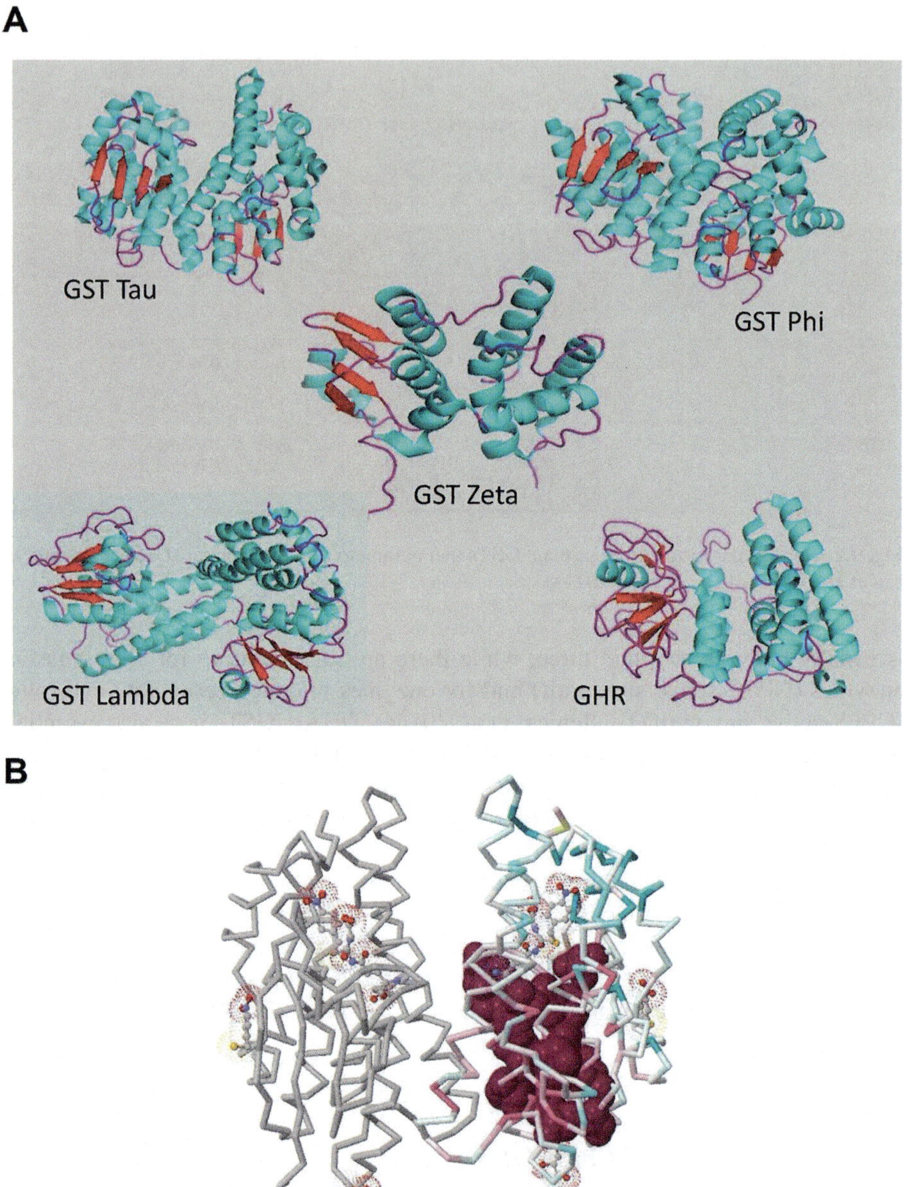

Fig. 9.3 (**a**) Structural representations of selected members of plant GSTs from tau (PDB code 2VO4), phi (PDB code 4RI6), zeta (PDB code 1E6B), lambda (PDB code 4PQH) and GHR (PDB code 4USS) classes. The GSTs are represented as cartoons with β-strands coloured red and α-helices coloured turquoise. The figures were produced using PyMOL (The PyMOL Molecular Graphics System, Version 1.8 Schrödinger, LLC). (**b**) ConSurf-HSSP mapping (http://bental.tau.ac.il/new_ConSurfDB/) of high evolutionary conservation, projected onto the van der Waals surface of PDB entry 2VO4

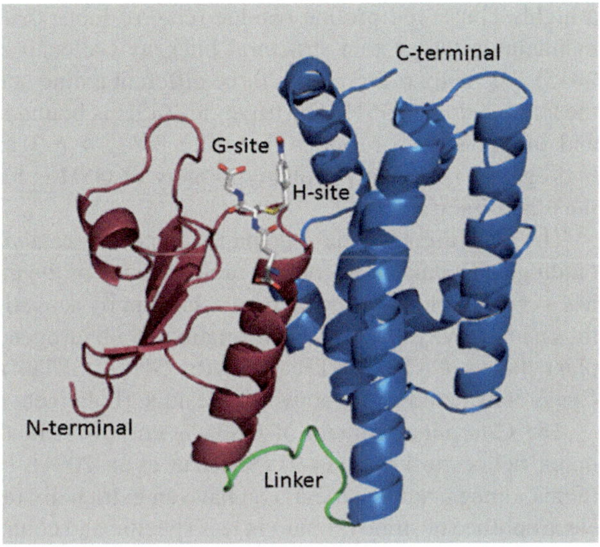

Fig. 9.4 Subunit structure of the tau class *Gm*GSTU4-4 (PDB code 2VO4). Secondary structure elements and the location of G- and H-sites are labelled. The bound inhibitor S-(p-nitrobenzyl)-glutathione (Nb-GSH) is shown in a stick representation. The figures were produced using PyMOL (The PyMOL Molecular Graphics System, Version 1.8 Schrödinger, LLC)

rice (Dixon et al. 1998, 2003), maize (Neuefeind et al. 1997a, b) and *Glycine max* (Axarli et al. 2009a, b; Skopelitou et al. 2015; Axarli et al. 2016); and the zeta class GST from *Arabidopsis* (Thom et al. 2001; Polekhina et al., 2001). Recently, the structure of a GST isoenzyme from *Populus trichocarpa* that belongs to GHR class (Lallement et al. 2015) and the structures of three isoenzymes that belong to GSTL class were characterized (Lallement et al. 2014b).

GSTs function as dimers, composed of two identical or different subunits of 23–30 kDa with an average length of 200–250 amino acids (Axarli et al. 2009a; Mohsenzadeh et al. 2011; Board and Menon 2013; Chronopoulou et al. 2014; Bathige et al. 2014; Labrou et al. 2015; Axarli et al. 2016; Skopelitou et al. 2015, 2017) (Fig. 9.3). However, the GSTL and DHAR enzymes are active as monomers (Fig. 9.3) (Kumar et al. 2013; Lallement et al. 2014a). The existence of a large inter-subunit cleft of a varying size is a typical structural feature of GSTs. Each subunit of dimeric GSTs has at least two ligand-binding sites: G-site and H-site. The G-site is specific for GSH and formed by a conserved group of amino acid residues at the N-terminal domain of the polypeptide (Lee et al., 1995; Board and Menon 2013; Lan et al., 2013) (Fig. 9.4). This domain has a thioredoxin-like fold that is formed by α/β structure (Axarli et al. 2009a; Cummins et al. 2011; Shaokui et al. 2012; Board and Menon 2013; Cho et al. 2015; Labrou et al. 2015; Jiang et al. 2016). This domain consists of the β-sheet sandwiched between α-helices (βαβαββα structural motif) (Sheehan et al. 2001; Parbhoo 2013). The former begins with an N-terminal β-strand (β-1), followed by an α-helix (α-1) and then a second β-strand (β-2) which is parallel to β-1. A loop region leads into a second α-helix (α-2), which is connected to the C-terminal motif (Fig. 9.4.). This motif consists of two sequential β-strands (β-3 and β-4), which are antiparallel and are followed by a third α-helix (α-3) at the C-terminus of the fold. The loop that connects α-2–β-3 features possesses a highly conserved proline residue (cis-Pro loop) that contributes significantly in maintaining the protein structural integrity (Allocati et al. 1999; Lallement et al. 2015). The helix α-2 provides three different amino acid residues that interact with the Gly residue of GSH. The tripeptide GSH is bound in an extended conformation and interacts with residues from the β-3–β-4–α-3 motif through a network of hydrogen bonds. The γ-glutamyl moiety of GSH is found pointing down towards the dimer interface.

The N-terminal domain contains a specific catalytic residue critical for GSH binding and catalysis. A conserved Ser residue of the theta, zeta, phi and tau classes has a crucial role in the catalytic mechanism by activating the GSH thiol ionization, thus its reactivity, through the formation of a hydrogen bond. However, in the other plant classes, GSTIs, GSTHs, DHARs, GHRs, GSTLs and mPGES-2s, a conserved Cys is considered as the active site residue (Lallement et al. 2014a).

The C-terminal domain of GSTs is entirely helical and formed by five or six major helices (α4-8 or α4-9) (Sheehan et al. 2001). However, some GST classes (alpha, omega, tau and theta) can have an extra helix (α-9). The binding site for the electrophilic substrate (H-site) is less specific and contributes to the ability of GSTs to display wide specificity towards electrophile substrates (Mannervik and Danielson 1988; Rushmore and Pickett 1993; Dirr et al. 1994; Armstrong 1997; Sheehan et al. 2001; Kapoli et al. 2008; Axarli et al., 2009a; Mohsenzadeh et al. 2011; Board and Menon 2013; Jacquot et al. 2013; Karpusas et al. 2013; Chronopoulou et al. 2014; Liu et al. 2015; Skopelitou et al. 2015).

Conformational changes and structural rearrangements of the active site upon ligand-binding are common features of many GST isoenzymes. Both the G- and H-sites can change conformations to accommodate the GSH and the electrophile substrate, giving rise to an induced-fit mechanism. Numerous biochemical and structural studies have confirmed the existence of this mechanism in maize GSTF1-1 and soybean GSTU4-4 (Neuefeind et al. 1997a; Axarli et al. 2009a, b).

4 GSH-Dependent Catalytic Functions

The sulphur atom of GSH is considered as a soft nucleophile and as a consequence prefers soft electrophile to form conjugates. Its nucleophilic character allows electron transfer to electrophilic centres, and in most cases, the thiolate anion $-S^-$ and the electrophile acceptor molecule share an electron pair (Ivarsson and Mannervik 2007). Therefore, the GSH/electrophile conjugative reactions can be grouped as substitution or addition reactions (Fig. 9.5) (Dixon et al. 1998; Chronopoulou et al. 2015). In substitution reactions, replacement of the leaving group of the electrophilic substrate is achieved by the $-S^-$ anion (Ivarsson and Mannervik 2007), such as with diphenylether herbicides (e.g. fluorodifen, fenoxaprop-ethyl), where the nitrobenzyl group is released (Schröder 2001). In addition reactions, the $-S^-$ anion reacts with an electrophile carbon–carbon double bond, an oxirane ring or similar structures usually neighboured by an electron-withdrawing group (Talalay et al.

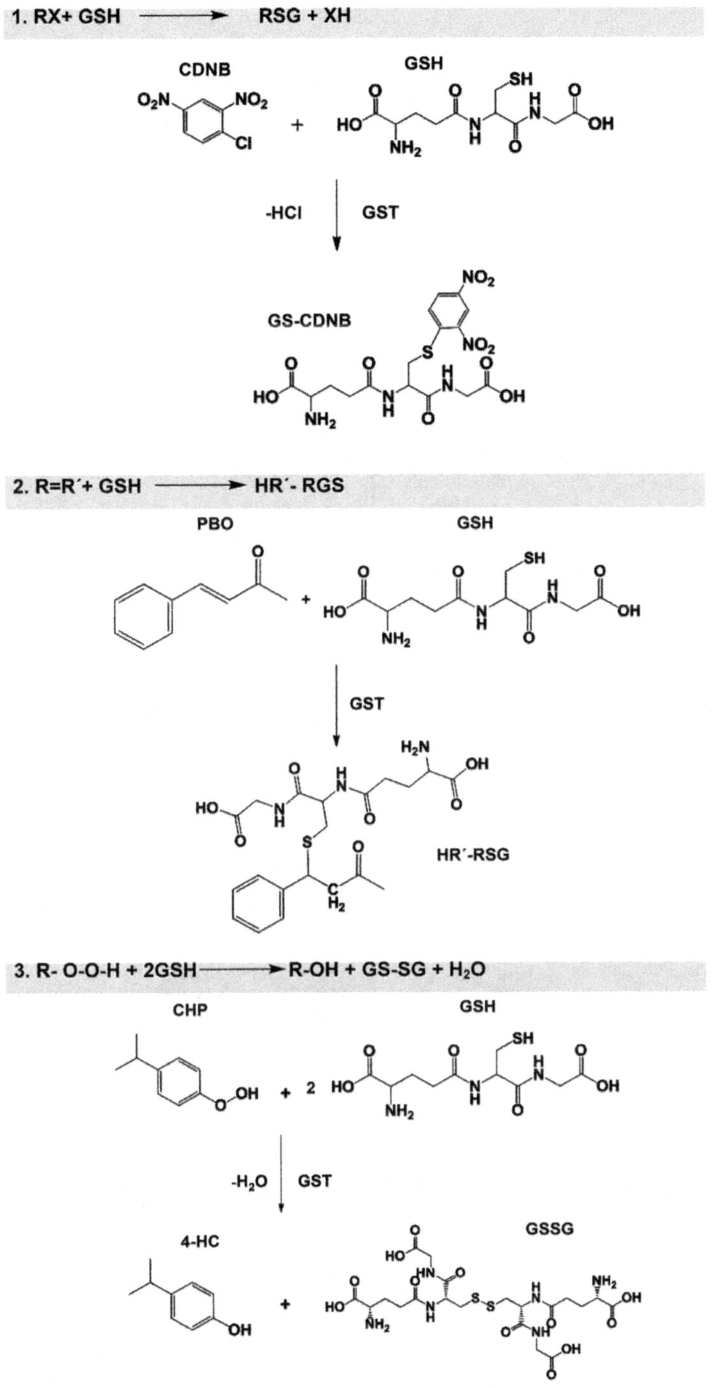

Fig. 9.5 Typical GST-catalysed reactions: (*1*) nucleophilic aromatic substitution with 1-chloro-2,4-dinitrobenzene (CDNB), (*2*) addition reaction with *trans*-4-phenyl-3-buten-2-one (PBO), (*3*) reduction of cumene hydroperoxide, (*4*) Michael addition reaction with allyl-isothiocyanate, (*5*) double-bond transition, (*6*) reduction reaction of oxidized thiol proteins, (*7*) dehydroascorbate reduction reaction, (*8*) dithiothreitol reduction reaction, (*9*) GS-quinone reduction. Reactions created with ChemBioDraw Ultra 13 (PerkinElmer)

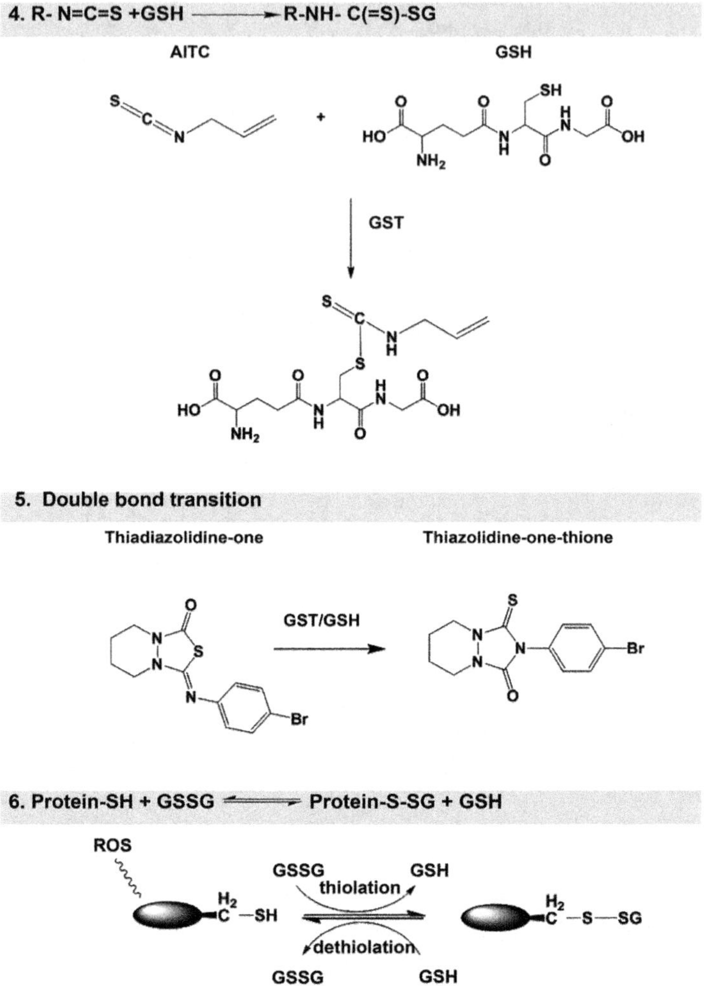

Fig. 9.5. (continued)

1988; Ivarsson and Mannervik 2007), like 2,4,6-trinitrotoluene (TNT), a global military environmental pollutant, which can be detoxified from the field through its denitration from GSTs (Gunning et al. 2014; Tzafestas et al. 2016).

GSTs have significant role in herbicide detoxification and selectivity. Phi and tau classes of GSTs are able to conjugate GSH to a number of electrophilic herbicides like triazines, thiocarbamates, chloroacetanilides, sulphonylureas and diphenyl ethers (Cummins et al. 1997; Edwards and Dixon 2000; Axarli et al. 2009a; Chronopoulou et al. 2012; Axarli et al. 2016). A majority of herbicides are detoxified through substitution reactions and, more rarely, with GSH addition reactions (Cummins et al. 2011). GST isomerase activity has been reported to function in

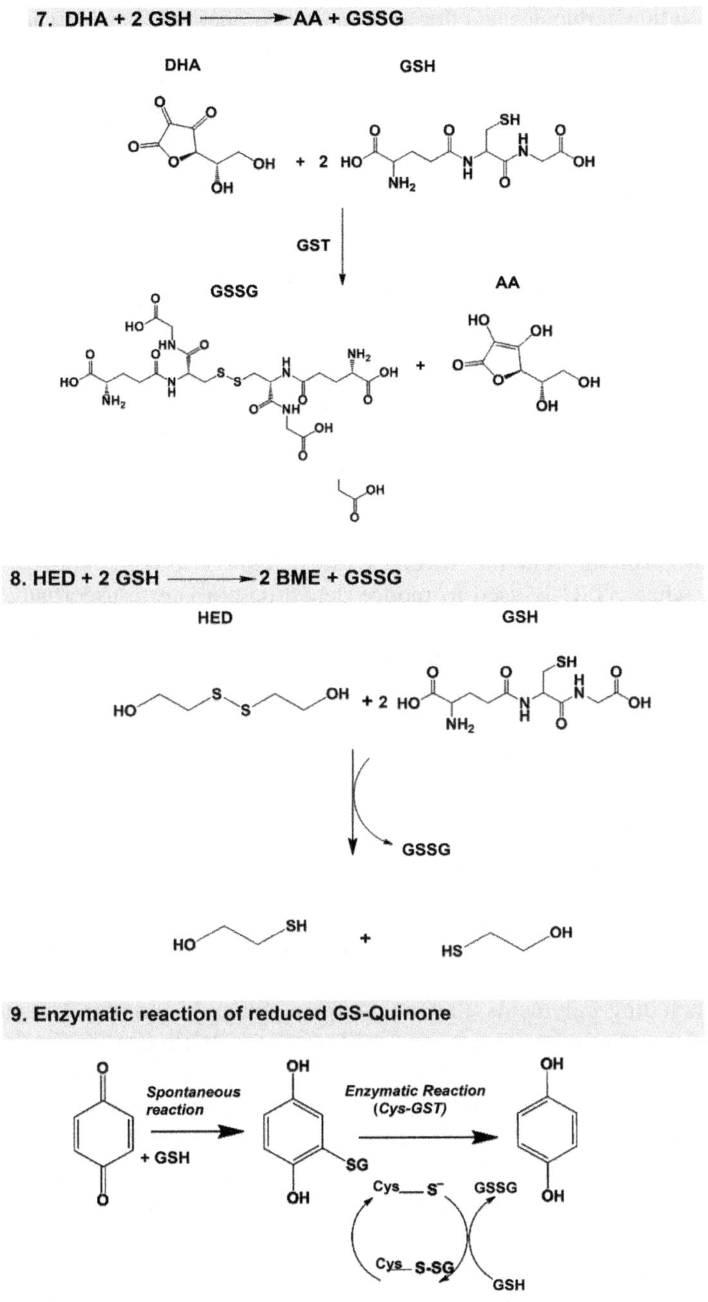

Fig. 9.5. (continued)

thiadiazolidine herbicides and fluthiacet-methyl proherbicides which are isomerized to triazolidine and urazol herbicides, respectively (Shimizu et al. 1995; Edwards et al. 2000; Edwards and Dixon 2000). Another GST isomerase activity in plants has been reported for the zeta class enzyme from *Arabidopsis* (*At*GSTZ1) that displays maleylacetone isomerase activity in tyrosine catabolism pathway (Dixon et al. 2000).

Plant GSTs also possess hydroperoxide scavenging activity (Dixon et al. 1998; Edwards et al. 2000; Axarli et al. 2009a; Chronopoulou et al. 2012; Chronopoulou et al. 2015). They catalyse the reduction of hydroperoxides to alcohols, which are less toxic. Hydroperoxides are formed by oxidative injury of membrane fatty acids (Cummins et al. 2011; Chronopoulou et al. 2015). Plant members of the theta, phi and tau classes exhibit high glutathione hydroperoxidase activity, contributing to lipid hydroperoxide detoxification (Dixon et al. 2009; Cummins et al. 2011; Chronopoulou et al. 2015). GSTs that belong to GHR class function as glutathionyl-hydroquinone reductases (GS-HQR) and catalyse the reduction of GS-trichloro-p-hydroquinone to trichloro-p-hydroquinone (Belchik and Xun 2011; Lam et al. 2012).

Another similar function involves the dehydroascorbate reductase (DHAR) activity, where GSH is used to reduce dehydroascorbate to ascorbate, producing GSH disulphide as by-product (Dixon et al. 2002a).The thioltransferase activity of GSTs protects the reduced thiol groups in proteins and regulates their function through a reverse S-thiolation/deglutathionylation process. GSTLs and DHARs from *Arabidopsis* were found to exhibit GSH-dependent thioltransferase activities with model substrates (Dixon et al. 2002a).

5 Non-catalytic Functions of GSTs

In addition to catalytic function, GSTs display a non-catalytic function (known as ligandin function) and are able to bind tightly a wide range of hydrophobic substrates including fatty acids, flavonoids (quercetin and kaempferol), anthocyanins, hormones (auxin and cytokinin), phytoalexins, porphyrinogens and other metabolites (Zettl et al. 1994; Gonneau et al. 1998; Smith et al. 2003; Dixon et al. 2008, 2010; Dixon and Edwards 2009, 2010b). The binding of such compounds is achieved in the ligandin-binding site (L-site) that is located in a distinct region or overlaps the G- and H-sites. The facilitation of metabolite transport between cellular compartments or their storage is presumably the central roles of ligandin function (Cummins et al. 2011; Lallement et al. 2014a).

Recently, the isoenzyme *At*GSTF2 from *A. thaliana* was reported to bind camalexin, flavonoids (quercetin, quercetin-3-O-rhamnoside and kaempferol) and other heterocyclic compounds such as harmane, norharmane, indole-3-aldehyde and lumichrome (Dixon et al. 2011). *At*GSTF2 does not exhibit GSH conjugation activity with these compounds, indicating that *At*GSTF2 plays a role as a carrier protein. In addition, the binding of these molecules to *At*GSTF2 reveals that either did not alter *At*GSTF2 catalytic activity or even increased it, suggesting the presence of multiple ligand/substrate-binding sites.

6 The Role of GSTs in Natural Product Metabolism

The involvement of GSTs in plant secondary metabolism is another key role that has been recently assigned (Cummins et al. 2011). Although there are not enough biochemical knowledge on the involvement of GSTs in plant secondary metabolism, some roles that have been proposed include (1) reversible S-glutathionylation to facilitate metabolite transport; (2) S-glutathionylation for the addition of sulphur into specific metabolites, such as glucosinolates; (3) cotransport of metabolites and GSH (e.g. anthocyanins); and (4) transient conjugation as in the case of isomerization reactions (Zettl et al. 1994; Gonneau et al. 1998; Smith et al. 2003; Kitamura et al., 2004; Dixon et al. 2008, 2010; Dixon and Edwards 2009, 2010b;Chronopoulou et al., 2015; Labrou et al. 2015).

Recent reported examples showed the involvement of GSTs in the metabolism of oxophytodienoic acid, an electrophilic intermediate in jasmonate synthesis, which is subjected to reversible addition reactions by certain phi and tau class GSTs (Dueckersho et al. 2008; Dixon and Edwards 2009). These reactions may provide protection to the oxophytodienoic acid during transport between cell compartments (Cummins et al. 2011). Another example includes the metabolic processing of isothiocyanates that are considered as potential natural substrates for GSTs. Isothiocyanates, which are produced from glucosinolates, are very good substrates for GSTs from different classes (e.g. phi, tau) (Meyer et al. 1995; Wagner et al. 2002; Chronopoulou et al. 2012, 2014). In this case, GSTs catalyse the addition of the –SH group of GSH to the electrophilic central carbon of the isothiocyanate group to form dithiocarbamates [R-NH-C(=S)-SG] (Fahey et al. 2001; Ivarsson and Mannervik 2007). Lambda GSTs from wheat and *A. thaliana* are able to catalyse the deglutathionylation of flavonol conjugates spontaneously (Dixon and Edwards 2010b). The isoenzyme *At*GSTF6-6 from *A. thaliana* catalyses the reaction of GSH with indole-3-acetonitrile (Su et al. 2011), leading to the synthesis of the defence compound camalexin. *At*GSTF2-2 from *A. thaliana* has been reported to bind with high-affinity camalexin that might be involved in its transport (Dixon et al. 2011). On the other hand, *A. thaliana At*GSTF8-8 conjugates GSH with two stress signalling molecules: prostaglandin 12-oxophytodienoic acids and A1-phytoprostanes (Mueller et al. 2008). In addition, the plant defence secondary metabolite coumarin was found to silence GST gene of *Rhynchophorus ferrugineus* larvae, which reinforces new aspects for pesticide development (AlJabr et al. 2017).

In conclusion, this chapter summarizes information into plant GSTs' functional, structural and catalytic diversity. GSTs are multifunctional enzymes, involved in different biological mechanisms such as cellular detoxification, stress response, cell proliferation, oxidative factor scavenging, redox homeostasis, cell death regulation and secondary metabolite biosynthesis.

Acknowledgements FSA and NEL extend their appreciation to the International Scientific Partnership Program (ISPP) at King Saud University for funding this research work through ISPP# 0071.

References

AlJabr AM, Hussain A, Rizwan-Ul-Haq M, Al-Ayedh H (2017) Toxicity of plant secondary metabolites modulating detoxification genes expression for natural red palm weevil pesticide development. Molecules 22:169

Allocati N, Casalone E, Masulli M, Ceccarelli I, Carletti E, Parker MW, Di Ilio C (1999) Functional analysis of the evolutionarily conserved proline 53 residues in *Proteus mirabilis* glutathione S-transferase B1-1. FEBS Lett 445:347–350

Allocati N, Federici L, Masulli M, Di Ilio C (2009) Glutathione transferases in bacteria. FEBS J 276:58–75

Armstrong RN (1997) Structure, catalytic mechanism, and evolution of the glutathione transferases. A useful review of GSTs from an evolutionary perspective. Chem Res Toxicol 10:2–18

Arora PK, Bae H (2014) Role of dehalogenases in aerobic bacterial degradation of chlorinated aromatic compounds. J Chem 2014:1–10

Axarli I, Dhavala P, Papageorgiou AC, Labrou NE (2009a) Crystallographic and functional characterization of the fluorodifen-inducible glutathione transferase from *Glycine max* reveals an active site topography suited for diphenylether herbicides and a novel L-site. J Mol Biol 385:984–1002

Axarli I, Dhavala P, Papageorgiou AC, Labrou NE (2009b) Crystal structure of Glycine max glutathione transferase in complex with glutathione: investigation of the mechanism operating by the tau class glutathione transferases. Biochem J 422:247–256

Axarli I, Muleta AW, Vlachakis D, Kossida S, Kotzia G, Maltezos A, Dhavala P, Papageorgiou AC, Labrou NE (2016) Directed evolution of tau class glutathione transferases reveals a site that regulates catalytic efficiency and masks co-operativity. Biochem J 473:559–570

Basantani M, Srivastava A (2007) Plant glutathione transferases: a decade falls short. Can J Bot 85:443–456

Bathige SDNK, Umasuthan N, Revathy KS, Lee Y, Kim S, Cho MY, Park M-A, Whang I, Lee J (2014) A mu class glutathione S-transferase from manila clam *Ruditapes philippinarum* (RpGSTμ): cloning, mRNA expression, and conjugation assays. Comp Biochem Physiol C Toxicol Pharmacol 162:85–95

Belchik M, Xun L (2011) S-glutathionyl-(chloro)hydroquinone reductases: a new class of glutathione transferases functioning as oxidoreductases. Drug Metab Rev 43:307–316

Board P, Menon D (2013) Glutathione transferases, regulators of cellular metabolism and physiology. Biochemica et Biophysica Acta 1830:3267–3288

Chan C, Lam H-M (2014) A putative lambda class glutathione S-transferase enhances plant survival under salinity stress. Plant Cell Physiol 55:570–579

Chen S, Chen X, Dou W, Wang L, Yin H, Guo S (2016) GST(phi) gene from macrophyte *Lemna minor* is involved in cadmium exposure responses. Chinese J Oceanol and Limnology 34:342–353

Cho HY, Kong KH (2007) Study on the biochemical characterization of herbicide detoxification enzyme, glutathione S-transferase. Biofactors 30:281–287

Cho HY, Maeng SJ, Cho HJ, Choi YS, Chung JM, Lee S, Kim HK, Kim JH, Eom C-Y, Kim Y-G, Guo M, Jung HS, Kang BS, Kim S (2015) Assembly of multi-tRNA synthetase complex via heterotetrameric glutathione transferase-homology domains. J Biol Chem 290:29313–29328

Chronopoulou E, Madesis P, Asimakopoulou B, Platis D, Tsaftaris A, Labrou NE (2012) Catalytical and structural diversity of the fluazifop-inducible glutathione transferases from *Phaseolus vulgaris*. Planta 235:1253–1269

Chronopoulou E, Madesis P, Tsaftaris A, Labrou NE (2014) Cloning and characterization of a biotic-stress-inducible glutathione transferase from *Phaseolus vulgaris*. Appl Biochem Biotechnol 172:595–609

Chronopoulou E, Kontouri K, Chantzikonstantinou M, Pouliou F, Perperopoulou F, Voulgari G, Bosmali E, Axarli I, Nianiou-Obeidat I, Madesis P, Tsaftaris A, Labrou EN (2015) Plant glutathione transferases: structure, antioxidant catalytic function and in planta protective role in biotic and abiotic stress. Curr Chem Biol 8:58–75

Csiszar J, Horvath E, Vary Z, Galle A, Bela K, Brunner S et al (2014) Glutathione transferase supergene family in tomato: salt stress-regulated expression of representative genes from distinct GST classes in plants primed with salicylic acid. Plant Physiol Biochem 78:15–26

Cummins I, Cole DJ, Edwards R (1997) Purification of multiple glutathione transferases involved in herbicide detoxification from wheat (*Triticum aestivum* L.) treated with the safener fenchlorazole-ethyl. Pestic Biochem Physiol 59:35–49

Cummins I, Bryant DN, Edwards R (2009) Safener responsiveness and multiple herbicide resistance in the weed black-grass (*Alopecurus myosuroides*). Plant Biotechnol J 7:807–820

Cummins I, Dixon DP, Freitag-Pohl S, Skipsey M, Edwards R (2011) Multiple roles for plant glutathione transferases in xenobiotic detoxification. Drug Metab Rev 43:266–280

Dirr H, Reinemer P, Huber R (1994) X-ray crystal structures of cytosolic glutathione S-transferases – implications for protein architecture, substrate recognition and catalytic function. Eur J Biochem 220:645–661

Dixon DP, Edwards R (2009) Selective binding of glutathione conjugates of fatty acid derivatives by plant glutathione transferases. J Biol Chem 284:21249–21256

Dixon DP, Edwards R (2010a) Glutathione transferases. The Arabidopsis Book 8:e0131. https://doi.org/10.1199/tab.0131

Dixon DP, Edwards R (2010b) Roles for stress-inducible lambda glutathione transferases in flavonoid metabolism in plants as identified by ligand fishing. J Biol Chem 285:36322–36329

Dixon DP, Cummins L, Cole DJ, Edwards R (1998) Glutathione-mediated detoxification system in plants. Curr Opin Plant Biol 1:258–266

Dixon DP, Cole DJ, Edwards R (2000) Characteristics of a zeta class GST from *Arabidopsis thaliana* with a putative role in tyrosine catabolism. Arch Biochem Biophys 384:407–412

Dixon DP, Davis BG, Edwards E (2002a) Functional divergence in the glutathione transferase superfamily in plants. J Biol Chem 277:30859–30869

Dixon DP, Lapthorn A, Edwards R (2002b) Plant glutathione transferases. Genome Biol 3:1–10

Dixon DP, McEwen AG, Lapthorn AJ, Edwards R (2003) Forced evolution of a herbicide detoxifying glutathione transferase. J Biol Chem 278:23930–23935

Dixon DP, Lapthorn A, Madesis P, Mudd EA, Day A, Edwards R (2008) Binding and glutathione conjugation of porphyrinogens by plant glutathione transferases. J Biol Chem 283:20268–20276

Dixon DP, Hawkins T, Hussey PJ, Edwards R (2009) Enzyme activities and subcellular localization of members of the *Arabidopsis* glutathione transferase superfamily. J Exp Bot 60:1207–1218

Dixon DP, Skipsey M, Edwards R (2010) Roles for glutathione transferases in plant secondary metabolism. Phytochemistry 71:338–350

Dixon DP, Sellars JD, Edwards R (2011) The *Arabidopsis* phi class glutathione transferase AtGSTF2: binding and regulation by biologically active heterocyclic ligands. Biochem J 438:63–70

Dueckersho K, Mueller S, Mueller MJ, Reinders J (2008) Impact of cyclopentenone-oxylipins on the proteome of *Arabidopsis thaliana*. Biochim Biophys Acta 1784:1975–1985

Edwards R, Dixon DP (2000) The role of glutathione transferases in herbicide metabolism. In: Cobb AH, Kirkwood RC (eds) Herbicides and their mechanisms of action. Sheffield Academic Press, Sheffield, pp 38–71

Edwards R, Dixon DP (2005) Plant glutathione transferases. Methods Enzymol 401:169–186

Edwards R, Dixon DP, Walbot V (2000) Plant glutathione S-transferases: enzymes with multiple functions in sickness and in health. Trends Plant Sci 5:193–198

Fahey JW, Zalcmann AT, Talalay P (2001) The chemical diversity and distribution of glucosinolates and isothiocyanates among plants. Phytochemistry 56:5–51

Frova C (2003) The plant glutathione transferase gene family: genomic structure, functions, expression and evolution. Physiol Plantarum 119:469–479

Frova C (2006) Glutathione transferases in the genomics era: new insights and perspectives. Biomol Eng 23:149–169

George S, Venkataraman G, Parida A (2010) A chloroplast-localized and auxin-induced glutathione S-transferase from phreatophyte *Prosopis juliflora* confer drought tolerance on tobacco. J Plant Physiol 167:311–318

Ginsberg G, Smolenski S, Hattis D, Guyton KZ, Johns DO, Sonawane B (2009) Genetic polymorphism in glutathione transferases (GST): population distribution of GSTM1, T1, and P1 conjugating activity. J Toxicol Environ Health B Crit Rev 12:389–439

Gonneau M, Mornet R, Laloue MA (1998) *Nicotiana plumbaginifolia* protein labeled with an azido cytokinin agonist is a glutathione S-transferase. Physiol Plant 103:114–124

Gunning V, Tzafestas K, Sparrow H, Johnston EJ, Brentnall AS, Potts JR, Rylott EL, Bruce NC (2014) *Arabidopsis* glutathione transferases U24 and U25 exhibit a range of detoxification activities with the environmental pollutant and explosive, 2,4,6-trinitrotoluene. Plant Physiol 165:854–865

Huberts DH, van der Klei IJ (2010) Moonlighting proteins: an intriguing mode of multitasking. Biochim Biophys Acta 1803:520–525

Ivarsson Y, Mannervik B (2007) Combinatorial protein chemistry in three dimensions: a paradigm for the evolution of glutathione transferases with novel activities. In Awasthi YC (ed) Toxicology of glutathione transferases, CRC Press, Boca Raton, Florida; pp 213-231. Please provide publisher name and location for Ivarsson and Mannervik (2007) and Jacquot et al. (2013).Were added

Jacquot JP, Dietz KJ, Rouhier N, Meux E, Lallement PA, Selles B, Hecker A. Redox regulation in plants: glutathione and "redoxin" related families. U. Jakob and D. Reichmann (eds.), Oxidative stress and redox regulation. 2013; Springer, Dordrecht, pp. 213–231

Jain M, Ghanashyam C, Bhattacharjee A (2010) Comprehensive expression analysis suggests overlapping and specific roles of rice glutathione S-transferase genes during development and stress responses. BMC Genomics 11:73

Jeffery CJ (2009) Moonlighting proteins- an update. Mol BioSyst 5:345–350

Jia B, Sun M, Sun X, Li R, Wang Z, Wu J, Wei Z, DuanMu H, Xiao J, Zhu Y (2016) Overexpression of GsGSTU13 and SCMRP in *Medicago sativa* confers increased salt-alkaline tolerance and methionine content. Physiol Plant 156:176–189

Jiang J, Ma S, Ye N, Jiang M, Cao J, Zhang J (2016) WRKY transcription factors in plant responses to stresses. J Integr Plant Biol 59:86–101

Kampranis SC, Damianova R, Atallah M et al (2000) A novel plant glutathione S-transferase/ peroxidase suppresses bax lethality in yeast. J Biol Chem 275:29207–29216

Kao CW, Bakshi M, Sherameti I, Dong S, Reichelt M, Oelmüller R, Yeh KW (2016) A Chinese cabbage (*Brassica campetris* subsp. Chinensis) τ-type glutathione-S-transferase stimulates *Arabidopsis* development and primes against abiotic and biotic stress. Plant Mol Biol 92:643–659

Kapoli P, Axarli IA, Platis D, Fragoulaki M, Paine M, Hemingwayb J, Vontas J, Labrou EN (2008) Engineering sensitive glutathione transferase for the detection of xenobiotics. Biosens Bioelectron 24:498–503

Karpusas M, Axarli I, Chiniadis L, Papakyriakou A, Bethanis K, Scopelitou K, Clonis YD, Labrou NE (2013) The interaction of the chemotherapeutic drug chlorambucil with human glutathione transferase A1-1: kinetic and structural analysis. PLoS One 8:e56337

Kearse M, Moir R, Wilson A, Stones-Havas S, Cheung M, Sturrock S, Buxton S, Cooper A, Markowitz S, Duran C, Thierer T, Ashton B, Mentjies P, Drummond A (2012) Geneious basic: an integrated and extendable desktop software platform for the organization and analysis of sequence data. Bioinformatics 28:1647–1649

Kitamura S, Shikazono N, Tanaka A (2004) TRANSPARENT TESTA 19 is involved in the accumulation of both anthocyanins and proanthocyanidins in *Arabidopsis*. Plant J 37:104–114
Kumar S, Asif MH, Chakrabarty D, Tripathi RD, Dubey RS, Trivedi PK (2013) Differential expression of rice lambda class GST gene family members during plant growth, development, and in response to stress conditions. Plant Mol Biol Rep 31:569–580
Labrou NE, Papageorgiou AC, Pavli O, Flemetakis E (2015) Plant GSTome: structure and functional role in xenome network and plant stress response. Curr Opin Biotechnol 32:186–194
Labudda M, Azam FMS (2014) Glutathione-dependent responses of plants to drought: a review. Acta Soc Bot Pol 83:3–12
Lallement P-A, Brouwer B, Keech O, Hecker A, Rouhier N (2014a) The still mysterious roles of cysteine-containing glutathione transferases in plants. Front Pharmacol 5:1–22
Lallement PA, Meux E, Gualberto JM, Prosper P, Didierjean C, Saul F, Haouz A, Rouhier N, Hecker A (2014b) Structural and enzymatic insights into lambda glutathione transferases from *Populus trichocarpa*, monomeric enzymes constituting an early divergent class specific to terrestrial plants. Biochem J 462:39–52
Lallement PA, Meux E, Gualberto JM, Dumarcay S, Favier F, Didierjean C, Saul F, Haouz A, Morel-Rouhier M, Gelhaye E, Rouhier N, Hecker A (2015) Glutathionyl-hydroquinone reductases from poplar are plastidial proteins that deglutathionylate both reduced and oxidized glutathionylated quinones. FEBS Lett 589:37–44
Lam LK, Zhang Z, Board PG, Xun L (2012) Reduction of benzoquinones to hydroquinones via spontaneous reaction with glutathione and enzymatic reaction by S-glutathionyl-hydroquinone reductases. Biochemistry 51:5014–5021
Lan T, Yang ZL, Yang X, Liu YJ, Wang XR, Zeng QY (2009) Extensive functional diversification of the Populus glutathione S-transferase supergene family. Plant Cell 21:3749–3766
Lan T, Wang XR, Zeng QY (2013) Structural and functional evolution of positively selected sites in pine glutathione S-transferase enzyme family. J Biol Chem 288:24441–24451
Lee H-C, Toung Y-PS TY-SL, Tu C-PD (1995) A molecular genetic approach for the identification of essential residues in human glutathione S-transferase function in *Escherichia coli*. J Biol Chem 270:99–109
Li D, Xu L, Pang S, Liu Z, Wang K, Wang C (2017) Variable levels of glutathione S-transferases are responsible for the differential tolerance to metolachlor between maize (*Zea mays*) shoots and roots. J Agric Food Chem 65:39–44
Liu YJ, Han XM, Ren LL, Yang HL, Zeng QY (2013) Functional divergence of the glutathione S-transferase supergene family in *Physcomitrella patens* reveals complex patterns of large gene family evolution in land plants. Plant Physiol 161:773–786
Liu H, He J, Zhao R, Chi C, Bao Y (2015) A novel biomarker for marine environmental pollution of pi-class glutathione S-transferase from *Mytilus coruscus*. Ecotoxicol Environ Saf 118:47–54
Lu WD, Atkins WA (2004) Novel antioxidant role for ligandin behavior of glutathione S-transferases: attenuation of the photodynamic effects of hypericin. Biochemistry 43:12761–12769
Mannervik B, Danielson UH (1988) Glutathione transferases: structure and catalytic activity. CRC Crit Rev Biochem 23:283–337
Mashiyama ST, Malabanan MM, Akiva E, Bhosle R, Branch MC et al (2014) Large-scale determination of sequence, structure, and function relationships in cytosolic glutathione transferases across the biosphere. PLoS Biol 12:1–19
Meyer DJ, Crease DJ, Ketterer B (1995) Forward and reverse catalysis and product sequestration by human glutathione S-transferases in the reaction of GSH with dietary aralkyl isothiocyanates. Biochem J 306:565–569
Mohsenzadeh S, Esmaeili M, Moosavi F, Shahrtash M, Saffari B, Mohabatkar H (2011) Plant glutathione S-transferase classification, structure and evolution. African J Biotech 10:8160–8165
Morel M, Meux E, Mathieu Y, Thuillier A, Chibani K, Harvengt L et al (2013) Xenomic networks variability and adaptation traits in wood decaying fungi. Microb Biotechnol 6:248–263
Mueller S, Hilbert B, Dueckershoff K, Roitsch T, Krischke M, Mueller MJ et al (2008) General detoxification and stress responses are mediated by oxidized lipids through TGA transcription factors in *Arabidopsis*. Plant Cell 20:768–785

Neuefeind T, Huber R, Dasenbrock H, Prade L, Bieseler B (1997a) Crystal structure of herbicide-detoxifying maize glutathione S-transferase-I in complex with lactoylglutathione: evidence for an induced- fit mechanism. J Mol Biol 274:446–453
Neuefeind T, Huber R, Reinemer P, Knäblein J, Prade L, Mann K, Bieseler B (1997b) Cloning, sequencing, crystallization and X-ray structure of glutathione S-transferase-III from *Zea mays* Var. mutin: a leading enzyme in detoxification of maize herbicides. J Mol Biol 274:577–587
Parbhoo N (2013) Topologically conserved hydrophobic residues of the thioredoxin C-subdomain stabilise GSTs. WIReDSpace
Pearson WR (2005) Phylogenies of glutathione transferase families. Methods Enzymol 401:186–204
Pickett CB, Lu AYH (1989) Glutathione S-transferases: gene structure, regulation, and biological function. Annu Rev Biochem 58:743–764
Polekhina G, Board PG, Blackburn AC, Parker MW (2001) Crystal structure of maleylacetoacetate isomerase/glutathione transferase zeta reveals the molecular basis for its remarkable catalytic promiscuity. Biochemistry 40:1567–1576
Reinemer P, Prade L, Hof P, Neuefeind T, Huber R, Zettl R, Palme K, Schell J, Jkoelln I, Bartunik H, Bieseler B (1996) An inducible glutathione S-transferase in soybean hypocotyl is localized in the apoplast. Plant Physiol 3:1185–1190
Rezaei MK, Shobbar ZS, Shahbazi M, Abedini R, Zare S (2013) Glutathione S-transferase (GST) family in barley: identification of members, enzyme activity, and gene expression pattern. J Plant Physiol 170:1277–1284
Roxas VP, Smith RK Jr, Allen ER, Allen RD (1997) Overexpression of glutathione S-transferase/glutathione peroxidase enhances the growth of transgenic tobacco seedlings during stress. Nat Biotechnol 15:988–991
Rushmore TH, Pickett CB (1993) Glutathione S-transferases, structure, regulation, and therapeutic implications. J Biol Chem 268:11475–11478
Schröder P (2001) The role of glutathione and glutathione S-transferase in plant reaction and adaptation to xenobiotics. Plant Ecophysiol 2:155–183
Shaokui W, Kun W, Qingbo Y, Xueying L, Zhengbin L, Xiaoyan L, Ruizhen Z, Haitao Z, Guojun D, Qian Q, Guiquan Z, Xiangdong F (2012) Control of grain size, shape and quality by *OsSPL16* in rice. Nat Genet 44:950–954
Sheehan D, Meade G, Foley VM, Dowd CA (2001) Structure, function and evolution of glutathione transferases: implications for classification of non- mammalian members of an ancient enzyme superfamily. Biochem J 360:1–16
Shimizu T, Hashimoto N, Nakayama I, Nakao T, Mizutani H, Unai T (1995) A novel isourazole herbicide, fluthiacet-methyl, is a potent inhibitor of protoporphyrinogen oxidase after isomerization by glutathione S-transferase. Plant Cell Physiol 36:625–632
Sievers F, Wilm A, Dineen D, Gibson TJ, Karplus K, Li W, Lopez R, McWilliam H, Remmert M, Söding J, Thompson JD, Higgins DG (2011) Fast, scalable generation of high-quality protein multiple sequence alignments using Clustal omega. Mol Syst Biol 7:539
Skopelitou K, Muleta AW, Papageorgiou AC, Chronopoulou E, Labrou NE (2015) Catalytic features and crystal structure of a tau class glutathione transferase from *Glycine max* specifically upregulated in response to soybean mosaic virus infections. Biochim Biophys Acta 1854:166–177
Skopelitou K, Muleta AW, Papageorgiou AC, Chronopoulou EG, Pavli O, Flemetakis E, Skaracis GN, Labrou NE (2017) Characterization and functional analysis of a recombinant tau class glutathione transferase GmGSTU2-2 from *Glycine max*. Int J Biol Macromol 94:802–812
Smith AP, Nourizadeh SD, Peer WA, Xu J, Bandyopadhyay A, Murphy AS, Goldsbrough PB (2003) *Arabidopsis* AtGSTF2 is regulated by ethylene and auxin, and encodes a glutathione S-transferase that interacts with flavonoids. Plant J 36:433–442
Su T, Xu J, Li Y, Lei L, Zhao L, Yang H, Feng J, Liu G, Ren D (2011) Glutathione-indole-3-acetonitrile is required for camalexin biosynthesis in *Arabidopsis thaliana*. Plant Cell 23:364–380

Talalay P, De Long MJ, Prochaska HJ (1988) Identification of a common chemical signal regulating the induction of enzymes that protect against chemical carcinogenesis. Proc Natl Acad Sci U S A 85:8261–8265

Thom R, Dixon DP, Edwards R, Cole DJ, Lapthorn A (2001) The structure of a zeta class glutathione S-transferase from *Arabidopsis thaliana*: characterization of a GST with novel active site architecture and a putative role in tyrosine catabolism. J Mol Biol 308:949–962

Thom R, Cummins I, Dixon DP, Edwards R, Cole DJ, Lapthorn AJ (2002) Structure of a tau class glutathione S-transferase from wheat active in herbicide detoxification. Biochemistry 41:7008–7020

Tzafestas K, Razalan MM, Gyulev I, Mazari AM, Mannervik B, Rylott EL, Bruce NC (2016) Expression of a drosophila glutathione transferase in *Arabidopsis* confers the ability to detoxify the environmental pollutant, and explosive, 2,4,6-trinitrotoluene. New Phytol. https://doi.org/10.1111/nph.14326

Wagner U, Edwards R, Dixon DP, Mauch F (2002) Probing the diversity of the *Arabidopsis* glutathione S-transferase gene family. Plant Mol Biol 49:515–532

Yang Q, Liu Y-J, Zeng Q-Y (2014) Biochemical functions of the glutathione transferase supergene family of *Larix kaempferi*. Plant Physiol Biochem 77:99–107

Zettl R, Schell J, Palme K (1994) Photoaffinity labelling of *Arabidopsis thaliana* plasma membrane vesicles by 5-azido-[7-3H]indole-3-acetic acid identification of a glutathione S-transferase. Proc Natl Acad Sci U S A 91:689–693

Chapter 10
Plant Glutathione Transferases in Abiotic Stress Response and Herbicide Resistance

Evangelia Chronopoulou, Nikolaos Georgakis, Irini Nianiou-Obeidat, Panagiotis Madesis, Fereniki Perperopoulou, Fotini Pouliou, Eleni Vasilopoulou, Elisavet Ioannou, Farid S. Ataya, and Nikolaos E. Labrou

Abstract Plant responses and adaptations to stress conditions are of great interest for both basic and applied science, and represent the key factors for the improvement of economically important crops worldwide. Glutathione *S*-transferases (GSTs, EC. 2.5.1.18) are multifunctional enzymes encoded by a highly divergent ancient gene family. GSTs catalyze the conjugation of tripeptide glutathione (GSH) with endogenous electrophilic compounds (secondary metabolites, hydroperoxides) and xenobiotics, such as herbicides, leading to their cellular detoxification. Therefore, GSTs are implicated in metabolism-based herbicide resistance in crop weeds. This chapter discusses the involvement of plant GSTs in abiotic stress response with focus on metabolism-based herbicide resistance and attempts to give an overview of their catalytic roles and *in planta* function.

Keywords Glutathione transferase • Herbicide resistance • Abiotic stress

---

E. Chronopoulou • N. Georgakis • F. Perperopoulou • F. Pouliou • E. Vasilopoulou
E. Ioannou • N.E. Labrou (✉)
Laboratory of Enzyme Technology, Department of Biotechnology, School of Food, Biotechnology and Development, Agricultural University of Athens,
75 Iera Odos Street, GR-11855 Athens, Greece
e-mail: lambrou@aua.gr

I. Nianiou-Obeidat
Laboratory of Genetics and Plant Breeding, School of Agriculture, Forestry and Natural Environment, Aristotle University of Thessaloniki,
P.O. Box 261, Thessaloniki GR-54124, Greece

P. Madesis
Institute of Applied Biosciences, CERTH,
6th km Charilaou-Thermis Road, P.O. Box 361, Thermi GR-57001 Thessaloniki, Greece

F.S. Ataya
Department of Biochemistry, College of Science, King Saud University,
P.O. Box 2455, Riyadh 11451, Saudi Arabia

M.A. Hossain et al. (eds.), *Glutathione in Plant Growth, Development, and Stress Tolerance*, https://doi.org/10.1007/978-3-319-66682-2_10

Contents

1 Introduction

Abiotic stress factors are known to impose oxidative stress in plants. In order to handle stress conditions, plants have evolved several mechanisms: physiological or enzymatic, which help them to withstand the stress. Variation does exist in these mechanisms among plants. Certain morphological features of some plants, make them to avoid stress factors. But it may not be the case in all plants, thus the only option for plants is to adapt their physiology, metabolic mechanisms, gene expression and developmental activities to cope with stress effects (Rao et al. 2006).

In nature, plants are continuously subjected to unfavorable environmental stress conditions, which lead to reactive oxygen species (ROS) production (Del Río 2015; Inupakutika et al. 2016). The alteration of production versus scavenging equilibrium is then disturbed, causing significant damage to cell structures (Dat et al. 2000; Mittler 2002; Mittler et al. 2004; Kärkönen and Kuchitsu 2015; Singh et al. 2016a, b). One of the common ROS targets is the different cellular membrane systems that perform several vital processes such as photosynthesis, respiration, nitrogen assimilation and protein synthesis. Thus, under stress conditions all these processes can be negatively affected, leading to the decrease of plant development and productivity (Kärkönen and Kuchitsu 2015; Singh et al. 2016a, b). Adaptation and tolerance of plants to biotic and abiotic stress conditions are of great interest in agriculture, because they are the major limiting factors for maximum plant production (Popelka et al. 2010; Gray and Brady 2016).

Some important abiotic stresses that are potentially harmful to the plants include heat, cold, freezing, drought, UV or other ionizing rays, high salinity or acidity of the soil and the presence of pollutants (xenobiotics), including herbicides (Nianiou-Obeidat et al. 2017). The effect of each abiotic stress on plant physiology depends on its severity, duration and developmental stages of the plant (Pandey et al. 2017). For example, in the case of herbicides, the response of plant to an individual herbicide is variable and can range from very injury to severe lethal impacts, since it depends on the plant species as well as on the herbicide's biochemical target. In plants, the most important steps in herbicide detoxification are catalyzed by cytochrome P450 mono-oxygenases (CYPs), which catalyze oxidation reactions, and glutathione transferases (GSTs), which conjugate electrophilic herbicides with the tripeptide glutathione (GSH), resulting to their detoxification (Gray and Brady 2016; Nianiou-Obeidat et al. 2017). GSTs are expressed both constitutively and induced in response to some herbicides as well as herbicide safeners, compounds that increase herbicide tolerance in cereals (Skipsey et al. 2011; Kissoudis et al. 2015b; Labrou et al. 2015). GSTs are encoded by a large and diverse gene family in plants, which differ in number, herbicide specificity and inducibility across different plants. In addition, plant GSTs provide protection against oxidative stress induced by other abiotic stresses by acting as glutathione peroxidases (GPX) and dehydroascorbate reductases (DHAR) (Labrou et al. 2015). GPX can catalyze the reduction of hydroperoxides, whereas DHAR contributes to redox homeostasis by regenerating ascorbic acid (AsA) from dehydroascorbate. In this chapter, we will review the function of GSTs in abiotic stress mechanism, with focus on metabolism-based herbicide resistance.

2 Enzymatic Systems for Preventing Oxidative Stress

Enzymatic and non-enzymatic systems play an important role in balancing and preventing oxidative damage (Bowler et al. 1994; Foyer et al. 1994; Saidi et al. 2014). The enzymatic defense system of plants (Fig. 10.1) includes enzymes, such as

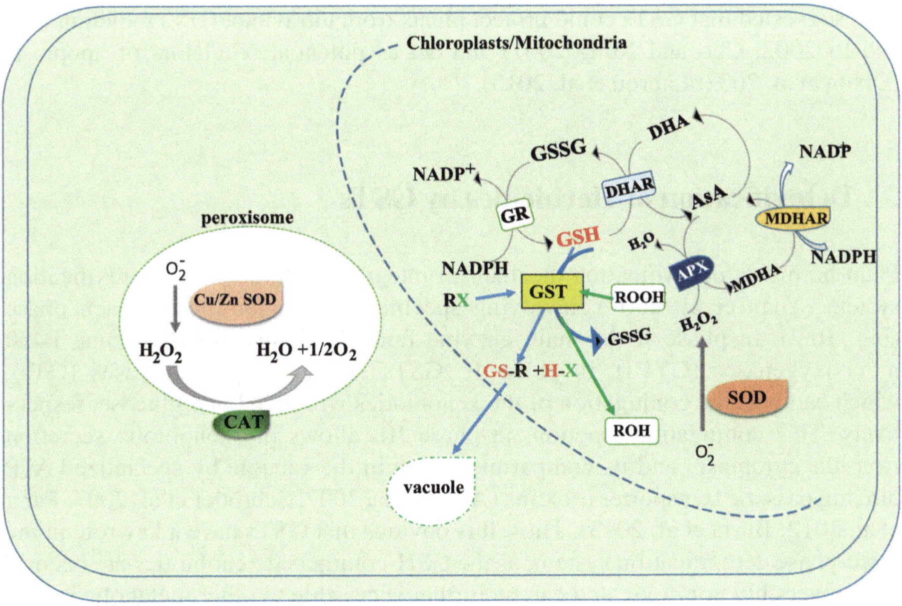

Fig. 10.1 The enzymatic defense system of plants. *SOD* superoxide dismutase, *CAT* catalase, *APX* ascorbate peroxidase, *MDHAR* monodehydroascorbate reductase, *DHAR* dehydroascorbate reductase, *GR* glutathione reductase, *GPX* glutathione peroxidase, *GST* glutathione transferase, *GSR* glutathione-disulfide reductase, *RX* haloalkane (e.g., xenobiotic compound), *HX* hydrohalogen

superoxide dismutases (SOD), catalases (CAT), ascorbate peroxidases (APX), GST and GPX, that catalyze the scavenging of ROS (Yang et al. 2014). The activities of APX, GST and GPX depend on the availability of AsA and GSH that are maintained by glutathione reductase (GR), DHAR and monodehydroascorbate reductase (MDHAR) using NAD(P)H as an electron donor (Park et al. 2016; Roxas et al. 2000). The non-enzymatic system comprises compounds, such as AsA, GSH, phenolic compounds, alkaloids, non-protein amino acids and α-tocopherols. Enzymatic and non-enzymatic systems work in concert to control oxidative damage and protect plant cells by scavenging ROS (Kasote et al. 2015; Mostofa et al. 2015).

GSTs are enzymes encoded by a highly divergent ancient gene family (Lallement et al. 2015; Nianiou-Obeidat et al. 2017). Their presence in plants was first detected in the 1970s, when GSTs activity in maize was shown to be responsible for the conjugation of atrazine with GSH, thereby protecting maize from injury by this herbicide (Frear and Swanson 1970; Li et al. 2016). The ability of GSTs to detoxify herbicides is well studied mainly because of their importance in determining herbicide selectivity (McGonigle et al. 2000). In addition, GSTs have been found to function in other cellular processes as for example in targeting numerous secondary metabolites to an appropriate cellular localization (Marrs 1996; Csiszar et al. 2014; Skopelitou et al. 2017). For instance, some GSTs function as reversible ligand binding proteins that play a regulatory role in hormone signaling (Bilang and Sturm 1995; Kitamura et al. 2004; Moons 2005; Axarli et al. 2016; Nianiou-Obeidat et al. 2017; Skopelitou et al. 2017; Banday and Nandi 2017). In addition, recent studies have suggested that GSTs could protect plants from ultraviolet (UV) radiation (Liu and Li 2002; Cho and Kong 2007) and act as potential regulators of apoptosis (Dixon et al. 2002; Labrou et al. 2015).

3 Detoxification of Herbicides by GSTs

Plant herbicide detoxification mechanism integrates a three-phase detoxification system (Yuan et al. 2007), involving specific enzyme families in each phase (Fig. 10.2). In phase I, the main enzyme family belongs to cytochrome P450 monooxygenases (CYPs), in phase II, GSTs and glycosyltransferases (GTs), which catalyze the conjugation of the xenobiotics with GSH and glucose, respectively. This conjugation reaction, in phase III, allows the xenobiotic secretion from the cytoplasm and its compartmentation in the vacuole by specialized ATP binding cassette transporter proteins (ABC) (Rea 2007; Schröder et al. 2007; Pang et al. 2012; Burla et al. 2013). Thus, it is obvious that GSTs have a key role in the three-phase detoxification system, as the GSH-conjugated xenobiotics are becoming irreversibly non-toxic and can be further accessible to other metabolic procedures (Schröder 2001; Duhoux et al. 2015, 2017; Li et al. 2017). Collectively, these xenobiotic detoxifying enzymes and transporters are termed "xenome" (Morel et al. 2013; Labrou et al. 2015). The xenome in plants is very large. For example, the relatively small genome of the well-studied model species

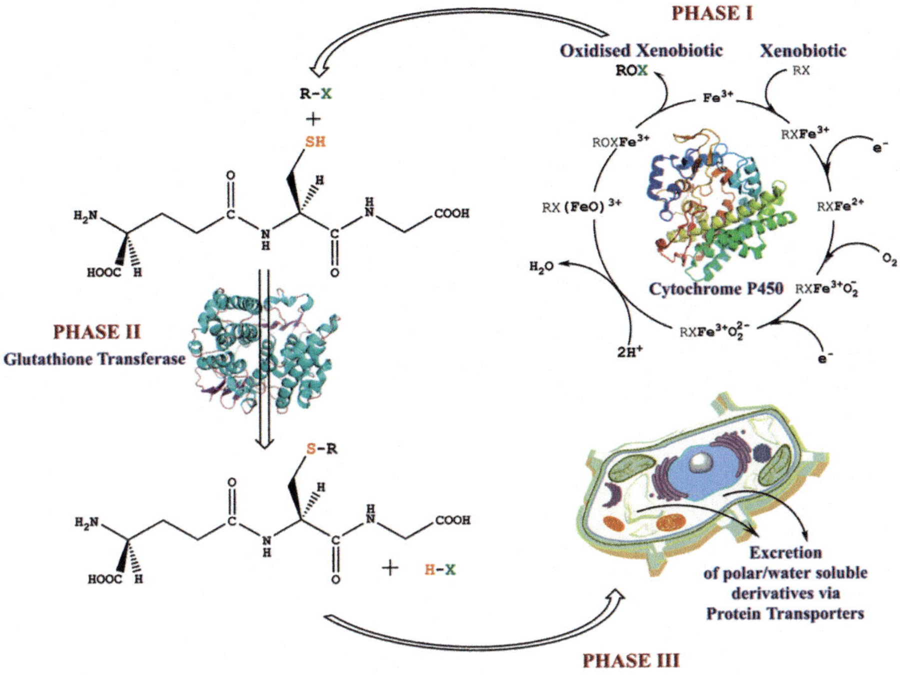

Fig. 10.2 Plant detoxification mechanisms integrate a three-phase detoxification system, involving specific enzyme families in each phase. The enzyme families in phase I and II belong to CytP450 monooxygonases and GSTs, respectively. In phase III, the GSH-conjugates are secreted from the cytoplasm by specialized ATP binding cassette transporters

Arabidopsis thaliana contains 273 CYPs, 107 family 1 UDP-glucose GTs, 55 GSTs and 120 ABCs.

The GST/GSH system has been found to play a major role in the detoxification of several classes of herbicides (Fig. 10.3) and therefore in herbicide-resistance mechanism of weeds (Cummins et al. 1997a, 2011; Chronopoulou et al. 2012; Axarli et al. 2016; Nianiou-Obeidat et al. 2017; Duhoux et al. 2017; Li et al. 2017). Several major classes of herbicides, including sulfonylureas (chlorimuronethyl, trisulfuron methyl and flupyrsulfuronmethyl), triazines (atrazine), chloroacetanilides (alachlor and metolachlor), thiocarbamate sulfoxides (S-ethyl dipropylthiocarbamatesulfoxide) and diphenylethers (fluorodifen) are found as a GSH conjugate (Rossini et al. 1996; Hatton et al. 1999; Dixon et al. 1997; Cummins et al. 2011; Chronopoulou et al. 2012; Axarli et al. 2016; Nianiou-Obeidat et al. 2017; Duhoux et al. 2015, 2017; Li et al. 2017).

Specific plants belonging to monocots have been found to be tolerant to herbicides in correlation with the expression of certain herbicide-detoxifying GSTs in their foliage (Dixon et al. 1997). Moreover, GSTs have a wide-range substrate specificity and an ability to produce homo-or heterodimers, rendering plants to tolerate a broad spectrum of herbicides (Cummins et al. 1997a; Dixon et al. 1999; Axarli

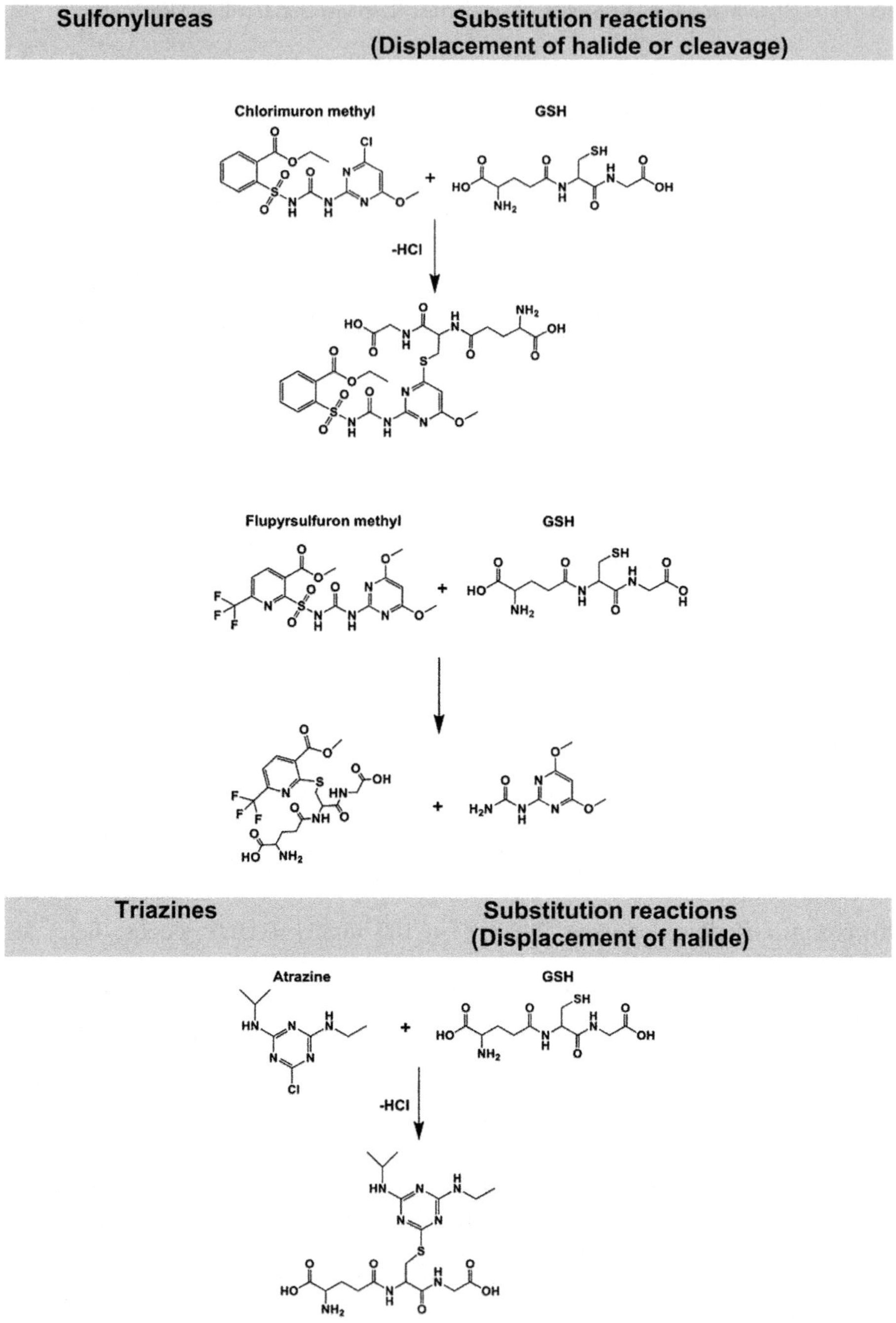

Fig. 10.3 Detoxification reactions of major classes of herbicides catalyzed by GSTs. The picture depicts the following herbicides: sulfonylureas (chlorimuronethyl and flupyrsulfuronmethyl), triazines (atrazine), chloroacetanilides (alachlor and metolachlor), thiocarbamate sulfoxides (S-ethyl dipropylthiocarbamatesulfoxide) and diphenylethers (fluorodifen) (The figure was created by ChemBio Draw Ultra 13 (Perkin Elmer))

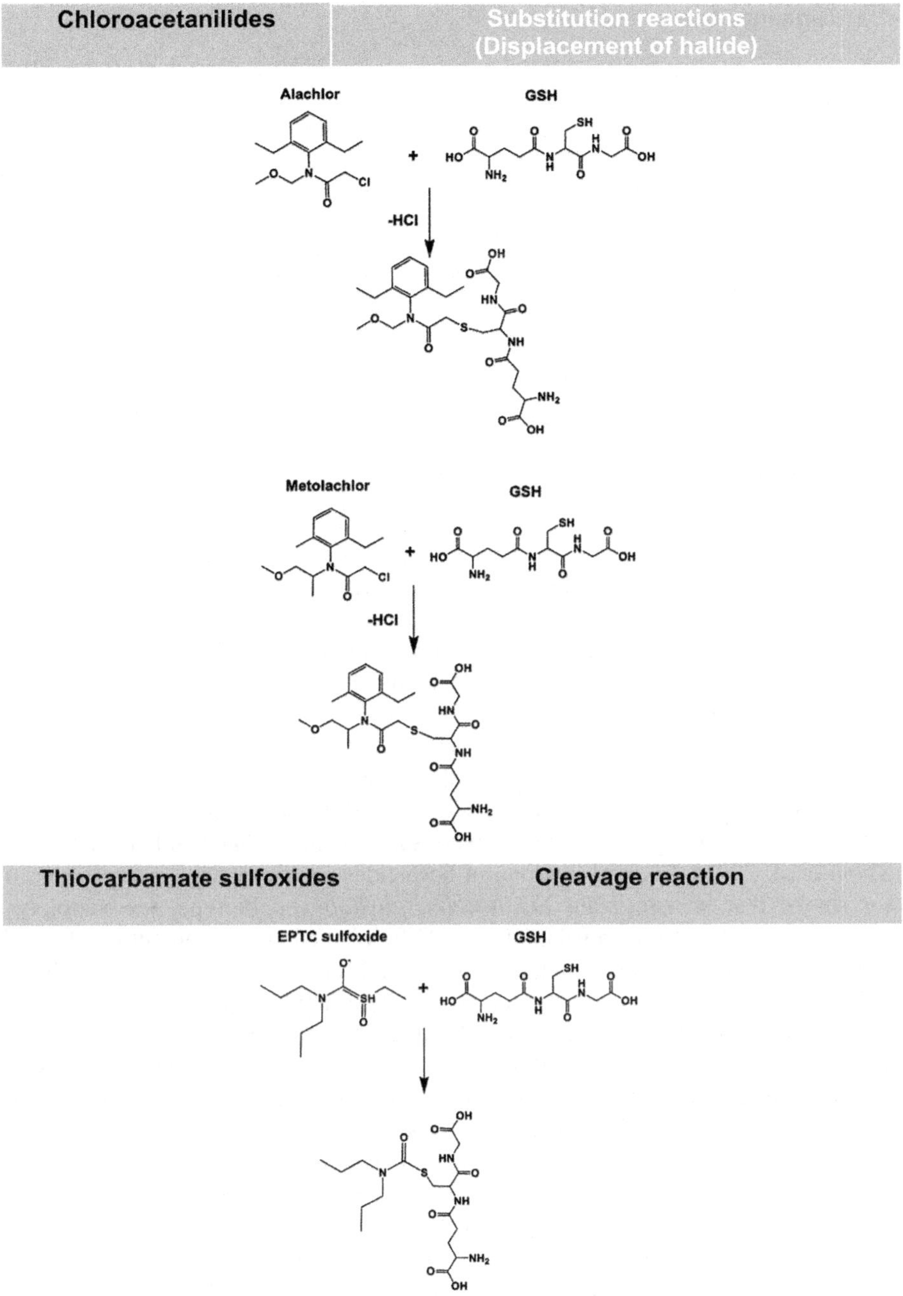

Fig. 10.3 (continued)

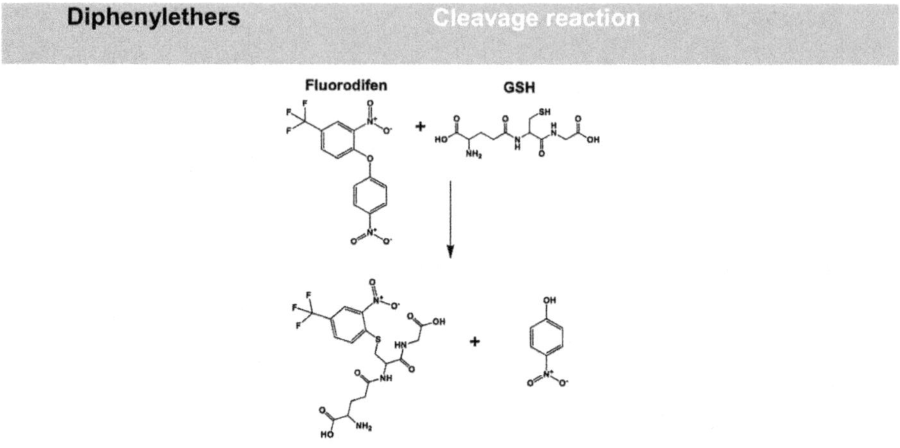

Fig. 10.3 (continued)

et al. 2016; Nianiou-Obeidat et al. 2017; Labrou et al. 2016; Skopelitou et al. 2017). The variety of GST homo- and heterodimers formed is subjected in transcriptional regulation of the individual subunits (Cummins et al. 1997a; Dixon et al. 1997).

The GSH conjugation with herbicides through the enzymatic action of GSTs is found to play an important role in the metabolism and detoxification of selective herbicides in other cereals, including the hexaploid bread wheat and maize (Cummins et al. 2003; Jiang et al. 2016; Li et al. 2017). Thus, rapid GSH conjugation of the chloroacetamide, dimethenamid and sulphonylurea flupyrsulfuronmethyl is a key factor in the detoxification and selectivity of these herbicides in wheat. The GSTs involved in the processes above were subsequently cloned and characterized (Thom et al. 2002). In another group of herbicides aryloxyphenoxy propionate, it was shown that in a resistant *Echinochloa phyllopogon* biotype, fenoxaprop-p-methyl tolerance might be attributed to GSH-herbicide conjugation, although GST activity was not determined (Bakkali et al. 2007). Moreover, when fluazifop-*p*-butyl of the aryloxyphenoxypropionic group of herbicides applied on *Phaseolus vulgaris* resulted in the induction of GST activity, three inducible GST isoenzymes were isolated showing high homology with GSTs that belong to phi and tau classes (Chronopoulou et al. 2012). In *Echinochloa cruss-galli*, the expression level of *Ec*GST1 was found to be higher to a ratio of six- to ten-fold in resistant biotypes as compared to susceptible ones, and moreover, the difference in expression level increased after herbicide treatment (Li et al. 2013).

GSTs have also been extensively studied in rice, either at biochemical or genomic level (Deng and Hatzios 2002a; Soranzo et al. 2004; Tripathi et al. 2014; Dubey et al. 2016). At least four constitutive and one safener inducible GST isozymes with activity toward pretilachlor have been reported in *Oryza sativa* variety Nipponbare (japonica type) (Deng et al. 1997). In addition, Deng and Hatzios have identified in Teqing rice two GST isozymes that detoxify pretilachlor and at the same time confer tolerance against oxidative stress (Deng and Hatzios 2002b). Another GST from

O. sativa, which belongs to the phi class, *Os*GSTF3–3, was found to detoxify metolachlor and alachlor by GSH conjugation (Cho and Kong 2005). Legumes like soybean and beans, unlike other plants, have as predominant thiol the homoglutathione (hGSH, γ-glutamyl-L-cysteinyl-L-alanine) rather than GSH. This difference has direct effect on differential catalytic efficiency of several soybean GSTs when using either of the thiol substrates (McGonigle et al. 1998). This has been demonstrated with soybean isoenzymes *Gm*GSTU1 and *Gm*GST2U6, which conjugate more efficiently diphenyl ether herbicide fomesafen with hGSH rather than GSH (Skipsey et al. 1997, 2005). Safeners could induce GST gene expression (Hatzios and Burgos 2004; Zhang and Riechers 2004; DeRidder and Goldsbrough 2006; Skipsey et al. 2011). For instance, crops like maize and sorghum exhibit low degree of tolerance to thiocarbamate (e.g., EPTC) and chloroacetamide herbicides (e.g., metolachlor, alachlor), but pretreatment with safeners like flurazole, dichlormid and benoxacor significantly improve the tolerance to these herbicides by inducing specific GST activity, thus enhancing the speed of herbicide detoxification via GSH conjugation (Fuerst and Gronwald 1986; Viger et al. 1991; Irzyk and Fuerst 1993; Jiang et al. 2016; Li et al. 2017).

4 Herbicide Resistance in Weeds

Herbicide resistance is a major problem, which poses a threat to the sustainability of agriculture (Green 2014; Heap 2014; Walsh and Powles 2014). Recent reports point out an increase in the frequency and diversity of herbicide-resistant weed biotypes over the past two decades. This fact can be attributed to the introduction of herbicide-resistant crops during the 1990s, which expanded farmer reliance on herbicides (Benbrook 2012; Green 2014; Perry et al. 2016). It is noteworthy that the management of herbicide resistance reached a total estimated cost of $53 millions (Lambert et al. 2017).

Multiple mechanisms are responsible for herbicide resistance in weeds. These can be classified in two broad classes: (i) target-site resistance and (ii) non-target-site resistance (or metabolism-based herbicide resistance, MHR) (Yu and Powles 2014; Duhoux et al. 2015). Target-site resistance mechanism has been extensively studied compared to the mechanisms that enable non-target-site resistance. The well-studied mechanism within the target-site resistance class is the evolutionary advantage of adaptive mutations at a site-of-action (Kaundun 2014; Sammons and Gaines 2014; Yang et al. 2016a, b; Kleinman and Rubin 2017). These mutations abolish the binding interaction between the site-of-action and the respective herbicide, which targets that site. Non-target site resistance refers to up-regulation mechanisms, resulting in amplification of metabolic and detoxification pathways that increase the herbicide detoxification and/or sequestration systems. This causes the limitation of the bioavailability of herbicides in the plant, in a manner, which makes them inadequate for binding to their site-of-action in sufficient level (Yang et al. 2016a, b; Kleinman and Rubin 2017).

The enzymes that are involved in metabolic herbicide resistance can confer cross-resistance to herbicides that belong to different chemical groups and sites of action, as they can automatically confer resistance to already existing, new or herbicides which have never been applied to plants before (Green 2014; Heap 2014; Walsh and Powles 2014). For example, in grass weeds, herbicide detoxification and development herbicide resistant is associated with elevated levels of Phase I, II and III proteins (Fig.10.2), including CYPs, UDP-glucose-dependent glycosyltransferases (UGTs), GSTs, as well as membrane-associated ABC transporter proteins (Green 2014; Heap 2014; Walsh and Powles 2014). Recent work has demonstrated that a GST isoenzyme, which belongs to the phi class (MHR-GSTF1), plays a significant role in detoxifying herbicides in resistant *Alopecurus myosuroides* and *Lolium rigidum* populations (Cummins et al. 1997b, 1999, 2009, 2013). MHR-GSTF1 is relevant to the phi class of GSTs that are induced in crops following treatment with herbicide safeners (Cummins et al. 2009). MHR-GSTF1 displays low catalytic activity to herbicides; however, it exhibits high activity toward organic hydroperoxides. This GST isoenzyme is believed to contribute to herbicide resistance by preventing the accumulation of cytotoxic hydroperoxides, which can be formed either directly or indirectly as an outcome of injury by herbicides (Cummins et al. 1999). In addition, the enzyme seems to play a regulatory role on cell metabolism and controls the accumulation of protective flavonoid compounds in the plant cell (Cummins et al. 2013). The significant role of MHR-GSTF1 in MHR was confirmed by verifying that the GST-inhibiting pharmacophore compounds (e.g., 4-chloro-7-nitro-benzoxadiazole) inhibit its activity in *Alopecurus myosuroides* and help restore herbicide control in MHR black-grass (Cummins et al. 2013).

5 Development of Engineered Plants to Overexpress Specific GSTs

In recent years, the advantages in recombinant DNA technology and in the methods for horizontal gene transfer from an organism to another have opened new ways to plant genetic engineering. The use of genetic engineering renders the targeted plant breeding possible beyond the limitation of natural hybridization and existing genetic variability. For example, one of the major contributions in the agricultural efficiency was the development of herbicide-resistant crops as for example the roundup ready *Glycine max* and *Zea mays* (Samsel and Seneff 2013; Bakshi and Dewan 2013). In this case, the transgenic plants were genetically engineered to overexpress a glyphosate tolerant 5-enolpyruvylshikimate 3-phosphate (CP4 EPSP) enzyme, thus being able to tolerate the herbicide, enabling more effective weed control (Padgette et al. 1995; Funke et al. 2006; Feng et al. 2010; Zhao et al. 2011; Duke 2011). GSTs consist an important enzyme family offered for efficient genetic breeding through the development of resistant varieties to multiple stresses, with maximum yield potential. The development of engineered plants to overexpress specific GSTs represents an area of intense research. Several reports over the last two decades have demonstrated the efficiency of this approach for the development of transgenic plants with increased herbicide-detoxifying capability and stress tolerance. Initial examples, showed that wheat overexpressing the maize *Zm*GSTF27 isoenzyme (Milligan et al. 2001) and tobacco plants overexpressing the maize *Zm*GSTF1 (Karavangeli et al. 2005), resulted in increased tolerance to alachlor. In another example, the combined co-expression of soybean *Gm*GSTU21 and a homoglutathione synthetase in tobacco plants conferred tolerance to the herbicide fomesafen (Skipsey et al. 2005).

Dixon et al. (2003) using in vitro directed evolution developed a chimeric form of maize *Zm*GSTU1-1 and *Zm*GSTU2-2, which displayed 27-fold enhanced fluorodifen detoxification activity, compared to the parent enzymes (Dixon et al. 2003). This enzyme when expressed in *Arabidopsis thaliana,* conferred enhanced tolerance to fluorodifen, compared to the wild-type plants. More recently, Cummins et al. (2013), investigated the biological function of MHR-GSTF1 isoenzyme from *Alopecurus myosuroides*, which was expressed in *A. thaliana* (Cummins et al. 2013). The work showed that transgenic *A. thaliana* acquired resistance to multiple herbicides and showed similar changes in their secondary, xenobiotic and antioxidant metabolism to those determined in MHR weeds.

Extensive work has been carried out over the last decade on the isolation, characterization and application of the herbicide-inducible GST isoenzyme *Gm*GSTU4-4 from *Glycine max*. *Gm*GSTU4-4 was used for the development of transgenic tobacco plants that displayed increased tolerance toward the herbicides fluorodifen, oxyfluorfen and alachlor, compared to wild-type plants (Benekos et al. 2010). In a similar work, Kissoudis et al. (2015a) studied the effects of the chloroacetanilide herbicide alachlor on the metabolome of wild-type and tobacco plants overexpressing *Gm*GSTU4. They showed that the increased metabolic capacity of *Gm*GSTU4 overexpressing plants is accompanied by pleiotropic metabolic alterations.

There are plenty of evidence, based on recent reports, indicating the contribution of GSTs in conferring tolerance toward a wide range of abiotic stresses, such as salt, drought, temperature and heavy metals (Table 10.1). For example, Le Martret et al. (2011) expressed in tobacco chloroplast three antioxidant enzymes: DHAR, GST and GR. Homoplasmic chloroplast transformants containing either DHAR or GST displayed enhanced salt and cold tolerance. Co-expression of DHAR:GR and GST:GR conferred methyl viologen-induced oxidative stress tolerance. Dixit et al. (2011) established that transgenic tobacco plants expressing a *Trichoderma virens* GST (*Tv*GST) are more tolerant to Cd, without enhancing its accumulation in the plant biomass. In another example, Liu et al. (2013) showed that the zeta class GST from *Pyrus pyrifolia* Nakai cv Huobali (*Pp*GST), when expressed in *Nicotiana tabacum* L. cv Xanthi, caused enhanced tolerance of transgenic tobacco line to oxidative stress triggered by drought, NaCl and Cd treatments.

Transgenic *A. thaliana* expressing the zeta class GST from *Tamarix hispida* (*Th*GSTZ1), showed enhanced resistance to salt and drought stress (Yang et al. 2014). Similarly, Sharma et al. (2014) showed that overexpression of a rice tau class GST gene improves tolerance to salinity and oxidative stresses in *Arabidopsis*. Chan and Lam (2014) showed that a putative lambda class GST from *Glycine max*

Table 10.1 Summary of the main studies reporting altered abiotic stress tolerance responses in GST transgenics

GST isoenzyme/class/origin	Transgenic plant	Improved trait	References
*Zm*GSTF27/Phi/*Zea mays*	*Triticum aestivum*	Increased tolerance to alachlor	Milligan et al. (2001)
Engineered form of maize *Zm*GSTU1-1 and *Zm*GSTU2-2/Tau/ *Zea mays*	*Arabidopsis thaliana*	Increased tolerance to fluorodifen	Dixon et al. (2003)
*Zm*GSTF1/Phi/*Zea mays*	*Nicotiana tabacum*	Increased tolerance to alachlor	Karavangeli et al. (2005)
*Gm*GSTU21/Tau/*Glycine max*	*Nicotiana tabacum*	Increased tolerance to fomesafen	Skipsey et al. 2005
*Gm*GSTU4-4/Tau/*Glycine max*	*Nicotiana tabacum*	Increased tolerance to fluorodifen, oxyfluorfen and alachlor	Benekos et al. (2010)
*Tv*GST/*Trichoderma virens*	*Nicotiana tabacum*	Increased tolerance to Cd	Dixit et al. (2011)
*Pp*GST/Zeta/*Pyrus pyrifolia* Nakai cv Huobali	*Nicotiana tabacum*	Increased tolerance to drought, NaCl, and Cd stresses	Liu et al. (2013)
MHR-GSTF1/Phi/*Alopecurus myosuroides*	*Arabidopsis thaliana*	Increased tolerance to multiple herbicides	Cummins et al. (2013)
ThGSTZ1/Zeta/*Tamarix hispida*	*Arabidopsis thaliana*	Increased tolerance to salt and drought stress	Yang et al. (2014)
*Os*GST/Tau/*Oryza sativa*	*Arabidopsis thaliana*	Increased tolerance to salinity and oxidative stresses	Sharma et al. (2014)
*Gm*GSTL1/Lambda/*Glycine max*	*A. thaliana* and tobacco BY-2 cells	Increased tolerance to salinity stress	Chan and Lam (2014)
*Cs*GST/Tau/*Citrus sinensis*	*Nicotiana tabacum*	Increased tolerance to fluorodifen, salt and drought stresses	Lo Cicero et al. (2015)
*Gm*GSTU4-4/Tau/*Glycine max*	*Nicotiana tabacum*	Increased tolerance to herbicide alachlor	Kissoudis et al. (2015b)
*Sl*GST/Tau/*Solanum lycopersicum*	*Arabidopsis thaliana*	Increased tolerance to salt and osmotic stress	Xu et al. (2015)
*Jr*GSTTau1/Tau/*Juglans regia*	*Nicotiana tabacum*	Increased tolerance to cold stress	Yang et al. (2016a, b)

(*Gm*GSTL1) when expressed in *A. thaliana* and tobacco BY-2 cells, enhances survival under salinity stress. They observed a marked reduction of ROS accumulation in transgenic *A. thaliana* plants under salt treatment. Kissoudis et al. (2015b) showed that *Gm*GSTU4-4 expression is highly induced following salt stress and atrazine treatment in *Glycine max*. Tobacco plants engineered to express *Gm*GSTU4-4 displayed high tolerance to salt stress (150 mM NaCl). Metabolomics analysis and comparison of transgenic and wild-type plants showed that under salt stress, transgenic tobacco plants maintained their cellular homeostasis in contrast to the wild-type plants that exhibited deregulated energy metabolism. Yang et al. (2016a, b) showed that a tau class GST gene from *Juglans regia* (*Jr*GSTTau1) is involved in chilling tolerance in transgenic tobacco plants. In a similar work, Xu et al. (2015) showed that a tau class GST isoenzyme from *Solanum lycopersicum* (*Sl*GST) when expressed in *A. thaliana* promoted resistance to salt and osmotic stress, induced by NaCl and mannitol, in the transgenic plant. The authors concluded that the increased tolerance was correlated with the changes in the levels of proline, malondialdehyde and other antioxidant enzyme activities. In a relevant work, Lo Cicero et al. (2015) reported that tobacco plants over-expressing a tau class isoenzyme from sweet orange *Citrus sinensis* (*Cs*GST) acquired tolerance to the diphenyl ether herbicide fluorodifen as well as to salt and drought stresses (Lo Cicero et al. 2015). Vijayakumar et al. (2016) using in silico bioinformatics analysis characterized the family of GSTs from *Brassica oleracea* and found the presence of 65 different isoenzymes. They proposed possible pathways in which GST genes were involved in cold stress.

6 Conclusion

GSTs can confer resistance to different herbicides, which have diverse biochemical mechanism of action, thus permitting the control of weeds with increased herbicide resistance or weeds that have developed resistance to a specific herbicide. In addition, GSTs function could be exploited for the development of plants tolerant to salt, drought, chilling or enhanced tolerance to herbicides like alachlor and possibly other chlorinated compounds (Karavangeli et al. 2005). In addition, engineering of GSTs through directed mutagenesis and/or directed evolution could lead to the development of improved enzyme forms (Axarli et al. 2016, 2017), with enhanced structural stability and catalytic activity toward xenobiotics, which may be explored in future research efforts.

Acknowledgement We acknowledge support by the research project for excellence IKY/SIEMENS; Grants4Targets 2016-2-25, funded by Bayer CropScience AG; the ISPP#0071 program, funded by King Saud University; and the IKY program "Strengthening Postdoctoral Researchers" (1st Cycle), funded by the National Strategic Reference Framework (ESPA) 2014-2020.

References

Axarli I, Muleta AW, Vlachakis D, Kossida S, Kotzia G, Maltezos A, Dhavala P, Papageorgiou AC, Labrou NE (2016) Directed evolution of tau class glutathione transferases reveals a site that regulates catalytic efficiency and masks co-operativity. Biochem J 473:559–570

Axarli I, Muleta AW, Chronopoulou EG, Papageorgiou AC, Labrou NE (2017) Directed evolution of glutathione transferases towards a selective glutathione-binding site and improved oxidative stability. BBA(General Subjects) 1861:3416–3428

Bakkali Y, Ruiz-Santaella JP, Osuna MD, Wagner J, Fischer AJ, De Prado R (2007) Late watergrass (*Echinochloa phyllopogon*): mechanisms involved in the resistance to fenoxaprop-p-ethyl. J Agr Food Chem 55:4052–4058

Bakshi S, Dewan D (2013) Status of transgenic cereal crops: a review. Clon Transgen 3:119

Banday ZZ, Nandi AK (2017) *Arabidopsis thaliana* GLUTATHIONE-S-TRANSFERASE THETA 2 interacts with RSI1/FLD to activate systemic acquired resistance. Mol Plant Pathol. https://doi.org/10.1111/mpp.12538

Benbrook CM (2012) Impacts of genetically engineered crops on pesticide usein the U.S. the first sixteen years. Environ Sci Eur 24:24

Benekos K, Kissoudis C, Nianiou-Obeidat I, Labrou N, Madesis P, Kalamaki M, Makris A, Tsaftaris A (2010) Overexpression of a specific soybean GmGSTU4 isoenzyme improves diphenyl ether and chloroacetanilide herbicide tolerance of transgenic tobacco plants. J Biotechnol 150:195–201

Bilang J, Sturm A (1995) Cloning and characterization of a glutathione S-transferase that can be photolabeled with 5-azido-indole-3-acetic acid. Plant Physiol 109:253–260

Bowler C, Van Camp W, Van Montagu M, Inze D (1994) Superoxide dismutase in plants. Crit Rev Plant Sci 13:199–218

Burla B, Pfrunder S, Nagy R, Francisco RM, Lee Y, Martinoia E (2013) Vacuolar transport of abscisic acid glucosyl ester is mediated by ATP-binding cassette and proton-antiport mechanisms in Arabidopsis. Plant Physiol 163:1446–1458

Chan C, Lam HM (2014) A putative lambda class glutathione S-transferase enhances plant survival under salinity stress. Plant Cell Physiol 55:570–579

Cho HY, Kong KH (2005) Molecular cloning, expression, and characterization of a phi-type glutathione S-transferase from *Oryza sativa*. Pestic Biochem Phys 83:29–36

Cho HY, Kong KH (2007) Study on the biochemical characterization of herbicide detoxification enzyme, glutathione S-transferase. Biofactors 30:281–287

Chronopoulou E, Madesis P, Asimakopoulou B, Platis D, Tsaftaris A, Labrou NE (2012) Catalytic and structural diversity of the fluazifop-inducible glutathione transferases from *Phaseolus vulgaris*. Planta 235:1253–1269

Csiszar J, Horvath E, Vary Z, Galle A, Bela K, Brunner S, Tari I (2014) Glutathione transferase supergene family in tomato: salt-stress-regulated expression of representative genes from distinct GST classes in plants primed with salicylic acid. Plant Physiol Biochem 78:15–26

Cummins I, Cole DJ, Edwards R (1997a) Purification of multiple glutathione transferases involved in herbicide detoxification from wheat (*Triticum aestivum* L.) treated with the safener fenchlorazole-ethyl. Pestic Biochem Phys 59:35–49

Cummins I, Moss S, Cole DJ, Edwards R (1997b) Glutathione transferases in herbicide-resistant and herbicide-susceptible black-grass (*Alopecurus myosuroides*). Pest Manag Sci 51:244–250

Cummins I, Cole DJ, Edwards R (1999) A role for glutathione transferases functioning as glutathione peroxidases in resistance to multiple herbicides in black-grass. Plant J 18:285–292

Cummins I, O'Hagan D, Jablonkai I, Cole DJ, Hehn A, Werck-Reichhart D, Edwards R (2003) Cloning, characterization and regulation of a family of phi class glutathione transferases from wheat. Plant Mol Biol 52:591–603

Cummins I, Bryant DN, Edwards R (2009) Safener responsiveness and multiple herbicide resistance in the weed black-grass (*Alopecurus myosuroides*). Plant Biotechnol J 7:807–820

Cummins I, Dixon DP, Freitag-Pohl S, Skipsey M, Edwards R (2011) Multiple roles for plant glutathione transferases in xenobiotic detoxification. Drug Metab Rev 43:266–280

Cummins I, Wortley DJ, Sabbadin F, He Z, Coxon CR, Straker HE, Sellars JD, Knight K, Edwards L, Hughes D, Kaundun SS, Hutchings SJ, Steel PG, Edwards R (2013) Key role for a glutathione transferase in multiple-herbicide resistance in grass weeds. Proc Nat Acad Sci USA 110:5812–5817

Dat J, Vandenabeele S, Vranova E, Van Montagu M, Inze D, Van Breusegem F (2000) Dual action of the active oxygen species during plant stress responses. CMLS 57:779–795

Del Río LA (2015) ROS and RNS in plant physiology: an overview. J Exp Bot 66:2827–2837

Deng F, Hatzios KK (2002a) Characterization and safener induction of multiple glutathione S-transferases in three genetic lines of rice. Pestic Biochem Phys 72:24–39

Deng F, Hatzios KK (2002b) Purification and characterization of two glutathione S-transferase isozymes from indica-type rice involved in herbicide detoxification. Pestic Biochem Phys 72:10–23

Deng F, Nagao A, Shim I, Usui K (1997) Induction of glutathione S-trasnferase isozymes in rice shoots treated with a combination of petilachlor and fenclorim. J Weed Sci Tech 42:277–288

DeRidder BP, Goldsbrough PB (2006) Organ-specific expression of glutathione S-transferases and the efficacy of herbicide safeners in Arabidopsis. Plant Physiol 140:167–175

Dixit P, Mukherjee PK, Ramachandran V, Eapen S (2011) Glutathione transferase from Trichoderma Virens enhances cadmium tolerance without enhancing its accumulation in transgenic *Nicotiana tabacum*. PLoS One 6:e16360

Dixon D, Cole DJ, Edwards R (1997) Characterisation of multiple glutathione transferases containing the GST I subunit with activities toward herbicide substrates in maize (*Zea mays*). Pestic Sci 50:72–82

Dixon DP, Cole DJ, Edwards R (1999) Dimerisation of maize glutathione transferases in recombinant bacteria. Plant Mol Biol 40:997–1008

Dixon DP, Laphtorn A, Edwards R (2002) Plant glutathione transferases. Genome Biol 3:3004.3001–3004.3010

Dixon DP, McEwen AG, Lapthorn AJ, Edwards R (2003) Forced evolution of a herbicide detoxifying glutathione transferase. J Biol Chem 278:23930–23935

Dubey AK, Kumar N, Sahu N, Verma PK, Ranjan R, Chakrabarty D, Behera SK, Mallick S (2016) Response of two rice cultivars differing in their sensitivity towards arsenic, differs in their expression of glutaredoxin and glutathione S-transferase genes and antioxidant usage. Ecotoxicol Environ Saf 124:393–405

Duhoux A, Carrère S, Gouzy J, Bonin L, Délye C (2015) RNA-Seq analysis of rye-grass transcriptomic response to an herbicide inhibiting acetolactate-synthase identifies transcripts linked to non-target-site-based resistance. Plant Mol Biol 87:473–487

Duhoux A, Carrère S, Duhoux A, Délye C (2017) Transcriptional markers enable identification of rye-grass (*Lolium sp.*) plants with non-target-site-based resistance to herbicides inhibiting acetolactate-synthase. Plant Sci 257:22–36

Duke SO (2011) Glyphosate degradation in glyphosate-resistant and-susceptible crops and weeds. J Agric Food Chem 59:5835–5841

Feng PCC, CaJacob CA, Martino-Catt SJ, Cerny RE, Elmore GA, Heck GR, Huang J, Kruger WM, Malven M, Miklos JA, Padgette SR (2010) Glyphosate-resistant crops: developing the next generation products. In: Nandula VK (ed) Glyphosate resistance in crops and weeds: history, development, and management. Wiley, Hoboken

Foyer CH, Descourvieres P, Kunert KJ, Psip PP (1994) Photooxidative stress in plants. Physiol Plant 92:696–717

Frear DS, Swanson HR (1970) Biosynthesis of s-(4-ethylamino-6-isopropylamino-2-s-triazino) glutathione: partial purification and properties of a glutathione S-transferase from corn. Phytochemistry 9:2123–2132

Fuerst EP, Gronwald JW (1986) Induction of rapid metabolism of metolachlor in sorghum (*Sorghum bicolor*) shoots by Cga-92194 and other antidotes. Weed Sci 34:354–361

Funke T, Han H, Healy-Fried ML, Fischer M, Schonbrunn E (2006) Molecular basis for the herbicide resistance of roundup ready crops. Proc Nat Acad Sci USA 103:13010–13015

Gray SB, Brady SM (2016) Plant developmental responses to climate change. Dev Biol 419:64–77

Green JM (2014) Current state of herbicides in herbicide-resistant crops. Pest Manag Sci 70:1351–1357

Hatton PJ, Cummins I, Cole DJ, Edwards R (1999) Glutathione transferases involved in herbicide detoxification in the leaves of *Setaria faberi* (giant foxtail). Physiol Plant 105:9–16

Hatzios KK, Burgos N (2004) Metabolism-based herbicide resistance: regulation by safeners. Weed Sci 52:454–467

Heap I (2014) Global perspective of herbicide-resistant weeds. Pest Manag Sci 70:1306–1315
Inupakutika MA, Sengupta S, Devireddy AR, Azad RK, Mittler R (2016) The evolution of reactive oxygen species metabolism. J Exp Bot 67:5933–5943
Irzyk GP, Fuerst EP (1993) Purification and characterization of a glutathione-s-transferase from benoxacor-treated maize (*Zea mays*). Plant Physiol 102:803–810
Jiang L, Yang Y, Jia LX, Lin JL, Liu Y, Pan B, Lin Y (2016) Biological responses of wheat (*Triticum aestivum*) plants to the herbicide simetryne in soils. Ecotoxicol Environ Saf 127:87–94
Karavangeli M, Labrou NE, Clonis YD, Tsaftaris A (2005) Development of transgenic tobacco plants overexpressing maize glutathione S-transferase I for chloroacetanilide herbicides phytoremediation. Biomol Eng 22:121–128
Kärkönen A, Kuchitsu K (2015) Reactive oxygen species in cell wall metabolism and development in plants. Phytochemistry 112:22–32
Kasote DM, Katyare SS, Hegde MV, Bae H (2015) Significance of antioxidant potential of plants and its relevance to therapeutic applications. Int J Biol Sci 11:982–991
Kaundun SS (2014) Resistance to acetyl-CoA carboxylase-inhibiting herbicides. Pest Manag Sci 70:1405–1417
Kissoudis C, Kalloniati C, Flemetakis E, Madesis P, Labrou NE, Tsaftaris A, Nianiou-Obeidat I (2015a) Maintenance of metabolic homeostasis and induction of cytoprotectants and secondary metabolites in alachlor treated GmGSTU4 overexpressing tobacco plants, as resolved by metabolomics. Plant Biotechnol Rep 9:287–296
Kissoudis C, Kalloniati C, Pavli O, Flemetakis E, Labrou NE, Madesis P, Skaracis G, Tsaftaris A, Nianiou-Obeidat I (2015b) Stress inducible GmGSTU4 shapes transgenic tobacco plants metabolome towards increased salinity tolerance. Acta Physiol Plant 37:102
Kitamura S, Shikazono N, Tanaka A (2004) TRANSPARENT TESTA 19 is involved in the accumulation of both anthocyanins and proanthocyanidins in *Arabidopsis*. Plant J 37:104–114
Kleinman Z, Rubin B (2017) Non-target-site glyphosate resistance in *Conyza bonariensis* is based on modified subcellular distribution of the herbicide. Pest Manag Sc 73:246–253
Labrou NE, Papageorgiou AC, Pavli O, Flemetakis E (2015) Plant GSTome: structure and functional role in xenome network and plant stress response. Curr Opin Biotechnol 32:186–194
Labrou NE, Muharram MM, Abdelkader MS (2016) Delineation of the structural and functional role of Arg111 in GSTU4-4 from Glycine max by chemical modification and site-directed mutagenesis. Biochim Biophys Acta 1864:1315–1321
Lallement PA, Meux E, Gualberto JM, Dumarcay S, Favier F, Didierjean C, Saul F, Haouz A, Morel-Rouhier M, Gelhaye E, Rouhier N, Hecker A (2015) Glutathionyl-hydroquinone reductases from poplar are plastidial proteins that deglutathionylate both reduced and oxidized glutathionylated quinones. FEBS Lett 589:37–44
Lambert DM, Larson JA, Roberts RK, English BC, Zhou XV, Falconer LL, Hogan RJ Jr, Johnson JL, Reeves JM (2017) Resistance is futile: estimating the costs of managing herbicide resistance as a first-order Markov process and the case of U.S. upland cotton producers. Agric Econ 48:1–10
Le Martret B, Poage M, Shiel K, Nugent GD, Dix PJ (2011) Tobacco chloroplast transformants expressing genes encoding dehydroascorbate reductase, glutathione reductase, and glutathione-S-transferase, exhibit altered anti-oxidant metabolism and improved abiotic stress tolerance. Plant Biotechnol J 9:661–673
Li G, Wu SG, Yu RX, Cang T, Chen LP, Zhao XP, Cai LM, Wu CX (2013) Identification and expression pattern of a glutathione S-transferase in *Echinochloa crus*-galli. Weed Res 53:314–321
Li D, Gao Q, Li X, Pang S, Liu Z, Wang C, Tan W (2016) Characterization of glutathione S-transferases in the detoxification of metolachlor in two maize cultivars of differing herbicide tolerance. Pestic Biochem Physiol. https://doi.org/10.1016/j.pestbp.2016.12.003
Li D, Xu L, Pang S, Liu Z, Wang K, Wang C (2017) Variable levels of glutathione S-transferases are responsible for the differential tolerance to metolachlor between maize (*Zea mays*) shoots and roots. J Agric Food Chem 65:39–44
Liu XF, Li JY (2002) Characterization of an ultra-violet inducible gene that encodes glutathione S-transferase in *Arabidopsis thaliana*. Acta Genet Sinica 29:458–465

Liu D, Liu Y, Rao J, Wang G, Li H, Ge F, Chen C (2013) Overexpression of the glutathione S-transferase gene from *Pyrus pyrifolia* fruit improves tolerance to abiotic stress in transgenic tobacco plants. Mol Biol (Mosk) 47:591–601

Lo Cicero L, Madesis P, Tsaftaris A, Lo Piero AR (2015) Tobacco plants over-expressing the sweet orange tau glutathione transferases (CsGSTUs) acquire tolerance to the diphenyl ether herbicide fluorodifen and to salt and drought stresses. Phytochemistry 116:69–77

Marrs KA (1996) The functions and regulation of glutathione S-Transferasesin plants. Annu Rev Plant Physiol Plant Mol Biol 47:127–158

McGonigle B, Lau SMC, Jennings LD, O'Keefe DP (1998) Homoglutathione selectivity by soybean glutathione S-transferases. Pestic Biochem Physiol 62:15–25

McGonigle B, Keeler SJ, Lan SMC, Koeppe MK, O'Keefe DP (2000) A genomics approach to the comprehensive analysis of the glutathione S-transferase gene family in soybean and maize. Plant Physiol 124:1105–1120

Milligan AS, Daly A, Parry MAJ, Lazzeri PA, Jepson I (2001) The expression of a maize glutathione S-transferase gene in transgenic wheat confers herbicide tolerance, both in planta and in vitro. Mol Breed 7:301–315

Mittler R (2002) Oxidative stress, antioxidants and stress tolerance. Trends Plant Sci 7:405–410

Mittler R, Vanderauwera S, Gollery M, Van Breusegem F (2004) Reactive oxygen gene network of plants. Trends Plant Sci 9:490–498

Moons A (2005) Regulatory and functional interactions of plant growth regulators and plant glutathione S-transferases (GSTs). Vitam Horm 72:155–202

Morel M, Meux E, Mathieu Y, Thuillier A, Chibani K, Harvengt L, Jacquot JP, Gelhaye E (2013) Xenomic networks variability and adaptation traits in wood decaying fungi. Microb Biotechnol 6:248–263

Mostofa MG, Hossain MA, Fujita M (2015) Trehalose pretreatment induces salt tolerance in rice (*Oryza sativa* L.) seedlings: oxidative damage and co-induction of antioxidant defense and glyoxalase systems. Protoplasma 252:461–475

Nianiou-Obeidat I, Madesis P, Kissoudis C, Voulgari G, Chronopoulou E, Tsaftaris A, Labrou NE (2017) Plant glutathione transferase-mediated stress tolerance: functions and biotechnological applications. Plant Cell Rep. https://doi.org/10.1007/s00299-017-2139-7

Padgette SR, Kolacz KH, Delannay X, Re DB, Lavallee BJ, Tinius CN, Rhodes WK, Otero YI, Barry GF, Eichholtz DA, Peschke VM, Nida DL, Taylor NB, Kishore GM (1995) Development, identification, and characterization of a glyphosate-tolerant soybean line. Crop Sci 35:1451–1461

Pandey S, Fartyal D, Agarwal A, Shukla T, James D, Kaul T, Negi YK, Arora S, Reddy MK (2017) Abiotic stress tolerance in plants: myriad roles of ascorbate peroxidase. Front Plant Sci 8:581

Pang S, Duan L, Liu Z, Song X, Li X, Wang C (2012) Co-induction of a glutathione-s-transferase, a glutathione transporter and an ABC transporter in maize by xenobiotics. PLoS One 7:e40712

Park AE, Kim IS, Do H, Jeon BW, Lee CW, Roh SJ, Shin SC, Park H, Kim YS, Kim YH, Yoon HS, Lee JH, Kim HW (2016) Structure and catalytic mechanism of monodehydroascorbate reductase, MDHAR, from *Oryza sativa* L. japonica. Sci Rep 6:33903

Perry ED, Moschini G, Hennessey DA (2016) Testing for complementarity: glyphosate tolerant soybeans and conservation tillage. Am J Agric Econ 98:765–784

Popelka M, Tuinstra M, Weil F (2010) Discovering genes for abiotic stress tolerance in crop plants. Clifford. Wiley-Blackwell Ames, Aiowa

Rao Madhava KV, Raghavendra AS, Reddy Janardhan K (2006) Physiology and molecular biology of stress tolerance in plants. Springer Netherlands, Dordrecht

Rea PA (2007) Plant ATP-binding cassette transporters. Annu Rev Plant Biol 58:347–375

Rossini L, Jepson I, Greenland AJ, Gorla MS (1996) Characterization of glutathione S-transferase isoforms in three maize inbred lines exhibiting differential sensitivity to alachlor. Plant Physiol 112:1595–1600

Roxas VP, Lodhi SA, Garrett DK, Mahan JR, Allen RD (2000) Stress tolerance in transgenic tobacco seedlings that overexpress glutathione S-transferase/glutathione peroxidase. Plant Cell Physiol 41:1229–1234

Saidi I, Chtourou Y, Djebali W (2014) Selenium alleviates cadmium toxicity by preventing oxidative stress in sunflower (*Helianthus annuus*) seedlings. J Plant Physiol 171:85–91
Sammons RD, Gaines TA (2014) Glyphosate resistance: state of knowledge. Pest Manag Sci 70:1367–1377
Samsel A, Seneff S (2013) Glyphosate's suppression of cytochrome P450 enzymes and amino acid biosynthesis by the gut microbiome: pathways to modern diseases. Entropy 15:1416–1463
Schröder P (2001) The role of glutathione and glutathione S-transferase in plant reaction and adaptation to xenobiotics. Kluwer, Dordrecht
Schröder P, Scheer CE, Diekmann F, Stampfl A (2007) How plants cope with foreign compounds – translocation of xenobiotic glutathione conjugates in roots of barley (*Hordeum vulgare*). Environ Sci Pollut R 14:114–122
Sharma R, Sahoo A, Devendran R, Jain M (2014) Overexpression of a rice tau class glutathione s-transferase gene improves tolerance to salinity and oxidative stresses in *Arabidopsis*. PLoS One 9:e92900
Singh S, Parihar P, Singh R, Singh VP, Prasad SM (2016a) Heavy metal tolerance in plants: role of transcriptomics, proteomics, metabolomics, and ionomics. Front Plant Sci 6:1143
Singh R, Singh S, Parihar P, Mishra RK, Tripathi DK, Singh VP, Chauhan DK, Prasad SM (2016b) Reactive oxygen species (ROS): beneficial companions of plants' developmental processes. Front Plant Sci 7:1299
Skipsey M, Andrews CJ, Townson JK, Jepson I, Edwards R (1997) Substrate and thiol specificity of a stress-inducible glutathione transferase from soybean. FEBS Lett 409:370–374
Skipsey M, Cummins I, Andrews CJ, Jepson I, Edwards R (2005) Manipulation of plant tolerance to herbicides through co-ordinated metabolic engineering of a detoxifying glutathione transferase and thiol cosubstrate. Plant Biotechnol J 3:409–420
Skipsey M, Knight KM, Brazier-Hicks M, Dixon DP, Steel PG, Edwards R (2011) Xenobiotic responsiveness of *Arabidopsis thaliana* to a chemical series derived from a herbicide safener. J Biol Chem 286:32268–32276
Skopelitou K, Muleta AW, Papageorgiou AC, Chronopoulou EG, Pavli O, Flemetakis E, Skaracis GN, Labrou NE (2017) Characterization and functional analysis of a recombinant tau class glutathione transferase GmGSTU2-2 from *Glycine max*. Int J Biol Macromol 94:802–812
Soranzo N, Gorla MS, Mizzi L, De Toma G, Frova C (2004) Organisation and structural evolution of the rice glutathione S-transferase gene family. Mol Gen Genomics 271:511–521
Thom R, Cummins I, Dixon DP, Edwards R, Cole DJ, Lapthorn AJ (2002) Structure of a tau class glutathione S-transferase from wheat active in herbicide detoxification. Biochemistry 41:7008–7020
Tripathi A, Indoliya Y, Tiwari M, Tiwari P, Srivastava D, Verma PK, Verma S, Gautam N, Chakrabarty D (2014) Transformed yeast (*Schizosaccharomyces pombe*) overexpressing rice tau class glutathione S-transferase (OsGSTU30 and OsGSTU41) shows enhanced resistance to hexavalent chromium. Metallomics 6:1549–1557
Viger PR, Eberlein CV, Fuerst EP, Gronwald JW (1991) Effects of Cga-154281 and temperature on metolachlor absorption and metabolism, glutathione content, and glutathione-S-transferase activity in corn (*Zea mays*). Weed Sci 39:324–328
Vijayakumar H, Thamilarasan SK, Shanmugam A, Natarajan S, Jung HJ, Park JI, Kim H, Chung MY, Nou IS (2016) Glutathione transferases superfamily: cold-inducible expression of distinct gst genes in *Brassica oleracea*. Int J Mol Sci 17:1211
Walsh MJ, Powles SB (2014) Management of herbicide resistance in wheat cropping systems: learning from the Australian experience. Pest Manag Sci 70:1324–1328
Xu J, Xing XJ, Tian YS, Peng RH, Xue Y, Zhao W, Yao QH (2015) Transgenic Arabidopsis plants expressing tomato glutathione s-transferase showed enhanced resistance to salt and drought stress. PLoS One 10:e0136960
Yang G, Wang Y, Xia D, Gao C, Wang C, Yang C (2014) Overexpression of a GST gene (ThGSTZ1) from *Tamarix hispida* improves drought and salinity tolerance by enhancing the ability to scavenge reactive oxygen species. Plant Cell Tissue Organ Cult 117:99–112

Yang Q, Deng W, Li X, Yu Q, Bai L, Zheng M (2016a) Target-site and non-target-site based resistance to the herbicide tribenuron-methyl in flixweed (*Descurainia sophia* L.) BMC Genomics 17:551

Yang G, Xu Z, Peng S, Sun Y, Jia C, Zhai M (2016b) In planta characterization of a tau class glutathione S-transferase gene from *Juglans regia* (JrGSTTau1) involved in chilling tolerance. Plant Cell Rep 35:681–692

Yu Q, Powles S (2014) Metabolism-based herbicide resistance and cross-resistance in crop weeds: a threat to herbicide sustainability and global crop production. Plant Physiol 166:1106–1118

Yuan JS, Tranel PJ, Stewart CN Jr (2007) Non-target-site herbicide resistance: a family business. Trends Plant Sci 12:6–13

Zhang O, Riechers DE (2004) Proteomic characterization of herbicide safener-induced proteins in the coleoptile of *Triticum tauschii* seedlings. Proteomics 4:2058–2071

Zhao T, Lin CY, Shen ZC (2011) Development of transgenic glyphosate-resistant rice with G6 gene encoding 5-enolpyruvylshikimate-3-phosphate synthase. Agri Sci China 10:1307–1312

Chapter 11
Glyoxalase System: A Glutathione-Dependent Pathway for Abiotic Stress Tolerance in Plants

Rituraj Batth, Muskan Jain, Sumita Kumari, and Ananda Mustafiz

Abstract Due to their sessile nature, plants have to go through various adverse environmental conditions. Abiotic stresses, such as salinity, drought, flooding, cold, heat, etc. have been the major environmental factors contributing to the decreased yield of important crop plants. Abiotic stress leads to an abrupt increase in the contents of methylglyoxal (MG) in plants. MG is a potent cytotoxin, and readily reacts with major macromolecules of the cell to form advanced glycation end-products (AGEs). MG detoxification is principally carried out by the glyoxalase (GLY) system, which consists of two enzymes, GLYI and GLYII. GLYI acts upon the non-enzymatically formed complex of MG and a molecule of reduced glutathione (GSH), leading to the production of S-D-lactoylglutathione (SLG). GLYII, then, catalyzes the conversion of SLG to D-lactate giving GSH back to the system, thereby maintaining GSH homeostasis. The glyoxalase pathway keeps a check on the elevation of the MG level and helps in maintaining a higher "reduced to oxidized" GSH ratio. The glyoxalase pathway has been directly correlated to abiotic stress tolerance. Overexpression of GLY enzymes confers improved abiotic stress tolerance in plants. This chapter provides insights into the importance of the glyoxalase pathway in stress response and sheds light on the dependence of the glyoxalase pathway on GSH as the key player in regulating the pathway.

Keywords Glyoxalase • Methylglyoxal • Gluthathione • Abiotic stress

---

Rituraj Batth and Muskan Jain contributed equally to this work.

R. Batth • M. Jain • A. Mustafiz (✉)
Faculty of Life Sciences and Biotechnology, Plant Molecular Biology Laboratory, South Asian University, New Delhi 110021, India
e-mail: amustafiz@sau.ac.in

S. Kumari
Sher-e-Kashmir University of Agriculture Sciences and Technology, Jammu 180009, India

M.A. Hossain et al. (eds.), *Glutathione in Plant Growth, Development, and Stress Tolerance*, https://doi.org/10.1007/978-3-319-66682-2_11

Contents

1 Introduction

Being immobile and sensitive organisms, plants have to encounter various abiotic stresses in nature. The abiotic stress decreases plant growth and productivity, resulting in reduced crop yields. The decrease in yields of crops can be due to various other factors too, but abiotic stress by far contributes most towards the reduced yield of crops (Boyer 1982). This scenario is going to worsen with the predicted forthcoming global changes in climate, bringing with them extreme environmental conditions, as well as the continuous increase of the world population, and ever-increasing deterioration of arable land and scarcity of fresh water (Turral et al. 2011). All this emphasizes the importance of developing stress-tolerant crops that are able to sustain growth and productivity in stressful environments (Wang et al. 2003). There is an increase by many-fold in methylglyoxal (MG) concentrations in response to various stress conditions, and this increased MG forms adducts with the major macromolecules of the cell, leading to cellular injury (Ahmed and Thornalley 2007; Yadav et al. 2005a, b). To combat such harmful conditions, plants have evolved a mechanism to detoxify excessive MG produced under adverse situations. Various enzymes are present in systems to help plants cope with this stress, such as glyoxalase system enzymes, MG reductase, aldo-keto reductases, etc. These enzymes convert the harmful metabolite MG to lactate or ultimately pyruvate, which undergoes the tricarboxylic acid (TCA) cycle, thus diverting the cell towards energy production. Engineering of glyoxalase pathway enzymes improves stress tolerance in plants. So, a detailed understanding of this pathway may help to build a new platform towards generation of stress-tolerant crops, thus addressing the problem of decreased yields, particularly under unfavorable environmental conditions.

2 Abiotic Stress and Its Effect on Plant Systems Through Methylgloxal (MG) Formation

Abiotic stress is defined as the impact of non-living factors on living organisms in a specific environment. Abiotic stress is basically the result of continuously changing environmental conditions, such as salinity, drought, heat, and cold. Plant response to abiotic stresses is dynamic and complex (Skirycz and Inzé 2010; Cramer 2010). The biology of cells is so complex that with any environmental stimulus, multiple signaling pathways are activated in response to it. The earliest metabolic responses to abiotic stresses include inhibition of protein synthesis and growth (Good and Zaplachinski 1994; Vincent et al. 2007; Ben-Zioni et al. 1967; Dhindsa and Cleland 1975), and an increase in protein folding and processing (Liu and Howell 2010). Energy metabolism is affected by the increasing severity of stress (Pinheiro and Chaves 2011; Cramer et al. 2007; Kilian et al. 2007). Thus, there are gradual and complex changes in metabolism in response to stress. The severity of stress and the genetic background of the plants are the basic determinants for the ultimate survival or death of plants. Plants when exposed to abiotic stresses (such as temperature, light, salinity, drought, and water and nutrient availability) show a drastic decrease in overall growth and yield performance (Wang et al. 2003). One of the reasons for the decrease in growth and metabolism of plants is due to the overproduction of MG, a compound that is produced during normal metabolism of glucose and lipid but has been found to increase by two to six fold in response to abiotic stresses (Yadav et al. 2005a, b). Stress-induced accumulation of MG has been observed in animals, mammals, yeast and bacterial systems (Cooper 1984; Kalapos et al. 1992), and in plants (Yadav et al. 2005a, b).

3 MG Detoxification and Its Role in Abiotic Stress Tolerance

3.1 Introduction to MG

MG is a three-carbon metabolite that exists ubiquitously across the genera from prokaryotes to eukaryotes. MG is produced as a by-product of various metabolic reactions, such as glycolysis, lipid peroxidation, protein degradation and photosynthesis. At higher concentrations, MG is known to be toxic for the cell, but at lower concentrations, MG can stimulate various stress responsive pathways (Kaur et al. 2015). Thus, MG can function as a signaling molecule in bacteria (Campbell et al. 2007), yeast (Maeta et al. 2004, 2005; Zuin et al. 2005; Takatsume et al. 2006; Nomura et al. 2008), animals (Akhand et al. 2001; Du et al. 2001; Fukunaga et al. 2005; Yamawaki et al. 2008; Riboulet-Chavey et al. 2006; Chang et al. 2011; Jia et al. 2012; Kevin and Anthony 1994; Laga et al. 2007) and plants (Hoque et al. 2012a-d; Sharma et al. 2012; Kaur et al. 2014). MG also acts as a stress signal molecule in plant systems and triggers a response by inducing several protein kinases and transcription factors, which in turn affect the expression of various downstream targets, thereby causing a global change in transcriptome (Kaur et al. 2015). In the same study, authors have suggested MAP kinase pathway genes are involved in MG signaling, since the transcript level of putative histidine kinase genes and six MAPK genes were found to be induced in response to MG stress in rice. MG in plants also modulates stress responses by interacting with other signaling molecules, including ROS and Ca^{2+}. MG regulates stomatal opening and closure, the production of ROS, cytosolic Ca^{2+} concentration, and expression of many stress responsive genes (Kaur et al. 2015). The signaling roles of MG in up-regulating stress-responsive pathways and its potential to activate multiple pathways have made MG a suitable marker for abiotic stress tolerance in plants (Hoque et al. 2012a-d). MG synthesis in the system occurs via both enzymatic and non-enzymatic pathways.

3.2 Enzymatic Methods of MG Formation

The only enzyme known to directly catalyze MG formation is MG synthase, which converts DHAP (dihydroxy acetone phosphate) into MG and inorganic phosphate (Hopper and Cooper 1972) (Fig. 11.1). This enzyme has been found only in bacteria (Cooper and Anderson 1970; Hopper and Cooper 1971, 1972; Cooper 1974, 1975) and some yeast species (Babel and Hofmann 1981; Murata et al. 1985). Although the isolation of MG synthase from the goat liver has been reported (Ray and Ray 1981), the presence of this enzyme in eukaryotic cells has not yet been detected (Phillips and Thornalley 1993a, b; Sato et al. 1980). MG is also produced through acetone via cytochrome P450, which catalyzes the reaction in two consecutive steps and consumes NADPH (Casazza et al. 1984; Koop and Casazza 1985). Amine oxidase utilizes aminoacetone obtained from threonine and glycine metabolism as a substrate to form MG (Elliott 1960; Urata and Granick 1963) (Fig. 11.1). The presence of an amine oxidase in mammals, yeast, and microbes has been demonstrated (Elliott 1959, 1960; Inoue and Kimura 1995).

3.3 Non-enzymatic Methods of MG Formation

The major route of MG formation in plants is the non-enzymatic breakdown of triose sugars (Thornalley 1990; Kalapos 1999). Under stress conditions, the increase in glycolytic activity leads to an imbalance in glycolysis pathway. Triose phosphates (DHAP and GAP) are believed to be quite unstable metabolites, which undergo spontaneous β-elimination of phosphate groups to form stable and reactive MG (Richard 1984, 1991, 1993; Phillips and Thornalley 1993b). Auto-oxidation of sugars, ketone bodies, Maillard reaction, and lipoperoxidation also lead to formation of MG (Turk et al. 2006; Thornalley et al. 1999; Esterbauer et al. 1982; Kalapos 1999) (Fig. 11.1). Thus, MG production is an inevitable outcome of glycolysis pathway in stress conditions.

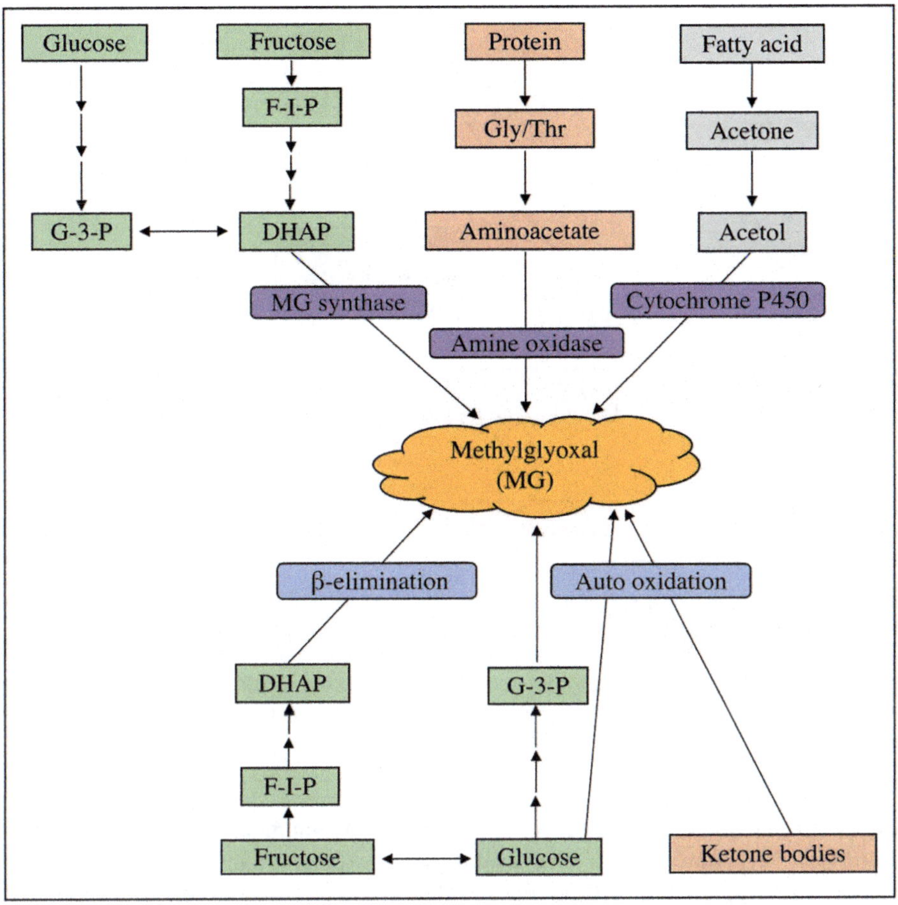

Fig. 11.1 A diagrammatic representation of MG synthesis. MG is synthesized from metabolism of sugars, proteins and fatty acids. Triose phosphate intermediates produced from glycolysis are the main source of MG. MG synthase can catalyze MG production from these intermediates or also they can undergo β-elimination of phosphate group forming MG. Similarly, actions of certain enzymes on protein and fatty acid metabolites can produce MG. Auto oxidation of sugars and ketone bodies is also known to form MG. The *upper panel* of the figure shows enzymatic pathways leading to production of MG and the *lower one* shows MG synthesis from non-enzymatic pathway

3.4 MG Accumulation

MG has two functional groups: a ketone group and an aldehyde group (Leoncini 1979). MG can react with the amino acids lysine, cysteine, and arginine to produce glycated proteins, referred to as advanced glycation end-products (AGEs) (Ahmed and Thornalley 2007), causing degradation of proteins and inactivation of the anti-oxidant defense system (Martins et al. 2001). Being a mutagen and a genotoxic agent, MG needs to be detoxified to maintain its homeostasis in the cell (Thornalley 2006).

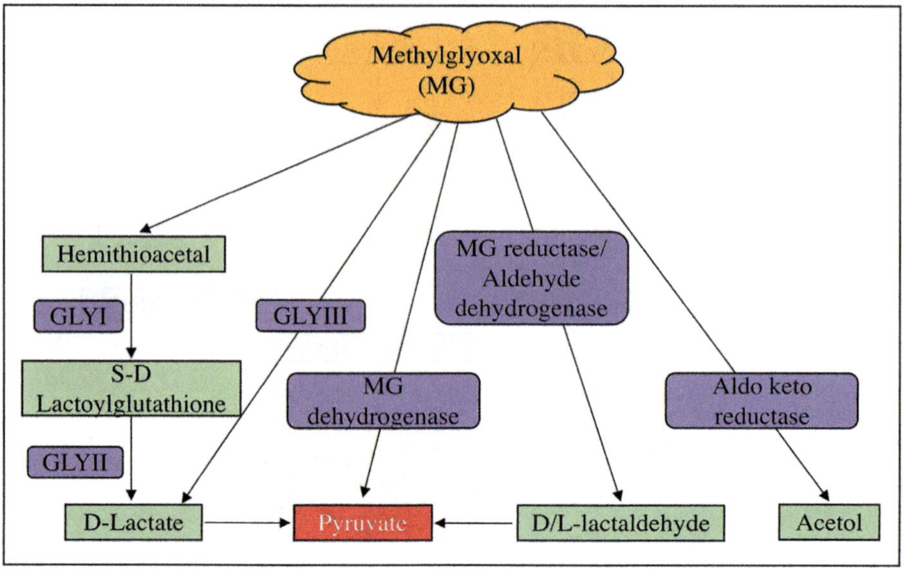

Fig. 11.2 Methylglyoxal detoxification system. MG acts as substrate for various enzymes that converts it into non-toxic molecules. Glyoxalase pathway converts MG into D-lactate via its two enzymes, GLYI and GLYII. GLYIII converts MG to D-lactate directly without using GSH or conceding any intermediate step. Other enzymes also act upon MG to detoxify it. MG dehydrogenase, MG reductase, aldehyde dehydrogenases, aldo keto reductases also act upon MG converting it to different products. Ultimately MG is converted to pyruvate, which enters TCA cycle of energy production

Under normal metabolic conditions, plants usually maintain a lower level (30–75 μM) of MG; however, an abrupt increase was observed in response to various abiotic stresses, including drought, salinity, cold, etc. (Yadav et al. 2005a, b; Hossain et al. 2009; Mostofa et al. 2015). The dicarbonyl group within MG reacts readily with the amine groups of proteins and nucleic acids. MG accumulation has been referred to as dicarbonyl stress, which has been implicated in tissue damage and aging (Rabbani and Thornalley 2014). This excessive MG accumulation in plant cells under stress can inhibit cell proliferation and cause the inactivation and degradation of proteins and inactivation of antioxidant defenses, leading to disruption of many cellular functions (Hoque et al. 2010).

3.5 *MG Detoxification*

MG detoxification involves conversion of toxic MG to less toxic molecules. The major pathway for MG detoxification is the glyoxalase pathway in which two enzymes (GLYI and GLYII) catalyze the conversion of MG to D-lactate using one molecule of GSH as a cofactor (Thornalley 1990) (Fig. 11.2). Apart from glyoxalases, several other enzymes also work to detoxify MG. A new enzyme, GLYIII, has recently been discovered which converts MG directly to D-lactate in a single step, without using GSH or any other cofactors (Misra et al. 1995). The ability of both the functional groups present in MG to be either reduced or oxidized makes MG a substrate for the enzymes involved in oxido-reduction reactions (Kalapos 1999). Various aldo-keto reductases and dehydrogenases have been identified in different species (Vander Jagt et al. 2001; Grant et al. 2003; Ko et al. 2005; Simpson et al. 2009). Aldo-keto reductases form acetol, lactaldehyde, and pyruvate from MG using NADH or NADPH (Yamauchi et al. 2011; Simpson et al. 2009; Narawongsanont et al. 2012; Kalapos 1999) (Fig. 11.2). MG dehydrogenase catalyzes conversion of MG to pyruvate, and MG reductase and aldehyde dehydrogenase converts MG to L/D-lactaldehyde, which in turn converted into pyruvate by L/D-lactate dehydrogenases. Ultimately MG is converted to pyruvate, which enters the TCA cycle of energy production.

4 The Glyoxalase System: Detoxification of Dicarbonyls

Initially, the glyoxalase system was thought to be a part of glycolysis, where glucose split into two molecules of triose phosphates, which were then converted to MG. In 1913, an enzymatic activity was discovered that converted this MG to lactic acid and this discovery paved the way for glucose metabolism from glucose to pyruvic acid or ethanol (Dakin and Dudley 1913; Neuberg 1913; Neuberg and Kobel 1928). Thus, MG and the glyoxalase system became the focal point of glycolysis (Neuberg and Kobel 1928). Later, various observations led to the dismissal of the glycolytic role of MG and glyoxalase. It was soon realised that DHAP and GAP are quite unstable molecules and MG is not a key metabolite but is produced non-enzymatically. When lactic acid formed from the glyoxalase pathway was found to be D-lactate, the exit of MG and GLY from glycolysis was confirmed (Racker 1954). Next the question about the physiological role of the glyoxalase pathway arose since it did not have any physiological substrate and led to a dead end-product. Many hypotheses were proposed explaining the role of glyoxalase enzymes but all of them were eventually dismissed (Szent-Györgyi 1974). Mannervik proposed the role of GLY in the elimination of dicarbonyl compounds (Mannervik et al. 1974), and his proposal remains the most convincing one to date. The hypothesis was that MG and other dicarbonyls are highly reactive compounds, and that the GSH-dependent enzymes play detoxification roles, thereby justifying the need for GSH-dependent MG removing the glyoxalase system.

5 The Glyoxalase System: A Glutathione (GSH)-Dependent Pathway

The two-step glyoxalase system catalyzes the conversion of MG to D-lactic acid via the intermediate S-D lactoylglutathione (SLG) (Carrington and Douglas 1986; Thornalley 1990). In the first step, GLYI (glyoxalase I; lactoylglutathione lyase) catalyzes the isomerization of the non-enzymatically formed hemithioacetal of MG and GSH to SLG. Thereafter, GLY II (glyoxalase II; S-2-hydroxyacylglutathione hydrolase) hydrolyses the product of the GLYI reaction to D-lactate releasing the molecule of GSH back to the system. Since the substrate for GLYI is the complex formed from MG and GSH, MG detoxification is strongly dependent on the availability of cellular GSH. A deficiency of GSH limits the production of hemithioacetal, leading to the accumulation of MG, which causes cellular damage.

6 The Glyoxalase Pathway in the Plant System

6.1 Glyoxalase I

GLYI is a primary enzyme of the glyoxalase pathway. In plants, its activity was first detected in Douglas fir needles (Smits and Jhonson 1981). Later GLYI activity was also detected in various monocots and dicots, for example *Lycopesicon esculentum* (Espartero et al. 1995), *Brassica oleracea* (Clugston et al. 1998), *Brassica juncea* (Veena et al. 1999), *Oryza sativa* (Usui et al. 2001), *Triticum aestivum* (Lin et al. 2010), *Arabidopsis thaliana* (Jain et al. 2016), and various other plants (Chen et al. 2004; Skipsey et al. 2000; Hossain and Fujita 2009; Hossain et al. 2009). GLYI has also been characterized in different organisms, like mammals (Aronsson et al. 1978; Cameron et al. 1997), yeast (Aronsson et al. 1978; Marmstal et al. 1979; Gomes et al. 2005; Martins et al. 2001), bacteria (He et al. 2000; Sukdeo et al. 2004; MacLean et al. 1998), and protozoan parasites (Akoachere et al. 2005; Ariza et al. 2006; Greig et al. 2006; Iozef et al. 2003; Sousa et al. 2005; Barata et al. 2010). Although GLYI is ubiquitously present, it is absent in some protozoa, like *Entamoeba histolytica*, *Trypanosoma brucei,* and *Giardia lamblia* (Sousa et al. 2012). It has recently been shown that degradation of the GLYI enzyme in stigma leads to a self-incompatibility response due to an increased level of MG and formation of MG-modified proteins in tissue (Subramanian et al. 2015). In this study, GLYI acts as a stigmatic compatibility factor and is required for pollination to occur, and its overexpression in *B. napus* lines was sufficient to partially break down pollen self-incompatibility (Subramanian et al. 2015).

GLYI catalyzes a virtually irreversible reaction with the formation of SLG from hemithioacetal, formed previously in a non-enzymatic step between MG and GSH (Martins et al. 2001; Ariza et al. 2005; Clugston et al. 2004; Thornalley 2003; Deponte et al. 2007). Other substrates of GLYI are glyoxals, phenylglyoxals, and hydroxyl-pyruvatealdehyde discovered first by Racker in 1952. The three chemical species MG, GSH, and hemimercaptal adducts (hemithoacetal) are simultaneously in equilibrium, making kinetic studies of GLYI quite difficult (Vander et al. 1972, 1975; Mannervik et al. 1973, 1974; Mannervik and Ridderstrom 1993). The reaction mechanisms involving hemithioacetal as the single substrate for GLYI was first proposed in 1961 and was universally accepted (Cliffe and Waley 1961). Later the mechanism of ordered reaction was discovered, which proposed binding of GSH followed by α-oxoaldehyde in the active site (Bartfai et al. 1973). By performing a steady-state kinetic analysis of GLYI from erythrocytes and yeast, the same authors proposed a random mechanism comprising two hypotheses: first GLYI can react with one substrate, the hemithioacetal; second, GLYI can display a two-substrate ordered mechanism, where GSH and MG act as the first and second substrates, respectively (Mannervik et al. 1973, 1974). These authors were the first to report a goodness-of-fit mechanism that differentiates between steady-state kinetics. According to these authors intracellular GLYI may already be complexed to thiol and then reacts with MG at the catalytic surface. These studies were rarely considered while doing GLYI enzyme kinetics and a single-substrate mechanism is usually followed (Akoachere et al. 2005; Sousa et al. 2005; Deponte et al. 2007; Vickers et al. 2004; Allen et al. 1993). A new approach of model discrimination was proposed in a study conducted with yeast GLYI (Lages et al. 2012). They optimized the initial substrate concentration of GLYI in the presence of the subsequent pathway enzyme GLYII. The method they used provided a design where extension of the Kullback-Leibler distance is maximized when computed over various time courses predicted by different models. With this, the best possible discrimination conditions were achieved. In this study, the single substrate model for yeast GLYI was not valid, rather two substrate mechanisms for the kinetics of yeast GLYI are reported (Lages et al. 2012).

GLYI is metalloenzymatic and requires divalent metal ions for its activity. It allows two metal ions in the active site (Ridderstorm et al. 1998). Metal ion is envisioned to play a catalytic role by directly coordinating with the *cis*-enediolate intermediate that forms along the reaction pathway of GLYI (Cameron et al. 1999). GLYI is broadly classified into two metal activation classes, i.e., Zn^{2+}-dependent and Zn^{2+}-independent (Neuberg 1913; Thornalley 2003). Previously, Zn^{2+}-dependent GLYIs were thought to have a eukaryotic origin and Zn^{2+}-independent GLYIs to have a prokaryotic origin. For example, the GLYI apoenzyme from *Homo sapiens* has been fully reactivated by Zn^{2+} and Mg^{2+} (Sellin et al. 1982, 1983). GLYI from *S. cerevisiae* is partially reactivated by the addition of Mg^{2+}, Ca^{2+}, and Mn^{2+}, but not by Fe^{2+}, Co^{2+}, and Ni^{2+} (Murata et al. 1986). The bacterial GLYI enzymes from *P. aeruginosa, N. meningitides*, and *Y. pestis* are not activated by Zn^{2+}, rather they are activated by Ni^{2+} or Co^{2+} (Sukdeo et al. 2004) and maximum activation of GLYI from *E. coli* is seen with the Ni^{2+} ion. This origin-based classification of GLYIs was later falsified with the discovery of newer glyoxalase genes belonging to both metal activation classes to co-exist in prokaryotes as well as eukaryotes. For example, multiple GLYI encoding genes from both metal activation classes are present in *P. aeruginosa,* a eubacterial species (Sukdeo and Honek 2007). Also, a unique Ni^{2+}-dependent and MG-inducible GLYI is reported in rice (Mustafiz et al. 2014). A higher eukaryote, such as plants, is known to consist of multiple isoforms of GLYI proteins (Mustafiz et al. 2011; Kaur et al. 2013; Ghosh and Islam 2016). As seen, a recent study in *Arabidopsis thaliana* reported the presence of both Zn^{2} -dependent and non-Zn^{2+}-dependent forms of GLYI proteins encoded by multiple GLYI genes (Jain et al. 2016). Out of three reported active GLYIs, AtGLYI2 is found to be Zn^{2+} dependent whereas AtGLYI3 and AtGLYI6 are Ni^{2+}-dependent GLYI enzymes.

Amino acid sequence comparison indicated the presence of a more extended peptide chain in Zn^{2+}-dependent GLYI [*A. thaliana* (AtGLYI2), *H. sapiens* (GlyI), *P. putida* (GlyI) *and P. aeruginosa* (GloA3)] compared to non-Zn^{2+}-dependent GLYI from *E. coli* (GlxI), *Y. pestis* (Gly I), *P. aeruginosa* (GloA2, GloA3), and *N. meningitidies* (Gly I), *A. thaliana* (AtGLYI3, AtGLYI6), and *O. sativa* (OsGly11.2) (Sukdeo and Honek 2007; Jain et al. 2016). The kinetic profile reveals AtGLYI2 to be 250 and 670 times more active than AtGLYI3 and AtGLYI6, respectively (Jain et al. 2016). Not only this, Arabidopsis AtGLYI2 can be considered the most active, with specific activity of 5,157 μmol/min/mg protein (Jain et al. 2016), in comparison to GLYIs of other plant species, such as rice [*O. sativa*: 120 μmol/min/mg (Mustafiz et al. 2014)], onion (*A. cepa:* 4.45 μmol/min/mg), radish (*R. sativus*: 2.19 μmol/min/mg), carrot (*D. carota*: 1.13 μmol/min/mg), and sweet potato (*I. batatas*: 1.04 μmol/min/mg) (Hossain et al. 2005). Also, AtGLYI2 shows the highest activity among other known Zn^{2+}-dependent GLYI enzymes belonging to different species, such as *P. aeruginosa* (Sukdeo and Honek 2007), *P. falciparum* (Vander and Han 1973), *S. cerevisiae* (Iozef et al. 2003), and *H. sapiens* (Ridderstrom and Mannervik 1996). AtGLYI3 and AtGLYI6 showed the highest activity (Jain et al. 2016) among other known Ni^{2+}-dependent GLYI enzymes from *O. sativa* (Mustafiz et al. 2014), *P. aeruginosa* (Sukdeo and Honek, 2007), *E. coli* (Clugston et al. 1998), and *C. acetobutylicum* (Suttisansanee et al. 2011).

So far structural characterization of GLYI enzymes has been done in six different species, namely *Homo sapiens* (Aronsson et al. 1978; Cameron et al. 1997), *E. coli* (He et al. 2000), *Leishmania major* (Ariza et al. 2006), *Mus musculus* (Kawatani et al. 2008), *Clostridium aycetobutylicum* (Suttisansanee et al. 2011), and *Pseudomonas aeruginosa (*Bythell-Douglas et al. 2015). GLYI from these species display a characteristic homodimeric quaternary structure, in which each monomer comprises a βαβββ domain that interacts to generate a continuous eight-stranded *β*-sheet with another βαβββ domain present in the opposite monomer. The two active sites are formed at the concavities of each interdomain *β*-sheet, which corresponds to the interface between the monomers. An exception to this is *C. acetobutylicum*, in which an eight-stranded *β*-sheet results from interaction with the βαβββ domain in the same monomer (Suttisansanee et al. 2011). Such versatile topology of a typical glyoxalase fold has been suggested to be a result of domain swapping and gene duplication (Cameron et al. 1997). Although these GLYI are homodimeric in nature, the existence of a monomeric form of GLYI has also been suggested based on the primary sequence comparison in some organisms, including yeast, *Plasmodium*, rice, and wheat (Ridderstrom and Mannervik 1996; Iozef et al. 2003; Mustafiz et al. 2014; Lin et al. 2010). Monomeric GLYI has been suggested to contain a four-βαβββ domain, even though their spatial organization in the protein quaternary structure has not been recognized. GLYI from yeast and *P. falciparum* are among the very few characterized enzymes with a single polypeptide with two active sites that catalyze the same reaction (Frickel et al. 2001; Deponte et al. 2007). Also, GLYI from rice appears to have only one active site (Mustafiz et al. 2014). The structural analysis of GLYI enzymes suggested that the glyoxalases in which the metal cofactor displays an octahedral coordination geometry were catalytically active (He et al. 2000; Suttisansanee et al. 2011, 2015). Such octahedral geometry arises from metal coordination of four conserved residues present in His/Glu/Gln/Glu or His/Glu/His/Glu metal binding motifs with two water molecules. Recently, a monomeric GLYI has been biochemically and structurally characterized from *Zea mays* (ZmGLX1) (Turra et al. 2015), reporting the first atomic model of GLYI from plants. The results indicate that the overall fold of ZmGLX1 constitutes an arrangement of four βαβββ motifs in two domains, as revealed by a high-resolution structure. However, instead of forming eight-stranded *β*-sheets between two monomers, ZmGLX1 is a single polypeptide comprising two interdomain eight-stranded *β*-sheets, out of which only the middle eight-stranded *β*-sheet, called site A, bears a functional active site and binds a single Ni^{2+} ion (Turra et al. 2015). The other site B seems to be able to bind GSH conjugates, presenting the possibility of ZmGLX1 being noncompetitively affected by such molecules, probably as part of a yet to be discovered regulatory mechanism or an altogether different catalytic activity, like the one that has recently been discovered for GLYI from *P. aeruginosa* (Bythell-Douglas et al. 2015). Such a characteristic, where the same enzyme shows different catalytic specificity or binds different substrates, is not uncommon in the GLYI protein family. For instance, homodimeric GLYI from *E. coli* displays different specificities in each of its two active sites (Su et al. 2008) and monomeric GLYI from *S. cerevisea* and *P. falciparum* contains two active sites with different substrate preferences and catalytic activity (Frickel et al. 2001; Deponte et al. 2007).

GLYI enzymes have long been considered to be cytosolic proteins (Thornalley 2003), until few reports have shown otherwise, for example, subcellular localization of GLYI proteins in apicoplast of *Plasmodium* (Urscher et al. 2010) and in the peroxisome of Arabidopsis (Quan et al. 2010). Also, it has been predicted that many plant GLYI can be compartmentalized into different organelles (Kaur et al. 2013). This suggests that intracellular compartmentalization of GLYI is another mode of MG detoxification in cell organelles. A genome-wide study carried out in rice reported the presence of 11 putative GLYI genes (Mustafiz et al. 2014), out of which only four were functionally active and were found to be in either cytoplasm or chloroplast (Kaur et al. 2013). In a recent study carried out by Kaur et al. (2016), it was found that out of four active GLYIs in rice, OsGLYI-8 is localized in nucleus. It is a dimeric enzyme that binds Zn^{2+}/Mn^{2+} ions and shows enzyme activity in the presence of a trace amount of metal ions, and exhibits unusual biphasic steady-state kinetics. Despite the presence of two active sites, most of the dimeric enzymes followed Michaelis-Menten kinetics (Sukdeo and Honek 2007). However, OsGLYI-8 was found to be biphasic with two apparent K_m and K_{cat} values. This biphasic kinetics is not a usual characteristic of GLYIs. Another GLYI that has been reported to show biphasic kinetics is the *Pf*GloI from *Plasmodium* (Deponte et al. 2007). However, unlike OsGLYI-8, it possesses two non-identical active sites. For homodimeric OsGLYI-8, the observed biphasic kinetics is rather intriguing (Kaur et al. 2016).

6.2 Glyoxalase II

GLYII is characterized in several eukaryotes and prokaryotes, including humans (Cameron et al. 1999; Ridderstrom et al. 1996), plants (Limphong et al. 2010; Zang et al. 2001), yeast (Gomes et al. 2005; Martins et al. 2001; Inoue et al. 2011), bacteria (Suttisansanee and Honek 2011; O'Young et al. 2007), and protozoan parasites (Sousa et al. 2005; Irsch and Krauth 2004; Silva et al. 2008; Urscher and Deponte 2009; Wendler et al. 2009). Like GLYI, this enzyme is also nearly ubiquitous in all living organisms, although some studies reported its absence in some mammals (Agar et al. 1984; Valentine et al. 1970). Another study shows the presence of more than one GLYII-encoding gene, for example protozoan parasites like *Trypanosoma brucei* lack GLYI altogether and have two GLYII encoding genes (Sousa et al. 2012).

GLYII catalyzes the conversion of SLG into D-Lactic acid. It is a member of the metallo-*β*-lactamase superfamily (Daiyasu et al. 2001). The proteins in this superfamily share a common αβ/βα fold. GLYII exists in the mitochondria as well as the cytosol of eukaryotes. In animals, multiple forms of GLYII are found in the mitochondria (both in the intermembrane space as well as in the matrix), and only one form in the cytosol, and this cytosolic form of GLYII appears to resemble the GLYII from inter-membraneous space (Principato et al. 1987; Talesa et al. 1988, 1989, 1990). Plants, on the other hand, appear to have a single mitochondrial GLYII form and multiple cytosolic forms (Talesa et al. 1990; Norton et al. 1990; Ridderstrom and Mannervik 1997). GLYII enzyme purification and characterization has been carried out in *Aloe vera*, spinach (Norton et al. 1990; Talesa et al. 1990), and *Arabidopsis thaliana* (Maiti et al. 1997; Ridderstrom and Mannervik 1997), where five different isoforms of GLYII have been identified (Maiti et al. 1997). Three of these isoforms are predicted to be mitochondrial (GLX2-1, GLX2-4 and GLX2-5), while GLX2-2 is found to be cytosolic. Yeast also has a cytosolic as well as a mitochondrial form of GLYII (Bito et al. 1997, 1999). The presence of GLYII in mitochondria suggests that the SLG is transported from cytosol to the mitochondria in eukaryotes (Talesa et al. 1990), where D-lactic acid is converted to pyruvate by D-α-hydroxyacid dehydrogenase (Talesa et al. 1990).

GLYII is very specific for the GSH moiety of the substrate, although it can interact with other thioesters too (Thornalley 1990, 1993; Vander 1993). A mutation in the GSH-binding domain at the C terminus of human GLYII resulted in a complete loss of its activity (Park et al. 2006). All species having GLYII, including human, yeast, and Arabidopsis, contain a highly conserved metal binding domain (THXHXDH), which is also present in the family of metallo-*β*-lactamases and is known to require Zn^{2+} ion (Maiti et al. 1997; Crowder et al. 1996, 1997; Concha et al. 1996). Based on its similarity to the metallo-β-lactamases, it is predicted that GLYII binds two Zn^{2+} ions by utilizing five histidines, two aspartic acids, and a bridging water molecule (Crowder et al. 1997). These predictions have also been shown by Melino et al. (1998) through determination of the human GLYII crystal structure (Cameron et al. 1999). GLYII has been characterized in more detail in

H. sapiens (Allen et al. 1993; Ridderstrom et al. 1996) and *A. thaliana* (Ridderstrom et al. 1997; Zang et al. 2001; Schilling et al. 2003; Wenzel et al. 2004). Metal binding analysis of the *H. sapiens* and the *A. thaliana* GLYII have indicated a metal stoichiometry of approximately two metal ions per monomeric enzyme, and that these ions are necessary for catalysis (Wenzel et al. 2004; Cameron et al. 1999). The nature of the metal ions present in the active site of GLYII enzyme has been studied in detail in *A. thaliana*. A mutant form of the *A. thaliana* cytosolic GLYII (R248W) contains two Zn^{2+} ions (Crowder et al. 1997); the wildtype enzyme and the mitochondrial enzyme have been shown to contain varying ratios of Zn^{2+}, Fe^{2+}, and Mn^{2+} (Zang et al. 2001; Schilling et al. 2003; Wenzel et al. 2004). The crystal structures of the *H. sapiens* and *A. thaliana* enzymes show that GLYII shares the same overall fold as the Zn^{2+}dependent metallo-β-lactamases (Cameron et al. 1999; Marasinghe et al. 2005).

A genome-wide study by Mustafiz et al. (2011) has shown that the plant genome contains multiple isoforms of GLYII enzymes, three in rice and five in Arabidopsis, based on the common Pfam id: PF00753. Recently a OsGLYII-2, a GLYII enzyme from *O. sativa* has been functionally characterized and found to be active (Ghosh et al. 2014). In a previous study, expression of the OsGLYII-2 transcript has been reported to up-regulate in response to salinity (Mustafiz et al. 2011). The K_m and k_{cat}/K_m values for OsGLYII-2 were found to be 254 μM and 2.00×10^6 M^{-1} sec^{-1}, respectively. These values were higher in comparison to prokaryotic GLYII and lower than eukaryotic GLYII, and were somewhat comparable with higher eukaryotes (Ghosh et al. 2014). As per the authors, the activity of OsGLYII-2 could be under the influence of reduced GSH produced or recycled as an end-product by the glyoxalase pathway. From the previous reports, it is already known that the activity of the enzymes can be affected by their products or substrates as a quick response towards cellular needs (Majumdar et al. 2013). GSH acts as a signaling molecule (Ghanta and Chattopadhyay 2011) and plays a significant role in maintaining cellular redox status, so it might influence GLYII activity as well. Under normal physiological conditions, the total activity of GLYI is much higher than GLYII and it occupies one molecule of GSH to form SLG. In addition to this, the total GSH in plants (~4.8 μmol g^{-1}) is relatively lower in comparison to animals (16–25 μmol g^{-1}; Newton et al. 1996). So, in this scenario, the product inhibition of OsGLYII-2 gains an important and tight correlation between OsGLYII-2 activity and cellular GSH level, thereby leading to maintenance of redox status. As the level of MG increases during stress, more GSH binds to MG to form SLG. Lower levels of GSH would increase the activity of OsGLYII-2, which will regenerate GSH from SLG and finally detoxify MG to D-lactate (Ghosh et al. 2014).

OsGLYII-2 is predicted to contain a binuclear Zn/Fe metal center, like its Arabidopsis counterpart, AtGLYII-2 (Zang et al. 2001). Interestingly, neither Zn^{2+} or Fe^{2+} could restore the activity of OsGLYII-2 when added externally to the metal-chelated form of the enzyme (Ghosh et al. 2014). The inability of Zn^{2+} to reactivate chelated OsGLYII-2 could be due to the presence of eight cysteine residues in the OsGLYII-2 protein, which could lead to de-metallation by thiolate ligation as observed in a previous study (O'Young et al. 2007). The activity of chelated

OsGLYII-2 protein could be restored by incubation with Co^{2+} or Mn^{2+} because of their simplistic nature, and inserted into the active site cavity (Ghosh et al. 2014). Similar observations have also been noted in the case of *E. coli* GLYII (O'Young et al. 2007).

6.3 Glyoxalase III

Discovery of GLYIII (EC 4.2.1.130) in 1995 in *E. coli* challenged the concept of the glyoxalase pathway. It catalyzes the conversion of methylglyoxal into D-lactate without requiring GSH or any other cofactors (Misra et al. 1995). This is an irreversible reaction with neither formation nor catabolism of SLG (Fig. 11.2). This enzyme exhibits a higher activity than GLYI and GLYII and represents the main system for methylglyoxal detoxification in *E. coli* cells in the stationary phase (Benov et al. 2004; Okado-Matsumoto and Fridovich 2000). However, its expression is not induced in the presence of methylglyoxal in growth medium (Matsumoto and Fridovich 2000) or in mutants lacking GLYI (MacLean et al. 1998). Also, it is not a very efficient enzyme with a K_{cat}/K_m value of $1.8 \times 10^3\ M^{-1}\ s^{-1}$ (Subedi et al. 2011), when compared with GLYI, which has a K_{cat}/K_m value of $1.2 \times 10^7\ M^{-1}\ s^{-1}$ (Clugston et al. 1998). Previously it had also been shown that a known Hsp31, a heat-inducible chaperon, showed GLYIII-like activity in *E. coli* (Subedi et al. 2011). Further analysis of *E. coli* GLYIII revealed that it is a member of the DJ-1/*Pfp*-I super family. DJ-1 family proteins are also known as PARK7 (Parkinson disease protein 7), a molecular chaperone associated with the early onset of Parkinson's disease. GLYIII activity was also found in its homolog from the mouse and *Caenorhabditis elegans* (Lee et al. 2012). It was proposed that DJ-1/GLYIII is a scavenger for reactive carbonyl species (Lee et al. 2012), but its relevance for methylglyoxal catabolism is still unclear.

GLYIII is a 23 kDa protein dimer with a putative active site close to the dimer interface, in which the residues Cys_{106}, His_{126}, and Glu_{18} may play an important role in catalysis and are highly conserved (Tao and Tong 2003). This catalytic triad, cysteine, histidine, and glutamic acid/aspartic acid, is identical to the one at the putative protease catalytic site in *E. coli* Hsp31, also known as GLYIII (Zhao et al. 2003), and the structures of both proteins are rather similar, sharing an evolutionarily conserved domain (Zhao et al. 2003). DJ-1/PfpI domain-containing proteins have been found to perform other physiological roles, such as regulation of mitochondrial function (Canet-Aviles et al. 2004), regulation of transcription (Clements et al. 2006), molecular chaperone (Subedi et al. 2011), and stimulation of antioxidant enzymes (Xu et al. 2010) and protease (Du et al. 2000) in various organisms. DJ-1 proteins play an important role in protection against MG-induced cell death in human, mouse and *C. albicans* (Lee et al. 2012). Moreover, mutation in human DJ-1 protein leads to Parkinson's disease (Tao and Tong 2003) and various forms of cancers (Hod 2004). In case of plants, loss of AtDJ-1a function causes cell death, whereas overexpression leads to increased oxidative stress tolerance (Xu et al. 2010). It has been also observed that expression of AtDJ-1a is induced by various stresses and interacts with several stress-inducible proteins, such as superoxide dismutase and GSH peroxidase (Xu et al. 2010).

Recently a sequence analysis of DJ-1 members in various plant species revealed the existence of DJ-1 genes in monocots, dicots, lycopods, gymnosperms, and bryophytes (Ghosh et al. 2016). The domain structure(s) (single or two-domain DJ-1/pfp-1) was conserved in both monocots and dicots, and a large number of plant species had preference for two-domain structures. Multiple members of DJ-1 proteins were found in both Arabidopsis and rice genomes. Six DJ-1 genes were found in both rice and Arabidopsis that encode for 12 and 11 proteins, respectively. Out of 12 DJ-1 genes in rice, upregulation of the OsDJ-1 gene has been observed in rice. Members of the DJ-1 gene family in rice showed upregulation in response to both biotic (4-hydroxyphenylacetic acid and linolenic acid) and abiotic factors (cold, osmosis, heat, heavy metals, etc.). In silico analysis of OsDJ-1 proteins revealed that OsDJ-1 proteins may interact with a total of 503 proteins of rice. Highly stress-responsive *cis*-acting elements, such as AERB, MTS, and LBR, were predicted to be present in the OsDJ-1 gene promoter, further emphasizing their role towards abiotic stress tolerance (Ghosh et al. 2016). Functionally GLYIII activity of DJ-1 proteins has been reported in various species, including *E. coli* (Subedi et al. 2011), *H. sapiens* (Lee et al. 2012), *C. albicans* (Hasim et al. 2014), *A. thaliana* (Kwon et al. 2013), and *S. pombe* (Zhao et al. 2014). Different roles, including molecular chaperones, amino peptidase, and proteases (Du et al. 2000; Malki et al. 2003), have been observed in these proteins. In rice, a member of the OsDJ-1 family, OsDJ-1C showed high consecutive expression at all developmental stages in most tissues of rice, and it can utilize MG as a substrate to produce D-lactate in a GSH-independent manner (Ghosh et al. 2016). Although the efficiency of the OsGLYIII enzyme has been found to be relatively low as compared to GLYI/II in terms of its kinetic parameters, it is still the highest among other GLYIIIs except the AtDJ-1d of Arabidopsis (Ghosh et al. 2016).

7 Role of the Glyoxalase Pathway in Abiotic Stress Tolerance

Adverse environmental conditions, such as drought, salinity, extreme temperatures, and heavy metal toxicity, are critical factors that vigorously reduce crop yields worldwide (Wang et al. 2003). Also, keeping in mind that global climate change adds to these stressful environments, efforts need to be put in for the development of sustainable agriculture to meet the demand of the food supply for a continuously increasing population (Howden et al. 2007). The glyoxalase system plays an important role because of its ability to combat adverse environmental situations by maintaining MG levels, which otherwise increase to toxic levels under various abiotic and biotic stresses (Yadav et al. 2005a; Chen et al. 2004). The genetic manipulation

Table 11.1 Overexpressing GLYI and GLYII genes in transgenic plants provides improved abiotic stress tolerance in different plant species

Gene name	Plant species	Observed phenotype	Reference
(B. juncea) GLYI	*Nicotiana tabacum*	Enhances salt stress tolerance	Veena et al. (1999)
(B. juncea) GLYI	*Vigna mungo*	Better salt stress tolerance	Bhomkar et al. (2008)
(B. juncea) GLYI	*Arabidopsis thaliana*	Provides salt stress tolerance	Roy et al. (2008)
(Sugar beat) *GLYI*	*Nicotiana tabacum*	Enhanced tolerance to MG, salt stress, excessive mannitol, and H_2O_2	Wu et al. (2013)
GLYI	*Nicotiana tabacum*	Better salt stress tolerance	Yadav et al. (2005a)
(T. aestivum) GLYI	*Nicotiana tabacum*	Confers zinc tolerance	Lin et al. (2010)
(O. sativa) GLYI	*Nicotiana tabacum*	Significant tolerance towards MG and salt stress	Mustafiz et al. (2014)
(B. juncea) GLYI	*Brassica juncea*	Tolerance towards salinity, heavy metal, and drought stress	Rajwanshi et al. (2016)
(B. juncea) GLYI	*Oryza sativa*	Better salinity tolerance	Verma et al. (2005)
(O. sativa) GLYII	*Oryza sativa*	Better salt stress tolerance	Singla-pareek et al. (2008)
(O sativa) GLYII	*Brassica juncea*	Provides salinity tolerance	Saxena et al. (2011)
(O. sativa) GLYII	*Oryza sativa*	Better salinity tolerance	Wani and Gosal (2011)
(O. sativa) GLYII	*Nicotiana tabacum*	Enhanced salinity tolerance	Ghosh et al. (2014)
(A. thaliana) GLYII	*Arabidopsis thaliana*	Tolerance towards anoxic stress	Devanathan et al. (2014)
(B. juncea) GLYI+ (O. sativa) GLYII	*Nicotiana tabacum*	Better salinity tolerance; set viable seeds in zinc spiked soils	Singla-Pareek et al. (2003, 2006), Yadav et al. (2005b)
(B. juncea) GLYI+ (P. glaucum) GLYII	*Solanum lycopersicum*	Confers salt stress tolerance	Alvarez-Viveros et al. (2013)
(B. juncea) GLYI+ (P. glaucum) GLYII	*Citrus sinensis* x *Poncirus trifoliata*	Confers salinity tolerance	Alvarez-Gerding et al. (2015)

of the glyoxalase system in plants has successfully contributed to improved tolerance to multiple abiotic factors, such as salinity, heavy metal, and MG treatments, etc., as shown in Table 11.1.

The importance of the glyoxalase pathway in plants has been investigated using various approaches, one of them being the transgenic approach, where overexpression of GLYI and GLYII genes individually or together in plants have led to enhanced tolerance towards abiotic stresses. For example, overexpressing GLYI gene from *B. juncea* resulted in transgenic tobacco plants that showed higher tolerance to salt and MG toxicity compared with untransformed plants (Veena et al. 1999). The same GLYI gene was later overexpressed in *Vigna mungo* and blackgram, and found to be effective in imparting salinity tolerance (Bhomkar et al. 2008). Furthermore, transgenic tobacco and rice plants overexpressing the rice GLYII gene have also been raised, and like GLYI transgenic plants, GLYII enhances tolerance to high MG and NaCl concentrations. The transgenic tobacco plants could grow and flower properly and set viable seeds under continuous salt stress conditions (Singla-Pareek et al. 2003, 2008; Wani and Gosal 2011). The expression of the same GLYII gene in *B. juncea* could also impart increased salinity tolerance by delaying senescence (Saxena et al. 2011). The double transgenic tobacco lines that overexpressed both GLYI and GLYII genes exhibited a better response than single-gene transformed lines and the untransformed plants under salinity (Singla-Pareek et al. 2003). Very less yield penalty was observed in double transgenic lines, with about 5% loss in total productivity at 200 mM NaCl concentration. Also, double transgenic tomato plants have been generated, which showed improved salinity tolerance, probably because of decreased oxidative stress (Alvarez-Viveros et al. 2013). The overexpression of GLYI and GLYII genes from *B. juncea* and *Pennisetum glaucum*, respectively, in *Cirrizo citrange* rootstock imparted salt stress tolerance (Alvarez-Gerding et al. 2015).

The transgenic plants overexpressing GLY genes have also imparted tolerance towards heavy metal stress (Singla-Pareek et al. 2006). The glyoxalase transgenic plants could grow normally in the presence of 5 mM $ZnCl_2$ without any cost to yield. Other than Zn, transgenic plants could tolerate toxic concentrations of other heavy metals, such as cadmium and lead. Double transgenics were again better in terms of survival, growth, and yield than either of the single-gene transformants (either GLYI or II). The double transgenic plants grown in the presence of high Zn could produce 95% of the total seeds obtained from wildtype plants grown in water. MG accumulation and lipid peroxidation were also reduced under high concentrations of Zn. However, no correlation between the levels of expression of TcGLX1, a GLYI from the zinc hyperaccumulator plant *Thlaspi caerulescens*, and the degrees of Zn tolerance have been observed. No phenotype was visible in *A. thaliana* T-DNA insertion line closest to *A. thaliana* homolog of TcGLX1 (Tuomainen et al. 2011). There can be serval reasons for this, firstly it is likely that overexpression of both GLYI and GLYII is responsible for Zn tolerance. Secondly, different glyoxalase isoforms may have different effects. It is known that Arabidopsis has at least ten potentially redundant GLYI-like genes, but this information is lacking for tobacco, *B. juncea, Triticum aestivum,* and *Thlaspi caerulescens*. Thirdly, localization of GLYI expression may also be different as GLYI and TcGLX1 genes are under the influence of different promoters (Tuomainen et al. 2011). On the other hand, GLYI from wheat in transgenic tobacco plants imparted tolerance to Zn, when compared to the control (Lin et al. 2010). In another study, a GLYI gene from sugar beet overexpressed in tobacco conferred significant tolerance to MG, salt, mannitol, and

H_2O_2 treatments (Wu et al. 2013). A recent study carried out in *A. thaliana* shows that expression of the AtGLX2-1 gene, an isoform of GLYI, increased in response to salinity, anoxia stress, and excess L-threonine. Additionally, mutation in AtGLX2-1 inhibits growth and survival of Arabidopsis plants in response to stress. In the same study, they have shown loss of function mutants and constitutively GLX2-1 overexpressing transgenic lines resembled wild type plants under normal growth conditions, suggesting AtG5LX2-1 is an essential protein during normal plant life, but is also required during specific stressful conditions (Devanathan et al. 2014). Overexpression of GLYI gene in *B. juncea* under constitutive (35S promoter) and stress-inducible promoter (rd29A) imparted tolerance towards salt, drought, and heavy metal stress as compared to untransformed control plants. However, constitutive transgenic lines showed a yield penalty under non-stressful conditions, and no such effects were seen in inducible lines. This suggests that GLYI overexpressing lines under the influence of stress-inducible promoter is a better option for improving the stress tolerance capacity of plants (Rajwanshi et al. 2016).

Thus, it can be said that overexpression of the glyoxalase pathway or individual genes has the potential to confer tolerance to multiple stresses. Other than the transgenic approach, various transcriptomic and proteomic studies have also contributed to defining the role of the glyoxalase pathway in abiotic stress tolerance. There have been various reports showing the induction in expression of glyoxalase genes and change in glyoxalase activities in response to various stimuli, such as hormonal treatment, xenobiotics, heavy metals, various pretreatments of seeds, etc., details of which were discussed in a report by Kaur et al. (2014).

8 Conclusion

Our knowledge regarding glyoxalase enzymes has notably expanded in recent years. At present, we can with certainty associate the glyoxalase pathway with plant abiotic stress tolerance. Numerous reports related to perturbation in the glyoxalase pool in response to stress conditions are available. Any change in cellular atmosphere caused by environmental factors upregulates these enzymes, hence glyoxalases can be called sirens of the plant defense system. The level of MG also increases in response to both biotic and abiotic stresses, and inhibits growth and developmental processes both in plants and in animals. Many researchers have tried to explore the mechanism of MG toxicity under various physiological and pathological conditions, and this has improved our knowledge to some extent, but still much more work remains to be done. MG detoxification is strongly dependent on the availability of cellular GSH. Deficiency of GSH limits the production of hemithioacetal, leading to the accumulation of MG and subsequent cellular toxicity. Recently, it has been predicted that the activity of GLYII enzymes can be influenced by GSH, formed as an end-product of glyoxalase pathway (Ghosh et al. 2014), eventually leading to maintenance of the redox state of the cell. Recent studies have also shown that Arabidopsis GLYI of different metal specificity have different levels of stress tolerance in *E. coli* and yeast when overexpressed (Jain et al. 2016). But in the plant system, a specific role of Zn^{2+}/Ni^{2+}-dependent GLYI is yet to be explored. Plants also maintain a very low level of MG even without stress (Yadav et al. 2005b), and this internal MG has been shown to play role as a stress signal in rice (Kaur et al. 2015). Even when both GLYI and GLYII enzymes are overexpressed, the MG level is not depleted from the plant system completely (Yadav et al. 2005b). Multiple GLY enzymes in the plant system might modulate the internal GSH and MG concentration, thus playing an important role in regulating the complex stress signaling cascade within the plant system. Already an engineered glyoxalase pathway has shown promising results in providing abiotic stress tolerance in plants. Detailed study is needed to understand this network and tweaking of this pathway to engineer enhanced multiple abiotic stress-tolerant crop plants for cultivation under adverse environmental threats.

Acknowledgement Funding from South Asian University (SAU) is duly acknowledged.

References

Agar NS, Board PG, Bell K (1984) Studies of erythrocyte glyoxalase-II in various domestic species: discovery of glyoxalase-II deficiency in the horse. Anim Blood Groups Biochem Genet 15:67–70

Ahmed N, Thornalley PJ (2007) Advanced glycation endproducts: what is their relevance to diabetic complications? Diabetes Obes Metab 9(3):233–245

Akhand AA, Hossain K, Mitsui H, Kato M, Miyata T, Inagi R, Suzuki H (2001) Glyoxal and methylglyoxal trigger distinct signals for map family kinases and caspase activation in human endothelial cells. Free Radic Biol Med 31(1):20–30

Akoachere M, Iozef R, Rahlfs S, Deponte M, Mannervik B, Creighton DJ, Schirmer H, Becker K (2005) Characterization of the glyoxalases of the malarial parasite *Plasmodium falciparum* and comparison with their human counterparts. Biol Chem 386:41–52

Allen RE, Lo TW, Thornalley PJ (1993) Purification and characterisation of glyoxalase II from human red blood cells. Eur J Biochem FEBS 213(3):1261–1267

Alvarez Viveros MF, Inostroza-Blancheteau C, Timmermann T, Gonzalez M, Arce-Johnson P (2013) Overexpression of GlyI and GlyII genes in transgenic tomato (*Solanum lycopersicum Mill.*) plants confers salt tolerance by decreasing oxidative stress. Mol Biol Rep 40:3281–3290

Alvarez-Gerding X, Cortés-Bullemore R, Medina C, Romero-Romero JL, Inostroza Blancheteau C, Aquea F et al (2015) Improved salinity tolerance in Carrizo citrange rootstock through overexpression of glyoxalase system genes. Bio Med Res Int 2015:827951

Ariza A, Vickers TJ, Greig N, Fairlamb AH, Bond CS (2005) Crystallization and preliminary X-ray analysis of *Leishmania major* glyoxalase I. Acta Crystallogr Sect F Struct Biol Cryst Commun 61(8):769–772

Ariza A, Vickers TJ, Greig N, Armour KA, Dixon MJ, Eggleston IM, Fairlamb AH, Bond CS (2006) Specificity of the trypanothione-dependent *Leishmania major* glyoxalase I: structure and biochemical comparison with the human enzyme. Mol Microbiol 59:1239–1248

Aronsson AC, Marmstål E, Mannervik B (1978) Glyoxalase I, a zinc metalloenzyme of mammals and yeast. Biochem Biophys Res Commun 81(4):1235–1240

Babel W, Hofmann KH (1981) The conversion of triosephosphate via methylglyoxal, a bypass to the glycolytic sequence in methylotrophic yeasts? FEMS Microbiol Lett 10(2):133–136

Barata L, Sousa Silva M, Schuldt L, Da Costa G, Tomás AM, Ferreira AEN, Cordeiro C (2010) Cloning, expression, purification, crystallization and preliminary X-ray diffraction analysis of glyoxalase I from *Leishmania infantum*. Acta Crystallogr Sect F Struct Biol Cryst Commun 66(5):571–574

Bartfai T, Ekwall K, Mannervik B (1973) Discrimination between steady-state kinetic models of the mechanism of action of yeast glyoxalase I. Biochemist 12:387–391

Benov L, Sequeira F, Beema AF (2004) Role of rpoS in the regulation of glyoxalase III in *Escherichia coli*. Acta Biochim Pol 51:857–860

Ben-Zioni A, Itai C, Vaadia Y (1967) Water and salt stresses, kinetin and protein synthesis in tobacco leaves. Plant Physiol 42(3):361–365

Bhomkar P, Upadhyay CP, Saxena M, Muthusamy A, Shiva Prakash N, Pooggin M, Sarin NB (2008) Salt stress alleviation in transgenic *Vigna mungo* L. Hepper (blackgram) by overexpression of the glyoxalase I gene using a novel cestrum yellow leaf curling virus (CmYLCV) promoter. Mol Breed 22(2):169–181

Bito A, Haider M, Hadler I, Breitenbach M (1997) Identifcation and phenotypic analysis of two glyoxalase II encoding genes from *Saccharomyces cerevisiae,* GLO2 and GLO4, and intracellular localization of the corresponding proteins. J Biol Chem 272:21509–21519

Bito A, Haider M, Briza P, Strasser P, Breitenbach M (1999) Heterologous expression, purification, and kinetic comparison of the cytoplasmic and mitochondrial glyoxalase II enzymes, Glo2p and Glo4p, from *Saccharomyces cerevisiae*. Protein Expr Purif 17(3):456–464

Boyer JS (1982) Plant productivity and environment. Science 218(4571):443–448

Bythell-Douglas R, Suttisansanee U, Flematti GR, Challenor M, Lee M, Panjikar S, Bond CS (2015) The crystal structure of a homodimeric pseudomonas glyoxalase I enzyme reveals asymmetric metallation commensurate with half-of-sites activity. Chemistry 21(2):541–544

Cameron AD, Olin B, Ridderström M, Mannervik B, Jones TA (1997) Crystal structure of human glyoxalase I-evidence for gene duplication and 3D domain swapping. EMBO J 16(12):3386–3395

Cameron AD, Ridderström M, Olin B, Kavarana MJ, Creighton DJ, Mannervik B (1999) Reaction mechanism of glyoxalase I explored by an X-ray crystallographic analysis of the human enzyme in complex with a transition state analogue. Biochemist 38:13480–13490

Campbell AK, Naseem R, Holland IB, Matthews SB, Wann KT (2007) Methylglyoxal and other carbohydrate metabolites induce lanthanum-sensitive Ca^{2+} transients and inhibit growth in *E. coli*. Arch Biochem Biophys 468(1):107–113

Canet-Avilés RM, Wilson MA, Miller DW, Ahmad R, McLendon C, Bandyopadhyay S, Cookson MR (2004) The Parkinson's disease protein DJ-1 is neuroprotective due to cysteine-sulfinic acid-driven mitochondrial localization. Proc Natl Acad Sci U S A 101(24):9103–9108

Carrington SJ, Douglas KT (1986) The glyoxalase enigma-the biological consequences of a ubiquitous enzyme. IRCS Med Sci-Biochemistry 14(8):763–768

Casazza JP, Felver ME, Veech RL (1984) The metabolism of acetone in rat. J Biol Chem 259(1):231–236

Chang T, Wang R, Olson DJH, Mousseau DD, Ross ARS, Wu L (2011) Modification of Akt1 by methylglyoxal promotes the proliferation of vascular smooth muscle cells. FASEB J 25(5):1746–1757

Chen Z-Y, Brown RL, Damann KE, Cleveland TE (2004) Identification of a maize kernel stress-related protein and its effect on aflatoxin accumulation. Phytopathology 94(13):938–945

Clements CM, McNally RS, Conti BJ, Mak TW, Ting JP (2006) DJ-1. A cancer- and Parkinson's disease-associated protein, stabilizes the antioxidant transcriptional master regulator Nrf2. Proc Natl Acad Sci U S A 103:15091–15096

Cliffe EE, Waley SG (1961) The mechanism of the glyoxalase I reaction, and the effect of ophthalmic acid as an inhibitor. Biochem J 79:475–482

Clugston SL, Daub E, Honek JF (1998) Identification of glyoxalase I sequences in *Brassica oleracea* and *Sporobolus stapfianus*: evidence for gene duplication events. J Mol Evol 47(2):230–234

Clugston SL, Yajima R, Honek JF (2004) Investigation of metal binding and activation of *Escherichia coli* glyoxalase I: kinetic, thermodynamic and mutagenesis studies. Biochem J 377(519):309–316

Concha NO, Rasmussen BA, Bush K, Herzberg O (1996) Crystal structure of the wide-spectrum binuclear zinc β -lactamase from Bacteroides fragilis. Structure 4(7):823–836

Cooper RA (1974) Methylglyoxal formation during glucose catabolism by *Pseudomonas saccharophila*. Eur J Biochem 44(1):81–86

Cooper RA (1975) Methylglyoxal synthase. Methods Enzymol 41(104):502–508

Cooper RA (1984) Metabolism of methylglyoxal in microorganisms. Annu Rev Microbiol 38(1):49–68

Cooper RA, Anderson A (1970) The formation and catabolism of methylglyoxal during glycolysis in *Escherichia coli*. FESB Lett 11(4):273–276

Cramer GR (2010) Abiotic stress and plant responses from the whole vine to the genes. Aust J Grape Wine Res 16(s1):86–93

Cramer GR, Ergül A, Grimplet J, Tillett RL, Tattersall EAR, Bohlman MC, Osborne C (2007) Water and salinity stress in grapevines: early and late changes in transcript and metabolite profiles. Funct Integr Genomics 7(2):111–134

Crowder MW, Wang Z, Franklin SL, Zovinka EP, Benkovic SJ (1996) Characterization of the metal-binding sites of the beta-lactamase from *Bacteroides fragilis*. Biochemist 35(37):12126–12132

Crowder MW, Maiti MK, Banovic L, Makaroff CA (1997) Glyoxalase II from *A. thialana* requires Zn(II) for catalytic activity. FEBS Lett 418(3):351–354

Daiyasu H, Osaka K, Ishino Y, Toh H (2001) Expansion of the zinc metallo-hydrolase family of the β -lactamase fold. FEBS Lett 503(1):1–6

Dakin HD, Dudley HW (1913) An enzyme concerned with the formation of hydroxy acids from ketonic aldehydes. J Biol Chem 14(2):155–157

Deponte M, Sturm N, Mittler S, Harner M, Mack H, Becker K (2007) Allosteric coupling of two different functional active sites in monomeric *Plasmodium falciparum* glyoxalase I. J Biol Chem 282:28419–28430

Devanathan S, Erban A, Perez-Torres R Jr, Kopka J, Makaroff CA (2014) Arabidopsis thaliana glyoxalase 2-1 is required during abiotic stress but is not essential under normal plant growth. PLoS One 9(4):e95971

Dhindsa RS, Cleland RE (1975) Water stress and protein synthesis I. Differential inhibition of protein synthesis. Plant Physiol 55(4):778–781

Du X, Choi IG, Kim R, Wang W, Jancarik J, Yokota H, Kim SH (2000) Crystal structure of an intracellular protease from *Pyrococcus horikoshii* at 2-a resolution. Proc Natl Acad Sci U S A 97(26):14079–14084

Du J, Suzuki H, Nagase F, Akhand AA, Ma X-Y, Yokoyama T, Nakashima I (2001) Superoxide-mediated early oxidation and activation of ASK1 are important for initiating methylglyoxal-induced apoptosis process. Free Radic Biol Med 31(4):469–478

Elliott WH (1959) Amino-acetone: its isolation and role in metabolism. Nature 183:1051–1052

Elliott WH (1960) Methylglyoxal formation from aminoacetone by ox plasma. Nature 185:467–468

Espartero J, Sánchez-Aguayo I, Pardo JM (1995) Molecular characterization of glyoxalase-I from a higher plant; upregulation by stress. Plant Mol Biol 29(6):1223–1233

Esterbauer H, Cheeseman KH, Dianzani MU, Poli G, Slater TF (1982) Separation and characterization of the aldehydic products of lipid peroxidation stimulated by ADP-Fe^{2+} in rat liver microsomes. Biochem J 208(1):129–140

Frickel EM, Jemth P, Widersten M, Mannervik B (2001) Yeast glyoxalase I is a monomeric enzyme with two active sites. J Biol Chem 276(3):1845–1849

Fukunaga M, Miyata S, Higo S, Hamada Y, Ueyama S, Kasuga M (2005) Methylglyoxal induces apoptosis through oxidative stress-mediated activation of p38 mitogen-activated protein kinase in rat Schwann cells. Ann N Y Acad Sci 1043(1):151–157

Ghanta S, Chattopadhyay S (2011) Glutathione as a signaling molecule. Plant Signal Behav 6:783–788

Ghosh A, Pareek A, Sopory SK, Singla-Pareek SL (2014) A glutathione responsive rice glyoxalase II, OsGLYII-2, function in salinity adaptation by maintaining better photosynthesis efficiency and anti-oxidant pool. Plant J 80:93–105

Ghosh A, Islam T (2016) Genome-wide analysis and expression profiling of glyoxalase gene families in soybean (*Glycine max*) indicate their development and abiotic stress specific response. BMC Plant Biol 16(1):87

Ghosh A, Kushwaha HR, Hasan MR, Pareek A, Sopory SK, Singla-Pareek SL (2016) Presence of unique glyoxalase III proteins in plants indicates the existence of shorter route for methylglyoxal detoxification. Sci Rep 6(November 2015):18358

Gomes RA, Silva MS, Vicente Miranda H, Ferreira AEN, Cordeiro CAA, Freire AP (2005) Protein glycation in *Saccharomyces cerevisiae*: argpyrimidine formation and methylglyoxal catabolism. FEBS J 272(17):4521–4531

Good AG, Zaplachinski ST (1994) The effects of drought stress on free amino acid accumulation and protein synthesis in *Brassica napus*. Physiol Plant 90(1):9–14

Grant AW, Steel G, Waugh H, Ellis EM (2003) A novel aldo-keto reductase from *Escherichia coli* can increase resistance to methylglyoxal toxicity. FEMS Microbiol Lett 218(1):93–99

Greig N, Wyllie S, Vickers TJ, Fairlamb AH (2006) Trypanothione-dependent glyoxalase I in Trypanosoma cruzi. Biochem J 400(2):217–223

Hasim S, Hussin NA, Alomar F, Bidasee KR, Nickerson KW, Wilson MA (2014) A glutathione-independent glyoxalase of the DJ-1 superfamily plays an important role in managing metabolically generated methylglyoxal in *Candida albicans*. J Biol Chem 289(3):1662–1674

He MM, Clugston SL, Honek JF, Matthews BW (2000) Determination of the structure of *Escherichia coli* glyoxalase I suggests a structural basis for differential metal activation. Biochemist 39:8719–8727

Hod Y (2004) Differential control of apoptosis by DJ-1 in prostate benign and cancer cells. J Cell Biochem 92(6):1221–1233

Hopper DJ, Cooper RA (1971) The regulation of *Escherichia coli* methylglyoxal synthase; a new control site in glycolysis? FEBS Lett 13(4):213–216

Hopper DJ, Cooper RA (1972) The purification and properties of *Escherichia coli* methylglyoxal synthase. Biochem J 128(2):321–329

Hoque MA, Uraji M, Banu MNA, Mori IC, Nakamura Y, Murata Y (2010) The effects of methylglyoxal on glutathione S-transferase from *Nicotiana tabacum*. Biosci, Biotech, Biochem 74(10):2124–2126

Hoque TS, Okuma E, Uraji M, Furuichi T, Sasaki T, Hoque MA, Murata Y (2012a) Inhibitory effects of methylglyoxal on light-induced stomatal opening and inward K^+ channel activity in Arabidopsis. Biosci, Biotech, Biochem 76(3):617–619

Hoque M, Uraji M, Torii A, Banu M, Akhter N, Mori IC, Murata Y (2012b) Methylglyoxal inhibition of cytosolic ascorbate peroxidase from *Nicotiana tabacum*. J Biochem Mol Toxicol 26(8):315–321

Hoque TS, Uraji M, Tuya A, Nakamura Y, Murata Y (2012c) Methylglyoxal inhibits seed germination and root elongation and up-regulates transcription of stress-responsive genes in ABA-dependent pathway in Arabidopsis. Plant Biol 14(5):854–858

Hoque TS, Uraji M, Ye W, Hossain MA, Nakamura Y, Murata Y (2012d) Methylglyoxal-induced stomatal closure accompanied by peroxidase-mediated ROS production in *Arabidopsis*. J Plant Physiol 169(10):979–986

Hossain D, Rohman M, Fujita M (2005) Comparative investigation of glutathione S-transferases, glyoxalase- I and alliinase activities in different vegetable crops. J Crop Sci Biotech 10:21–28

Hossain MA, Fujita M (2009) Purification of glyoxalase I from onion bulbs and molecular cloning of its cDNA. Biosci Biotechnol Biochem 73(9):2007–2013

Hossain MA, Hossain MZ, Fujita M (2009) Stress-induced changes of methylglyoxal level and glyoxalase I activity in pumpkin seedlings and cDNA cloning of glyoxalase I gene. Aust J Crop Sci 3(2):53–64

Howden SM, Soussana JF, Tubiello FN, Chhetri N, Dunlop M, Meinke H (2007) Adapting agriculture to climate change. Proc Natl Acad Sci U S A 104:19691–19696

Inoue Y, Kimura A (1995) Methylglyoxal and regulation of its metabolism in microorganisms. Adv Microb Physiol 37:177–227

Inoue Y, Maeta K, Nomura W (2011) Glyoxalase system in yeasts: structure, function, and physiology. Semin Cell Dev Biol 22(3):278–284

Iozef R, Rahlfs S, Chang T, Schirmer H, Becker K (2003) Glyoxalase I of the malarial parasite *Plasmodium falciparum*: evidence for subunit fusion. FEBS Lett 554(3):284–288

Irsch T, Krauth-Siegel RL (2004) Glyoxalase II of African trypanosomes is trypanothione-dependent. J Biol Chem 279(21):22209–22217

Jain M, Batth R, Kumari S, Mustafiz A (2016) *Arabidopsis thaliana* contains both Ni^{2+} and Zn^{2+} dependent glyoxalase I enzymes and ectopic expression of the latter contributes more towards abiotic stress tolerance in *E. coli*. PLoS One 11(7):e0159348

Jia X, Chang T, Wilson TW, Wu L (2012) Methylglyoxal mediates adipocyte proliferation by increasing phosphorylation of Akt1. PLoS One 7(5):e36610

Kalapos MP (1999) Methylglyoxal in living organisms: chemistry, biochemistry, toxicology and biological implications. Toxicol Lett 110(3):145–175

Kalapos MP, Garzó T, Antoni F, Mandl J (1992) Accumulation of S-D-lactoylglutathione and transient decrease of glutathione level caused by methylglyoxal load in isolated hepatocytes. Biochim Biophys Acta: BBA-Mol Cell Res 1135(2):159–164

Kaur C, Vishnoi A, Ariyadasa TU, Bhattacharya A, Singla-Pareek SL, Sopory SK (2013) Episodes of horizontal gene-transfer and gene-fusion led to co-existence of different metal-ion specific glyoxalase I. Sci Rep 3(Ii):3076

Kaur C, Mustafiz A, Sarkar AK, Ariyadasa TU, Singla-Pareek SL, Sopory SK (2014) Expression of abiotic stress inducible ETHE1-like protein from rice is higher in roots and is regulated by calcium. Physiol Plant 152(1):1–16

Kaur C, Kushwaha HR, Mustafiz A, Pareek A, Sopory SK, Singla-Pareek SL (2015) Analysis of global gene expression profile of rice in response to methylglyoxal indicates its possible role as a stress signal molecule. Front Plant Sci 6:682

Kaur C, Sharma S, Singla-Pareek SL, Sopory SK (2016) Methylglyoxal detoxification in plants: role of glyoxalase pathway. Indian J Plant Physiol 21(4):377–390

Kawatani M, Okumura H, Honda K, Kanoh N, Muroi M, Dohmae N, Osada H (2008) The identification of an osteoclastogenesis inhibitor through the inhibition of glyoxalase I. Proc Natl Acad Sci U S A 105(33):11691–11696

Kevin J Tracey MD, Cerami Anthony PD (1994) Tumor necrosis factor: a pleiotropic cytokine and therapuetic target. Annu Rev Med 45(1):491–503

Kilian J, Whitehead D, Horak J, Wanke D, Weinl S, Batistic O, Harter K (2007) The AtGenExpress global stress expression data set: protocols, evaluation and model data analysis of UV-B light, drought and cold stress responses. Plant J 50(2):347–363

Ko J, Kim I, Yoo S, Min B, Kim K, Park C (2005) Conversion of methylglyoxal to acetol by *Escherichia coli* aldo-keto reductases. J Bacteriol 187(16):5782–5789

Koop DR, Casazza JP (1985) Identification of ethanol-inducible P-450 isozyme 3a as the acetone and acetol monooxygenase of rabbit microsomes. J Biol Chem 260(25):13607–13612

Kwon K, Choi D, Hyun JK, Jung HS, Baek K, Park C (2013) Novel glyoxalases from *Arabidopsis thaliana*. FEBS J 280(14):3328–3339

Laga M, Cottyn A, Van Herreweghe F, Berghe WV, Haegeman G, Van Oostveldt P, Vancompernolle K (2007) Methylglyoxal suppresses TNF-α-induced NF-κB activation by inhibiting NF-κB DNA-binding. Biochem Pharmacol 74(4):579–589

Lages NF, Cordeiro C, Sousa Silva M, Ponces Freire A, Ferreira AE (2012) Optimization of time-course experiments for kinetic model discrimination. PLoS One 7:e32749

Lee JY, Song J, Kwon K, Jang S, Kim C, Baek K, Park C (2012) Human DJ-1 and its homologs are novel glyoxalases. Hum Mol Genet 21(14):3215–3225

Leoncini G (1979) The role of alpha-ketoaldehydes in biological systems. Ital J Biochem 28(4):285

Limphong P, Adams NE, Rouhier MF, McKinney RM, Naylor M, Bennett B, Crowder MW (2010) Converting GLX2-1 into an active glyoxalase II. Biochemist 49(37):8228–8236

Lin F, Xu J, Shi J, Li H, Li B (2010) Molecular cloning and characterization of a novel glyoxalase I gene TaGly I in wheat (*Triticum aestivum* L.) Mol Biol Rep 37(2):729–735

Liu J-X, Howell SH (2010) Endoplasmic reticulum protein quality control and its relationship to environmental stress responses in plants. Plant Cell 22(9):2930–2942

MacLean MJ, Ness LS, Ferguson GP, Booth IR (1998) The role of glyoxalase I in the detoxification of methylglyoxal and in the activation of the KefB K^+ efflux system in *Escherichia coli*. Mol Microbiol 27(3):563–571

Maeta K, Izawa S, Okazaki S, Kuge S, Inoue Y (2004) Activity of the Yap1 transcription factor in *Saccharomyces cerevisiae* is modulated by methylglyoxal, a metabolite derived from glycolysis. Mol Cell Biol 24(19):8753–8764

Maeta K, Izawa S, Inoue Y (2005) Methylglyoxal, a metabolite derived from glycolysis, functions as a signal initiator of the high osmolarity glycerol-mitogen-activated protein kinase cascade and calcineurin/Crz1-mediated pathway in *Saccharomyces cerevisiae*. J Biol Chem 280(1):253–260

Maiti MK, Krishnasamy S, Owen HA, Makaroff CA (1997) Molecular characterization of glyoxalase II from *Arabidopsis thaliana*. Plant Mol Biol 35(4):471–481

Majumdar R, Shao L, Minocha R, Long S, Minocha SC (2013) Ornithine: the overlooked molecule in the regulation of polyamine metabolism3. Plant Cell Physiol 54(6):990–1004

Malki A, Kern R, Abdallah J, Richarme G (2003) Characterization of the *Escherichia coli* YedU protein as a molecular chaperone. Biochem Biophys Res Commun 301(2):430–436

Mannervik B, Ridderstrom M (1993) Catalytic and molecular properties ofglyoxalase I. Biochem Soc Trans 21:515–517

Mannervik B, Gorna-Hall B, Bartfai T (1973) The steady-state kinetics of glyoxalase I from porcine erythrocytes. Evidence for a random-pathway mechanism involving one-and two-substrate branches. Eur J Biochem 37:270–281

Mannervik B, Bartfai T, Gorna-Hall B (1974) Random pathway mechanism involving parallel one- and two-substrate branches for glyoxalase I from yeast. J Biol Chem 249:901–903

Marasinghe GPK, Sander IM, Bennett B, Periyannan G, Yang KW, Makaroff CA, Crowder MW (2005) Structural studies on a mitochondrial glyoxalase II. J Biol Chem 280(49):40668–40675

Marmstål E, Aronsson AC, Mannervik B (1979) Comparison of glyoxalase I purified from yeast (*Saccharomyces cerevisiae*) with the enzyme from mammalian sources. Biochem J 183:23–30

Martins AM, Mendes P, Cordeiro C, Freire AP (2001) In situ kinetic analysis of glyoxalase I and glyoxalase II in *Saccharomyces cerevisiae*. Eur J of Biochem 268(14):3930–3936

Melino S, Capo C, Dragani B, Aceto A, Petruzzelli R (1998) A zinc-binding motif conserved in glyoxalase II, β-lactamase and arylsulfatases. Trends Biochem Sci 23(10):381–382

Misra K, Banerjee AB, Ray S, Ray M (1995) Glyoxalase III from *Escherichia coli*: a single novel enzyme for the conversion of methylglyoxal into D-lactate without reduced glutathione. Biochem J 305:999–1003

Mostofa MG, Hossain MA, Fujita M (2015) Trehalose pretreatment induces salt tolerance in rice (*Oryza sativa* L.) seedlings: oxidative damage and co-induction of antioxidant defense and glyoxalase systems. Protoplasma 252(2):461–475

Murata K, Fukuda Y, Watanabe K, Saikusa T, Shimosaka M, Kimura A (1985) Characterization of methylglyoxal synthase in *Saccharomyces cerevisiae*. Biochem Biophys Res Commun 131(1):190–198

Murata K, Saikusa T, Fukuda Y, Watnabe K, Inoue Y, Shimosaka M, Kimura A (1986) Metabolism of 2-oxoaldehydes in yeasts: possible role of glycolytic bypath as a detoxification system in L-threonine catabolism by *Saccharomyces cerevisiae*. Eur J Biochem 157(2):297–301

Mustafiz A, Singh AK, Pareek A, Sopory SK, Singla-Pareek SL (2011) Genome-wide analysis of rice and Arabidopsis identifies two glyoxalase genes that are highly expressed in abiotic stresses. Funct Integr Genomics 11(2):293–305

Mustafiz A, Ghosh A, Tripathi AK, Kaur C, Ganguly AK, Bhavesh NS, Singla-Pareek SL (2014) A unique Ni^{2+}-dependent and methylglyoxal-inducible rice glyoxalase i possesses a single active site and functions in abiotic stress response. Plant J 78(6):951–963

Narawongsanont R, Kabinpong S, Auiyawong B, Tantitadapitak C (2012) Cloning and characterization of AKR4C14, a rice aldo–keto reductase, from Thai jasmine rice. Protein J 31(1):35–42

Neuberg C (1913) The destruction of lactic aldehyde and methylglyoxal by animal organs. Biochem Z 49:502–506

Neuberg C, Kobel M (1928) On the question of evidence of methyl-glyoxal as an intermediate product of glycolysis. Biochem Z 193:464–467

Newton G, Arnold K, Price M, Sherrill C, Delcardayre S, Aharonowitz Y, Cohen G, Davies J, Fahey R, Davis C (1996) Distribution of thiols in microorganisms: mycothiol is a major thiol in most actinomycetes. J Bacteriol 178:1990–1995

Nomura W, Maeta K, Kita K, Izawa S, Inoue Y (2008) Role of Gcn4 for adaptation to methylglyoxal in *Saccharomyces cerevisiae*: methylglyoxal attenuates protein synthesis through phosphorylation of eIF2α. Biochem Biophys Res Commun 376(4):738–742

Norton SJ, Talesa V, Yuan WJ, Principato GB (1990) Glyoxalase I and glyoxalase II from Aloe Vera: purification, characterization and comparison with animal glyoxalases. Biochem Int 22:411–418

O'Young J, Sukdeo N, Honek JF (2007) *Escherichia coli* glyoxalase II is a binuclear zinc-dependent metalloenzyme. Arch Biochem Biophys 459(1):20–26

Okado-Matsumoto A, Fridovich I (2000) The role of α,β-dicarbonyl compounds in the toxicity of short chain sugars. J Biol Chem 275(45):34853–34857

Park HS, Nam SH, Lee JK, Yoon CN, Mannervik B, Benkovic SJ, Kim HS (2006) Design and evolution of new catalytic activity with an existing protein scaffold. Science (Washington, DC, United States) 311(5760):535–538

Phillips SA, Thornalley PJ (1993a) Formation of methylglyoxal and D-lactate in human red blood cells in vitro. Bioche Soc Trans 21(2):163S–163S

Phillips SA, Thornalley PJ (1993b) The formation of methylglyoxal from triose phosphates. Eu J Biochem 212(1):101–105

Pinheiro C, Chaves MM (2011) Photosynthesis and drought: can we make metabolic connections from available data? J Exp Bot 62(3):869–882

Principato GB, Rosi G, Talesa V, Giovannini E, Uotila L (1987) Purification and charac- terization of two forms of glyoxalase II from the liver and brain of Wistar rats. Biochim Biophys Acta 911:349–355

Quan S, Switzenberg R, Reumann S, Hu J (2010) In vivo subcellular targeting analysis validates a novel peroxisome targeting signal type 2 and the peroxisomal localization of two proteins with putative functions in defense in Arabidopsis. Plant Signal Behav 5:151–153

Rabbani N, Thornalley PJ (2014) Dicarbonyl proteome and genome damage in metabolic and vascular disease. Biochem Soc Trans 42(2):425–432

Racker E (1954) Glutathione as a coenzyme in intermediary metabolism. In: Glutathione, a symposium. Academic Press, Ridgefield, pp 165–183

Rajwanshi R, Kumar D, Yusuf MA, DebRoy S, Sarin NB (2016) Stress-inducible overexpression of glyoxalase I is preferable to its constitutive overexpression for abiotic stress tolerance in transgenic *Brassica juncea*. Mol Breed 36(6):1–15

Ray S, Ray M (1981) Isolation of methylglyoxal synthase from goat liver. J Biol Chem 256(12): 6230–6233

Riboulet-Chavey A, Pierron A, Durand I, Murdaca J, Giudicelli J, Van Obberghen E (2006) Methylglyoxal impairs the insulin signaling pathways independently of the formation of intracellular reactive oxygen species. Diabetes 55(5):1289–1212

Richard JP (1984) Acid-base catalysis of the elimination and isomerization reactions of triose phosphates. J Am Chem Soc 106(17):4926–4936

Richard JP (1991) Kinetic parameters for the elimination reaction catalyzed by triosephosphate isomerase and an estimation of the reaction's physiological significance. Biochemist 30(18):4581–4585

Richard JP (1993) Mechanism for the formation of methylglyoxal from triosephosphates. Biochem Soc Trans 21(2):549–553

Ridderstrom M, Mannervik B (1996a) The primary structure of monomeric yeast glyoxalase I indicates a gene duplication resulting in two similar segments homologous with the subunit of dimeric human glyoxalase I. Biochem J 316:1005–1006

Ridderstrom M, Mannervik B (1996b) Optimized heterologous expression of the human zinc enzyme glyoxalase I. Biochem J 314(2):463–467

Ridderstrom M, Mannervik B (1997) Molecular cloning and characterization of the thiolesterase glyoxalase II from *Arabidopsis thaliana*. Biochem J 322:449–454

Ridderström M, Saccucci F, Hellman U, Bergman T, Principato G, Mannervik B (1996) Molecular cloning, heterologous expression, and characterization of human glyoxalase II. J Biol Chem 271(1):319–323

Ridderström M, Cameron AD, Jones TA, Mannervik B (1998) Involvement of an active-site Zn^{2+} ligand in the catalytic mechanism of human glyoxalase I. J Biol Chem 273:21623–21628

Roy SD, Saxena M, Bhomkar PS, Pooggin M, Hohn T, Bhalla-Sarin N (2008) Generation of marker free salt tolerant transgenic plants of *Arabidopsis thaliana* using the gly I gene and cre gene under inducible promoters. Plant Cell Tiss Org 95(1):1–11

Sankaranarayanan S, Jamshed M, Samuel MA (2015) Degradation of glyoxalase I in *Brassica napus* stigma leads to self-incompatibility response. Nat Plants 1:15185

Sato J, Wang YM, Van Eys J (1980) Methylglyoxal formation in rat liver cells. J Biol Chem 255(5):2046–2050

Saxena M, Roy SB, Singla-Pareek SL, Sopory SK, Bhalla-Sarin N (2011) Overexpression of the glyoxalase II gene leads to enhanced salinity tolerance in *Brassica juncea*. Open Plant Sci J 5:23–28

Schilling O, Wenzel N, Naylor M, Vogel A, Crowder M, Makaroff C, Meyer-Klaucke W (2003) Flexible metal binding of the metallo-β-lactamase domain: glyoxalase II incorporates iron, manganese, and zinc in vivo. Biochemist 42(40):11777–11786

Sellin S, Eriksson LE, Mannervik B (1982) Fluorescence and nuclear relaxation enhancement studies of the binding of glutathione derivatives to manganese-reconstituted glyoxalase I from human erythrocytes. A model for the catalytic mechanism of the enzyme involving a hydrated metal ion. Biochemist 21:4850–4857

Sellin S, Eriksson LEG, Aronsson AC, Mannervik B (1983) Octahedral metal coordination in the active site of glyoxalase I as evidenced by the properties of co(II)-glyoxalase I. J Biol Chem 258:2091–2093

Sharma S, Mustafiz A, Singla-Pareek SL, Shankar Srivastava P, Sopory SK (2012) Characterization of stress and methylglyoxal inducible triose phosphate isomerase (OscTPI) from rice. Plant Signal Behav 7(10):1337–1345

Silva MS, Barata L, Ferreira AEN, Romão S, Tomás AM, Freire AP, Cordeiro C (2008) Catalysis and structural properties of *Leishmania infantum* glyoxalase II: Trypanothione specificity and phylogeny. Biochemist 47(1):195–204

Simpson PJ, Tantitadapitak C, Reed AM, Mather OC, Bunce CM, White SA, Ride JP (2009) Characterization of two novel aldo–keto reductases from Arabidopsis: expression patterns, broad substrate specificity, and an open active-site structure suggest a role in toxicant metabolism following stress. J Mol Biol 392(2):465–480

Singla-Pareek SL, Reddy MK, Sopory SK (2003) Genetic engineering of the glyoxalase pathway in tobacco leads to enhanced salinity tolerance. Proc Natl Acad Sci U S A 100:14672–14677

Singla-Pareek SL, Yadav SK, Pareek A, Reddy MK, Sopory SK (2006) Transgenic tobacco overexpressing glyoxalase pathway enzymes grow and set viable seeds in zinc-spiked soils. Plant Physiol 140:613–623

Singla-Pareek SL, Yadav SK, Pareek A, Reddy MK, Sopory SK (2008) Enhancing salt tolerance in a crop plant by overexpression of glyoxalase II. Transgenic Res 17:171–180

Skipsey M, Andrews CJ, Townson JK, Jepson I, Edwards R (2000) Cloning and characterization of glyoxalase I from soybean. Arch Biochem Biophys 374(2):261–268

Skirycz A, Inzé D (2010) More from less: plant growth under limited water. Curr Opin Biotech 21(2):197–203

Smits MM, Johnson MA (1981) Methylglyoxal: Enzyme distributions relative to its presence in Douglas-fir needles and absence in Douglas-fir needle callus. Arch Biochem Biophys 208(2):431–439

Sousa Silva M, Ferreira AE, Tomas AM, Cordeiro C, Ponces Freire A (2005) Quantitative assessment of the glyoxalase pathway in *Leishmania infantum* as a therapeutic target by modelling and computer simulation. FEBS J 272:2388–2398

Sousa Silva M, Ferreira AEN, Gomes R, Tomás AM, Ponces Freire A, Cordeiro C (2012) The glyoxalase pathway in protozoan parasites. Int J Med Microbiol 302(4–5):225–229

Su Z, Sukdeo N, Honek JF (2008) 15N-1H HSQC NMR evidence for distinct specificity of two active sites in *Escherichia coli* glyoxalase I. Biochemist 47(50):13232–13241

Subedi KP, Choi D, Kim I, Min B, Park C (2011) Hsp31 of *Escherichia coli* K-12 is glyoxalase III. Mol Microbiol 81(4):926–936

Sukdeo N, Clugston SL, Daub E, Honek JF (2004) Distinct classes of glyoxalase I: metal specificity of the *Yersinia pestis*, *Pseudomonas aeruginosa* and *Neisseria meningitidis* enzymes. Biochem J 384(Pt 1):111–117

Sukdeo N, Honek JF (2007) *Pseudomonas aeruginosa* contains multiple glyoxalase I-encoding genes from both metal activation classes. Biochim Biophys Acta, Proteins Proteomics 1774(6):756–763

Suttisansanee U, Honek JF (2011) Bacterial glyoxalase enzymes. Semin Cell Dev Biol 22:285–292

Suttisansanee U, Lau K, Lagishetty S, Rao KN, Swaminathan S, Sauder JM, Honek JF (2011) Structural variation in bacterial glyoxalase I enzymes: investigation of the metalloenzyme glyoxalase I from *Clostridium acetobutylicum*. J Biol Chem 286(44):38367–38374

Suttisansanee U, Ran Y, Mullings KY, Sukdeo N, Honek JF, Booth IR, Geourjon C (2015) Modulating glyoxalase I metal selectivity by deletional mutagenesis: underlying structural factors contributing to nickel activation profiles. Metallomics 7(4):605–612

Szent-Györgyi A (1974) Electronic biology and its relation to cancer. Life Sci 15(5):863–875

Takatsume Y, Izawa S, Inoue Y (2006) Methylglyoxal as a signal initiator for activation of the stress-activated protein kinase cascade in the fission yeast *Schizosaccharomyces pombe*. J Biol Chem 281(14):9086–9092

Talesa V, Uotila L, Koivusalo M, Principato G, Giovannini E, Rosi G (1988) Demonstration of glyoxalase II in rat liver mitochondria. Partial purification and occurrence in multiple forms. Biochim Biophys Acta 955:103–110

Talesa V, Uotila L, Koivusalo M, Principato G, Giovannini E, Rosi G (1989) Isolation of glyoxalase II from two different compartments of rat liver mitochondria. Kinetic and immunochemical characterization of the enzymes. BBA – General Subjects 993(1):7–11

Talesa V, Principato GB, Norton SJ, Contenti S, Mangiabene C, Rosi G (1990) Isolation of glyoxalase II from bovine liver mitochondria. Biochem Int 20:53–58

Tao X, Tong L (2003) Crystal structure of human DJ-1, a protein associated with early onset Parkinson's disease. J Biol Chem 278(33):31372–31379

Thornalley PJ (1990) The glyoxalase system: new developments towards functional characterization of a metabolic pathway fundamental to biological life. Biochem J 269:1–11

Thornalley PJ (1993) The glyoxalase system in health and disease. Mol Asp Med 14:287–371

Thornalley PJ (2003) Glyoxalase I – structure, function and a critical role in the enzymatic defense against glycation. Biochem Soc Trans 31(8):1343–1348

Thornalley PJ (2006) Quantitative screening of protein glycation, oxidation, and nitration adducts by LC-MS/MS: protein damage in diabetes, uremia, cirrhosis, and alzheimer's disease. In: Redox proteomics: from protein modifications to cellular dysfunction and diseases. John Wiley & Sons, Inc., Hoboken, pp 681–727

Thornalley PJ, Langborg A, Minhas HS (1999) Formation of glyoxal, methylglyoxal and 3-deoxyglucosone in the glycation of proteins by glucose. Biochem J 344(1):109–116

Tuomainen M, Ahonen V, Kärenlampi SO, Schat H, Paasela T, Svanys A, Tuohimetsä S, Peräniemi S, Tervahauta A (2011) Characterization of the glyoxalase 1 gene TcGLX1 in the metal hyperaccumulator plant *Thlaspi caerulescens*. Planta 233:1173–1184

Turk Z, Nemet I, Varga-Defteardarović L, Car N (2006) Elevated level of methylglyoxal during diabetic ketoacidosis and its recovery phase. Diabetes Metab 32(2):176–180

Turra GL, Agostini RB, Fauguel CM, Daniel A, Andreo CS, Gonz JM, Valeria A (2015) Structure of the novel monomeric glyoxalase I from *Zea mays* structure of the novel monomeric glyoxalase I from. Acta Crystallogr D Biol Crystallogr 71(10):2009–2020

Turral H, Burke J, Faurès JM (2011) Climate change, water and food security. FAO water reports. ISSN 1020–1203

Urata G, Granick S (1963) Biosynthesis of α-aminoketones and the metabolism of aminoacetone. J Biol Chem 238(2):811–820

Urscher M, Deponte M (2009) *Plasmodium falciparum* glyoxalase II: Theorell–chance product inhibition patterns, rate-limiting substrate binding via Arg257/Lys260, and unmasking of acid-base catalysis. Biol Chem 390:1171–1183

Urscher M, Przyborski J, Imoto M, Deponte M (2010) Distinct subcellular localization in the cytosol and apicoplast, unexpected dimerization and inhibition of *Plasmodium falciparum* glyoxalases. Mol Microbiol 76:92–195

Usui Y, Nakase M, Hotta H, Urisu A, Aoki N, Kitajima K, Matsuda T (2001) A 33-kDa allergen from rice (*Oryza sativa* L. Japonica): cDNA cloning, expression, and identification as a novel glyoxalase I. J Biol Chem 276(14):11376–11381

Valentine WN, Paglia DE, Neerhout RC, Konrad PN (1970) Erythrocyte glyoxalase II deficiency with coincidental hereditary elliptocytosis. Blood 36:797–808

Vander Jagt DL (1993) Glyoxalase II: molecular characteristics, kinetics and mechanism. Biochem Soc Trans 21(2):522–527

Vander Jagt DL, Han LP, Lehman CH (1972) Kinetic evaluation of substrate specificity in the glyoxalase-I-catalyzed disproportionation of α-ketoaldehydes. Biochemist 11:3735–3740

Vander Jagt DL, Han LPB (1973) Deuterium isotope effects and chemically modified coenzymes as mechanism probes of yeast glyoxalase-I. Biochemistry 12(25):5161–5167

Vander Jagt DL, Daub E, Krohn JA, Han LP (1975) Effects of pH and thiols on the kinetics of yeast glyoxalase I. An evaluation of the random pathway mechanism. Biochemist 14:3669–3675

Vander Jagt DL, Hassebrook RK, Hunsaker LA, Brown WM, Royer RE (2001) Metabolism of the 2-oxoaldehyde methylglyoxal by aldose reductase and by glyoxalase-I: roles for glutathione in both enzymes and implications for diabetic complications. Chem Biol Interac 130:549–562

Veena, Reddy VS, Sopory SK (1999) Glyoxalase I from *Brassica juncea*: molecular cloning, regulation and its over-expression confer tolerance in transgenic tobacco under stress. Plant J 17(4):385–395

Verma M, Verma D, Jain RK, Sopory SK, Wu R (2005) Overexpression of glyoxalaseI gene confers salinity tolerance in transgenic japonica and indica rice plants. News Lett 22:58–62

Vickers TJ, Greig N, Fairlamb AH (2004) A trypanothione-dependent glyoxalase I with a prokaryotic ancestry in *Leishmania major*. Proc Natl Acad Sci USA 101:13186–13191

Vincent D, Ergül A, Bohlman MC, Tattersall EAR, Tillett RL, Wheatley MD, Schlauch K (2007) Proteomic analysis reveals differences between *Vitis Vinifera* L. cv. Chardonnay and cv. Cabernet sauvignon and their responses to water deficit and salinity. J Exp Bot 58(7):1873–1892

Wang W, Vinocur B, Altman A (2003) Plant responses to drought, salinity and extreme temperatures: towards genetic engineering for stress tolerance. Planta 218(1):1–14

Wani SH, Gosal SS (2011) Introduction of OsglyII gene into *Oryza sativa* for increasing salinity tolerance. Biol Plantarum 55:536–540

Wendler A, Irsch T, Rabbani N, Thornalley PJ, Krauth-Siegel RL (2009) Glyoxalase II does not support methylglyoxal detoxification but serves as a general trypanothione thioesterase in African trypanosomes. Mol Biochem Parasitol 163(1):19–27

Wenzel NF, Carenbauer AL, Pfiester MP, Schilling O, Meyer-Klaucke W, Makaroff CA, Crowder MW (2004) The binding of iron and zinc to glyoxalase II occurs exclusively as di-metal centers and is unique within the metallo-β-lactamase family. J Biol Inorg Chem 9(4):429–438

Wu C, Ma C, Pan Y, Gong S, Zhao C, Chen S, Li H (2013) Sugar beet M14 glyoxalase I gene can enhance plant tolerance to abiotic stresses. J Plant Res 126:415–425

Xu XM, Lin H, Maple J, Björkblom B, Alves G, Larsen JP, Møller SG (2010) The Arabidopsis DJ-1a protein confers stress protection through cytosolic SOD activation. J Cell Sci 123(10):1644–1651

Yadav SK, Singla-Pareek SL, Ray M, Reddy MK, Sopory SK (2005a) Methylglyoxal levels in plants under salinity stress are dependent on glyoxalase I and glutathione. Biochem Biophys Res Commun 337(1):61–67

Yadav SK, Singla-Pareek SL, Ray M, Reddy MK, Sopory SK (2005b) Transgenic tobacco plants overexpressing glyoxalase enzymes resist an increase in methylglyoxal and maintain higher reduced glutathione levels under salinity stress. FEBS Lett 579:6265–6271

Yamauchi Y, Hasegawa A, Taninaka A, Mizutani M, Sugimoto Y (2011) NADPH-dependent reductases involved in the detoxification of reactive carbonyls in plants. J Biol Chem 286(9):6999–7009

Yamawaki H, Saito K, Okada M, Hara Y (2008) Methylglyoxal mediates vascular inflammation via JNK and p38 in human endothelial cells. Am J Cell Physiol 295(6):C1510–C1517

Zang TM, Hollman DA, Crawford PA, Crowder MW, Makaroff CA (2001) Arabidopsis glyoxalase II contains a zinc/iron binuclear metal center that is essential for substrate binding and catalysis. J Biol Chem 276:4788–4795

Zhao Y, Liu D, Kaluarachchi WD, Bellamy HD, White MA, Fox RO (2003) The crystal structure of *Escherichia coli* heat shock protein YedU reveals three potential catalytic active sites. Protein Sci 12(10):2303–2311

Zhao Q, Su Y, Wang Z, Chen C, Wu T, Huang Y (2014) Identification of glutathione (GSH)-independent glyoxalase III from *Schizosaccharomyces pombe*. BMC Evol Biol 14(1):86

Zuin A, Vivancos AP, Sansó M, Takatsume Y, Ayté J, Inoue Y, Hidalgo E (2005) The glycolytic metabolite methylglyoxal activates Pap1 and Sty1 stress responses in *Schizosaccharomyces pombe*. J Biol Chem 280(44):36708–36713

Chapter 12
Glutathione Reductase and Abiotic Stress Tolerance in Plants

Vokkaliga T. Harshavardhan, Tsung-Meng Wu, and Chwan-Yang Hong

Abstract Abiotic stress is a major factor impeding crop productivity globally. Almost all abiotic stresses induce the accumulation of reactive oxygen species and consequently cause oxidative stress. Glutathione (GSH) and glutathione reductase (GR) are important components of the antioxidant machinery that plants use to respond to abiotic stress. GR catalyzes the reduction of glutathione disulfide (GSSG) to GSH with the accompanying oxidation of NADPH, which plays a pivotal role in maintaining the cellular redox balance of GSH/GSSG. Recently, GR was found to play a positive role in tolerance to abiotic stress. In this chapter, we review this recent information on the subcellular localization of GR between monocots and eudicots, detection of the redox state of GSH, and the expression, signaling and physiological role of GR genes in response to abiotic stress in plants.

Keywords Abiotic stress • Glutathione • Glutathione disulfide • Glutathione reductase • Redox status

Contents

V.T. Harshavardhan • C.-Y. Hong (✉)
Department of Agricultural Chemistry, College of Bioresources and Agriculture, National Taiwan University, Taipei 10617, Taiwan
e-mail: cyhong@ntu.edu.tw

T.-M. Wu
Department of Aquaculture, National Pingtung University of Science and Technology, Pingtung 91201, Taiwan

M.A. Hossain et al. (eds.), *Glutathione in Plant Growth, Development, and Stress Tolerance*, https://doi.org/10.1007/978-3-319-66682-2_12

1 Introduction

All kinds of abiotic and biotic stresses trigger a generalized stress response called oxidative stress, caused by the accumulation of activated oxygen molecules – reactive oxygen species (ROS). In addition, reactive nitrogen species contribute to oxidative stress by oxidation of cell compounds. In general, oxidative stress can be defined as a physiological state in which oxidation exceeds oxygen reduction, thereby resulting in oxidative damage to cell compounds. Hence, oxidative stress is an imbalance in the reduction/oxidation (redox) state of the cell caused by lack of electrons.

ROS are alternatively also known as active oxygen species or reactive oxygen intermediates. They occur in all aerobic organisms and are considered an unavoidable byproduct of aerobic metabolism in cellular compartments with strong electron flow (Choudhury et al. 2013; Bose et al. 2014). In plants, the main sites of ROS production are the cellular compartments such as chloroplasts, peroxisomes, mitochondria, plasma membrane and apoplast (Corpas et al. 2015). In light, chloroplasts and peroxisomes are the major source of ROS production; whereas in the dark, the source is mitochondria (Foyer and Noctor 2003; Moller 2001).

Changes in the cellular redox environment play a pivotal role in integrating external stimuli and stress signaling network in plants (Fujita et al. 2006; Spoel and Loake 2011; Suzuki et al. 2012; Scheibe and Dietz 2012). The cells must tightly regulate ROS levels to avoid cellular damage. The two major components that regulate cellular redox homeostasis are production of ROS and presence of antioxidant machinery. The cells have evolved a complex battery of antioxidant machinery usually close to the site of ROS production to tightly regulate redox homeostasis and avoid cellular damage. Uncontrolled accumulation of ROS causes oxidation of DNA and RNA, protein denaturation and decreased enzyme activity as well as lipid peroxidation and carbohydrate oxidation (Scandalios 1993; Noctor and Foyer 1998). Cellular redox homeostasis plays an important role in regulating the plant response to development and environmental stimuli. Changes in the redox state are sensed and used to trigger different signaling pathways, which leads to redox-dependent reprograming in the cell. Hence, to utilize ROS as signaling molecules, ROS levels must be maintained below the threshold level of damage by controlling ROS production and the counter process of ROS scavenging.

2 Glutathione Reductase (GR) and Other Antioxidant Pathways

The level of most redox active compounds depends on the plant's growth and developmental stage, subcellular location and type of stress exposed to. The redox state of each of the redox-active molecules can be determined by the proportion of reduced molecule relative to its total pool size or the ratio of reduced to oxidized molecules within a pool. In addition, the different redox-active compounds interact with each other (Potters et al. 2010).

The harmful effects of increased ROS levels are regulated or controlled by ROS scavenging mechanisms. In plants, these can be broadly divided into enzymatic and non-enzymatic mechanisms. The enzymatic scavenging system includes superoxide dismutase, catalase (CAT), ascorbate peroxidase (APX), monodehydroascorbate reductase (MDHAR), dihydroascorbate reductase (DHAR), glutathione reductase (GR), glutathione peroxidase (GPX) and glutathione S-transferase (GST). The non-enzymatic system includes antioxidant molecules such as ascorbic acid/ascorbate (AsA), dehydroascorbate (DHA), glutathione (GSH), tocopherols, tocotrienols, proline, flavonoids, carotenoids, phenolics, quercetin, kaempferol-glycosides, cytochromes, polyamines and proteins carrying redox-active *S*-groups (thioredoxins, peroxiredoxins and sulfiredoxins). The balance of these redox-active compounds contributes to the general redox homeostasis in plant cells (Potters et al. 2010).

Within the antioxidant system in plants, the ascorbate-GSH (AsA-GSH) pathway plays a major and central role in regulating ROS. This pathway combines both enzymatic (APX, MDHAR, DHAR, GR) and non-enzymatic (AsA and GSH) antioxidant components (Foyer and Noctor 2011). In the first step of the cycle, H_2O_2 is reduced to water by APX with AsA as the electron donor. During this reduction, AsA is oxidized to monodehydroascorbate, which is reduced back to AsA by the action of MDHAR or spontaneously converted to DHA. DHA is reduced to AsA by the action of DHAR at the expense of GSH yielding glutathione disulfide (GSSG). GR, regenerates the GSH from GSSG, using NADPH as electron donor. In the whole pathway, the net electron flow occurs from NADPH to H_2O_2. This cycle requires the maintenance of high levels of both ascorbate and GSH pools (Foyer and Noctor 2011; Noctor et al. 2011).

GR, also known as glutathione disulfide reductase (GSR), is a flavoprotein belonging to the family of NADPH-dependent oxidoreductases. It catalyzes the reduction of GSSG to GSH and plays an essential central role in cell defense against ROS. The catalytic activity of GR is a two-step process. In the first step, NADPH reduces flavin moiety, then flavin is oxidized and a redox-active disulfide bridge is reduced, thereby producing a thiolate anion and a cysteine. In the second step, GSSG is reduced via the thiol-disulfide interchange reaction. The reduced enzyme must be re-oxidized by GSSG if not will result in reversible inactivation (Gill et al. 2013; Rao and Reddy 2008).

The pool of GSH is reduced by its synthesis or recycling by GR with NADPH as cofactor and electron donor. GSH is a multifunctional low-molecular-weight water-soluble thiol tripeptide (γ-Glu-Cys-Gly) containing a sulfhydryl group (−SH) and forms a substrate for DHAR in the AsA-GSH pathway. In plants, GSH is synthesized by γ-glutamyl-cystein synthetase (γ-ECS) and glutathione synthetase (GS). γ-ECS is the rate-limiting enzyme of GSH biosynthesis localized in the chloroplast; whereas, GS is localized in both the chloroplast and cytosol (Noctor et al. 2011). GSH has a multifunctional role in various biological processes (cell growth/division, sulfate transport, signal transduction, protein and nucleic acid synthesis, detoxification of xenobiotics, etc.) because of its high reducing power. The central component in maintaining cellular redox homeostasis is the balance between GSH and GSSG (Potters et al. 2010; Foyer and Noctor 2011).

GR occurs in both prokaryotes and eukaryotes. Plant GRs are from 60 to 190 kDa (Mullineaux and Creissen 1997) and have different quaternary structure in different organisms. In most plants, GRs are homodimers (Wingsle 1989; Anderson et al. 1990; Edwards et al. 1990; Madamanchi et al. 1992), but they are also found as monomers in *Chlamydomonas* (Takeda et al. 1993) and as heterodimers in *Pisum sativum* and *Zea maize* (Kalt-Torres et al. 1984; Mahan and Burke 1987). The assembly of GR dimers into a higher-order state may be one of the regulation mechanisms of GR activity. The assembly into tetramers or higher aggregative states depends on temperature and pH (Mullineaux and Creissen 1997).

3 Subcellular Localization of GR in Monocots and Dicots

GR is located in the chloroplasts, cytosol, mitochondria and peroxisomes (Edwards et al. 1990; Jimenez et al. 1997; Kataya and Reumann 2010). However, it is predominantly found in the chloroplasts, and 80% of GR activity in photosynthetic tissue occurs by the chloroplastic isoform (Edwards et al. 1990). Recent phylogenetic analyses of different plant species revealed a clear divergence of GR isoforms between monocot and eudicot plants. The eudicots have two GR isoforms, one localized at the cytosol and the other localized at both the chloroplasts and mitochondria (Creissen et al. 1995; Chew et al. 2003). However, monocots such as barley, brachypodium, maize, rice and sorghum all contain three GR isoforms: one cytosolic GR and two chloroplast/mitochondria dual-localized GRs (Wu et al. 2013). GR targets different cell compartments because of the different signal peptides found in GR. For example, a 74-amino acid sequence at the N-terminus of Arabidopsis GR2 rich in Ser and Thr directs targets to the chloroplast (Kubo et al. 1993). Similarly, cowpea and soybean feature a 52-amino acid GR signal peptide, whereas pea features a 60-amino acid GR signal peptide at the N-terminus (Contour-Ansel et al. 2006). Nevertheless, the chloroplastidic transit peptide is rich in Ser (15%), Thr (15%) and Ala (10%) (von Heijne et al. 1989). The first GR found to have dual targeting – chloroplast and mitochondria – was reported from pea (Romero-Puertas et al. 2006). The signal peptides are more hydrophobic for such dual targeting than chloroplast or mitochondrial targeting alone (Peeters and Small 2001).

In addition, GRs targeted to the same cellular organelle are genetically closer among different plants than from the same plant. For example, cowpea cytosolic GR showed 89.2%, 80.2% and 74.6% identities with pea, Arabidopsis and rice cytosolic GR, respectively (Contour-Ansel et al. 2006), but the chloroplastic and cytosolic GR of Arabidopsis showed only 55% identity. The subcellular isoforms of GR, including the cytosolic, chloroplastic, mitochondrial and peroxisome forms, have been characterized and reported from various plant species and are summarized in Table 12.1.

Table 12.1 Localization of glutathione reductase (GR) isoforms in different cell organelles in monocots and eudicots

Plants	Species	GR isoforms	Gene symbol	Subcellular localization	References
Arabidopsis	*A. thaliana*	2	GR1	Cytosol and peroxisomes	Ding et al. (2016)
			GR2	Chloroplast and mitochondria	
Tobacco	*N. tabacum*	2	GRT1	Cytosol	Creissen et al. (1995)
			GRT2	Chloroplast	
Cowpea	*V. unguiculata*	2	cGR	Cytosol	Contour-Ansel et al. (2006)
			dtGR	Chloroplast and mitochondria	
Common bean	*P. vulgaris*	2	cGR	Cytosol	Torres-Franklin et al. (2007)
			dtGR	Chloroplast and mitochondria	
Pea	*P. sativum*	2	GOR1	Cytosol	Creissen et al. (1992), Stevens et al. (1997)
			cGR	Cytosol	
Mustard	*B. campestris*	2	GR	Cytosol	Lee et al. (1998)
			GR	Chloroplast	
Rice	*O. sativa*	3	OsGR1	Chloroplast and mitochondria	Rouhier et al. (2006), Wu et al. (2013)
			OsGR2	Cytosol	
			OsGR3	Chloroplast and mitochondria	
Wheat	*T. aestivium*	3	GR1	Chloroplast and mitochondria	Lascano et al. (2001)
			GR2	Cytosol	
			GR3	Chloroplast and mitochondria	
Barley	*H. vulgare*	3	HvGR1	Chloroplast and mitochondria	Bashir et al. (2007)
			HvGR2	Cytosol	
			HvGR2	Chloroplast and mitochondria	
Maize	*Z. mays*	3	GR1	Chloroplast and mitochondria	Pastori et al. (2000)
			GR2	Cytosol	
			GR3	Chloroplast and mitochondria	
			GOR2	Chloroplast and mitochondria	

4 Expression and Signaling of GR Under Abiotic Stress

Under various environmental stress conditions, GR is differentially regulated at the mRNA or protein (amount or activity) level.

4.1 Drought

Under drought stress, the non-availability of the electron acceptor results in generation of free radicals (Asada 2000; Sairam and Saxena 2000; Reddy et al. 2004). GR activity was increased under drought in plant species such as barley (Smirnoff and Colombe 1988), maize (Jiang and Zhang 2002; Pastori and Trippi 1992), tobacco (van Rensburg and Kruger 1994), wheat (Chen et al. 2004; Sairam et al. 1997), rice (Selote and Khanna-Chopra 2004; Sharma and Dubey 2005; Srivalli et al. 2003), pea (Gogorcena et al. 1995) and *Lotus japonicus* (Signorelli et al. 2013). GR activity was increased with short-term drought treatment in leaves of wheat (Bartoli et al. 1999) and mung bean (*Vigna radiata*) (Nahar et al. 2015). Total GR activity was increased in the drought-tolerant sugarcane genotype under severe water stress but not mild stress, but was increased even under mild stress in non–drought-tolerant cultivars (Boaretto et al. 2014). GR activity was increased in a drought-resistant wheat cultivar subjected to 100% oxygen and water stress (Pastori and Trippi 1993).

In addition, GR activity was increased under drought stress in cotton (Ratnayaka et al. 2003), wheat (Bartoli et al. 1999), beans (Torres-Franklin et al. 2007), rice (Sharma and Dubey 2005) and cucumber (Liu et al. 2009). The transcript levels of GRs fluctuate with environmental variables. In rice seedlings, cytosolic GR was strongly induced by drought (Kaminaka et al. 1998). Similarly, drought stress induced the expression of cytosolic GR both in drought-resistant ('EPACE-1') and drought-sensitive ('1183') cowpea cultivars; however, under continuous drought, dual-targeted GR transcripts were upregulated in the drought-sensitive cultivar but downregulated in the resistant cultivar (Contour-Ansel et al. 2006).

4.2 Salinity

Salinity, characterized by a high concentration of soluble salts (electrical conductivity of 4 dS/m; NaCl concentration of 40 mM; osmotic pressure ~0.2 MPa) is one of the most widespread soil toxicity problems (Munns and Tester 2008). GR activity was increased or enhanced under salinity in rice (Dionisio-Sese and Tobita 1998; Vaidyanathan et al. 2003; Demiral and Turkan 2005; Tsai et al. 2005), wheat (Sairam et al. 2005), maize seedlings (AbdElgawad et al. 2016), foxtail millet (Sreenivasulu et al. 2000), Arabidopsis (Huang et al. 2005), pea (Hernández et al. 1993, 1995, 2000), soybean (Comba et al. 1998), chickpea (Kukreja et al. 2005; Eyidogan and

Oz 2007), tomato (Shalata et al. 2001; Molina et al. 2002; Mittova et al. 2003), mung bean (Sumithra et al. 2006), sunflower (Davenport et al. 2003) and citrus (Gueta-Dahan et al. 1997). The transcript levels of cytosolic, chloroplastic and mitochondrial GR were increased in rice seedlings under salinity (Kaminaka et al. 1998; Kim et al. 2007; Wu et al. 2013; Hong et al. 2009) and with additional treatment with exogenous H_2O_2 (Tsai et al. 2005; Wu et al. 2013). In pea, cytosolic GR was induced in a NaCl-tolerant but not NaCl-sensitive variety under salt stress (Hernández et al. 2000).

4.3 Temperature (High and Low)

High temperature inhibits the antioxidant enzymes of the AsA-GSH pathway. With high temperature, GR activity was decreased in tomato (Rivero et al. 2004), lentil (Chakraborty and Pradhan 2011) and wheat genotypes (Almeselmani et al. 2006), but enhanced in maize (Hu et al. 2010), rice seedlings (Chao et al. 2009), wheat (Hasanuzzaman et al. 2012), cucumber (Dai et al. 2012), tobacco (Tan et al. 2011) and mung bean (Kumar et al. 2011). Pretreating *Dolichos lablab* with NaCl and high temperature increased GR activity (D'Souza and Devaraj 2013). In heat-resistant wheat cultivars, GR activity was increased after an initial decrease, whereas in heat-sensitive cultivars, GR activity decreased with heat stress (Wang et al. 2014). Under non-freezing temperatures, GR activity was enhanced in wheat (Kocsy et al. 2000), maize (Kocsy et al. 2001), rice (Guo et al. 2006) and other cereals (Janda et al. 2003), soybean (Sun et al. 2011), watermelon (Rivero et al. 2002), cucumber (Kuk and Shin 2007) and *Populus suaveolens* (Lei et al. 2007) but was not changed in wheat (Yordanova and Popova 2007) and was decreased in rice (Huang and Guo 2005). Under low temperature, cytosolic GR expression was enhanced in rice (Kaminaka et al. 1998), wheat (Baek and Skinner 2003) and pea (Romero-Puertas et al. 2006), but was repressed under high temperature in pea (Romero-Puertas et al. 2006).

4.4 Other Abiotic Stresses

In pea, high light and continuous dark decreased the transcript abundance of cytosolic and chloroplastic GR (Romero-Puertas et al. 2006). In addition, paraquat treatment decreased the transcript level of chloroplastic GR (Donahue et al. 1997). The expression and activity of chloroplastic and cytosolic GR increased in response to Fe-deficient conditions in barley and wheat (Bashir et al. 2007). Waterlogging differentially modulated GR activity in Citrumelo CPB 4475 (Arbona et al. 2008), pigeon pea (Kumutha et al. 2009) and rice (Damanik et al. 2010), and ozone (O_3) differentially modulated GR activity in *Beta vulgaris* (Kumari et al. 2013). Under UV radiation, GR activity was elevated in peanut (Tang et al. 2010). GR activity was increased in weed tall fleabane (*Conyza sumatrensis*) in response to paraquat (Chiang et al. 2008) and in maize in response to the herbicides, pretilachlor and metribuzin (Alla et al. 2008). GR activity was increased with Cadmium treatment in pea (Dixit et al. 2001), wheat (Yannarelli et al. 2007), *Capsicum annuum* (León et al. 2002), *Vigna mungo* (Singh et al. 2008) and *Brassica juncea* (Mobin and Khan 2007). It was increased in alfalfa treated with Cu (Wang et al. 2011) and in both tolerant and sensitive cultivars of *B. juncea* trea/ted with As (V) and As (III) (Srivastava et al. 2010).

4.5 Signaling

Within the same plant, GR responds differentially to environmental and biotic stresses (Lascano et al. 2001; Mullineaux and Creissen 1997; Pastori and Trippi 1992; Romero-Puertas et al. 2006). Abscisic acid (ABA) levels increase in plants on exposure to different environmental cues, which is the key in the signal transduction pathway leading to ROS scavenging (Anderson et al. 1994; Bueno et al. 1998; Gong et al. 1998; Hung and Kao 2003; Jiang and Zhang 2001, 2002, Sreenivasulu et al. 2012). GR activity was increased in maize seedlings and rice roots with increased ABA concentration, and H_2O_2 is known to be involved in ABA-induced GR activity in plant tissues (Anderson et al. 1994; Bueno et al. 1998; Gong et al. 1998; Jiang and Zhang 2001, 2002; Hung and Kao 2003; Tsai and Kao 2004). ABA-mediated signal transduction was responsible for the expression of GR genes during drought (Contour-Ansel et al. 2006). Nevertheless, in pea, no ABA-responsive elements have been found for GR. The presence of ABA-responsive elements in rice cytosolic GR was responsible for its regulation by ABA-mediated signal transduction (Kaminaka et al. 1998). In contrast, salinity increased expressions of GR2 and GR3 in rice seedlings were mediated by H_2O_2 but not ABA (Hong et al. 2009).

5 Role of GR in Plant Abiotic Stress Tolerance

In general, APX and GR activities increase in various plant species under different stress conditions. Studies of transgenic plants have shown that GR plays important role in providing resistance to oxidative stress caused by different environmental cues. In addition to conferring stress tolerance, increased GR activity can alter the redox state of electron transport chain components (Gill et al. 2013). GR plays a central role in maintaining the reduced GSH pool and regulating the ROS scavenging pathway in the cell under stress (Noctor et al. 2011; Gill et al. 2013). Numerous reviews have covered the role of GR in abiotic stress responses (Rao and Reddy 2008; Noctor et al. 2011; Yousuf et al. 2012).

Transgenic tobacco and tomato plants with suppressed GR activity showed increased sensitivity to oxidative stresses, paraquat, methyl-viologen and chilling stress (Aono et al. 1995; Tyystjärvi et al. 1999; Ding et al. 2009, 2012; Shu et al. 2011). In contrast, transgenic tobacco overexpressing chloroplastic or cytosolic GR showed enhanced tolerance to photo-oxidative stress caused by paraquat, the air pollutant sulfur dioxide, ozone, methyl-viologen, salt and cold stresses (Aono et al. 1991; Broadbent et al. 1995; Le Martret et al. 2011). Enhanced chloroplastic GR activity in transgenic poplar and cotton resulted in increased resistance to oxidative stress caused by ozone, paraquat, high light, or chilling stress (Foyer et al. 1995; Payton et al. 2001; Kornyeyev et al. 2003). Insertional mutagenesis and functional complementation study of rice revealed a crucial role of chloroplastic/mitochondrial GR3 in imparting tolerance to methyl viologen and salinity (Wu et al. 2015). A study of the *cat2gr1* double mutant showed that Arabidopsis GR1 plays a specific role in the H_2O_2 response during stress (Mhamdi et al. 2010). A number of recent studies have shown that elevated GR activity in GR-overexpressing transgenic plants increases stress tolerance to various abiotic stresses (Table 12.2).

Transgenic wheat overexpressing *Haynaldia villosa* plastidial GR was tolerant to powdery mildew (Chen et al. 2007). Moreover, GR plays important roles in plant development. The Arabidopsis chloroplastic/mitochondrial *gr2* knockout mutant was embryo-lethal (Tzafrir et al. 2004; Marty et al. 2009), and a weak mutant of *gr2* showed defective root apical meristem and inhibited root growth (Diaz-Vivancos et al. 2015; Yu et al. 2013).

6 Detection of Redox State of GSH Under Abiotic Stress

Reduced GSH is the prevalent form in a cell; however, various stress conditions induce characteristic changes in the amount and redox state of GSH. Under stress, in addition to DHAR, free radicals and reactive aldehydes oxidize the thiol moiety, thereby resulting in GSSG, the oxidized form of GSH. Therefore, GSH functions as an important redox buffer system in maintaining a balanced GSH/GSSG ratio. Hence, increases in GSSG relative to GSH are a useful indicator of oxidative stress and an effective marker of cellular redox homeostasis. Under non-stress conditions, leaf tissue usually maintain GSH/GSSG ratios of at least 20:1 (Mhamdi et al. 2010), however, ratios of GSH/GSSG maybe higher or lower in different subcellular compartments of leaves (Queval et al. 2011).

Initially, the redox state of the GSH pool was assessed by the less-sensitive technique of spectrophotometric detection of GSH and GSSG simultaneously (Mergel et al. 1979). With advances in technology, a high-performance liquid chromatography (HPLC) method was used to detect the fluorescent derivatives of thiols for detecting the levels of total and reduced thiols (Cys, GSH and their homologs) after reduction (Kranner and Grill 1996). Furthermore, sensitive HPLC methods were used to detect GSH, GSSG and their precursors and homologs (Kranner and Grill 1996; Potesil et al. 2005; Rellan-Alvarez et al. 2006). Combination with electrochemical detection methods widened the simultaneous detection of Cys, GSH, GSSG and phytochelatins (Potesil et al. 2005). In addition, combination with mass

Table 12.2 Role of GR in stress tolerance analyzed by overexpression (OE) or downregulation (DR) or mutants by use of inducible or constitutive promoters in model and crop plants

Transgenic	Source	Approach	Promoter	Organelle specific	Total GSH	GSH	GSSG	GSH/GSSG	Stress tolerance	References
N. tabacum	*P. sativum*	OE	Constitutive	Cytosol, Mito./Chloro.	Increased	NA	NA	No difference	Reduced sensitivity to paraquat and ozone	Broadbent et al. (1995)
Poplar (hybrid)	*E. coli*	OE	Constitutive	Chloroplast, Cytosol	Increased in chloroplast GR OE not in cytosol	NA	NA	NA	Increased resistance to photoinhibition and oxidative stress	Foyer et al. (1995)
E. coli	*B. campestris*	OE	Inducible	NA	NA	NA	NA	NA	Tolerance to oxidative stress	Yoon et al. (2005)
N. tabacum	*N. tabacum*	DR	NA	Chloroplast	NA	Increased under stress	Increased under stress	Decreased under stress	Increased sensitivity to oxidative stress	Ding et al. (2009)
L. esculentum	*L. esculentum*	DR	Constitutive	Chloroplast	No difference	Decreased under stress	Increased under stress	Decreased under stress	Increased sensitivity to chilling stress	Shu et al. (2011)
N. tabacum	*S. oleracea*	DR	Rubisco	NA	NA	NA	NA	NA	Increased sensitivity to paraquat	Aono et al. (1995)
A. thaliana	*A. thaliana*	DR	NA	Chloroplast	Decreased	Decreased	Increased	Decreased	Onset of early senescence	Ding et al. (2016)
N. tabacum	Bacteria	OE	Constitutive	Cytosol	No difference	No difference	No difference	NA	NA	Foyer et al. (1991)
O. sativa	*B. campestris*	OE	Constitutive	Cytosol	NA	NA	NA	NA	Increased resistance to MV and heat stress	Kouřil et al. (2003)

B. juncea	Bacteria	OE	NA	Cytosol plastid	Increased in plastid targeted GR	NA	NA	NA	Plastid targeted GR transgenic showed tolerance to cadmium	Pilon-Smits et al. (2000)
G. hirsutum	*A. thaliana*	OE	NA	NA	NA	NA	NA	NA	No improvement in emergence rate under temperature stress	Mahan et al. (2009)
O. sativa	*O. sativa*	Mutant and OE	Ubiquitin1	Chloroplast and mitochondria	Decreased in mutantunder salt stress	Decreased under salt stress in mutant	Decreased in mutant under salt stress	Decreased in mutant under salt stress	Increased sensitivity to salt stress in mutant.	Wu et al. (2015)
N. tabacum	*N. tabacum*	DR	NA	Chloroplast	NA	Decreased under stress	Increased under stress	Decreased under stress	Highly sensitive to chilling stress	Ding et al. (2012)
G. hirsutum	*A. thaliana*	OE	NA	NA	NA	NA	NA	NA	Photoinhibition of growth during chilling stress	Logan et al. (2003)
N. tabacum	*E. coli*	OE	Constitutive	NA	NA	NA	NA	NA	Reduced sensitivity to paraquat	Aono et al. (1991)
N. tabacum	*E. coli*	OE	Constitutive	NA	NA	NA	NA	NA	Resistance to photooxidative stress caused by paraquat or sulfur dioxide.	Aono et al. (1993)

spectrometry (electrospray ionization) allowed for the simultaneous detection of GSH, GSSG, homo-GSH (hGSH) and homo-GSSG (hGSSG) (Rellan-Alvarez et al. 2006). Moreover, capillary-zone electrophoresis ensured simultaneous analysis of GSH and GSSG (Mendoza et al. 2004). Nevertheless, as compared with fluorescent detection, the latter three methods could not determine the reduced and oxidized forms of GSH precursors.

GSH/GSSG ratio was found higher in a freeze-tolerant than freeze-sensitive wheat genotype (Kocsy et al. 2001). Because of high GR activity, ratios of GSH/GSSG were increased in cold-tolerant wheat, maize and spruce under stress (Esterbauer and Grill 1978; Leipner et al. 1999; Kocsy et al. 2001). At high temperatures, ratios of GSH/GSSG were increased in wheat and maize and associated with heat tolerance (Kocsy et al. 2004). Similarly, with induction of GR by cold stress, a high reduction of GSSG was maintained in cold-hardened spruce (Hausladen and Alscher 1994). In addition, high GSH/GSSG ratio and GR activity are involved in salt and drought tolerances of tomato, *Myrothamnus flabellifolia* and wheat, respectively (Shalata et al. 2001; Kranner et al. 2002; Kocsy et al. 2004). In contrast, the GSH/GSSG ratio was decreased (more oxidized state) under Cu stress in *Silene cucubalus*, probably because of use of GSH for phytochelatin synthesis (de Vos et al. 1992). However, such a decrease in GSH/GSSG ratio was prevented in soybean under Cd stress because of the induction of GR (Ferreira et al. 2002). A high GSH/GSSG ratio, maintained by increased GSH synthesis and/or GSSG reduction, is necessary for efficient protection against abiotic stress-induced accumulation of ROS.

Over the past decades, most GSH redox research has been based on the chemical analysis of whole-cell extracts, which unavoidably destroys subcellular compartment-specific information. In recent years, a GFP-based ratiometric redox-sensitive GFP (roGFP) protein was generated and used to detect redox changes in GSH within different organelles (Jiang et al. 2006; Meyer et al. 2007). The roGFP protein can be targeted to distinct organelles by adding a signal sequence and then transfected into plants (Jiang et al. 2006; Meyer et al. 2007; Schwarzlander et al. 2009). The redox status of GSH can be determined by measuring the changes in GFP fluorescence before and after stress treatment by confocal microscopy. The use of roGFP provides the precise compartment-specific redox status of GSH and prevents underestimation of the redox status of GSH.

The half-cell reduction potential ($E_{GSSG/2GSH}$) of a cell or organelle can be calculated from concentrations of GSH and GSSG by using the Nernst equation and is closely associated with the biological status of the cell (Schafer and Buettner 2001). Hence, the redox potential of GSH (E_{GSH}) depends on and can be calculated by GSH concentration and ratio of GSH/GSSG (Meyer and Hell 2005). Under normal conditions, cytosolic E_{GSH} is highly reduced and hence more negative (about −310 mV) (Aller et al. 2013). Changes in the E_{GSH} buffer results in drastic changes. For example, an increase in E_{GSH} (decrease in reduction) to −260 mV prevented cell cycle progression and induced changes in transcript profiles, and an increase to −170 mV induced apoptosis in Arabidopsis (Aller et al. 2013; Schnaubelt et al. 2015). Since the E_{GSH} factor is associated with cellular status, it is used as a marker to monitor stress-induced damages and cell viability (Schafer and Buettner 2001; Kranner et al. 2006; Szalai et al. 2009) and the half-cell reduction potential of the GSH/GSSG couple is associated with the level of stress tolerance (Soltesz et al. 2011). Maize plants under various stresses showed induced changes in GSH concentration but minor changes in $E_{GSSG/2GSH}$ value. Lines with different levels of stress tolerance had different levels of GSH, but the difference in half-cell reduction potential was small. Hence, plants are able to maintain $E_{GSSG/2GSH}$ value in cells under moderate stress conditions (Szalai et al. 2009).

7 Conclusions and Future Perspectives

Oxidative stress is an unavoidable consequence of environmental stresses. As one of enzymes involved in the antioxidant machinery in plants, GR plays a crucial role in regulating the redox state of GSH, thereby protecting plants against abiotic stress-induced oxidative damage. The involvement of GR in response to different stresses has been reported from many plants, but the regulatory role of different isoforms under different environmental stresses has not been well studied. Understanding the role of compartment-specific GR in different organelles is helpful for understanding the precise regulation of the redox status of the thiol pool. In addition, the function of GR in modulating plant development is just beginning to be understood. Both the total amount and redox state of GSH might affect plant development and stress tolerance. Future studies to clarify the relative importance of the levels and redox state of GSH in tolerance to different abiotic stresses as well as in development might provide more information for the design of climate change-resilient crops.

Acknowledgments The project was supported by the Ministry of Science and Technology (MOST) of Taiwan to C.-Y. Hong. (grant no. MOST 104-2313-B-002-013-MY3 and 105-2628-B-002-036-MY3).

References

AbdElgawad H, Zinta G, Hegab MM, Pandey R, Asard H, Abuelsoud W (2016) High salinity induces different oxidative stress and antioxidant responses in maize seedlings organs. Front Plant Sci 7:276

Alla MMN, Badawi A-HM, Hassan NM, El-Bastawisy ZM, Badran EG (2008) Herbicide tolerance in maize is related to increased levels of glutathione and glutathione-associated enzymes. Acta Physiol Plant 30(3):371–379

Aller I, Rouhier N, Meyer AJ (2013) Development of roGFP2-derived redox probes for measurement of the glutathione redox potential in the cytosol of severely glutathione-deficient rml1 seedlings. Front Plant Sci 4:506

Almeselmani M, Deshmukh PS, Sairam RK, Kushwaha SR, Singh TP (2006) Protective role of antioxidant enzymes under high temperature stress. Plant Sci 171(3):382–388

Anderson JV, Hess JL, Chevone BI (1990) Purification, characterization, and immunological properties for two isoforms of glutathione reductase from eastern white pine needles. Plant Physiol 94(3):1402–1409

Anderson MD, Prasad TK, Martin BA, Stewart CR (1994) Differential gene expression in chilling-acclimated maize seedlings and evidence for the involvement of abscisic acid in chilling tolerance. Plant Physiol 105(1):331–339

Aono M, Kubo A, Saji H, Natori T, Tanaka K, Kondo N (1991) Resistance to active oxygen toxicity of transgenic *Nicotiana tabacum* that expresses the gene for glutathione reductase from *Escherichia coli*. Plant Cell Physiol 32(5):691–697

Aono M, Kubo A, Saji H, Tanaka K, Kondo N (1993) Enhanced tolerance to photo-oxidative stress of transgenic *Nicotiana tabacum* with high chloroplastic glutathione reductase activity. Plant Cell Physiol 34(1):129–135

Aono M, Saji H, Fujiyama K, Sugita M, Kondo N, Tanaka K (1995) Decrease in activity of glutathione reductase enhances paraquat sensitivity in transgenic *Nicotiana tabacum*. Plant Physiol 107(2):645–648

Arbona V, Hossain Z, López-Climent MF, Pérez-Clemente RM, Gómez-Cadenas A (2008) Antioxidant enzymatic activity is linked to waterlogging stress tolerance in citrus. Physiol Plant 132(4):452–466

Asada K (2000) The water–water cycle as alternative photon and electron sinks. Philos Trans R Soc Lond Ser B 355(1402):1419

Baek K-H, Skinner DZ (2003) Alteration of antioxidant enzyme gene expression during cold acclimation of near-isogenic wheat lines. Plant Sci 165(6):1221–1227

Bartoli CG, Simontacchi M, Tambussi E, Beltrano J, Montaldi E, Puntarulo S (1999) Drought and watering-dependent oxidative stress: effect on antioxidant content in *Triticum aestivum* L. leaves. J Exp Bot 50(332):375–383

Bashir K, Nagasaka S, Itai RN, Kobayashi T, Takahashi M, Nakanishi H, Mori S, Nishizawa NK (2007) Expression and enzyme activity of glutathione reductase is upregulated by Fe-deficiency in graminaceous plants. Plant Mol Biol 65(3):277–284

Boaretto LF, Carvalho G, Borgo L, Creste S, Landell MG, Mazzafera P, Azevedo RA (2014) Water stress reveals differential antioxidant responses of tolerant and non-tolerant sugarcane genotypes. Plant Physiol Biochem 74:165–175

Bose J, Rodrigo-Moreno A, Shabala S (2014) ROS homeostasis in halophytes in the context of salinity stress tolerance. J Exp Bot 65(5):1241–1257

Broadbent P, Creissen GP, Kular B, Wellburn AR, Mullineaux PM (1995) Oxidative stress responses in transgenic tobacco containing altered levels of glutathione reductase activity. Plant J 8(2):247–255

Bueno P, Piqueras A, Kurepa J, Savouré A, Verbruggen N, Van Montagu M, Inzé D (1998) Expression of antioxidant enzymes in response to abscisic acid and high osmoticum in tobacco BY-2 cell cultures. Plant Sci 138(1):27–34

Chakraborty U, Pradhan D (2011) High temperature-induced oxidative stress in *Lens culinaris*, role of antioxidants and amelioration of stress by chemical pre-treatments. J Plant Interact 6(1):43–52

Chao Y-Y, Hsu Y, Kao C (2009) Involvement of glutathione in heat shock–and hydrogen peroxide–induced cadmium tolerance of rice (*Oryza sativa* L.) seedlings. Plant Soil 318(1–2):37–45

Chen KM, Gong HJ, Chen GC, Wang SM, Zhang CL (2004) Gradual drought under field conditions influences the glutathione metabolism, redox balance and energy supply in spring wheat. J Plant Growth Regul 23(1):20–28

Chen YP, Xing LP, Wu GJ, Wang HZ, Wang XE, Cao AZ, Chen PD (2007) Plastidial glutathione reductase from *Haynaldia villosa* is an enhancer of powdery mildew resistance in wheat (*Triticum aestivum*). Plant Cell Physiol 48(12):1702–1712

Chew O, Rudhe C, Glaser E, Whelan J (2003) Characterization of the targeting signal of dual-targeted pea glutathione reductase. Plant Mol Biol 53(3):341–356

Chiang YJ, Wu YX, Chiang MY, Wang CY (2008) Role of antioxidative system in paraquat resistance of tall fleabane (*Conyza sumatrensis*). Weed Sci 56(3):350–355

Choudhury S, Panda P, Sahoo L, Panda SK (2013) Reactive oxygen species signaling in plants under abiotic stress. Plant Signal Behav 8(4):e23681

Comba ME, Benavides MP, Tomaro ML (1998) Effect of salt stress on antioxidant defence system in soybean root nodules. Funct Plant Biol 25(6):665–671

Contour-Ansel D, Torres-Franklin ML, Cruz DECMH, D'Arcy-Lameta A, Zuily-Fodil Y (2006) Glutathione reductase in leaves of cowpea: cloning of two cDNAs, expression and enzymatic activity under progressive drought stress, desiccation and abscisic acid treatment. Ann Bot 98(6):1279–1287

Corpas FJ, Gupta DK, Palma JM (2015) Production sites of reactive oxygen species (ROS) in organelles from plant cells. In: Reactive oxygen species and oxidative damage in plants under stress. Springer, Cham, pp 1–22

Creissen G, Edwards EA, Enard C, Wellburn A, Mullineaux P (1992) Molecular characterization of glutathione reductase cDNAs from pea (*Pisum sativum* L.) Plant J 2(1):129–131

Creissen G, Reynolds H, Xue Y, Mullineaux P (1995) Simultaneous targeting of pea glutathione reductase and of a bacterial fusion protein to chloroplasts and mitochondria in transgenic tobacco. Plant J 8(2):167–175

D'Souza MR, Devaraj V (2013) Oxidative stress biomarkers and metabolic changes associated with cadmium stress in hyacinth bean (*Lablab Purpureus*). Afr J Biotechnol 12(29):4670–4682

Dai A-H, Nie Y-X, Yu B, Li Q, Lu L-Y, Bai J-G (2012) Cinnamic acid pretreatment enhances heat tolerance of cucumber leaves through modulating antioxidant enzyme activity. Environ Exp Bot 79:1–10

Damanik RI, Maziah M, Ismail MR, Ahmad S, Zain AM (2010) Responses of the antioxidative enzymes in Malaysian rice (*Oryza sativa* L.) cultivars under submergence condition. Acta Physiol Plant 32(4):739–747

Davenport SB, Gallego SM, Benavides MP, Tomaro ML (2003) Behaviour of antioxidant defense system in the adaptive response to salt stress in *Helianthus annuus* L. cells. Plant Growth Regul 40(1):81–88

De Vos CH, Vonk MJ, Vooijs R, Schat H (1992) Glutathione depletion due to copper-induced phytochelatin synthesis causes oxidative stress in *Silene cucubalus*. Plant Physiol 98(3):853–858

Demiral T, Türkan İ (2005) Comparative lipid peroxidation, antioxidant defense systems and proline content in roots of two rice cultivars differing in salt tolerance. Environ Exp Bot 53(3):247–257

Diaz-Vivancos P, de Simone A, Kiddle G, Foyer CH (2015) Glutathione-linking cell proliferation to oxidative stress. Free Radic Biol Med 89:1154–1164

Ding S, Lu Q, Zhang Y, Yang Z, Wen X, Zhang L, Lu C (2009) Enhanced sensitivity to oxidative stress in transgenic tobacco plants with decreased glutathione reductase activity leads to a decrease in ascorbate pool and ascorbate redox state. Plant Mol Biol 69(5):577–592

Ding S, Lei M, Lu Q, Zhang A, Yin Y, Wen X, Zhang L, Lu C (2012) Enhanced sensitivity and characterization of photosystem II in transgenic tobacco plants with decreased chloroplast glutathione reductase under chilling stress. Biochim Biophys Acta 1817(11):1979–1991

Ding S, Jiang R, Lu Q, Wen X, Lu C (2016) Glutathione reductase 2 maintains the function of photosystem II in *Arabidopsis* under excess light. Biochim Biophys Acta 1857(6):665–677

Dionisio-Sese ML, Tobita S (1998) Antioxidant responses of rice seedlings to salinity stress. Plant Sci 135(1):1–9

Dixit V, Pandey V, Shyam R (2001) Differential antioxidative responses to cadmium in roots and leaves of pea (*Pisum sativum* L. cv. Azad). J Exp Bot 52(358):1101–1109

Donahue JL, Okpodu CM, Cramer CL, Grabau EA, Alscher RG (1997) Responses of antioxidants to paraquat in pea leaves (relationships to resistance). Plant Physiol 113(1):249–257

Edwards EA, Rawsthorne S, Mullineaux PM (1990) Subcellular distribution of multiple forms of glutathione reductase in leaves of pea (*Pisum sativum* L.) Planta 180(2):278–284

Esterbauer H, Grill D (1978) Seasonal variation of glutathione and glutathione reductase in needles of *Picea abies*. Plant Physiol 61(1):119–121
Eyidogan F, Öz MT (2007) Effect of salinity on antioxidant responses of chickpea seedlings. Acta Physiol Plant 29(5):485
Ferreira RR, Fornazier RF, Vitória AP, Lea PJ, Azevedo RA (2002) Changes in antioxidant enzyme activities in soybean under cadmium stress. J Plant Nutr 25(2):327–342
Foyer CH, Noctor G (2003) Redox sensing and signalling associated with reactive oxygen in chloroplasts, peroxisomes and mitochondria. Physiol Plant 119(3):355–364
Foyer CH, Noctor G (2011) Ascorbate and glutathione: the heart of the redox hub. Plant Physiol 155(1):2–18
Foyer CH, Souriau N, Perret S, Lelandais M, Kunert KJ, Pruvost C, Jouanin L (1995) Overexpression of glutathione reductase but not glutathione synthetase leads to increases in antioxidant capacity and resistance to photoinhibition in poplar trees. Plant Physiol 109(3):1047–1057
Fujita M, Fujita Y, Noutoshi Y, Takahashi F, Narusaka Y, Yamaguchi-Shinozaki K, Shinozaki K (2006) Crosstalk between abiotic and biotic stress responses: a current view from the points of convergence in the stress signaling networks. Curr Opin Plant Biol 9(4):436–442
Gill SS, Anjum NA, Hasanuzzaman M, Gill R, Trivedi DK, Ahmad I, Pereira E, Tuteja N (2013) Glutathione and glutathione reductase: a boon in disguise for plant abiotic stress defense operations. Plant Physiol Biochem 70:204–212
Gogorcena Y, Iturbe-Ormaetxe I, Escuredo PR, Becana M (1995) Antioxidant defenses against activated oxygen in pea nodules subjected to water stress. Plant Physiol 108(2):753–759
Gong M, Li YJ, Chen SZ (1998) Abscisic acid-induced thermotolerance in maize seedlings is mediated by calcium and associated with antioxidant systems. J Plant Physiol 153(3):488–496
Gueta-Dahan Y, Yaniv Z, Zilinskas BA, Ben-Hayyim G (1997) Salt and oxidative stress: similar and specific responses and their relation to salt tolerance in citrus. Planta 203(4):460–469
Guo Z, Ou W, Lu S, Zhong Q (2006) Differential responses of antioxidative system to chilling and drought in four rice cultivars differing in sensitivity. Plant Physiol Biochem 44(11–12):828–836
Hasanuzzaman M, Nahar K, Alam MM, Fujita M (2012) Exogenous nitric oxide alleviates high temperature induced oxidative stress in wheat (*Triticum aestivum* L.) seedlings by modulating the antioxidant defense and glyoxalase system. Australian J Crop Sci 6(8):1314
Hausladen A, Alscher RG (1994) Cold-hardiness-specific glutathione reductase isozymes in red spruce. Thermal dependence of kinetic parameters and possible regulatory mechanisms. Plant Physiol 105(1):215–223
Hernández JA, Corpas FJ, Gómez M, del Río LA, Sevilla F (1993) Salt-induced oxidative stress mediated by activated oxygen species in pea leaf mitochondria. Physiol Plant 89(1):103–110
Hernández JA, Olmos E, Corpas FJ, Sevilla F, del Río LA (1995) Salt-induced oxidative stress in chloroplasts of pea plants. Plant Sci 105(2):151–167
Hernández JA, Jiménez A, Mullineaux P, Sevilia F (2000) Tolerance of pea (*Pisum sativum* L.) to long-term salt stress is associated with induction of antioxidant defences. Plant Cell Environ 23(8):853–862
Hong CY, Chao YY, Yang MY, Cho SC, Kao CH (2009) Na^+ but not Cl^- or osmotic stress is involved in NaCl-induced expression of glutathione reductase in roots of rice seedlings. J Plant Physiol 166(15):1598–1606
Hu X, Liu R, Li Y, Wang W, Tai F, Xue R, Li C (2010) Heat shock protein 70 regulates the abscisic acid-induced antioxidant response of maize to combined drought and heat stress. Plant Growth Regul 60(3):225–235
Huang M, Guo Z (2005) Responses of antioxidative system to chilling stress in two rice cultivars differing in sensitivity. Biol Plant 49(1):81–84
Huang C, He W, Guo J, Chang X, Su P, Zhang L (2005) Increased sensitivity to salt stress in an ascorbate-deficient *Arabidopsis* mutant. J Exp Bot 56(422):3041–3049
Hung KT, Kao CH (2003) Nitric oxide counteracts the senescence of rice leaves induced by abscisic acid. J Plant Physiol 160(8):871–879

Janda T, Szalai G, Rios-Gonzalez K, Veisz O, Páldi E (2003) Comparative study of frost tolerance and antioxidant activity in cereals. Plant Sci 164(2):301–306

Jiang M, Zhang J (2001) Effect of abscisic acid on active oxygen species, antioxidative defence system and oxidative damage in leaves of maize seedlings. Plant Cell Physiol 42(11):1265–1273

Jiang M, Zhang J (2002) Water stress-induced abscisic acid accumulation triggers the increased generation of reactive oxygen species and up-regulates the activities of antioxidant enzymes in maize leaves. J Exp Bot 53(379):2401–2410

Jiang K, Schwarzer C, Lally E, Lally E, Zhang S, Ruzin S, Machen T, Remington J, Feldman L (2006) Expression and characterization of a redox-sensing green fluorescent protein (reduction-oxidation-sensitive green fluorescent protein) in Arabidopsis. Plant Physiol 141(2):397–403

Jimenez A, Hernandez JA, Del Rio LA, Sevilla F (1997) Evidence for the presence of the ascorbate-glutathione cycle in mitochondria and peroxisomes of pea leaves. Plant Physiol 114(1):275–284

Kalt-Torres W, Burke JJ, Anderson JM (1984) Chloroplast glutathione reductase: purification and properties. Physiol Plant 61(2):271–278

Kaminaka H, Morita S, Nakajima M, Masumura T, Tanaka K (1998) Gene cloning and expression of cytosolic glutathione reductase in rice (*Oryza sativa* L.) Plant Cell Physiol 39(12):1269–1280

Kataya AR, Reumann S (2010) Arabidopsis glutathione reductase 1 is dually targeted to peroxisomes and the cytosol. Plant Signal Behav 5(2):171–175

Kim DW, Shibato J, Agrawal GK, Fujihara S, Iwahashi H, Kim d H, Shim IS, Rakwal R (2007) Gene transcription in the leaves of rice undergoing salt-induced morphological changes (*Oryza sativa* L.) Mol Cells 24(1):45–59

Kocsy G, von Ballmoos P, Suter M, Ruegsegger A, Galli U, Szalai G, Galiba G, Brunold C (2000) Inhibition of glutathione synthesis reduces chilling tolerance in maize. Planta 211(4):528–536

Kocsy G, Galiba G, Brunold C (2001) Role of glutathione in adaptation and signalling during chilling and cold acclimation in plants. Physiol Plant 113(2):158–164

Kocsy G, Kobrehel K, Szalai G, Duviau M-P, Buzás Z, Galiba G (2004) Abiotic stress-induced changes in glutathione and thioredoxin h levels in maize. Environ Exp Bot 52(2):101–112

Kornyeyev D, Logan BA, Payton PR, Allen RD, Holaday AS (2003) Elevated chloroplastic glutathione reductase activities decrease chilling-induced photoinhibition by increasing rates of photochemistry, but not thermal energy dissipation, in transgenic cotton. Funct Plant Biol 30(1):101–110

Kouřil R, Lazár D, Lee H, Jo J, Nauš J (2003) Moderately elevated temperature eliminates resistance of rice plants with enhanced expression of glutathione reductase to intensive photooxidative stress. Photosynthetica 41(4):571–578

Kranner I, Grill D (1996) Determination of glutathione and glutathione disulphide in lichens: a comparison of frequently used methods. Phytochem Anal 7(1):24–28

Kranner I, Beckett RP, Wornik S, Zorn M, Pfeifhofer HW (2002) Revival of a resurrection plant correlates with its antioxidant status. Plant J 31(1):13–24

Kranner I, Birtic S, Anderson KM, Pritchard HW (2006) Glutathione half-cell reduction potential: a universal stress marker and modulator of programmed cell death? Free Radic Biol Med 40(12):2155–2165

Kubo A, Sano T, Saji H, Tanaka K, Kondo N, Tanaka K (1993) Primary structure and properties of glutathione reductase from *Arabidopsis thaliana*. Plant Cell Physiol 34(8):1259–1266

Kuk YI, Shin JS (2007) Mechanisms of low-temperature tolerance in cucumber leaves of various ages. J American Soc Hortic Sci 132(3):294–301

Kukreja S, Nandwal AS, Kumar N, Sharma SK, Sharma SK, Unvi V, Sharma PK (2005) Plant water status, H_2O_2 scavenging enzymes, ethylene evolution and membrane integrity of *Cicer arietinum* roots as affected by salinity. Biol Plant 49(2):305–308

Kumar S, Kaur R, Kaur N, Bhandhari K, Kaushal N, Gupta K, Bains TS, Nayyar H (2011) Heat-stress induced inhibition in growth and chlorosis in mungbean (*Phaseolus aureus* Roxb.) is partly mitigated by ascorbic acid application and is related to reduction in oxidative stress. Acta Physiol Plant 33(6):2091

Kumari S, Agrawal M, Tiwari S (2013) Impact of elevated CO_2 and elevated O_3 on *Beta vulgaris* L.: pigments, metabolites, antioxidants, growth and yield. Environ Pollut 174:279–288

Kumutha D, Ezhilmathi K, Sairam RK, Srivastava GC, Deshmukh PS, Meena RC (2009) Waterlogging induced oxidative stress and antioxidant activity in pigeon pea genotypes. Biol Plant 53(1):75–84

Lascano HR, Casano LM, Melchiorre MN, Trippi VS (2001) Biochemical and molecular characterisation of wheat chloroplastic glutathione reductase. Biol Plant 44(4):509–516

Le Martret B, Poage M, Shiel K, Nugent GD, Dix PJ (2011) Tobacco chloroplast transformants expressing genes encoding dehydroascorbate reductase, glutathione reductase, and glutathione-S-transferase, exhibit altered anti-oxidant metabolism and improved abiotic stress tolerance. Plant Biotechnol J 9(6):661–673

Lee H, Jo J, Son D (1998) Molecular cloning and characterization of the gene encoding glutathione reductase in *Brassica campestris*. Biochim Biophys Acta 1395(3):309–314

Lei L, Lin S, Zheng H, Lei Y, Zhang Q, Zhang Z (2007) The role of antioxidant system in freezing acclimation-induced freezing resistance of *Populus suaveolens* cuttings. Forest Stud China 9(2):107–113

Leipner J, Fracheboud Y, Stamp P (1999) Effect of growing season on the photosynthetic apparatus and leaf antioxidative defenses in two maize genotypes of different chilling tolerance. Environ Exp Bot 42(2):129–139

León AM, Palma JM, Corpas FJ, Gómez M, Romero-Puertas MC, Chatterjee D, Mateos RM, del Río LA, Sandalio LM (2002) Antioxidative enzymes in cultivars of pepper plants with different sensitivity to cadmium. Plant Physiol Biochem 40(10):813–820

Liu ZJ, Zhang XL, Bai JG, Suo BX, Xu PL, Wang L (2009) Exogenous paraquat changes antioxidant enzyme activities and lipid peroxidation in drought-stressed cucumber leaves. Sci Hortic 121(2):138–143

Logan BA, Monteiro G, Kornyeyev D, Payton P, Allen RD, Holaday AS (2003) Transgenic overproduction of glutathione reductase does not protect cotton, *Gossypium hirsutum* (Malvaceae), from photoinhibition during growth under chilling conditions. Am J Bot 90(9):1400–1403

Madamanchi NR, Anderson JV, Alscher RG, Cramer CL, Hess JL (1992) Purification of multiple forms of glutathione reductase from pea (*Pisum sativum* L.) seedlings and enzyme levels in ozone-fumigated pea leaves. Plant Physiol 100(1):138–145

Mahan JR, Burke JJ (1987) Purification and characterization of glutathione reductase from corn mesophyll chloroplasts. Physiol Plant 71(3):352–358

Mahan JR, Gitz DC, Payton PR, Allen R (2009) Overexpression of glutathione reductase in cotton does not alter emergence rates under temperature stress. Crop Sci 49(1):272–280

Marty L, Siala W, Schwarzlander M, Fricker MD, Wirtz M, Sweetlove LJ, Meyer Y, Meyer AJ, Reichheld JP, Hell R (2009) The NADPH-dependent thioredoxin system constitutes a functional backup for cytosolic glutathione reductase in Arabidopsis. Proc Natl Acad Sci U S A 106(22):9109–9114

Mendoza J, Soto P, Ahumada I, Garrido T (2004) Determination of oxidized and reduced glutathione, by capillary zone electrophoresis, in *Brassica juncea* plants treated with copper and cadmium. Electrophoresis 25(6):890–896

Mergel D, Andermann G, Andermann C (1979) Simultaneous spectrophotometric determination of oxidized and reduced glutathione in human and rabbit red cells. Methods Find Exp Clin Pharmacol 1(5):277–283

Meyer AJ, Hell R (2005) Glutathione homeostasis and redox-regulation by sulfhydryl groups. Photosynth Res 86(3):435–457

Meyer AJ, Brach T, Marty L, Kreye S, Rouhier N, Jacquot J-P, Hell R (2007) Redox-sensitive GFP in *Arabidopsis thaliana* is a quantitative biosensor for the redox potential of the cellular glutathione redox buffer. Plant J 52(5):973–986

Mhamdi A, Hager J, Chaouch S, Queval G, Han Y, Taconnat L, Saindrenan P, Gouia H, Issakidis-Bourguet E, Renou J-P, Noctor G (2010) Arabidopsis GLUTATHIONE REDUCTASE1 plays a crucial role in leaf responses to intracellular hydrogen peroxide and in ensuring appropriate gene expression through both salicylic acid and jasmonic acid signaling pathways. Plant Physiol 153(3):1144–1160

Mittova V, Tal M, Volokita M, Guy M (2003) Up-regulation of the leaf mitochondrial and peroxisomal antioxidative systems in response to salt-induced oxidative stress in the wild salt-tolerant tomato species *Lycopersicon pennellii*. Plant Cell Environ 26(6):845–856

Mobin M, Khan NA (2007) Photosynthetic activity, pigment composition and antioxidative response of two mustard (*Brassica juncea*) cultivars differing in photosynthetic capacity subjected to cadmium stress. J Plant Physiol 164(5):601–610

Molina A, Bueno P, Marín MC, Rodríguez-Rosales MP, Belver A, Venema K, Donaire JP (2002) Involvement of endogenous salicylic acid content, lipoxygenase and antioxidant enzyme activities in the response of tomato cell suspension cultures to NaCl. New Phytol 156(3):409–415

Moller IM (2001) PLANT MITOCHONDRIA AND OXIDATIVE STRESS: electron transport, NADPH turnover, and metabolism of reactive oxygen species. Ann Rev Plant Physiol Plant Mol Biol 52:561–591

Mullineaux PM, Creissen GP (1997) Glutathione reductase: regulation and role in oxidative stress. Cold Spring Harbor Monogr Archive 34:667–713

Munns R, Tester M (2008) Mechanisms of salinity tolerance. Ann Rev Plant Biol 59:651–681

Nahar K, Hasanuzzaman M, Alam MM, Fujita M (2015) Glutathione-induced drought stress tolerance in mung bean: coordinated roles of the antioxidant defence and methylglyoxal detoxification systems. AoB Plants 7:plv069

Noctor G, Foyer CH (1998) Simultaneous measurement of foliar glutathione, gamma-glutamylcysteine, and amino acids by high-performance liquid chromatography: comparison with two other assay methods for glutathione. Anal Biochem 264(1):98–110

Noctor G, Queval G, Mhamdi A, Chaouch S, Foyer CH (2011) Glutathione. Arabidopsis Book 9:e0142

Pastori GM, Trippi VS (1992) Oxidative stress induces high rate of glutathione reductase synthesis in a drought-resistant maize strain. Plant Cell Physiol 33(7):957–961

Pastori GM, Trippi VS (1993) Cross resistance between water and oxidative stresses in wheat leaves. J Agri Sci 120(3):289–294

Pastori GM, Mullineaux PM, Foyer CH (2000) Post-transcriptional regulation prevents accumulation of glutathione reductase protein and activity in the bundle sheath cells of maize. Plant Physiol 122(3):667–676

Payton P, Webb R, Kornyeyev D, Allen R, Holaday AS (2001) Protecting cotton photosynthesis during moderate chilling at high light intensity by increasing chloroplastic antioxidant enzyme activity. J Exp Bot 52(365):2345–2354

Peeters N, Small I (2001) Dual targeting to mitochondria and chloroplasts. Biochim Biophys Acta 1541(1–2):54–63

Pilon-Smits EA, Zhu YL, Sears T, Terry N (2000) Overexpression of glutathione reductase in Brassica juncea: effects on cadmium accumulation and tolerance. Physiol Plant 110(4):455–460

Potesil D, Petrlova J, Adam V, Vacek J, Klejdus B, Zehnalek J, Trnkova L, Havel L, Kizek R (2005) Simultaneous femtomole determination of cysteine, reduced and oxidized glutathione, and phytochelatin in maize (*Zea mays* L.) kernels using high-performance liquid chromatography with electrochemical detection. J Chromatogr A 1084(1–2):134–144

Potters G, Horemans N, Jansen MAK (2010) The cellular redox state in plant stress biology – a charging concept. Plant Physiol Biochem 48(5):292–300

Queval G, Jaillard D, Zechmann B, Noctor G (2011) Increased intracellular H_2O_2 availability preferentially drives glutathione accumulation in vacuoles and chloroplasts. Plant Cell Environ 34(1):21–32

Rao ASVC, Reddy AR (2008) Glutathione reductase: a putative redox regulatory system in plant cells. In: Khan NA, Singh S, Umar S (eds) Sulfur assimilation and abiotic stress in plants. Springer, Berlin/Heidelberg, pp 111–147

Ratnayaka HH, Molin WT, Sterling TM (2003) Physiological and antioxidant responses of cotton and spurred anoda under interference and mild drought. J Exp Bot 54(391):2293–2305

Reddy AR, Chaitanya KV, Vivekanandan M (2004) Drought-induced responses of photosynthesis and antioxidant metabolism in higher plants. J Plant Physiol 161(11):1189–1202
Rellan-Alvarez R, Hernandez LE, Abadia J, Alvarez-Fernandez A (2006) Direct and simultaneous determination of reduced and oxidized glutathione and homoglutathione by liquid chromatography-electrospray/mass spectrometry in plant tissue extracts. Anal Biochem 356(2):254–264
Rivero RM, Ruiz JM, García PC, López-Lefebre LR, Sánchez E, Romero L (2002) Response of oxidative metabolism in watermelon plants subjected to cold stress. Funct Plant Biol 29(5):643–648
Rivero R, Ruiz J, Romero L (2004) Oxidative metabolism in tomato plants subjected to heat stress. J Hortic Sci Biotechnol 79(4):560–564
Romero-Puertas MC, Corpas FJ, Sandalio LM, Leterrier M, Rodriguez-Serrano M, Del Rio LA, Palma JM (2006) Glutathione reductase from pea leaves: response to abiotic stress and characterization of the peroxisomal isozyme. New Phytol 170(1):43–52
Rouhier N, Couturier J, Jacquot J-P (2006) Genome-wide analysis of plant glutaredoxin systems. J Exp Bot 57(8):1685–1696
Sairam RK, Saxena DC (2000) Oxidative stress and antioxidants in wheat genotypes: possible mechanism of water stress tolerance. J Agron Crop Sci 184(1):55–61
Sairam RK, Shukla DS, Saxena DC (1997) Stress induced injury and antioxidant enzymes in relation to drought tolerance in wheat genotypes. Biol Plant 40(3):357–364
Sairam RK, Srivastava GC, Agarwal S, Meena RC (2005) Differences in antioxidant activity in response to salinity stress in tolerant and susceptible wheat genotypes. Biol Plant 49(1):85
Scandalios JG (1993) Oxygen stress and superoxide dismutases. Plant Physiol 101(1):7–12
Schafer FQ, Buettner GR (2001) Redox environment of the cell as viewed through the redox state of the glutathione disulfide/glutathione couple. Free Radic Biol Med 30(11):1191–1212
Scheibe R, Dietz K-J (2012) Reduction–oxidation network for flexible adjustment of cellular metabolism in photoautotrophic cells. Plant Cell Environ 35(2):202–216
Schnaubelt D, Queval G, Dong Y, Diaz-Vivancos P, Makgopa ME, Howell G, De Simone A, Bai J, Hannah MA, Foyer CH (2015) Low glutathione regulates gene expression and the redox potentials of the nucleus and cytosol in *Arabidopsis thaliana*. Plant Cell Environ 38(2):266–279
Schwarzlander M, Fricker MD, Sweetlove LJ (2009) Monitoring the in vivo redox state of plant mitochondria: effect of respiratory inhibitors, abiotic stress and assessment of recovery from oxidative challenge. Biochim Biophy Acta Bioenerg 1787(5):468–475
Selote DS, Khanna-Chopra R (2004) Drought-induced spikelet sterility is associated with an inefficient antioxidant defence in rice panicles. Physiol Plant 121(3):462–471
Shalata A, Mittova V, Volokita M, Guy M, Tal M (2001) Response of the cultivated tomato and its wild salt-tolerant relative *Lycopersicon pennellii* to salt-dependent oxidative stress: the root antioxidative system. Physiol Plant 112(4):487–494
Sharma P, Dubey RS (2005) Drought induces oxidative stress and enhances the activities of antioxidant enzymes in growing rice seedlings. Plant Growth Regul 46(3):209–221
Shu DF, Wang LY, Duan M, Deng YS, Meng QW (2011) Antisense-mediated depletion of tomato chloroplast glutathione reductase enhances susceptibility to chilling stress. Plant Physiol Biochem 49(10):1228–1237
Signorelli S, Casaretto E, Sainz M, Diaz P, Monza J, Borsani O (2013) Antioxidant and photosystem II responses contribute to explain the drought-heat contrasting tolerance of two forage legumes. Plant Physiol Biochem 70:195–203
Singh S, Khan NA, Nazar R, Anjum NA (2008) Photosynthetic traits and activities of antioxidant enzymes in blackgram (*Vigna mungo* L. Hepper) under cadmium stress. Am J Plant Physiol 3:25–32
Smirnoff N, Colombé SV (1988) Drought influences the activity of enzymes of the chloroplast hydrogen peroxide scavenging system. J Exp Bot 39(8):1097–1108
Soltesz A, Timar I, Vashegyi I, Toth B, Kellos T, Szalai G, Vagujfalvi A, Kocsy G, Galiba G (2011) Redox changes during cold acclimation affect freezing tolerance but not the vegetative/reproductive transition of the shoot apex in wheat. Plant Biol 13(5):757–766

Spoel SH, Loake GJ (2011) Redox-based protein modifications: the missing link in plant immune signaling. Curr Opin Plant Biol 14(4):358–364

Sreenivasulu N, Grimm B, Wobus U, Weschke W (2000) Differential response of antioxidant compounds to salinity stress in salt-tolerant and salt-sensitive seedlings of foxtail millet (*Setaria italica*). Physiol Plant 109(4):435–442

Sreenivasulu N, Harshavardhan VT, Govind G, Seiler C, Kohli A (2012) Contrapuntal role of ABA: does it mediate stress tolerance or plant growth retardation under long-term drought stress? Gene 506(2):265–273

Srivalli B, Sharma G, Khanna-Chopra R (2003) Antioxidative defense system in an upland rice cultivar subjected to increasing intensity of water stress followed by recovery. Physiol Plant 119(4):503–512

Srivastava S, Srivastava AK, Suprasanna P, D'Souza SF (2010) Comparative antioxidant profiling of tolerant and sensitive varieties of *Brassica juncea* L. to arsenate and arsenite exposure. Bull Environ Contam Toxicol 84(3):342–346

Stevens RG, Creissen GP, Mullineaux PM (1997) Cloning and characterization of a cytosolic glutathione reductase cDNA from pea (*Pisum sativum* L.) and its expression in response to stress. Plant Mol Biol 35(5):641–654

Sumithra K, Jutur PP, Carmel BD, Reddy AR (2006) Salinity-induced changes in two cultivars of *Vigna radiata*: responses of antioxidative and proline metabolism. Plant Growth Regul 50(1):11–22

Sun H, Li L, Wang X, Wu S, Wang X (2011) Ascorbate–glutathione cycle of mitochondria in osmoprimed soybean cotyledons in response to imbibitional chilling injury. J Plant Physiol 168(3):226–232

Suzuki N, Koussevitzky S, Mittler R, Miller G (2012) ROS and redox signaling in the response of plants to abiotic stress. Plant Cell Environ 35(2):259–270

Szalai G, Kellős T, Galiba G, Kocsy G (2009) Glutathione as an antioxidant and regulatory molecule in plants under abiotic stress conditions. J Plant Growth Regul 28(1):66–80

Takeda T, Ishikawa T, Shigeoka S, Hirayama O, Mitsunaga T (1993) Purification and characterization of glutathione reductase from *Chlamydomonas reinhardtii*. Microbiology 139(9):2233–2238

Tan W, Meng QW, Brestic M, Olsovska K, Yang X (2011) Photosynthesis is improved by exogenous calcium in heat-stressed tobacco plants. J Plant Physiol 168(17):2063–2071

Tang K, Zhan JC, Yang HR, Huang WD (2010) Changes of resveratrol and antioxidant enzymes during UV-induced plant defense response in peanut seedlings. J Plant Physiol 167(2):95–102

Torres-Franklin ML, Gigon A, de Melo DF, Zuily-Fodil Y, Pham-Thi AT (2007) Drought stress and rehydration affect the balance between MGDG and DGDG synthesis in cowpea leaves. Physiol Plant 131(2):201–210

Tsai YC, Kao CH (2004) The involvement of hydrogen peroxide in abscisic acid-induced activities of ascorbate peroxidase and glutathione reductase in rice roots. Plant Growth Regul 43(3):207–212

Tsai YC, Hong CY, Liu LF, Kao CH (2005) Expression of ascorbate peroxidase and glutathione reductase in roots of rice seedlings in response to NaCl and H_2O_2. J Plant Physiol 162(3):291–299

Tyystjärvi E, Riikonen M, Arisi A-CM, Kettunen R, Jouanin L, Foyer CH (1999) Photoinhibition of photosystem II in tobacco plants overexpressing glutathione reductase and poplars overexpressing superoxide dismutase. Physiol Plant 105(3):409–416

Tzafrir I, Pena-Muralla R, Dickerman A, Berg M, Rogers R, Hutchens S, Sweeney TC, McElver J, Aux G, Patton D, Meinke D (2004) Identification of genes required for embryo development in Arabidopsis. Plant Physiol 135(3):1206–1220

Vaidyanathan H, Sivakumar P, Chakrabarty R, Thomas G (2003) Scavenging of reactive oxygen species in NaCl-stressed rice (*Oryza sativa* L.)—differential response in salt-tolerant and sensitive varieties. Plant Sci 165(6):1411–1418 van Rensburg L, Krüger GHJ (1994) Evaluation of components of oxidative stress metabolism for use in selection of drought tolerant cultivars of *Nicotiana tabacum* L. J Plant Physiol 143(6):730–737

von Heijne G, Steppuhn J, Herrmann RG (1989) Domain structure of mitochondrial and chloroplast targeting peptides. Eur J Biochem 180(3):535–545

Wang SH, Zhang H, Zhang Q, Jin GM, Jiang SJ, Jiang D, He QY, Li ZP (2011) Copper-induced oxidative stress and responses of the antioxidant system in roots of *Medicago sativa*. J Agron Crop Sci 197(6):418–429

Wang X, Cai J, Liu F, Dai T, Cao W, Wollenweber B, Jiang D (2014) Multiple heat priming enhances thermo-tolerance to a later high temperature stress via improving subcellular antioxidant activities in wheat seedlings. Plant Physiol Biochem 74:185–192

Wingsle G (1989) Purification and characterization of glutathione reductase from Scots pine needles. Physiol Plant 76(1):24–30

Wu TM, Lin WR, Kao YT, Hsu YT, Yeh CH, Hong CY, Kao CH (2013) Identification and characterization of a novel chloroplast/mitochondria co-localized glutathione reductase 3 involved in salt stress response in rice. Plant Mol Biol 83(4–5):379–390

Wu TM, Lin WR, Kao CH, Hong CY (2015) Gene knockout of glutathione reductase 3 results in increased sensitivity to salt stress in rice. Plant Mol Biol 87:555–564

Yannarelli GG, Fernandez-Alvarez AJ, Santa-Cruz DM, Tomaro ML (2007) Glutathione reductase activity and isoforms in leaves and roots of wheat plants subjected to cadmium stress. Phytochemistry 68(4):505–512

Yoon HS, Lee IA, Lee H, Lee BH, Jo J (2005) Overexpression of a eukaryotic glutathione reductase gene from *Brassica campestris* improved resistance to oxidative stress in *Escherichia coli*. Biochem Biophys Res Commun 326(3):618–623

Yordanova R, Popova L (2007) Effect of exogenous treatment with salicylic acid on photosynthetic activity and antioxidant capacity of chilled wheat plants. Gen Appl Plant Physiol 33(3–4):155–170

Yousuf PY, Hakeem KUR, Chandna R, Ahmad P (2012) Role of glutathione reductase in plant abiotic stress. In: Ahmad P, Prasad MNV (eds) Abiotic stress responses in plants: metabolism, productivity and sustainability. Springer, New York, pp 149–158

Yu X, Pasternak T, Eiblmeier M, Ditengou F, Kochersperger P, Sun J, Wang H, Rennenberg H, Teale W, Paponov I, Zhou W, Li C, Li X, Palme K (2013) Plastid-localized glutathione reductase 2-regulated glutathione redox status is essential for *Arabidopsis* root apical meristem maintenance. Plant Cell 25(11):4451–4468

Chapter 13
Sulfur Assimilation and Glutathione Metabolism in Plants

Akiko Maruyama-Nakashita and Naoko Ohkama-Ohtsu

Abstract Sulfur is an essential element for all organisms. Plants utilize soil sulfate to synthesize an amino acid, cysteine, which is used for a variety of sulfur-containing compounds such as glutathione (GSH), methionine, proteins, lipids, coenzymes, and various secondary metabolites. Since animals cannot synthesize organic sulfur compounds from inorganic ones, sulfate assimilation in plants is important for the global sulfur cycle.

GSH is a tripeptide synthesized from the amino acids cysteine, glutamic acid, and glycine. By controlling the redox states of proteins and chemicals, GSH functions in many biological processes including enzymatic activity, detoxification of toxic agents, and eventually influences plant growth, development, and stress management in response to both abiotic and biotic factors. Maintaining an appropriate redox environment, for which GSH levels are crucial, is thus important for plant life.

GSH levels in plant cells are controlled by both synthesis and degradation processes. GSH is synthesized from cysteine by two-step reactions in plastids and cytosol. Since cysteine levels are relatively low in the cells, the sulfate assimilation pathway composed of sulfate uptake, sulfate reduction, and assimilation into cysteine, is a rate-limiting step in GSH synthesis. In this chapter, we review the molecular machineries and regulatory aspects of the sulfur assimilation pathway and GSH metabolism in plants.

Keywords Sulfate reduction • Cysteine synthesis • Transport • Glutathione

A. Maruyama-Nakashita (✉)
Graduate School of Agricultural Science, Kyushu University,
6-10-1 Hakozaki, Higashi-ku, Fukuoka 812-8581, Japan
e-mail: amaru@agr.kyushu-u.ac.jp

N. Ohkama-Ohtsu
Institute of Agriculture, Tokyo University of Agriculture and Technology,
3-5-8 Saiwai-cho, Fuchu-shi, Tokyo 183-8509, Japan

M.A. Hossain et al. (eds.), *Glutathione in Plant Growth, Development, and Stress Tolerance*, https://doi.org/10.1007/978-3-319-66682-2_13

Contents

1 Introduction

Sulfur is an essential macronutrient for all living organisms. It is found in various biomolecules such as amino acids, cysteine and methionine, tripeptide glutathione (GSH), proteins, lipids, vitamins, coenzymes, and various secondary metabolites, and these sulfur-containing compounds function in many biological processes such as redox control, detoxification, and regulation of protein activities. Since animals cannot synthesize organic sulfur compounds from inorganic ones, sulfate assimilation in plants is important for the global sulfur cycle (Long et al. 2015). Plants take up sulfate (SO_4^{2-}) from the soil environment, which is the most common form of sulfur in nature, and assimilate it into GSH through several steps. Here we summarize the metabolic process of sulfur assimilation and GSH synthesis as well as the regulatory aspects of these processes.

2 Sulfur in Plants

Plants contain a wide variety of sulfur compounds with primary and secondary functions. Cysteine and methionine are the two sulfur-containing amino acids used for protein synthesis. Cysteine is the primary product of sulfate assimilation, and also the starting compound for methionine and glutathione synthesis. The thiol groups of two cysteine residues in proteins can be oxidized to form a disulfide bond, and this can be reversed by reduction, which is the important linkage in protein folding and protein-protein interaction, and sometimes important for controlling protein activities. Several prosthetic groups, coenzymes and vitamins, including iron-sulfur centers, coenzyme A, *S*-adenosyl-L-methionine (SAM), thiamine, biotin, lipoic acid, and *S*-methylmethionine contain a sulfur moiety that is important for their functions. In addition, the sulfolipid sulfoquinovosyldiacylglycerol is a major component in chloroplast membranes.

The physiological roles of sulfur-containing compounds in abiotic and biotic stress responses are well documented. GSH functions in the detoxification of reactive oxygen species (ROS) and as a substrate for synthesis of phytochelatin, which is important for heavy metal detoxification. Camalexin and glucosinolates in *Brassicaceae* and alliins in onion and garlic function as defense compounds against herbivores and pathogens, and are also the causative compounds for flavor and smell, with health benefits for humans such as cancer prevention (Long et al. 2015).

In addition, several signaling compounds contain sulfur as a key component. The sulfated oligosaccharides function as nodulation (Nod) factors. Turgorin, a sulfated compound responsible for thigmotactic movement in leaves, and several peptide hormones, such as phytosulfokines, require sulfation of their tyrosine residues for biological function. Hydrogen sulfide is also known as a signaling compound in animals and plants (Calderwood and Kopriva 2014).

Sulfur deficiency causes growth retardation in plants, including the chlorosis of young leaves, increase in lateral roots, and decrease in the shoot-to-root ratio (Hell et al. 2010; Long et al. 2015). Besides the growth and developmental processes, many metabolic processes including photosynthesis are affected by sulfur deficiency. These influences of sulfur deficiency on developmental and metabolic processes sometimes result in reductions of crop yield and quality. Optimized sulfur fertilization is thus of great concern in agriculture.

3 Sulfate Assimilation

GSH accumulation in plants is significantly influenced by sulfur availability and assimilation of sulfur. Plants take up sulfur from soil in the form of sulfate (SO_4^{2-}) through the activity of sulfate transporters (SULTRs; Davidian and Kopriva 2010; Takahashi et al. 2011a, b). Sulfate absorbed into root cells is transported to the plastid and activated by ATP sulfurylase, forming 5′-adenylylsulfate (APS) (Leustek et al. 2000; Saito 2004). APS is reduced to sulfide by two reaction steps catalyzed by APS reductase (APR) and sulfite reductase (Leustek et al. 2000; Saito 2004). Cysteine, the first organic form of sulfur, is then synthesized from sulfide and *O*-acetyl-L-serine (Leustek et al. 2000; Saito 2004). In this section, we describe the molecular functions of components involved in the sulfate transport and assimilation in plants (Fig. 13.1).

3.1 Sulfate Uptake and Distribution

Sulfate uptake is a primary step in the sulfur assimilation pathway. Following uptake of sulfate from the soil, sulfate horizontally moves through the apoplast and symplast, and is loaded to the xylem to be transported to aerial parts. Both the uptake and internal mobilization of sulfate are mediated by the activity of sulfate transporters (Fig. 13.2a). The mechanism for sulfate transport is suggested to be a coupled H^+ co-transport with a molar ratio of 3:1 for H^+ and SO_4^{2-}, which is driven by a proton gradient across biological membranes. In general, sulfate transporters are proteins with 500–700 amino acids, 10–12 predicted trans-membrane domains, and

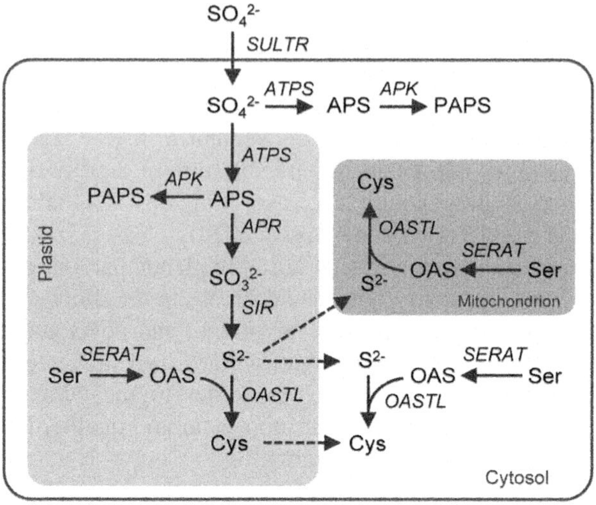

Fig. 13.1 Sulfur assimilatory pathway in plants. Sulfate absorbed by plant cells is activated, and reduced to sulfide (S^{2-}) in plastid. The cysteine is then synthesized in cytosol, plastid, and mitochondrion. *Abbreviations*: *SULTR* sulfate transporter, *ATPS* ATP sulfurylase, *APR* APS reductase, *APK* APS kinase, *SiR* sulfite reductase, *SERAT* serine acetyl-transferase, *OAS-TL* OAS(thiol)-lyase; *sulfate* SO_4^{2-}, *sulfite* SO_3^{2-}, *sulfide* S^{2-}, *APS* Adenosine-5′-phosphosulfate, *PAPS* 3′-Phosphoadenosine-5′-phosphosulfate, *Ser* Serine, *OAS* *O*-acetylserine, *Cys* Cysteine

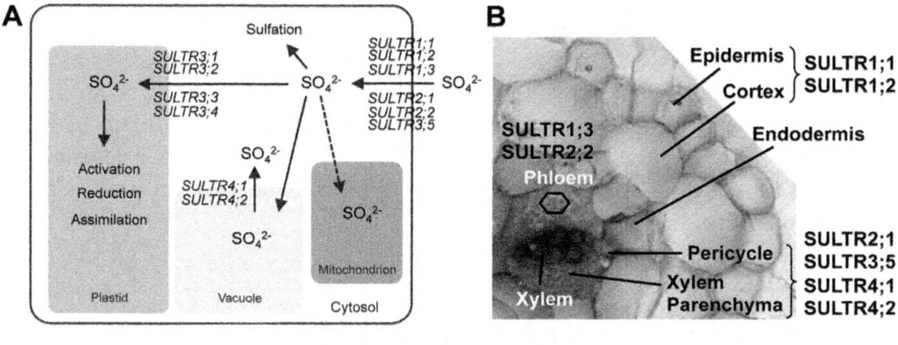

Fig. 13.2 Sulfate transport in plant cell and roots. Distribution of sulfate transporters in plant cell (**a**) and root tissues (**b**)

a short C-terminal domain similar to bacterial anti-sigma factor antagonists called the STAS (sulfate transporter and anti-sigma factor antagonists) domain. Plant sulfate transporters have been isolated from a number of species and analyzed for their functions (Buchner et al. 2004; Miller et al. 2009; Takahashi et al. 2011a, b). Among them, sulfate transporters (SULTRs) from *Arabidopsis thaliana* are relatively well understood with regard to their biochemical properties, tissue localization, and functions in plants (Davidian and Kopriva 2010; Takahashi et al. 2011a, b).

In *Arabidopsis*, there are 12 members of SULTRs, classified into four groups (SULTR1, SULTR2, SULTR3, and SULTR4) based on the similarity of their protein sequences (Fig. 13.2b; Takahashi et al. 2012). The group 1 sulfate transporters consist of high-affinity transporters, SULTR1;1, SULTR1;2 and SULTR1;3. SULTR1;1 and SULTR1;2 are expressed in the epidermis and cortex of roots and facilitate the initial uptake of sulfate from the soil (Takahashi et al. 2000; Vidmar et al. 2000; Shibagaki et al. 2002; Yoshimoto et al. 2002, 2007; Maruyama-Nakashita et al. 2003). Transcript levels of *SULTR1;2* were higher than those of *SULTR1;1* when sulfur was supplied in sufficient levels (Yoshimoto et al. 2002, 2007). When environmental sulfur was deficient, the transcript levels of both *SULTR*s were increased, but the level of induction was lower in *SULTR1;2* than in *SULTR1;1* (Shibagaki et al. 2002; Yoshimoto et al. 2002, 2007; Maruyama-Nakashita et al. 2003; Rouached et al. 2008). In addition, characterization of sulfate uptake, sulfate, cysteine, and GSH levels in *SULTR1;1* and *SULTR1;2* knockout lines indicated that SULTR1;2 mainly contributed to sulfate uptake under both sulfur sufficient (+S) and deficient (−S) conditions (Shibagaki et al. 2002; Yoshimoto et al. 2002, 2007; Maruyama-Nakashita et al. 2003). SULTR1;3 is localized in the phloem and mediates sulfate translocation from cotyledon to roots (Yoshimoto et al. 2003). The group 4 sulfate transporters, SULTR4;1 and SULTR4;2, are expressed in vascular tissues of roots and shoots and localized to the tonoplast in the cells (Kataoka et al. 2004a). Their transcript levels are higher in roots and are inducible by sulfur deficiency. Analysis of double disruption plants demonstrated their involvement in remobilization of the vacuolar sulfate pool.

In contrast to groups 1 and 4 sulfate transporters, the physiological functions of groups 2 and 3 sulfate transporters are only partially understood. The group 2 sulfate transporters, SULTR2;1 and SULTR2;2, exhibit low-affinity sulfate transport activity in yeast (Takahashi et al. 2000). SULTR2;1 is expressed in the xylem and phloem parenchyma cells of leaves and xylem parenchyma and pericycle cells of roots in *Arabidopsis* (Takahashi et al. 1997, 2000). Based on the tissue-specific localization and the analysis of the disruption lines for *SULTR2;1*, SULTR2;1 has been suggested to mediate the uptake of sulfate from the apoplast within the vascular tissues in roots (Takahashi et al. 2000; Kataoka et al. 2004b; Kawashima et al. 2011; Maruyama-Nakashita et al. 2015), and mediate the translocation of sulfate from old to young tissues through phloem transport in shoots (Awazuhara et al. 2005; Liang et al. 2010). SULTR2;2 is expressed in the phloem cells in roots, suggesting its contribution to phloem-mediated sulfate transport in roots (Takahashi et al. 2000). Although the group 3 sulfate transporters, SULTR3;1, SULTR3;2, SULTR3;3, SULTR3;4, and SULTR3;5, do not exhibit any sulfate transport activity in yeast, SULTR3;5 has been considered to function as a component of the sulfate transport system that interplays with SULTR2;1 in mediating the root-to-shoot transport of sulfate in *Arabidopsis* (Kataoka et al. 2004b). Simultaneous expression of SULTR2;1 and SULTR3;5 enhances sulfate uptake capacity compared to the expression of SULTR2;1 in yeast (Kataoka et al. 2004b). Both are expressed in the xylem parenchyma and pericycle cells of roots, which suggests a synergistic contribution of SULTR2;1 and SULTR3;5 to root-to-shoot transport of sulfate (Takahashi et al. 2000; Kataoka et al. 2004b). Because the co-expression of SULTR2;1 with SULTR3;5 increases sulfate uptake activity in yeast, the presence of *SULTR2;1* by inducible expression in roots can be a key component for increasing root-to-shoot transport of sulfate under –S (Kataoka et al. 2004b). Whether sulfate transporters facilitate xylem loading of sulfate has not been determined.

The sulfate import into plastids is necessary for assimilatory reduction of sulfate because the whole set of enzymes involved in sulfate reduction is only found in plastids. Recently, SULTR3;1 was identified as a plastid-localized sulfate transporter (Cao et al. 2013). Sulfate uptake by isolated chloroplasts was decreased by the disruption of *SULTR3;1,* and could be complemented by *SULTR3;1* expression (Cao et al. 2013). Although the reduction of sulfate uptake was moderated, similar tendencies were observed in isolated chloroplasts from the disruption lines of *SULTR3;2, SULTR3;3*, and *SULTR3;4*, which suggested that these SULTRs are also involved in sulfate import into chloroplasts. Preferential expression of *SULTR3;1, SULTR3;2*, and *SULTR3;3* in leaves also supports this conclusion. In addition, the regulatory role of the SULTR3 family in controlling sulfate translocation to developing seeds was indicated by a physiological analysis of *SULTR3* disruption lines (Zuber et al. 2010).

3.2 Sulfate Reduction

Assimilatory reduction of sulfate is predominantly a plastid-localized process since the full set of enzymes involved in sulfate reduction is only found in plastids. Sulfate reduction involves sulfate activation by ATP sulfurylase to 5′-adenylylsulfate (APS), and the reduction of APS to sulfide by two reaction steps catalyzed by APS reductase (APR) and sulfite reductase (Leustek et al. 2000; Saito 2004). Here we describe the function and cellular localization of the enzymes involved in sulfate reduction.

3.2.1 ATP Sulfurylase (ATPS)

ATP sulfurylase (ATPS) catalyzes the reaction between ATP and sulfate to yield adenosine 5′-phosphosulfate (APS) and pyrophosphate. ATPS activity can be detected in cytosol and chloroplasts in plants (Lunn et al. 1990; Rotte and Leustek 2000).

There are four *ATPS* genes (*ATPS1, ATPS2, ATPS3, ATPS4*) in the *Arabidopsis* genome (Leustek et al. 2000; Hatzfeld et al. 2000a). The four *ATPS* genes translate the proteins with N-terminal transit peptide for plastid localization, indicating that all ATPSs function within the plastids. Among them, ATPS1 was determined to be the causal gene of a natural variation of *Arabidopsis* with lower sulfur assimilation capacity, indicating that ATPS1 is the major isoform in sulfate activation (Koprivova et al. 2013). Cytosolic ATPS activity is considered to be due to ATPS2 because

ATPS2 is alternatively translated into two different isoforms that localize in plastids and cytosol in *Arabidopsis* (Hatzfeld et al. 2000a; Bohrer et al. 2014).

3.2.2 APS Reductase (APR)

APS reductase (APR) catalyzes the two-electron reduction of APS to sulfite (SO_3^{2-}) using 2 molecules of GSH as a reductant (Gutierrez-Marcos et al. 1996; Setya et al. 1996; Saito 2004). Since APS is also used for the synthesis of 3′-phosphoadenosine-5′-phosphosulfate (PAPS) catalyzed by APS kinase (APK), APK and APR compete for the substrate APS in plastids. PAPS is then used as a donor for sulfation reactions. Thus, the ratio of these two reactions controls the balance of reductive sulfur assimilation and secondary sulfur metabolism (Leustek et al. 2000; Saito 2004; Hell et al. 2010).

APR activity is predominantly detected in plastids (Leustek et al. 2000; Saito 2004). There are three *APR* genes (*APR1, APR2, APR3*) in the *Arabidopsis* genome, and the translated peptides contain plastid-localizing signals in their N-terminal region. The regulatory role of APR in the sulfate reduction pathway in Arabidopsis was demonstrated by several experimental results such as: (1) flux analysis using ^{35}S–sulfate (Vauclare et al. 2002), (2) responsibility of APR2 for genetic variation of sulfate content in Arabidopsis ecotypes (Loudet et al. 2007), and (3) strong transcriptional induction of *APR* genes under –S conditions (Takahashi et al. 1997; Koprivova et al. 2000).

In addition to the induction by –S, transcript levels of *APRs* are regulated by various environmental factors in the plants' surroundings. *APRs* are inhibited by reduced sulfur such as sulfur dioxide, hydrogen sulfide, cysteine, or glutathione (Kopriva and Koprivova 2004). APR activity and the transcript levels are influenced by nitrogen and carbon status and by diurnal rhythm, and are induced by abiotic stresses such as heavy metals, salinity, bright light, and cold (Davidian and Kopriva 2010). The activity of APR seems to follow the demand for reduced sulfur, because of the plant requirements for GSH and other sulfur-containing compounds involved in stress tolerances.

3.2.3 Sulfite Reductase (SiR)

Sulfite reductase (SiR) catalyzes the six-electron reduction of sulfite to sulfide (S^{2-}) using ferredoxin as a reductant. SiR is exclusively localized in plastids and consists of a homo-oligomer containing a siroheme and an iron-sulfur cluster per subunit. In non-photosynthetic cells, electrons can be supplied from NADPH (Yonekura-Sakakibara et al. 2000). Unlike the other enzymes working in the sulfur assimilation pathway, *SiR* is coded by one locus of the *Arabidopsis* genome. The low expression of *SiR* greatly reduces the accumulation of sulfur-containing compounds and results in severe growth retardation of plants (Khan et al. 2010), indicating the essential role of SiR in sulfur assimilation and plant growth.

3.3 Cysteine Synthesis

Cysteine synthesis, the final step of reductive sulfate assimilation into an organic sulfur compound, consists of two reaction steps. One is the synthesis of the carbon and nitrogen backbone for cysteine synthesis, and the other is the incorporation of reduced sulfur sulfide into the organic backbone. These steps are catalyzed by serine acetyltransferase (Serat) and *O*-acetylserine(thiol)lyase, respectively. Unlike the sulfate reduction pathway, both enzymes occur in the plastids, mitochondria, and cytosol in plant cells, and the catalytic activities are post-transcriptionally regulated (Hell et al. 2002; Saito 2004; Kawashima et al. 2005; Hell et al. 2010).

3.3.1 Serine Acetyltransferase (SERAT)

Serine acetyltransferase (SERAT) catalyzes the synthesis of *O*-acetylserine (OAS) from serine and acetyl-CoA (Hell et al. 2002; Droux 2004). Five genes encoding serine acetyltransferase, *SERAT1;1, SERAT2;1, SERAT2;2, SERAT3;1, SERAT3;2*, are found in the Arabidopsis genome (Hell et al. 2002; Kawashima et al. 2005). Each SERAT protein has a distinctive character to its cellular compartment and kinetic property (Noji et al. 1998; Kawashima et al. 2005). SERAT1;1, SERAT3;1, and SERAT3;2 are localized to cytosol, SERAT2;1 is localized to plastids, and SERAT2;2 is localized to mitochondria. The activities of SERAT1;1 and SERAT3;2 are regulated by allosteric feedback inhibition by L-cysteine, while those of the other three isoforms are not inhibited by cysteine, suggesting the existence of an isoform-specific regulatory mechanism.

Internal transport of OAS between each cellular compartment is suggested by the fact that none of the single knockout lines for SERAT shows a lethal phenotype (Haas et al. 2008; Watanabe et al. 2008). However, the analysis using knockout lines revealed that 80% of total SERAT activity is associated with mitochondria, while the residual activity is equally distributed in cytosol and plastids (Watanabe et al. 2008). The mitochondrial SERAT, SERAT2;2, seems to have a regulatory role in OAS net synthesis and limits total cysteine synthesis and growth of Arabidopsis (Haas et al. 2008).

3.3.2 *O*-Acetylserine(thiol)lyase (OAS-TL)

O-Acetylserine(thiol)lyase (OAS-TL) catalyzes the formation of cysteine from OAS and sulfide (Hell et al. 2010; Saito 2004). The activity is detected in the same three subcellular compartments as serine acetyltransferase. It belongs to a large family of enzymes catalyzing the ß-substitution of amino acids, which requires pyridoxal phosphate as a cofactor.

Nine genes encoding OAS-TL, *BSAS1;1, BSAS2;1, BSAS2;2, BSAS3;1, BSAS4;1, BSAS4;2, BSAS4;3, BSAS5;1*, are found in the Arabidopsis genome (Hatzfeld et al., 2000b; Watanabe et al. 2008). Among them, BSAS1;1, BSAS2;1 and BSAS2;2

function as OAS-TL in cytosol, plastids, and mitochondria, respectively. The analysis using knock out lines revealed that cytosolic OAS-TL BSAS1;1 is the major isoform in cysteine synthesis, while the other two contribute moderately but significantly to cysteine synthesis (Heeg et al. 2008; Watanabe et al. 2008). Interestingly, BSAS1;1 interacts with SULTR1;2 and stimulates sulfate uptake under –S (Shibagaki and Grossman 2010).

BSAS3;1, BSAS4;1, BSAS4;2, BSAS4;3, and BSAS5;1 show lower OAS-TL activities in vitro, but their relevance in cysteine synthesis is not proven. Several isoforms have their own activity and function in other metabolic processes, e.g., BSAS3;1 functions as ß-cyanoalanine synthase (Hatzfeld et al. 2000b; Watanabe et al. 2008), BSAS4;3 functions as cysteine desulfhydrase (Alvarez et al. 2010), and BSAS5;1 functions as S-sulfocysteine synthase (Bermudez et al. 2010).

3.3.3 Post-transcriptional Regulation of OAS and Cysteine Synthesis

SERAT and OAS-TL isoforms are not highly regulated at their transcript levels in response to sulfur status or exogenous application of OAS (Hirai et al. 2003; Maruyama-Nakashita et al. 2003; Nikiforova et al. 2004; Kawashima et al. 2005). However, SERAT and OAS-TL form a multi-enzyme complex, and the protein–protein interactions regulate the activities of both enzymes in the complex (Droux et al. 1998; Wirtz et al. 2001). As the concentration of OAS-TL is about 300 times higher than that of SERAT, only a part of the OAS-TL forms a complex with SERAT (Droux et al. 1998). Most of the OAS-TL existing in a free form is responsible for the catalytic function of cysteine synthesis. In a complexed form with SERAT, OAS-TL shows dramatically lower catalytic activity, and acts as a regulatory subunit to stimulate the activity of SERAT. The complex is stabilized by sulfide, resulting in the stimulation of OAS synthesis under +S. In contrast, OAS accumulated under –S promotes dissociation of the complex, resulting in reduced OAS synthesis. This system coordinately regulates OAS synthesis from serine and sulfate reduction for the efficient production of cysteine.

Cysteine synthesis is also influenced by a protein–protein interaction between OAS-TL and SULTR1;2 (Shibagaki and Grossman 2010). Cytosolic OAS-TL BSAS1;1 interacts with the C-terminus STAS domain of SULTR1;2 that extends into the cytosol from the plasma membrane. The interaction stimulates OAS-TL activity but reduces sulfate uptake activity of SULTR1;2, indicating that sulfate uptake and cysteine synthesis are coordinately regulated by the sulfur status in plant roots.

3.4 Regulation of Sulfate Acquisition and Assimilation

The sulfate assimilation process is highly regulated in both a transcriptional and a post-transcriptional manner, due to the importance of sulfur-containing compounds in plants. The regulation is generally demand-driven, i.e., repressed by reduced sulfur like cysteine and glutathione, and activated by the demand for reduced sulfur

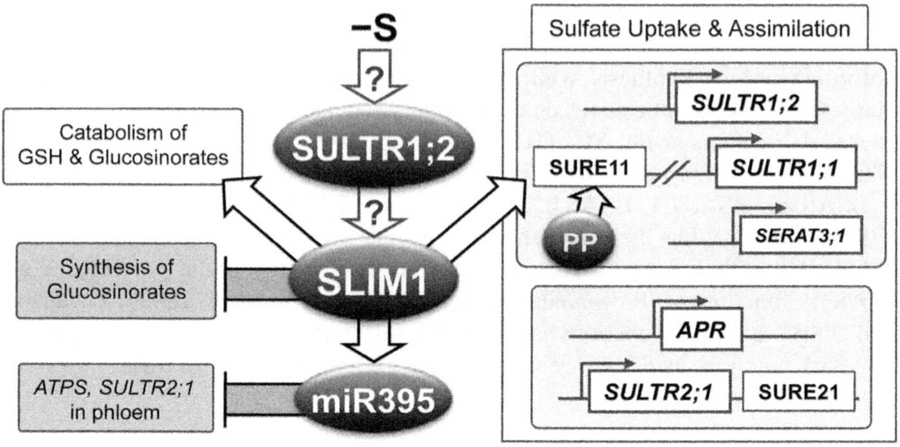

Fig. 13.3 Transcriptional regulation of sulfur assimilatory pathway under −S

under stressed conditions such as heavy metal exposure (Lappartient and Touraine 1996; Vauclare et al. 2002). As most of the post-transcriptional regulations were described in the sections above, here we mainly focus on the transcriptional regulation of the sulfate assimilation process, especially as regards sulfate uptake (Fig. 13.3).

The most well investigated mechanisms in the transcript levels of sulfur assimilatory genes are the plant responses to −S. Sulfur deficiency decreases the levels of sulfur-containing compounds such as sulfate, cysteine, GSH, and glucosinolates in *Brassicales*, etc. and enhances the concentrations of OAS. At the same time, −S increases the transcript levels of sulfur assimilatory genes, genes involved in glucosinolates and GSH catabolism, and decreases those of genes involved in glucosinolate synthesis (Maruyama-Nakashita et al. 2003; Hirai et al. 2005; Nikiforova et al. 2005). An ethylene-insensitive3-like (EIL) family transcription factor, sulfur limitation1 (SLIM1), has been identified as a regulator of plant −S responses associated with the upregulation of sulfate uptake, glucosinolate and GSH catabolism, and the downregulation of glucosinolate synthesis (Maruyama-Nakashita et al. 2006). The broad range of metabolic pathways regulated by SLIM1 suggests the existence of different factors mediating −S signals specific to each metabolic pathway.

Several *cis*-acting elements or regions responsive to –S, such as SURE11, SURE21, UPE-box, −S-responsive region of *NIT3*, and β-subunit gene promoter of β-conglycinin, have been reported (Awazuhara et al. 2002; Kutz et al. 2002; Maruyama-Nakashita et al. 2005, 2015; Wawrzynska et al. 2010). The −S-induced expressions of *SULTR1;1* and *SULTR1;2*, which facilitate the initial uptake of sulfate from the root surface, depend on the promoter activities of their 5′-upstream regions and SLIM1 activity (Maruyama-Nakashita et al. 2004a, b, 2006). A sulfur-responsive *cis*-acting element, called sulfur-responsive element 11 (SURE11), comprising 16 base pairs of a DNA sequence, has been identified in the 5′-region of *SULTR1;1* to induce its gene expression in response to –S (Maruyama-Nakashita et al. 2005), while corresponding sequences have not been identified in the *SULTR1;2* promoter. The transcript levels of *SULTR1;1* have been correlated with internal sulfate content and induced by local sulfate deficiency, whereas the levels of *SULTR1;2* are thought to be regulated by general metabolic demands (Rouached et al. 2008). Induction of *SULTR1;1* by –S involves protein phosphorylation/dephosphorylation as a regulatory process, but this is not the case for *SULTR1;2* (Maruyama-Nakashita et al. 2004a). These findings indicate different regulatory pathways of *SULTR1;1* and *SULTR1;2*, although both are regulated by SLIM1. In addition to the transcriptional regulation, unknown post-transcriptional mechanisms are important for the maintenance of SULTR1;1 and SULTR1;2 protein abundance under −S (Yoshimoto et al. 2007). In contrast to *SULTR1;1* and *SULTR1;2*, the –S responsive induction of *SULTR2;1* transcript in roots is mediated by the *cis*-acting element, SURE21, in the 3′-non-transcribed region (Maruyama-Nakashita et al. 2015), which is not regulated by SLIM1.

Post-transcriptional regulation mediated by microRNAs is also involved in the regulation of the sulfate assimilation pathway. Transcript levels of microRNA395 (miR395) are induced by –S in a SLIM1-dependent manner (Jones-Rhoades and Bartel 2004; Allen et al. 2005; Kawashima et al. 2009). Three ATPS isoforms, ATPS1, ATPS3, ATPS4, and SULTR2;1 are targeted by miR395 in response to -S, such that the miRNA395 levels and the transcript levels of these genes were negatively correlated in response to sulfur availability. Since miR395 is expressed predominantly in the phloem, *SULTR2;1* mRNA is considered to be targeted only in shoots where *SULTR2;1* is expressed in the phloem but not in the phloem of roots (Kawashima et al. 2009). The post-transcriptional regulations mediated by miR395 function in sulfate translocation from old to young leaves (Liang et al. 2010).

Interestingly, a sensor-like function of SULTR1;2 toward sulfur status in plants was reported (Zhang et al. 2014). The molecular mechanisms connecting each of the protein factors involved in –S regulation, such as SULTR1;2, SLIM1, and miR395 and the exploration of the signal transduction pathway between –S and the induction of SLIM1-independent genes like *APR* genes would further unravel the regulatory complex of sulfate assimilation. Other than –S, sulfate assimilation processes are modulated by phytohormone, biotic and abiotic stresses, nitrogen and carbon status, and the diurnal rhythms (Davidian and Kopriva 2010; Takahashi et al. 2011a, b). The interplay among these environmental factors and the regulatory protein factors of sulfur assimilation is another interesting feature of the sulfur assimilatory process and its regulatory process.

4 Glutathione Metabolism

Cysteine released from the sulfur assimilatory pathway is then used for the production of a variety of metabolites containing reduced sulfur, including methionine, glutathione, phytochelatins, and glucosinolates. In this section, we focus on the glutathione metabolism in plants (Fig. 13.4).

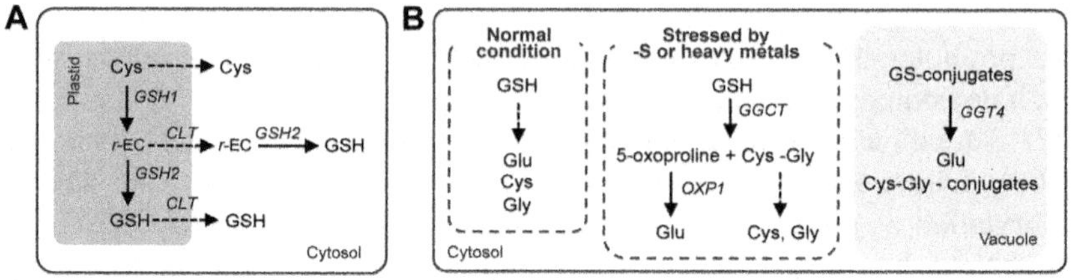

Fig. 13.4 Glutathione (GSH) metabolism in plants. (**a**) GSH synthesis. (**b**) GSH catabolism. *Abbreviations*: *GSH1* γ-glutamylcysteine synthase, *GSH2* glutathione synthase, *CLT* CRT-like transporter, *GGCT* γ-glutamyl cyclotransferase, *OXP1* cystathionine β-lyase, *GGT* γ-glutamyl transpeptidase, *Cys* cysteine, *r-EC* γ-glutamylcysteine, *GSH* glutathione, *Glu* glutamic acid, *Gly* glycine, *Cys-Gly* cysteinylglycine

4.1 Glutathione Synthesis

Glutathione, a tripeptide of γ-Glu-Cys-Gly, is synthesized in a two-step ATP-dependent reaction from these constituent amino acids. First Glu and Cys are ligated to generate γ-glutamylcysteine (γ-EC), which is catalyzed by γ-EC synthetase (Hell and Bergmann 1990). Next Gly is incorporated at the carboxy terminus to generate glutathione (GSH), which is catalyzed by GSH synthetase (Wang and Oliver 1996). In plants, γ-EC synthetase and GSH synthetase are each encoded by a single gene, *GSH1* (May and Leaver 1994) and *GSH2* (Wang and Oliver 1996), respectively.

As described in the above section, Cys is synthesized not only in chloroplasts, but also in cytosol and mitochondria, whereas synthesis of γ-EC from Glu and Cys is restricted to chloroplasts. Wachter et al. (2005) found in Arabidospsis and *Brassica juncea* two mRNA populations with long and short 5′-UTRs transcribed from the *GSH1* gene, and both mRNA sequences contained the entire transit peptide sequence for targeting to plastid. They experimentally revealed that plastidic targeting of GSH1 protein was not affected by the different length of 5′-UTRs in *GSH1* transcript and exclusively detected GSH1 protein in chloroplasts with immunological and immunocytochemical analysis.

In contrast to γ-EC synthesis, GSH synthesis from γ-EC and Gly takes place in both chloroplasts and cytosol. Wachter et al. (2005) also observed multiple mRNA populations transcribed from the *GSH2* gene, the longest of which carried the transit peptide sequence while others lacked it. They experimentally demonstrated that the protein translated from the longest *GSH2* mRNA targeted to plastid and those that lacked the transit peptide expressed in cytosol. Although GSH2 protein is expressed in both plastid and cytosol, it was shown that restricting cytosolic glutathione synthesis is sufficient for normal plant development (Pasternak et al. 2008). They observed that a seedling-lethal phenotype of *gsh2* insertion mutant of Arabidopsis was complemented by *GSH2* cDNA lacking the plastid transit peptide whose expression was restricted to cytosol.

As biosynthesis of γ-EC is restricted in plastid, it has to be transported into cytosol for use as the substrate in GSH synthesis. In Arabidopsis, γ-EC is transported from plastid into cytosol by transporters designated as CRT-like transporter (CLT) 1, 2, and

3, which are homologues to the *Plasmodium falciparum* chloroquine-resistance transporter, PfCRT (Maughan et al. 2010). Maughan et al. (2010) experimentally confirmed the localization of CLT 1, 2, and 3 to plastids and their transport activity for γ-EC and GSH using *Xenopus oocytes* expressing CLT genes. The cytosolic concentration of GSH was lower in *clt* mutants than in wild type plants and the GSH redox potential was more oxidized in the cytosol of *clt1clt2clt3* triple mutants compared to wild type although it was similar between mutants and wild type in the chloroplasts. They also observed that knockout mutants of these transporter genes were hypersensitive to heavy metal and pathogen infection. Thus it was considered that transport of γ-EC and GSH by CLTs from plastids into cytosol is required to maintain GSH levels and its redox balance, which is important for stress responses (Maughan et al. 2010).

4.2 Glutathione in Metabolic Pathways

Under heavy metal stressed conditions, GSH is polymerized for the synthesis of phytochelatins, which detoxify heavy metals by chelating them (Cobbett 2000; Mendoza-Cózatl 2011). GSH also functions as a molecule to provide reduced sulfur to other sulfur-containing metabolites, such as camalexin and glucosinolates, both of which are important secondary metabolites for defense responses against insects or pathogens. Cys has long been considered as the direct sulfur donor for them, but the finding by Parisy et al. (2007) that the Arabidopsis *pad 2–1* mutant with less camalexin was caused by a mutation in *GSH1* gene raised GSH as the candidate for sulfur donor to camalexin. The *pad2–1* mutant also showed reduced amounts of glucosinolates upon insect feeding (Schlaeppi et al. 2008), suggesting that GSH is the sulfur donor to glucosinolates. More certain evidence was provided by engendering of benzylflucosinolate in *Nicotiana benthamiana* (Geu-Flores et al. 2009). They observed accumulation of a related GSH conjugate, S-[(Z)-phenyl-acetohydroximoly] –L-glutathione, when genes for biosynthesis benzylflucosinolate were transiently coexpressed in leaves of *N. benthamiana,* and the GSH conjugates were diminished by coexpressing a *γ-glutamyl peptidase 1* (*GGP1*) gene. It was also shown that GGP1 heterologously expressed activity to hydrolyze γ-glutamyl residue from the GSH conjugates (Geu-Flores et al. 2009). Su et al. (2011) showed that indole-3-acetonitrile conjugated with GSH is the precursor for camalexin biosynthesis. Furthermore, a study with mutants for *ggp* genes demonstrated that GGP1 was the major enzyme to hydrolyze γ-glutamyl residue from the precursors in glucosinolates and camalexin biosynthesis (Geu-Flores et al. 2011).

4.3 Glutathione Catabolism

GSH is stable and exists at almost mM order in plant cells, which is more than 10 times higher than Cys, and thus GSH is considered to be major storage form of organic sulfur. After Cys is incorporated into GSH, it is stored in the organ synthesized or transported through phloem to sink organs. Kuzuhara et al. (2000) showed that the GSH concentration of phloem sap from rice taken with the insect laser method was high at around 5 mM, which was several times higher than sulfate, indicating that GSH is a major form of sulfur transported through phloem. Furthermore, phloem-transported GSH was shown to regulate genes for sulfate transport and assimilation (Lappartient et al. 1999).

After transport to a sink organ such as a sink where reduced sulfur is required, part of GSH is considered be degraded to Cys for further incorporation into other molecules such as proteins and coenzymes. When GSH synthesis was inhibited in Arabidopsis using buthionine sulfoxisimine, the inhibitor for γ-EC synthetase, the GSH concentration dropped to 20% in one day (Ohkama-Ohtsu et al. 2008). This means that turnover of GSH is rapid and its catabolism contributes to Cys availability in cells for other S-containing metabolites.

Compared to the GSH synthesis pathway, study of the pathway for its catabolism lags behind. In mammals, GSH has long been considered to be degraded in the γ-glutamyl cycle (Orlowski and Meister 1970). In this cycle, GSH is exported to the outside of cells where it is degraded by γ-glutamyl transpeptidase (GGT). GGT hydrolases the γ-bond between Glu and Cys. GGT also has transpeptidase activity to transfer the γ-glutamyl bond from GSH to other amino acids to produce γ-glutamyl amino acids. Then the γ-glutamyl amino acids are taken up into the cells and converted to 5-oxoproline by γ-glutamyl cyclotransferase (GGCT); 5-oxoproine is metabolized to Glu by oxoprolinase, while cystenylglycine is degraded into Cys and Gly. Glu, Cys, and Gly are used for other metabolites or again used for GSH synthesis.

In Arabidopsis, there are four homologues of mammal: GGT1, 2, 3, and 4. Homology of GGT1 and GGT2 is high and both proteins are shown to localize to the apoplast (Strozhenko et al. 2002; Ohkama-Ohtsu et al. 2007a; Martin et al. 2007). As the knockout mutant of *ggt1* showed yellowing in leaves by oxidative stress and accumulated oxidized glutathione GSSG in the apoplast, it was considered that GGT1 functions to alleviate oxidative stress by degrading GSSG in the apoplast (Ohkama-Ohtsu et al. 2007). Expression of GGT1 was in the vascular tissues and that of GGT2 was restricted to the young siliques. Especially strong expression of GGT2 was observed in funiculus, the tissue for transporting nutrients into seeds. From these expression patterns, it was considered possible that GGT1 and GGT2 function in GSH transport from leaves to seeds (Ohkama-Ohtsu et al. 2007). GGT3 is encoded by a short fragment of a gene that is probably derived from the C-terminal portion of GGT1 or GGT2, and expression of *GGT3* was not detected by RT-PCR (Ohkama-Ohtsu et al. 2007), thus it is unclear if GGT3 is a functional protein or not.

GGT4 has a unique N-terminal sequence not observed in other GGTs, and this sequence was predicted to be the signal peptide for vacuolar localization. Vacuolar localization of GGT4 was experimentally verified and its function was revealed to degrade glutathione conjugates in the vacuole (Ohkama-Ohtsu et al. 2007; Grzam et al. 2007). Xenobiotics such as herbicides have long been demonstrated to be conjugated with GSH for detoxification, but little is understood about how endogenous GSH conjugates other than the precursors of glucosinolates and camalexins described above. Glutathione S-transferases (GSTs) are the enzymes catalyzing conjugation of electrophiles with GSH. As there are 55 genes coding GSTs in the Arabidopsis genome (Labrou et al. 2015), it is considered that various endogenous metabolites are conjugated with GSH in plant cells. Metabolomics analysis with *ggt4* knockout mutants was found to be a good strategy to identify endogenous glutathione conjugates. This mutant is defective in degradation of GSH conjugates in the vacuole, so it is possible that metabolites accumulated in *ggt4* mutants are glutathione conjugates compared to wild type. With this strategy, Ohkama-Ohtsu et al. (2011) searched metabolites accumulated in both allelic *ggt4–1* and *ggt4–2* mutants compared to their corresponding wild type. Seven metabolites whose molecular weights corresponded to the sum of the molecular weight of GSH and that of Arabidopsis metabolites registered in the KNApSAcK database (http://kanaya.naist.jp/KNApSAcK/) were accumulated in both *ggt4* mutants. Among them, one metabolite whose molecular weight corresponded to 12-oxo-phytodienoic acid (OPDA) conjugated with GSH was verified using synthetic standards. The $[m/z]^+$ value and retention time in capillary electrophoresis–mass spectrometry of OPDA-GSH coincided with the peaks accumulated in *ggt4* mutants, confirming that OPDA is conjugated with GSH and transported into the vacuole followed by degradation by GGT4. As OPDA is the substrate of jasmonate, it was considered possible that conjugation of OPDA with GSH is involved in the regulation of jasmonate synthesis (Ohkama-Ohtsu et al. 2011). Using a similar strategy, Gläser et al. (2014) identified several candidate GSH conjugates accumulated in *ggt4* mutants, most of them previously being unassigned sulfur compounds.

As described above, GGTs have specific roles in the apoplast or vacuole. However, most of the GSH exists inside of the cells. Thus it was considered that an enzyme other than GGT is responsible for GSH breakdown in the cells. In fact, the degradation rate of GSH in cells, as determined by applying BSO, the inhibitor for GSH synthesis pathway, to Arabidopsis seedling culture, was not changed in *ggt1/ggt4* double mutan,t with no detectable GGT activity in seedlings (Ohkama-Ohtsu et al. 2008). Ohkama-Ohtsu et al. (2008) suggested with physiological analysis using oxoprolinase mutant that GSH is degraded in the cells via 5-oxoproine to Glu. The enzyme converting GSH to 5-oxoproine is γ-glutamyl cyclotransferase (GGCT), and it was found that the ChaC1 family, which are known as mammalian proapoptic proteins, functions as GGCT acting specifically to degrade GSH (Kumar et al. 2012). There are three homologues for the mammal ChaC1 family in Arabidopsis, named GGCT2;1, GGCT2;2, and GGCT2;3 (Kumar et al. 2015). Paulose et al. (2013) suggested using the Arabidopsis knockout mutant of *ggct2;1* that Arabidopsis GGCT2;1 functions in recycling of Glu by degrading GSH under heavy stress conditions. Expression of *GGCT2;1* is inducible by heavy metals (Paulose et al. 2013) or sulfur deficiency (Maruyama-Nakashita et al. 2006), and thus it is considered that catabolism of GSH is accelerated by this enzyme in response to the demand of constituent amino acids such as Cys and Glu. But it is possible that GSH catabolism also contributes to providing Cys or Glu in normal growth conditions especially in developing organs and another enzyme constitutively expressed other than GGCT2;1 would be responsible for the activity under normal growth conditions.

4.4 Regulation of Glutathione Metabolism

Glutathione is one of the signals to regulate sulfate uptake and assimilation in plants. Lappartient and Touraine (1996) and Lappartient et al. (1999) demonstrated with split roots experiments that GSH is the signal to transmit sulfur status in plants through phloem and regulate sulfate uptake and assimilation at transcriptional levels where GSH represses transcription of them. The GSH level is affected by availability of sulfate and its assimilation into Cys as suggested by an Arabidopsis mutant defective in the *sultr1;2* gene having decreased levels of GSH (Maruyama-Nakashita et al. 2003). Overexpression of the sulfate assimilation pathway into Cys increased GSH contents in Arabidopsis, tobacco and potato (Harms et al. 2000; Noji and Saito 2002; Wirtz and Hell 2007). It is also known that uptake and assimilation of sulfate into Cys is upregulated when GSH is demanded, such as in oxidative stressed or heavy metal stressed conditions (Queval et al. 2009, Hernández et al. 2015).

GSH is regulated not only by sulfur assimilation steps, but also at the step of GSH synthesis from its constituent amino acids, Glu, Cys, and Gly. The γ-EC synthetase is at the regulatory step for GSH synthesis and is feedback-inhibited by GSH (Hell and Bergmann 1990; Noctor et al. 2011). It was shown at protein structural basis that γ-EC synthetase is redox regulated at intermolecular disulfide bonds that link feedback regulation of this enzyme by GSH levels in cells (Hothorn et al. 2006; Hicks et al. 2007).

The γ-EC synthetase is regulated not only by post-transcriptional regulation but also at the transcriptional level. Expression of the *GSH1* gene encoding this enzyme in addition to that of the *GSH2* gene encoding GSH synthetase and the *GR1* encoding glutathione reductase in Arabidopsis was upregulated by copper or cadmium exposure (Xiang and Oliver 1998). As application of jasmonic acid also increased expression of these three genes, it was suggested that this plant hormone is involved in signaling of heavy metal stress by regulating genes for GSH synthesis (Xiang and Oliver 1998). Recently it was found in Arabidopsis that the Zinc-Finger transcription factor ZAT6, which is inducible by cadmium, positively regulates transcription of *GSH1*, *GSH2,* and genes for phytochelatin synthesis (Chen et al. 2016).

The GSH concentration in cells is considered to be regulated by both its synthesis and its catabolism. As mentioned above, GSH in cells was decreased to 20% in one day when its synthesis was chemically inhibited in Arabidopsis (Ohkama-Ohtsu et al. 2008), demonstrating that turnover of GSH is rapid. Compared to GSH synthesis, the pathway for its catabolism has just started to be unraveled in plants and its regulation is still unclear. To comprehensively understand homeostasis of GSH in plants, we need to take into account all of sulfur assimilation, GSH synthesis, and its catabolism.

Acknowledgement This work was supported by Japan Society for the Promotion of Science KAKENHI grant Number 15KT0028 (for N.O.O. and A.M.N.) and 24380040 (for A.M.N.).

References

Allen E, Xie Z, Gustafson AM, Carrington JC (2005) microRNA-directed phasing during trans-acting siRNA biogenesis in plants. Cell 121:207–221

Alvarez C, Calo L, Romero LC, Garcia I, Gotor C (2010) An O-acetylserine(thiol)lyase homolog with L-cysteine desulfhydrase activity regulates cysteine homeostasis in Arabidopsis. Plant Physiol 152:656–669

Awazuhara M, Kim H, Hayashi H, Chino M, Kim SG, Fujiwara T (2002) Composition of seed storage proteins changed by glutathione treatment of soybeans. Biosci Biotechnol Biochem 66:1751–1754

Awazuhara M, Fujiwara T, Hayashi H, Watanabe-Takahashi A, Takahashi H, Saito K (2005) The function of SULTR2;1 sulfate transporter during seed development in *Arabidopsis thaliana*. Physiol Plant 125:95–105

Bermudez MA, Paez-Ochoa MA, Gotor C, Romero LC (2010) Arabidopsis S-sulfocysteine synthase activity is essential for chloroplast function and long-day light-dependent redox control. Plant Cell 22:403–416

Bohrer AS, Yoshimoto N, Sekiguchi A, Rykulski N, Saito K, Takahashi H (2014) Alternative translational initiation of ATP sulfurylase underlying dual localization of sulfate assimilation pathways in plastids and cytosol in *Arabidopsis thaliana*. Front Plant Sci 5:750

Buchner P, Takahashi H, Hawkesford MJ (2004) Plant sulphate transporters: co-ordination of uptake, intracellular and long-distance transport. J Exp Bot 55:1765–1773

Calderwood A, Kopriva S (2014) Hydrogen sulfide in plants: from dissipation of excess sulfur to signaling molecule. Nitric Oxide 41:72–78

Cao MJ, Wang Z, Wirtz M, Hell R, Oliver DJ, Xiang CB (2013) SULTR3;1 is a chloroplast-localized sulfate transporter in *Arabidopsis thaliana*. Plant J 73:607–616

Chen J, Yang L, Yan X, Liu Y, Wang R, Fan T, Ren Y, Tang X, Xiao F, Cao S (2016) Zinc-finger transcription factor ZAT6 positively regulates cadmium tolerance through the glutathione-dependent pathway in Arabidopsis. Plant Physiol 171:707–719

Cobbett CS (2000) Phytochelatin biosynthesis and function in heavy-metal detoxification. Curr Opin Plant Biol 3:211–216

Davidian JC, Kopriva S (2010) Regulation of sulfate uptake and assimilation–the same or not the same? Mol Plant 3:314–325

Droux M (2004) Sulfur assimilation and the role of sulfur in plant metabolism: a survey. Photosynth Res 79:331–348

Droux M, Ruffet ML, Douce R, Job D (1998) Interactions between serine acetyltransferase and O-acetylserine (thiol) lyase in higher plants – structural and kinetic properties of the free and bound enzymes. Eur J Biochem 255:235–245

Geu-Flores F, Nielsen MT, Nafisi M, Moldrup ME, Olsen CE, Motawia MS, Halkier BA (2009) Glucosinolate engineering identifies a γ-glutamyl peptidase. Nat Chem Biol 5:575–577

Geu-Flores F, Moldrup ME, Bottcher C, Olsen CE, Scheel D, Halkier BA (2011) Cytosolic γ-glutamyl peptidases process glutathione conjugates in the biosynthesis of glucosinolates and camalexin in Arabidopsis. Plant Cell 23:2456–2469

Glaser K, Kanawati B, Kubo T, Schmitt-Kopplin P, Grill E (2014) Exploring the Arabidopsis sulfur metabolome. Plant J 77:31–45

Grzam A, Martin MN, Hell R, Meyer AJ (2007) γ-Glutamyl transpeptidase GGT4 initiates vacuolar degradation of glutathione S-conjugates in Arabidopsis. FEBS Lett 581:3131–3138

Gutierrez-Marcos JF, Roberts MA, Campbell EI, Wray JL (1996) Three members of a novel small gene-family from *Arabidopsis thaliana* able to complement functionally an Escherichia Coli mutant defective in PAPS reductase activity encode proteins with a thioredoxin-like domain and "APS reductase" activity. Proc Natl Acad Sci U S A 93:13377–13382

Haas FH, Heeg C, Queiroz R, Bauer A, Wirtz M, Hell R (2008) Mitochondrial serine acetyltransferase functions as a pacemaker of cysteine synthesis in plant cells. Plant Physiol 148:1055–1067

Harms K, von Ballmoos P, Brunold C, Hofgen R, Hesse H (2000) Expression of a bacterial serine acetyltransferase in transgenic potato plants leads to increased levels of cysteine and glutathione. Plant J 22:335–343

Hatzfeld Y, Lee S, Lee M, Leustek T, Saito K (2000a) Functional characterization of a gene encoding a fourth ATP sulfurylase isoform from *Arabidopsis thaliana.* Gene 248:51–58

Hatzfeld Y, Maruyama A, Schmidt A, Noji M, Ishizawa K, Saito K (2000b) β-Cyanoalanine synthase is a mitochondrial cysteine synthase-like protein in spinach and Arabidopsis. Plant Physiol 123:1163–1171

Heeg C, Kruse C, Jost R, Gutensohn M, Ruppert T, Wirtz M, Hell R (2008) Analysis of the Arabidopsis O-acetylserine(thiol)lyase gene family demonstrates compartment-specific differences in the regulation of cysteine synthesis. Plant Cell 20:168–185

Hell R, Bergmann L (1990) γ-Glutamylcysteine synthetase in higher-plants – catalytic properties and subcellular-localization. Planta 180:603–612

Hell R, Jost R, Berkowitz O, Wirtz M (2002) Molecular and biochemical analysis of the enzymes of cysteine biosynthesis in the plant *Arabidopsis thaliana.* Amino Acids 22:245–257

Hell R, Khan MS, Wirtz M (2010) Cellular biology of sulfur and its functions in plants. In: Hell R, Mendel RR (eds) Cell biology of metals and nutrients, plant cell monographs 17. Springer, Berlin/Heidelberg, pp 243–279

Hernandez LE, Sobrino-Plata J, Montero-Palmero MB, Carrasco-Gil S, Flores-Caceres ML, Ortega-Villasante C, Escobar C (2015) Contribution of glutathione to the control of cellular redox homeostasis under toxic metal and metalloid stress. J Exp Bot 66:2901–2911

Hicks LM, Cahoon RE, Bonner ER, Rivard RS, Sheffield J, Jez JM (2007) Thiol-based regulation of redox-active glutamate-cysteine ligase from *Arabidopsis thaliana.* Plant Cell 19:2653–2661

Hirai MY, Fujiwara T, Awazuhara M, Kimura T, Noji M, Saito K (2003) Global expression profiling of sulfur-starved Arabidopsis by DNA macroarray reveals the role of O-acetyl-l-serine as a general regulator of gene expression in response to sulfur nutrition. Plant J 33:651–663

Hirai MY, Klein M, Fujikawa Y, Yano M, Goodenowe DB, Yamazaki Y, Kanaya S, Nakamura Y, Kitayama M, Suzuki H, Sakurai N, Shibata D, Tokuhisa J, Reichelt M, Gershenzon J, Papenbrock J, Saito K (2005) Elucidation of gene-to-gene and metabolite-to-gene networks in arabidopsis by integration of metabolomics and transcriptomics. J Biol Chem 280:25590–25595

Hothorn M, Wachter A, Gromes R, Stuwe T, Rausch T, Scheffzek K (2006) Structural basis for the redox control of plant glutamate cysteine ligase. J Biol Chem 281:27557–27565

Jones-Rhoades MW, Bartel DP (2004) Computational identification of plant microRNAs and their targets, including a stress-induced miRNA. Mol Cell 14:787–799

Kataoka T, Watanabe-Takahashi A, Hayashi N, Ohnishi M, Mimura T, Buchner P, Hawkesford MJ, Yamaya T, Takahashi H (2004a) Vacuolar sulfate transporters are essential determinants controlling internal distribution of sulfate in Arabidopsis. Plant Cell 16:2693–2704

Kataoka T, Hayashi N, Yamaya T, Takahashi H (2004b) Root-to-shoot transport of sulfate in Arabidopsis: evidence for the role of SULTR3;5 as a component of low-affinity sulfate transport system in the root vasculature. Plant Physiol 136:4198–4204

Kawashima CG, Berkowitz O, Hell R, Noji M, Saito K (2005) Characterization and expression analysis of a serine acetyltransferase gene family involved in a key step of the sulfur assimilation pathway in Arabidopsis. Plant Physiol 137:220–230

Kawashima CG, Yoshimoto N, Maruyama-Nakashita A, Tsuchiya YN, Saito K, Takahashi H, Dalmay T (2009) Sulphur starvation induces the expression of microRNA-395 and one of its target genes but in different cell types. Plant J 57:313–321

Kawashima CG, Matthewman CA, Huang S, Lee BR, Yoshimoto N, Koprivova A, Rubio-Somoza I, Todesco M, Rathjen T, Saito K, Takahashi H, Dalamay T, Kopriva S (2011) Interplay of SLIM1 and miR395 in the regulation of sulfate assimilation in Arabidopsis. Plant J 66:863–876

Khan MS, Haas FH, Samami AA, Gholami AM, Bauer A, Fellenberg K, Reichelt M, Hansch R, Mendel RR, Meyer AJ, Wirtz M, Hell R (2010) Sulfite reductase defines a newly discovered bottleneck for assimilatory sulfate reduction and is essential for growth and development in *Arabidopsis thaliana*. Plant Cell 22:1216–1231

Kumar A, Tikoo S, Maity S, Sengupta S, Kaur A, Bachhawat AK (2012) Mammalian proapoptotic factor ChaC1 and its homologues function as γ-glutamyl cyclotransferases acting specifically on glutathione. EMBO Rep 13:1095–1101

Kumar S, Kaur A, Chattopadhyay B, Bachhawat AK (2015) Defining the cytosolic pathway of glutathione degradation in *Arabidopsis thaliana*: role of the ChaC/GCG family of γ-glutamyl cyclotransferases as glutathione-degrading enzymes and AtLAP1 as the Cys-Gly peptidase. Biochem J 468:73–85

Koprivova A, Suter M, Op den Camp R, Brunold C, Kopriva S (2000) Regulation of sulfate assimilation by nitrogen in Arabidopsis. Plant Physiol 122:737–746

Kopriva S, Koprivova A (2004) Plant adenosine 50-phosphosulphate reductase: the past, the present, and the future. J Exp Bot 55:1775–1783

Koprivova A, Giovannetti M, Baraniecka P, Lee BR, Grondin C, Loudet O, Kopriva S (2013) Natural variation in the ATPS1 isoform of ATP sulfurylase contributes to the control of sulfate levels in Arabidopsis. Plant Physiol 163:1133–1141

Kutz A, Muller A, Hennig P, Kaiser WM, Piotrowski M, Weiler EW (2002) A role for nitrilase 3 in the regulation of root morphology in sulphur-starving *Arabidopsis thaliana*. Plant J 30:95–106

Kuzuhara Y, Isobe A, Awazuhara M, Fujiwara T, Hayashi H (2000) Glutathione levels in phloem sap of rice plants under sulfur deficient conditions. Soil Sci Plant Nutr 46:265–270

Labrou NE, Papageorgiou AC, Pavli O, Flemetakis E (2015) Plant GSTome: structure and functional role in xenome network and plant stress response. Curr Opin Biotechnol 32:186–194

Lappartient AG, Touraine B (1996) Demand-driven control of root ATP sulfurylase activity and SO42- uptake in intact canola (the role of phloem-translocated glutathione). Plant Physiol 111:147–157

Lappartient AG, Vidmar JJ, Leustek T, Glass AD, Touraine B (1999) Inter-organ signaling in plants: regulation of ATP sulfurylase and sulfate transporter genes expression in roots mediated by phloem-translocated compound. Plant J 18:89–95

Leustek T, Martin MN, Bick JA, Davies JP (2000) Pathways and regulation of sulfur metabolism revealed through molecular and genetic studies. Annu Rev Plant Physiol Plant Mol Biol 51:141–165

Liang G, Yang F, Yu D (2010) MicroRNA395 mediates regulation of sulfate accumulation and allocation in *Arabidopsis thaliana*. Plant J 62:1046–1057

Long SR, Kahn M, Seefeldt L, Tsay YF, Kopriva S (2015) Chapter 16: Nitrogen and sulfur. In: Buchana BB, Gruissem W, Jones RL (eds) Biochemistry and molecular biology of plants. Wiley Blackwell, Oxford, pp 746–768

Loudet O, Saliba-Colombani V, Camilleri C, Calenge F, Gaudon V, Koprivova A, North K, Kopriva S, Daniel-Vedele F (2007) Natural variation for sulfate content in Arabidopsis thaliana is highly controlled by APR2. Nat Genet 39:896–900

Lunn JE, Droux M, Martin J, Douce R (1990) Localization of ATP-sulfurylase and O-acetylserine (thiol)lyase in spinach leaves. Plant Physiol 94:1345–1352

Martin MN, Saladores PH, Lambert E, Hudson AO, Leustek T (2007) Localization of members of the γ-glutamyl transpeptidase family identifies sites of glutathione and glutathione S-conjugate hydrolysis. Plant Physiol 144:1715–1732

Maruyama-Nakashita A, Inoue E, Watanabe-Takahashi A, Yamaya T, Takahashi H (2003) Transcriptome profiling of sulfur-responsive genes in Arabidopsis reveals global effects of sulfur nutrition on multiple metabolic pathways. Plant Physiol 132:597–605

Maruyama-Nakashita A, Nakamura Y, Yamaya T, Takahashi H (2004a) A novel regulatory pathway of sulfate uptake in Arabidopsis roots: implication of CRE1/WOL/AHK4-mediated cytokinin-dependent regulation. Plant J 38:779–789

Maruyama-Nakashita A, Nakamura Y, Yamaya T, Takahashi H (2004b) Regulation of high-affinity sulphate transporters in plants: towards systematic analysis of sulphur signaling and regulation. J Exp Bot 55:1843–1849

Maruyama-Nakashita A, Nakamura Y, Watanabe-Takahashi A, Inoue E, Yamaya T, Takahashi H (2005) Identification of a novel cis-acting element conferring sulfur deficiency response in Arabidopsis roots. Plant J 42:305–314

Maruyama-Nakashita A, Nakamura Y, Tohge T, Saito K, Takahashi H (2006) Arabidopsis SLIM1 is a central transcriptional regulator of plant sulfur response and metabolism. Plant Cell 18:3235–3251

Maruyama-Nakashita A, Watanabe-Takahashi A, Inoue E, Yamaya T, Saito K, Takahashi H (2015) Sulfurresponsive elements in the 3'-nontranscribed intergenic region are essential for the induction of SULFATE TRANSPORTER 2;1 gene expression in Arabidopsis roots under sulfur deficiency. Plant Cell 27:1279–1296

Maughan SC, Pasternak M, Cairns N, Kiddle G, Brach T, Jarvis R, Haas F, Nieuwland J, Lim B, Muller C, Salcedo-Sora E, Kruse C, Orsel M, Hell R, Miller AJ, Bray P, Foyer CH, Murray JA, Meyer AJ, Cobbett CS (2010) Plant homologs of the plasmodium falciparum chloroquine-resistance transporter, PfCRT, are required for glutathione homeostasis and stress responses. Proc Natl Acad Sci U S A 107:2331–2336

May MJ, Leaver CJ (1994) *Arabidopsis thaliana* γ-glutamylcysteine synthetase is structurally unrelated to mammalian, yeast, and Escherichia coli homologs. Proc Natl Acad Sci U S A 91:10059–10063

Mendoza-Cozatl DG, Jobe TO, Hauser F, Schroeder JI (2011) Long-distance transport, vacuolar sequestration, tolerance, and transcriptional responses induced by cadmium and arsenic. Curr Opin Plant Biol 14:554–562

Miller AJ, Shen Q, Xu G (2009) Freeways in the plant: transporters for N, P and S and their regulation. Curr Opin Plant Biol 12:284–290

Nikiforova VJ, Gakiere B, Kempa S, Adamik M, Willmitzer L, Hesse H, Hoefgen R (2004) Towards dissecting nutrient metabolism in plants: a systems biology case study on sulphur metabolism. J Exp Bot 55:1861–1870

Nikiforova VJ, Kopka J, Tolstikov V, Fiehn O, Hopkins L, Hawkesford MJ, Hesse H, Hoefgen R (2005) Systems rebalancing of metabolism in response to sulfur deprivation, as revealed by metabolome analysis of Arabidopsis plants. Plant Physiol 138:304–318

Noctor G, Queval G, Mhamdi A, Chaouch S, Foyer CH (2011) Glutathione. The Arabidopsis book. Am Soc Plant Biol 9: 2–32. URL: http://www.bioone.org/doi/full/10.1199/tab.0142

Noji M, Inoue K, Kimura N, Gouda A, Saito K (1998) Isoform-dependent differences in feedback regulation and subcellular localization of serine acetyltransferase involved in cysteine biosynthesis from *Arabidopsis thaliana*. J Biol Chem 273:32739–32745

Noji M, Saito K (2002) Molecular and biochemical analysis of serine acetyltransferase and cysteine synthase towards sulfur metabolic engineering in plants. Amino Acids 22:231–243

Ohkama-Ohtsu N, Radwan S, Peterson A, Zhao P, Badr AF, Xiang C, Oliver DJ (2007a) Characterization of the extracellular γ-glutamyl transpeptidases, GGT1 and GGT2, in Arabidopsis. Plant J 49:865–877

Ohkama-Ohtsu N, Zhao P, Xiang C, Oliver DJ (2007b) Glutathione conjugates in the vacuole are degraded by γ-glutamyl transpeptidase GGT3 in Arabidopsis. Plant J 49:878–888

Ohkama-Ohtsu N, Oikawa A, Zhao P, Xiang C, Saito K, Oliver DJ (2008) A γ-glutamyl transpeptidase-independent pathway of glutathione catabolism to glutamate via 5-oxoproline in Arabidopsis. Plant Physiol 148:1603–1613

Ohkama-Ohtsu N, Sasaki-Sekimoto Y, Oikawa A, Jikumaru Y, Shinoda S, Inoue E, Kamide Y, Yokoyama T, Hirai MY, Shirasu K, Kamiya Y, Oliver DJ, Saito K (2011) 12-oxo-phytodienoic acid-glutathione conjugate is transported into the vacuole in Arabidopsis. Plant Cell Physiol 52:205–209

Orlowski M, Meister A (1970) The γ-glutamyl cycle: a possible transport system for amino acids. Proc Natl Acad Sci U S A 67:1248–1255

Parisy V, Poinssot B, Owsianowski L, Buchala A, Glazebrook J, Mauch F (2007) Identification of PAD2 as a γ-glutamylcysteine synthetase highlights the importance of glutathione in disease resistance of Arabidopsis. Plant J 49:159–172

Pasternak M, Lim B, Wirtz M, Hell R, Cobbett CS, Meyer AJ (2008) Restricting glutathione biosynthesis to the cytosol is sufficient for normal plant development. Plant J 53:999–1012

Paulose B, Chhikara S, Coomey J, Jung HI, Vatamaniuk O, Dhankher OP (2013) A γ-glutamyl cyclotransferase protects Arabidopsis plants from heavy metal toxicity by recycling glutamate to maintain glutathione homeostasis. Plant Cell 25:4580–4595

Queval G, Thominet D, Vanacker H, Miginiac-Maslow M, Gakiere B, Noctor G (2009) H_2O_2-activated up-regulation of glutathione in Arabidopsis involves induction of genes encoding enzymes involved in cysteine synthesis in the chloroplast. Mol Plant 2:344–356

Rotte C, Leustek T (2000) Differential subcellular localization and expression of ATP sulfurylase and 5'-adenylylsulfate reductase during ontogenesis of Arabidopsis leaves indicates that cytosolic and plastid forms of ATP sulfurylase may have specialized functions. Plant Physiol 124:715–724

Rouached H, Wirtz M, Alary R, Hell R, Arpat AB, Davidian JC, Fourcroy P, Berthomieu P (2008) Differential regulation of the expression of two high-affinity sulfate transporters, SULTR1.1 and SULTR1.2, in Arabidopsis. Plant Physiol 147:897–911

Saito K (2004) Sulfur assimilatory metabolism. The long and smelling road. Plant Physiol 136:2443–2450

Schlaeppi K, Bodenhausen N, Buchala A, Mauch F, Reymond P (2008) The glutathione-deficient mutant pad2-1 accumulates lower amounts of glucosinolates and is more susceptible to the insect herbivore Spodoptera Littoralis. Plant J 55:774–786

Setya A, Murillo M, Leustek T (1996) Sulfate reduction in higher plants: molecular evidence for a novel 5′-adenylylsulfate reductase. Proc Natl Acad Sci U S A 93:13383–13388

Shibagaki N, Grossman AR (2010) Binding of cysteine synthase to the STAS domain of sulfate transporter and its regulatory consequences. J Biol Chem 285:25094–25102

Shibagaki N, Rose A, McDermott JP, Fujiwara T, Hayashi H, Yoneyama T, Davies JP (2002) Selenate-resistant mutants of *Arabidopsis thaliana* identify Sultr1;2, a sulfate transporter required for efficient transport of sulfate into roots. Plant J 29:475–486

Storozhenko S, Belles-Boix E, Babiychuk E, Herouart D, Davey MW, Slooten L, Van Montagu M, Inze D, Kushnir S (2002) γ-glutamyl transpeptidase in transgenic tobacco plants. Cellular localization, processing, and biochemical properties. Plant Physiol 128:1109–1119

Su T, Xu J, Li Y, Lei L, Zhao L, Yang H, Feng J, Liu G, Ren D (2011) Glutathione-indole-3-acetonitrile is required for camalexin biosynthesis in *Arabidopsis thaliana*. Plant Cell 23:364–380

Takahashi H, Yamazaki M, Sasakura N, Watanabe A, Leustek T, Engler JD, Engler G, VanMontagu M, Saito K (1997) Regulation of sulfur assimilation in higher plants: a sulfate transporter induced in sulfate-starved roots plays a central role in *Arabidopsis thaliana*. Proc Natl Acad Sci USA 94:11102–11107

Takahashi H, Watanabe-Takahashi A, Smith FW, Blake-Kalff M, Hawkesford MJ, Saito K (2000) The roles of three functional sulphate transporters involved in uptake and translocation of sulphate in *Arabidopsis thaliana*. Plant J 23:171–182

Takahashi H, Buchner P, Yoshimoto N, Hawkesford MJ, Shiu SH (2011a) Evolutionary relationships and functional diversity of plant sulfate transporters. Front Plant Sci 2:119

Takahashi H, Kopriva S, Giordano M, Saito K, Hell R (2011b) Sulfur assimilation in photosynthetic organisms: molecular functions and regulations of transporters and assimilatory enzymes. Annu Rev Plant Biol 62:157–184

Takahashi H, Buchner P, Yoshimoto N, Hawkesford MJ, Shiu SH (2012) Evolutionary relationships and functional diversity of plant sulfate transporters. Front Plant Sci 2:a119

Vauclare P, Kopriva S, Fell D, Suter M, Sticher L, von Ballmoos P, Krahenbuhl U, den Camp RO, Brunold C (2002) Flux control of sulphate assimilation in *Arabidopsis thaliana*: adenosine 5'-phosphosulphate reductase is more susceptible than ATP sulphurylase to negative control by thiols. Plant J 31:729–740

Vidmar JJ, Tagmount A, Cathala N, Touraine B, Davidian JCE (2000) Cloning and characterization of a root specific high-affinity sulfate transporter from *Arabidopsis thaliana*. FEBS Lett 475:65–69

Wachter A, Wolf S, Steininger H, Bogs J, Rausch T (2005) Differential targeting of GSH1 and GSH2 is achieved by multiple transcription initiation: implications for the compartmentation of glutathione biosynthesis in the *Brassicaceae*. Plant J 41:15–30

Wang CL, Oliver DJ (1996) Cloning of the cDNA and genomic clones for glutathione synthetase from *Arabidopsis thaliana* and complementation of a gsh2 mutant in fission yeast. Plant Mol Biol 31:1093–1104

Watanabe M, Kusano M, Oikawa A, Fukushima A, Noji M, Saito K (2008a) Physiological roles of the beta-substituted alanine synthase gene family in Arabidopsis. Plant Physiol 146:310–320
Watanabe M, Mochida K, Kato T, Tabata S, Yoshimoto N, Noji M, Saito K (2008b) Comparative genomics and reverse genetics analysis reveal indispensable functions of the serine acetyltransferase gene family in Arabidopsis. Plant Cell 20:2484–2496
Wawrzynska A, Lewandowska M, Sirko A (2010) Nicotiana tabacum EIL2 directly regulates expression of at least one tobacco gene induced by sulphur starvation. J Exp Bot 61:889–900
Wirtz M, Hell R (2007) Dominant-negative modification reveals the regulatory function of the multimeric cysteine synthase protein complex in transgenic tobacco. Plant Cell 19:625–639
Wirtz M, Berkowitz O, Droux M, Hell R (2001) The cysteine synthase complex from plants – mitochondrial serine acetyltransferase from *Arabidopsis thaliana* carries a bifunctional domain for catalysis and protein-protein interaction. Eur J Biochem 268:686–693
Xiang C, Oliver DJ (1998) Glutathione metabolic genes coordinately respond to heavy metals and jasmonic acid in Arabidopsis. Plant Cell 10:1539–1550
Yonekura-Sakakibara K, Onda Y, Ashikari T, Tanaka Y, Kusumi T, Hase T (2000) Analysis of reductant supply systems for ferredoxin-dependent sulfite reductase in photosynthetic and non-photosynthetic organs of maize. Plant Physiol 122:887–894
Yoshimoto N, Takahashi H, Smith FW, Yamaya T, Saito K (2002) Two distinct high-affinity sulfate transporters with different inducibilities mediate uptake of sulfate in Arabidopsis roots. Plant J 29:465–473
Yoshimoto N, Inoue E, Saito K, Yamaya T, Takahashi H (2003) Phloem-localizing sulfate transporter, Sultr1;3, mediates re-distribution of sulfur from source to sink organs in Arabidopsis. Plant Physiol 131:1511–1517
Yoshimoto N, Inoue E, Watanabe-Takahashi A, Saito K, Takahashi H (2007) Posttranscriptional regulation of high-affinity sulfate transporters in Arabidopsis by sulfur nutrition. Plant Physiol 145:378–388
Zhang B, Pasini R, Dan H, Joshi N, Zhao YH, Leustek T, Zheng ZL (2014) Aberrant gene expression in the Arabidopsis *SULTR1;2* mutants suggests a possible regulatory role for this sulfate transporter in response to sulfur nutrient status. Plant J 77:185–197
Zuber H, Davidian JC, Aubert G, Aime D, Belghazi M, Lugan R, Heintz D, Wirtz M, Hell R, Thompson R, Gallardo K (2010) The seed composition of Arabidopsis mutants for the group 3 sulfate transporters indicates a role in sulfate translocation within developing seeds. Plant Physiol 154:913–926

Chapter 14
Glutathione-Mediated Biotic Stress Tolerance in Plants

J.A. Hernández, G. Barba-Espín, and P. Diaz-Vivancos

Abstract Glutathione, along with ascorbate, is the main non-enzymatic antioxidant and redox buffers in plant cells. The reduced form of glutathione (GSH) is involved in the protection of cells from the oxidative damage induced by environmental challenges. GSH plays an important role in the recycling of reduced ascorbate in the reaction catalyzed by the enzyme dehydroascorbate reductase in the so-called ascorbate–glutathione cycle. Several studies reported that glutathione is involved in the induction of plant defense genes, and the increase in GSH and/or GSH-related enzymes is correlated with the resistance to different biotic challenges, including plant virus, bacteria, and fungi. Also, different works evidenced that decreases in GSH can be responsible for pathogen-elicited symptom development in susceptible plants. In that respect, it is important to mention that treatments leading to an increase in GSH and/or the redox state of glutathione can reduce the virus contents and/or the symptoms even during compatible plant–virus interactions. In addition, subcellular glutathione contents, reactive oxygen species production, and the antioxidative metabolism are considered valuable biotic stress indicators within plants during situations of pathogen attack.

Keywords Bacteria • Dehydroascorbate reductase • Fungi • Glutathione • Oxidative stress • Redox state • Virus

J.A. Hernández • P. Diaz-Vivancos (✉)
Department of Plant Breeding, CEBAS-CSIC, 30100 Murcia, Spain
e-mail: pdv@cebas.csic.es

G. Barba-Espín
Department of Plant and Environmental Sciences, University of Copenhagen, 1017 Copenhagen, Denmark

M.A. Hossain et al. (eds.), *Glutathione in Plant Growth, Development, and Stress Tolerance*, https://doi.org/10.1007/978-3-319-66682-2_14

Contents

1 Introduction

In plant cells, the reactive oxygen species (ROS) scavenging is dependent on ascorbate (AsA) and glutathione (GSH), the two main non-enzymatic hydrophilic antioxidants and redox buffers (Noctor and Foyer 1998). Both antioxidant molecules act as signaling molecules in many cellular processes and are involved in basic metabolic reactions as well as in the protection of plant cells under environmental stress situations. Ascorbate and glutathione take part in the important ascorbate–glutathione (AsA–GSH) cycle, which plays an important role in the scavenging of H_2O_2 and in the recycling to the reduced forms of ascorbate and glutathione (AsA and GSH). These functions are catalyzed by APX (ascorbate peroxidase), DHAR (dehydroascorbate reductase), MDHAR (monodehydroascorbate reductase), and GR (glutathione reductase). DHAR is the enzyme that links both antioxidants in the AsA–GSH cycle (Fig. 14.1). APX activity depends not only on the AsA availability but also on GSH through the DHAR activity, a GSH-dependent enzyme that can regenerate AsA from DHA (dehydroascorbate). It is also true that the reaction catalyzed by MDHAR for AsA recycling is more economical for the plant cell, but

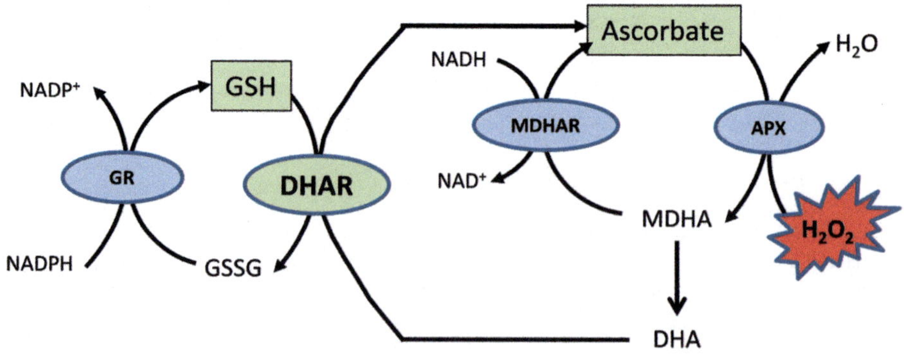

Fig. 14.1 Ascorbate–glutathione (AsA–GSH) cycle in plants. This cycle consists of a series of redox-coupled reactions whose main function is the scavenging of H_2O_2 and the recycling to the reduced forms of ascorbate and glutathione

under certain conditions, DHAR activation can be a good strategy to maintain ascorbate redox state.

Glutathione is the major low-molecular-weight thiol in plants, and its role in plant defense and tolerance against abiotic and biotic stresses has been widely described. In 1998, Wingate et al. reported the importance of GSH for local resistance responses. Treatment of suspension-cultured cells of bean (*Phaseolus vulgaris* L.) with GSH resulted in a massive and selective induction of the transcription of defense genes encoding enzymes related to phytoalexin and lignin biosynthesis, as well as stimulation of genes encoding cell wall hydroxyproline-rich glycoproteins. GSH is considered the most important thiol antioxidant in plant–pathogen interactions (Kuźniak 2010), and a clear relation between GSH increases and pathogen resistance has been reported in different studies (Gullner et al. 1999; Kużniak and Sklodowska 2004; Zechmann et al. 2007a; Clemente-Moreno et al. 2010, 2013).

The redox state of glutathione plays an important role regulating the expression of defense genes. One of the most typical examples corresponds to the activation of NPR1 proteins as well as *NPR1* gene expression. The reduction of NPR1 requires an increase in GSH contents, the NPR1 protein conformation being sensitive to cellular redox changes (Mou et al. 2003). Treatments with L-2-oxo-4-thiazolidine-carboxylic acid (OTC) increased GSH content and glutathione redox state, inducing the expression of the *NPR1* gene in both healthy and *Plum pox virus* (PPV)-infected peach plantlets (Clemente-Moreno et al. 2012). In that regard, GSH seems to be the main antioxidant involved in the activation of plant defense genes (Wingate et al. 1988; Ghanta et al. 2011).

2 Plant–Virus Interaction

Different authors observed the importance of the glutathione content as well as the GSH-related enzymes in the physiological and biochemical responses of plants against virus infection (Hernández et al. 2016). The GSH contents, and therefore the redox state of glutathione, have been often associated with a resistance response to plant viruses (Table 14.1). In addition, the availability of the GSH precursors, cysteine, glycine, and glutamate, which can limit GSH synthesis, is also important in the plant–virus interaction (Zechmann et al. 2007a). In this sense, treatments aimed at increasing GSH levels, or the redox state of glutathione, also induced some kind of resistance in different plant–virus interactions. In 1999, Gullner et al. reported that the exposure of tobacco (*Nicotiana tabacum*) leaf discs to the cysteine precursor L-2-oxo-4-thiazolidine-carboxylic acid (OTC) led to a massive accumulation of GSH, as well as reduced *tobacco mosaic virus* (TMV) coat protein contents and decreased number of necrotic lesions and virus contents in TMV-inoculated tobacco leaf discs (Gullner et al. 1999). Similar results were also recorded by Zechmann et al. (2007a) in pumpkin (*Cucurbita pepo*) seedlings. These authors treated pumpkin seedlings with 1 mM OTC for 48 h previous to the infection with *Zucchini*

Table 14.1 Response of glutathione and GSH-related enzymes in some plant–virus interactions

Interaction	GSH	GSH-related enzymes	Response	References
Tobacco (*Nicotiana tabacum*)–TMV (OTC-treated)	Increase	nd	Reduced necrotic lesions and viral coat proteins	Gullner et al. (1999)
Pumpkin (*Cucurbita pepo*)–ZYMV (OTC-treated)	Increase in roots and cotyledons	nd	Reduced, delayed, or complete suppression of symptoms	Zechmann et al. (2007a, b)
Pea (*Pisum sativum*)–PPV (OTC- or BTH-treated)	Increased redox state of glutathione	Increase in sDHAR, sGR, cDHAR, cGPX	Reduction of percentage of infected leaves	Clemente-Moreno et al. (2010)
In vitro peach (*Prunus persica*)–PPV	Increased redox state of glutathione	Decrease in GST (infected plantlets) Decrease in GR (healthy plantlets)	No protection	Clemente-Moreno et al. (2012)
Peach–PPV (OTC-treated)	Increase	Increase in cGPX	Reduced percentage of infected leaves Chloroplast protection	Clemente-Moreno et al. (2013)
Tobacco–TMV (sulfate-treated)	Increase	Increase in GST	Reduced necrotic lesions	Király et al. (2012)
Apricot (*Prunus armeniaca*) seeds–PNRSV	nd	Decrease in DHAR and GR	Low germination rate	Amari et al. (2007)
N. benthamiana–PMMoV-I	No change	Decrease GR (up to 21 dpi)	Symptom recovery	Hakmaoui et al. (2012)
N. benthamiana–PMMoV-S	Decreased levels Reduced GSH/GSSG ratio	Decrease GR (up to 21 dpi)	Severe symptoms	Hakmaoui et al. (2012)
Cucumber (*Cucumis sativus*)–CMV	nd	Increase in mDHAR, sGR Decreased sGR	Reduced Pn Reduced mitochondrial complex I and II	Song et al. (2009)
Tomato (*Lycopersicon esculentum*)–CMV	nd	Increase in sDHAR, cDHAR, mDHAR, sGR	Reduced Pn (less affected) Reduced mitochondrial complex I	Song et al. (2009)

s soluble fraction, *c* chloroplastic, *m* mitochondrial, *nd* not determined, *dpi* days post-inoculation *yellow mosaic virus* (ZYMV) and evaluated the symptoms of ZYMV disease. They observed a general delay, reduction, or complete suppression of symptom development in OTC-treated plants, depending on the time of infection and the severity of symptoms (Zechmann et al. 2007a).

Glutathione content is influenced by sulfate nutrition, and high sulfate supply correlated with increased levels of Cys (cysteine) and GSH (Kopriva and Rennenberg 2004; Király et al. 2012). In a recent work, Király et al. (2012) reported that TMV-resistant tobacco plants grown with sufficient sulfate developed less necrotic lesions response when compared with plants grown with sulfate deficiency. This enhanced virus resistance correlated with elevated levels of Cys and glutathione as well as the induction of glutathione S-transferase (GST) (Király et al. 2012).

In compatible interactions PPV–*Prunus* and PPV–pea, an imbalance in the antioxidant machinery at subcellular level is produced (Hernández et al. 2004, 2006, Diaz-Vivancos et al. 2008), and in all cases an inhibitory effect on some GSH-dependent enzymes was observed. For example, a decrease in soluble and chloroplastic DHAR, GR, or glutathione peroxidase (GPX) was noticed (Hernández et al. 2004, 2006; Diaz-Vivancos et al. 2008).

The treatment of pea and peach plants with 1 mM OTC, previous to the infection with PPV, resulted in a partial resistance response, in terms of PPV symptoms as well as in the percentage of leaves showing PPV symptoms (Clemente-Moreno et al. 2010, 2013). In pea plants, this response was correlated with a higher redox state of glutathione as well as with an increase in APX, POX, and GSH-related enzymes at the subcellular level (Clemente-Moreno et al. 2010). Interestingly, asymptomatic leaves of infected plants displayed a higher redox state of glutathione, a fact that could also play a role in the reduction of symptoms. In peach plants, OTC treatment, in addition to an increased plant growth, provides protection to the photosynthetic machinery and/or the chloroplast metabolism in PPV-infected plants (Clemente-Moreno et al. 2013). However, when OTC was applied to PPV-infected peach plantlets, the OTC treatments did not reduce the virus contents, although GSH levels were increased (Clemente-Moreno et al. 2012). In OTC-treated plantlets, an induction of *NPR1* gene expression took place, mainly in PPV-infected plants, suggesting that GSH could play an important role in the *NPR1* induction under a viral infection (Clemente-Moreno et al. 2012). Accordingly, a similar conclusion was reported by Ghanta et al. (2011) in tobacco plants.

The infection of apricot seeds by *Prunus necrotic ringspot virus* (PNRSV) induced an oxidative stress that was parallel with a general decrease of all AsA–GSH cycle enzymes, including GSH-dependent enzymes such as DHAR and GR. This response was correlated with a more limited ability to eliminate H_2O_2 but also in the recycling of AsA and GSH in infected seeds (Amari et al. 2007).

Hakmaoui et al. (2012) studied the response of *Nicotiana benthamiana* plants against two different strains of *Pepper mild mottle virus*, the Italian (PMMoV-I) and the Spanish (PMMoV-S) strains. The PMMoV-S was the most virulent, inducing dramatic symptoms, whereas plants infected with PMMoV-I were able to recover from their symptoms. This response was linked, among other factors, with the maintenance of GSH levels during the infection phase (up to 21 days). However, the plants infected with PMMoV-S strain showed a dramatic decrease in both total and reduced glutathione (Hakmaoui et al. 2012).

Cucumber mosaic virus (CMV) infection disrupted electron transport in chloroplast and mitochondria from cucumber and tomato plants. However, the net photosynthesis rate as well as the rate of mitochondrial complex I electron transport in mitochondria was less affected in tomato than in cucumber leaves (Song et al. 2009). This fact could be related with a better response of the antioxidant defenses in CMV-infected tomato plants than in cucumber, including an induction of DHAR in chloroplasts, mitochondria, and soluble fractions and GR in soluble fractions from tomato leaves. Nevertheless, in cucumber leaves, DHAR and GR only increased in mitochondria, whereas in chloroplasts a decrease in GR was recorded (Song et al. 2009).

3 Plant–Bacteria Interaction

A role for GSH and GSH-utilizing enzymes in the resistance against bacteria has been also suggested in different plant–bacteria interactions (Table 14.2). GSH mediates plant–bacteria interaction in both pathogenesis and symbiosis establishment.

The *Arabidopsis* mutant *pad2*–1 showed an increased susceptibility to the bacterial pathogen *Pseudomonas syringae* (Glazebrook et al. 1996) as well as enhanced susceptibility to additional pathogens (Parisy et al. 2007). PAD2 encodes γ-glutamylcysteine synthetase (GSH1), the enzyme that catalyzes the first step of de novo GSH biosynthesis, suggesting that in *Arabidopsis* the maintenance of adequate levels of GSH is an important factor during *P. syringae* and other pathogen infections (Parisy et al. 2007). In addition, *rax1–1*, an *Arabidopsis* GSH1 mutant which accumulates less than 50% of wild-type GSH content, was shown to be more susceptible to avirulent strains of *P. syringae* (Ball et al. 2004). However, the *cad2–1* mutant, with approximately about 30% of wild-type amounts of GSH, showed an unaltered disease resistance phenotype to virulent and avirulent strains of *P. syringae* (May et al. 1996).

The response of two different tomato cultivars against *P.syringae* pv. tomato is related with the glutathione levels and its redox state as well as the behavior of GSH-dependent enzymes (Kuźniak and Sklodowska 2004). In that regard, the tomato A100 cultivar, susceptible to *P. syringae*, responded to the infection with a decrease in GSH and an accumulation of oxidized glutathione (GSSG) leading to a decrease in the redox state of glutathione. Under the same conditions, increases in GPX and GR were produced that was insufficient to keep the glutathione pool reduced (Kuźniak and Sklodowska 2004). In contrast, the resistant tomato cultivar (Ontario), in addition to showing higher constitutive GSH levels than the susceptible cultivar, also maintained the glutathione pool homeostasis in response to *P. syringae*. Moreover, in response to the infection, the Ontario cultivar progressively increased GST activity (Kuźniak and Sklodowska 2004). This enzyme not only plays an important role in the detoxification of organic hydroperoxides but also

Table 14.2 Response of GSH and GSH-related enzymes in some plant–bacteria interactions

Interaction	GSH	GSH-related enzymes	Response	References
Arabidopsis pad2–1–P. syringae	Decrease	nd	Increased susceptibility	Glazebrook et al. (1996)
Arabidopsis rax1–1–P. syringae	Decrease	Increased MDHAR	Increased susceptibility to avirulent strains	Ball et al. (2004)
Arabidopsis cad2–1–P. syringae	Decrease	nd	Unaltered disease resistance phenotype	May et al. (1996)
Tomato cv. A100–*P. syringae*	Decrease (GSSG accumulation)	Increased GPX and GR	Susceptible	Kuzniak and Sklodowska (2004)
Tomato cv. Ontario–*P. syringae*	Increase	Increased GST	Resistant	Kuzniak and Sklodowska (2004)
Transgenic tobacco–*P. syringae*	nd	Increased GR	Delayed necrosis	Faize et al. (2012)
Transgenic tobacco–*P. syringae*	Increase	nd	Improved defense response	Matern et al. (2015)
Soybean (*Glycine max*)–*Bradyrhizobium japonicum*	Increase	Increased DHAR and GR	Defense against oxidative stress	Dalton et al. (1986, 1991)
Medicago truncatula–Sinorhizobium meliloti	Decrease by BSO or antisense glutathione synthetase genes	nd	Decreased nodulation	Frendo et al. (2005)

nd not determined

displays DHAR activity (Dixon et al. 2002). In that sense, the increase in GST was parallel with a decrease in the concentration of DHA, maintaining the AsA levels and thus increasing the redox state of ascorbate (Kuźniak and Sklodowska 2004).

The treatment of apple plants with benzothiadiazole (BTH), a SA analogue, limited the infection by *Erwinia amylovora* (Sklodowska et al. 2011). At short term (2 days post-inoculation, dpi), this response was correlated with an increase in GSH and GSSG, leading to a decrease in the GSH/GSSH ratio. After 7 dpi, BTH-treated plants showed a decline in GSH but a low increase in GSSG, about 20%, in relation to control plants. However, non-treated plants displayed a threefold increase in GSSG. In both cases, a decrease in GSH/GSSG ratio was produced, especially in non-treated plants, where the reduction in GSH/GSSG ratio was near four times. At long term (14 dpi), no significant changes in the redox state of glutathione occurred in any case. However, in this case, the BTH-induced resistance against the bacterial infection was not correlated with increases in GSH-dependent enzymes, such as GPX or GST (Sklodowska et al. 2011).

Additional evidence of the GSH role in plant–bacteria interaction comes from studies using transgenic plants. Tobacco plants overexpressing cytosolic Cu, Zn-superoxide dismutase (cyt*sod*), and/or ascorbate peroxidase (cyt*apx*) genes displayed a disease tolerance phenotype, with various levels of resistance, against bacterial wildfire caused by *P. syringae* pv. tabaci (Faize et al. 2012). Transgenic lines harboring cyt*apx* and both transgenes showed the best response in terms of resistance. Inoculated transgenic lines displayed increased levels of GR activity when compared with wild-type inoculated plants (Faize et al. 2012), suggesting that the maintenance of adequate glutathione redox state could be an important factor during *P. syringae* infection. More recently, it has been reported that transgenic high-glutathione *Nicotiana tabacum* lines showed also an improved defense response against *P. syringae*, this response being modulated by the GSH redox potential (Matern et al. 2015).

The causal agent of bacterial wilt disease of plants is the bacteria *Ralstonia solanacearum*. This pathogen uses virulence effector proteins leading to the suppression of disease resistance responses to succeed in infection. Mukaihara et al. (2016) have described that the *R. solanacearum* effector protein RipAY is able to degrade GSH. This protein displays γ-glutamylcyclotransferase activity and due to its high GSH degradation activity could be considered as an effective mechanism to overcome pathogen plant defenses. Moreover, because GSH is also important for bacterial environmental stress tolerance and growth, RipAY displays a very interesting safety mechanism to avoid unwanted activation, and it is specifically activated by host eukaryotic thioredoxins (Mukaihara et al. 2016).

On the other hand, rhizobial bacteria interact with legume root to establish a symbiotic relationship leading to the formation of a new specialized organ, the nodule, which is capable of fixing atmospheric nitrogen. In root nodules, high level of GSH or homoglutathione (hGSH, GSH homolog present instead of or in addition to GSH in certain legumes) and the presence of an active AsA–GSH cycle have been reported (Becana et al. 2000; Dalton et al. 1986; Frendo et al. 1999). Thus, it has been suggested that GSH and hGSH protect nitrogen-fixing nodules against toxic oxygen species resulting from the active nodule metabolism and from varying physiological conditions, as well as from environmental challenges. In this sense, Dalton et al. (1991) reported an increase in ascorbate peroxidase and ascorbate-recycling enzymes (especially DHAR) and GR activities, as well as an increase in ascorbate and glutathione content in soybean nodules exposed to elevated ambient pO_2, linking N_2 fixation and antioxidative metabolism.

Moreover, during the nodulation process, an active root cell division is triggered in order to establish nodule primordia. Through the characterization of different *A. thaliana* GSH1 mutants, it has been showed that GSH plays a key role during root development (Diaz-Vivancos et al. 2015). The *root meristemless1* (*rml1*) mutant, having only about 5% of the wild-type GSH levels (Schnaubelt et al. 2015), is not able to maintain cell division following germination. The cell cycle in *rml1* is arrested in G1 phase of the cell cycle, being GSH the required factor to reactivate the cell division in the root apical cells (Vernoux et al. 2000). In addition, buthionine sulfoximine (BSO), a GSH synthesis inhibitor, caused the arrest of root but not shoot development in wild-type seedlings (Schnaubelt et al. 2015). In *Medicago truncatula*, both BSO treatment and antisense glutathione synthetase genes in roots resulted in a decrease of (1) the average number of nodules in inoculated roots and (2) the expression of genes involved in the nodulation process, suggesting an important role for GSH in the symbiotic plant–bacteria interaction during the nodulation process (Frendo et al. 2005).

Taking together, all the reported evidences suggest that the maintenance of adequate levels of GSH is important for both the establishment of pathogen bacteria disease resistance and symbiotic plant–bacteria interactions.

4 Plant–Fungi Interaction

The effects of fungal infection on the glutathione metabolism in different cell compartments have been scarcely studied. Most of the information available has been obtained using crude extracts. In addition, the majority of information corresponds to interactions with a low range of fungi, including *Botrytis cinerea*, *Fusarium oxysporum*, or *Trichoderma harzianum* (Kuźniak and Skłodowska 1999, 2001, 2005; García-Limones et al. 2002; Bernal-Vicente et al. 2015) (Table 14.3).

During a plant–fungi interaction, ROS can be generated by both the pathogen and/or the host plant. In the case of a necrotrophic fungus, ROS overproduction can be a strategy to kill the host tissue in the initial phase of infection (Tiedemann 1997). In such conditions, GSH seems to be the limiting factor for a proper functioning of the AsA–GSH cycle during the progression of the infection. In tomato plants, the infection by the necrotrophic fungus *B. cinerea* caused a progressive decrease in GSH contents, whereas GSSG was barely affected (Kuźniak and Skłodowska 1999). The fungal infection also affected GSH-dependent enzymes. In that regard, an increase in GR activity was produced in tomato leaves in order to try to maintain the redox state of the glutathione. However, a decrease in other GSH-dependent enzymes, such as DHAR, occurred (Kuźniak and Skłodowska 1999). The same authors studied the effect of *B. cinerea* in the antioxidative mechanisms in chloroplasts from tomato plants (Kuźniak and Skłodowska 2001). These authors reported that *B. cinerea* infection promoted senescence symptoms, the chloroplasts being one of the earliest cell compartments affected, as indicated by the loss of chlorophyll observed even after 1 dpi. This effect was correlated with a decrease in chloroplastic GSH and total glutathione pools as well as a decrease in GPX, an enzyme that participates in the reduction of lipid hydroperoxides by using GSH as reducing power (Asada 1999).

Years later, Kuzniak and Sklodowska (2005) studied the changes of the AsA–GSH cycle in different cell compartments (chloroplasts, mitochondria, and peroxisomes) in *B.cinerea*-infected tomato leaves. The oxidative stress caused by the fungal infection affected all cellular compartments, although the authors observed organelle-specific

Table 14.3 Response of glutathione and GSH-related enzymes in some plant–fungi interactions

Interaction	GSH	GSH-related enzymes	Response	References
Tomato–*B. cinerea*	Decrease	Decrease in DHAR Increase in GR	Visible symptoms at 3 dpi Gray mold in oldest leaves	Kuźniak and Skłodowska (1999)
Tomato–*B. cinerea*	Decrease	Decreased cGPX (4–5 dpi). Increased cGR (4–5 dpi) and cGST (3 dpi)	Visible symptoms at 3 dpi Gray mold in oldest leaves	Kuzniak and Sklodowska (2001)
Tomato–*B. cinerea*	Decrease Increased mGSSG and pGSSG	Decreased total DHAR, mDHAR, pDHAR, mGR and pGR. Increased cDHAR, total GR	Dark necrotic lesions (2–3 dpi)	Kuzniak and Sklodowska (2001)
Chickpea (*Cicer arietinum* cv. JG62)–*F. oxysporum*	nd	Increased root GR	Susceptible response Vascular infection (20–22 dpi)	García-Limones et al. (2002)
Chickpea (cv. Ontario)–*F. oxysporum*	nd	Higher constitutive stem GR levels	Resistant response	García-Limones et al. (2002)
Olea europaea–Glomus claroideum	nd	Increased DHAR	Increased FW	Alguacil et al. (2003)
Retama sphaerocarpa–Glomus claroideum	nd	Increased DHAR Increased GR	Increased FW	Alguacil et al. (2003)
Rhamnus lycioides–Glomus claroideum	nd	Increased DHAR Increased GR	Increased FW	Alguacil et al. (2003)
Soybean–*Glomus mosseae*	nd	Increased root GR)	Increased plant biomass	Porcel et al. (2003)
Melon–*T. harzianum*	nd	Increased DHAR and GST	Increased FW	Bernal-Vicente et al. (2015)

S soluble fraction, *c* chloroplastic, *m* mitochondrial, *p* peroxisomal, *nd* not determined, *dpi* days post-inoculation

changes, such variations being masked when data were analyzed in whole-leaf extract. A general decrease in glutathione concentration occurred by the infection in different cell compartments from tomato leaves, mitochondria and peroxisomes being the most affected organelles. In chloroplasts and mitochondria, the total glutathione contents declined after 2–3 dpi, but in peroxisomes, this decrease started earlier, only 1 dpi (Kuzniak and Sklodowska 2005). The reduction in total glutathione was parallel with a significant increase in GSSG, especially in mitochondria and peroxisomes. As a result, in all cell compartments, the infection produced an important decrease in the GSH/GSSG ratios, appearing earlier in mitochondria and peroxisomes than in chloroplasts, showing lower GSH/GSSG ratio at the initial state of the infection phase (Kuzniak and Sklodowska 2005). GSH-dependent enzymes were also affected by *B. cinerea* infection in the different cell compartments studied. In this sense, the infection induced a decrease in DHAR activity in whole-leaf extracts, mainly from 2–4 dpi. The response of chloroplastic DHAR was somewhat different to that observed in mitochondria or peroxisomes. No important effect was produced in chloroplast, and even an important increase after 3 dpi occurred. However, mitochondrial and peroxisomal DHAR activities decreased after 3 dpi (Kuzniak and Sklodowska 2005). Regarding GR activity, an initial increase at 1 dpi was maintained during the infection period in whole-leaf extracts as well as in chloroplasts. In contrast, mitochondrial GR peaked at 3 dpi and then progressively decreased until the end of the infection period (5 dpi), whereas peroxisomal GR showed a decline only at 2 dpi. The authors suggested that the increases in GR can be an effective protection to avoid an excessive GSSG accumulation in order to maintain the redox state of glutathione (Kuzniak and Sklodowska 2005).

Fusarium oxysporum is another necrotrophic fungus that produces the fusarium wilt (a vascular wilt fungal disease) in many plant species, including tomato, pepper, melon, or legumes, among others. The information about the effect of *F. oxysporum* infection on the glutathione metabolism of higher plants is very scarce. In 2002, García-Limones et al. studied the possible role of ROS production in the development of the fusarium wilt disease in chickpea in compatible and incompatible interactions. However, these authors did not measure glutathione contents but only GR activity (among other antioxidant enzymes). The authors observed that the first symptoms appeared 15–17 dpi in the susceptible cultivar (cv. JG62). During this period, about 50% of plants were systemically infected. At the end of the disease development phase (20–22 dpi), more than 90% of susceptible plants showed vascular infection. In the case of the resistant cultivar (cv. WR315), no evidences of infection were observed (García-Limones et al. 2002). The authors found a constitutive GR activity much higher in stems than in roots in both chickpea cultivars. In infected plants, a transient increase in GR occurred only in roots from susceptible plants, and at the end of the disease development (20–22 dpi), a correlation among GR increase, H_2O_2-scavenging enzymes (APX, CAT), and H_2O_2-producing enzymes (SOD) took place. Although GR activity did not show significant changes in stems during the development of the fungal disease, in the resistant cultivar, GR activity levels were higher than the susceptible cultivar. In stems, APX, CAT, and SOD activities were induced only in susceptible plants during the disease development. All these responses led the authors to suggest that the lack of induction of antioxidant enzymes in the stem of resistant plants can denote a less efficient ROS scavenging defense and thus a higher ROS level accumulation that could be related with the resistance mechanism against *F. oxysporum* infection (García-Limones et al. 2002).

The addition of specific fungus such as *T. harzianum* to plant growth substrates can increase plant yield and also reduce plant diseases produced by other plant pathogens present in soils. However, the mechanisms of action of these biostimulant and biocontrol effects are not fully understood and knowledge about their influence in the antioxidative metabolism is very scarce. The inoculation with *T. harzianum* increased FW of melon plants grown in different organic substrates (Bernal-Vicente et al. 2015). This response was parallel with the increase of some GSH-dependent enzymes, such as DHAR and GST (Bernal-Vicente et al. 2015). More specifically, the combination of *T. harzianum* with either citrus or bentonite compost stimulated DHAR activity in melon leaves, whereas the combination of *T. harzianum* with either peat substrate or bentonite compost increased leaf GST activity (Bernal-Vicente et al. 2015). The increase in DHAR involves a higher AsA-recycling capacity, and according to Gong et al. (2005), ascorbate and GST seem to play key roles in plant growth and development.

Mycorrhizae may help plants to grow in semiarid ecosystems improving their response to the environmental changes that involve the progressive increase in atmospheric CO_2 concentration in a climate change context (Terrer et al. 2016). Mycorrhizal inoculation increased the plant biomass in three Mediterranean shrubs, *Olea europaea* L. ssp. sylvestris, *Retama sphaerocarpa* (L.) Boissier, and *Rhamnus lycioides* L. This stimulant effect in plant growth was related with increased mineral content (N, P, K, Mg, Fe, Ca, etc.) as well as with increased antioxidant capacity, including GSH-dependent enzymes. The inoculation with the allochthonous arbuscular mycorrhizal (AM) fungus *Glomus claroideum* strongly increased DHAR in the three mentioned shrubs. In contrast, the inoculation with a mixture of native AM fungi produced a lower stimulation of DHAR activity (Alguacil et al. 2003). The presence of *G. claroideum* also increased GR activity in *R. sphaerocarpa* and *R. lycioides*, whereas the inoculation with the mixture of native AM fungi only raised GR activity in *R. lycioides* plants, producing even higher increases than *G.claroideum*. However, no effect in GR was observed in the shrub *O. europaea* (Alguacil et al. 2003).

Since the abovementioned experiment was carried out in semiarid conditions, the mycorrhizal-induced increases in antioxidant enzymes could be a strategy used by such shrubs to face the ROS overproduction under the environmental conditions assayed. Specifically, and as far as GSH-enzymes are concerned, the increase in DHAR can involve a better capability to recycle AsA, whereas GR, in addition to GSH recycling, may serve to provide $NADP^+$ availability to accept electrons from the photosynthetic electron chain in order to minimize the reduction of O_2 to O_2^-. In conclusion, these authors suggested that the increase in antioxidant enzymes could be involved, at least partially, in the beneficial effect of mycorrhizal colonization on the performance of shrubs species grown under semiarid conditions (Alguacil et al. 2003).

Oxidative damage to biomolecules is one of the most important mechanisms triggering nodule senescence in stressed nodules (Becana et al. 2000). Mycorrhizal symbiosis can also protect plants against premature nodule senescence induced by drought situations, as observed in soybean plants (Porcel et al. 2003). This response seemed to be linked to a higher GR activity in roots and nodules in mycorrhizal plants. These authors proposed that the higher GR activity in roots and nodules of mycorrhizal plants has contributed to the protection of nodules from premature senescence (Porcel et al. 2003).

5 Changes in the Subcellular Compartmentalization of Glutathione Under Biotic Stress Conditions

As an antioxidant, glutathione is involved in detoxifying ROS through the ascorbate–glutathione cycle (Foyer and Noctor 2009, 2013). Changes in the subcellular contents of glutathione during biotic stimuli reflect the occurrence of compartment-specific defense mechanisms, which is associated with compartment-specific ROS accumulation and oxidative stress. Since virtually all plant pathogens cause ROS generation and oxidative stress, changes in subcellular glutathione contents are therefore valuable biotic stress indicators within plants during situations of pathogen attack. Whereas the role of glutathione and, by extension, of the antioxidative metabolism in different organelles to abiotic stress has been well reported, the data on the glutathione compartment-specific role during plant–pathogen interaction is poorly understood. The following lines summarize the findings on this subject, discussing the existing connection between subcellular accumulation of glutathione and ROS, and the documented functional differences of the subcellular compartments during biotic stress situations.

5.1 Apoplast

Glutathione concentrations in the apoplast constitute a minor portion of its total pool (Vanacker et al. 1998, 2000; Zechmann et al. 2008; Tolin et al. 2013). In leaf cells, the apoplast contains only 1–2% of the total cell glutathione (Foyer et al. 2001), although higher values has been reported in pea leaves (Hernández et al. 2001). Moreover, the glutathione homeostasis is easily alterable in the apoplast due to the absence of systems to regenerate the reduced glutathione form. Consequently the capacity of redox buffering in the apoplast is weaker than that found inside the cell (Horemans et al. 2000). These facts make the apoplast a sensor of changes in the environmental conditions (Tolin et al. 2013). In response to biotic stimuli, the low buffering capacity of the apoplast favors a rapid accumulation of ROS (Mittler 2002). Herein, ROS overproduction is one of the early events following pathogen attack, occurring mainly in the apoplast via plasma membrane NADPH oxidases (Torres et al. 2002; Suzuki et al. 2011) and cell wall peroxidases (Bindschedler et al. 2006). This has been reported in numerous plant–pathogen interactions as common event of plants' hypersensitive response (HR) leading to programmed cell death (Lamb and Dixon 1997; Wojtaszek 1997; Jones and Dangl 2006).

5.2 *Chloroplasts and Peroxisomes*

Changes in glutathione contents in chloroplasts and peroxisomes are involved in plant defense against pathogens. *P. syringae* and *B. cinerea* caused enhanced accumulation of glutathione in *Arabidopsis* at early stages of infection, reaching 73% and 450% of control levels in chloroplast and peroxisomes, respectively. At later stages of infection, a pronounced decrease of glutathione in both cell compartments was accompanied by increased ROS accumulation (Großkinsky et al. 2012). This highlights the importance of glutathione during stress signaling. Similar results have been achieved in peroxisomes of tomato plants during *B. cinerea* infection, where the initial increase of glutathione was followed by a pronounced decrease and the disruption of the antioxidative system in peroxisomes (Kuźniak and Skłodowska 2005). Some authors have pointed out the existence of a connection between apoplastic and chloroplastic ROS signaling during the biotic response, in which the chloroplast may act as an amplifier of the signal from the apoplast (Joo et al. 2005; Vahisalu et al. 2010). As antioxidants protect the organelles by counteracting the level of ROS (Green et al. 2006), their contents in chloroplast can be decisive in the tuning of ROS signaling.

5.3 *Mitochondria*

Mitochondria possess a strong antioxidant system to protect them against the constant generation of ROS, in which GSH is of particular importance (Zechmann et al. 2008). In this sense, the drop of glutathione contents is associated with ROS accumulation and oxidative stress in this compartment, leading to the induction of programmed cell death (Vianello et al. 2007). *Arabidopsis* plants infected with *B. cinerea* displayed a strong drop of total glutathione content in mitochondria 48-h post-inoculation, which correlated with a strong increase of H_2O_2 in this compartment and the development of necrosis symptoms (Simon et al. 2013). In a similar way, the infection of *N. tabacum* with an incompatible strain of *tobacco mosaic virus* (TMV) provoked a depletion of glutathione contents in mitochondria and the development of necrotic spots (Király et al. 2012). In tomato plants, *B. cinerea* produced a decrease of glutathione contents mainly in mitochondria, coupled with the accumulation of oxidized glutathione 48-h post-inoculation (Kuźniak and Skłodowska 2005).

5.4 *Nucleus*

Glutathione in nucleus plays essential roles in protection of DNA against oxidative damage, cell proliferation, DNA synthesis, and regulation of the nuclear matrix organization and proteins (Green et al. 2006; Díaz-Vivancos et al. 2010b; Go and Jones 2010). Glutathione also regulates the expression of genes involved in the activation of plant defense mechanisms (Han et al. 2013). The roles of glutathione in nuclei during pathogen attack are not fully understood. However it has been reported that there is a notable accumulation of glutathione in nuclei during early stages of viral, fungal, and bacterial infection. Such is the case of *Arabidopsis* plants infected with *P. syringae* (Großkinsky et al. 2012) and *B. cinerea* (Simon et al. 2013), leaves of *Cucurbita pepo* infected with ZYMV (Zechmann et al. 2005), and TMV-infected *N. tabacum* plants (Király et al. 2012). In *Arabidopsis*, increased GSH accumulation in the nuclei has been reported to be concomitant with decreased levels in the cytosol, followed by enhanced levels in the whole cell (Diaz-Vivancos et al. 2010a). Similarly, the increase of total glutathione in the nuclei of *Arabidopsis* plants infected with *P. syringae* (Großkinsky et al. 2012) and *B. cinerea* (Simon et al. 2013) was followed by a rapid accumulation of glutathione in the chloroplasts and cytosol. It was hypothesized that the initial accumulation of GSH in nuclei is perceived as a signal in order to increase its levels in the whole cell under stress conditions (Diaz-Vivancos et al. 2010a).

5.5 An Outlook of the Methods to Determine Subcellular Glutathione Concentrations

Determination of GSH and GSSG on the subcellular levels is technically challenging as the sample preparation itself can be perceived as a stress, thus altering the GSH levels. There are two major approaches to study the compartment distribution of glutathione in plants, presenting advantages and disadvantages inherent in both types of methods: (1) biochemical determination after subcellular fractionation of plant tissues and (2) microscopical visualization following glutathione labeling. Large amount of starting plant material and cross-contamination among fractions during organelle isolation are the major constraints of the biochemical determination. Moreover, the equivalence of the results obtained in vitro with the actual glutathione levels in vivo is, often, uncertain. Nevertheless, these methods allow the determination of glutathione in millimolar range, and glutathione redox state can be calculated through the measurement of both reduced and oxidized forms (Jiménez et al. 1997; Vanacker et al. 1998; Kuźniak and Skłodowska 2001; Ohkama-Ohtsu et al. 2007; Krueger et al. 2009). Regarding the microscopic approaches, they can be separated into light microscopical methods, in which glutathione is labeled with specific antibodies or dyes (Meyer and Fricker 2000; Müller et al. 2005; Zechmann and Müller 2010), fluorescence microscopy determination following labeling with redox-sensitive green fluorescent protein (GFP)(Meyer et al. 2007; Gutscher et al. 2008), and electron microscopy following immunogold labeling (Gao et al. 2012). Unfortunately, the antibody that is currently used for detecting glutathione cannot differentiate between the reduced and oxidized form (Zechmann et al. 2005).These methods allow determining the localization and concentration of glutathione in vivo in the different cell compartments in a more accurate way, which opens up new prospects in the study of glutathione dynamics in plant defense mechanisms. Main limitations of the microscopic techniques are intrinsic to the sample preparation and visualization, as mechanical separation of cells and tissues and exposure to light and dehydration under microscope can be perceived as a stress to the sample.

6 Conclusion and Future Perspectives

Biotic environmental stress situations lead to considerable yield drop causing important economic losses worldwide. One common consequence of exposure to stress conditions is the increased production of ROS. The most potentially deleterious effect of ROS is that at high concentrations they trigger genetically programmed cell suicide events. Far from being only damaging agents, ROS are also used by plants as second messengers in signal transduction cascades in a variety of process, their accumulation being crucial for plant development as well as defense (Foyer and Noctor 2013). In plants, ROS production and scavenging are intimately linked, and the balance between them will determine defense responses. Thus, the major low-molecular-weight antioxidants ascorbate and glutathione determine the specificity of this oxidative signaling. The tripeptide glutathione exerts a strong influence on plant responses against pathogens, not only as an antioxidant but also as a defense signaling compound. Due to its relatively high cellular concentration, GSH acts as ROS scavenger or sacrificial nucleophile. Although many other secondary metabolites can function similarly, GSH is distinguished from most of these compounds by three main characteristics: (1) the presence of specific enzymes that couple GSH to the oxidative metabolism, (2) the existence of relatively stable oxidizing form, and (3) the recycling of GSSG to GSH by high-capacity enzyme-based system.

Biological systems adapt to changing environments by reorganizing their cellular and physiological program with metabolites representing one important response level. Glutathione can thus be considered as multifunctional metabolite that is important in redox homeostasis and signaling as well as in developmental and defense reactions (Foyer and Noctor 2011). Changes in redox metabolism will inevitably modify much larger signaling network that integrates information from many pathways regulating plant growth and defense responses. The importance of GSH and GSH-related enzymes in reducing the incidence of plant pathogens and symptom development has been widely reported. In this sense, treatments inducing increases in GSH and/or the redox state of glutathione can be beneficial during plant–pathogen interactions. Nevertheless, how pathogens alter plant metabolism and biochemistry is not fully understood yet. New knowledge on this topic and the discovery of new products stimulating GSH redox homeostasis would lead to develop new strategies to achieve a durable tolerance against pathogens.

Acknowledgments PDV thanks CSIC and the Spanish Ministry of Economy and Competitiveness for their "Ramon &Cajal" research contract, co-financed by FEDER funds.

References

Alguacil MM, Hernandez JA, Caravaca F, Portillo B, Roldán A (2003) Antioxidant enzyme activities in shoots from three mycorrhizal shrub species afforested in a degraded semiarid soil. Physiol Plant 118:562–570

Amari K, Díaz-Vivancos P, Pallás V, Sánchez-Pina MA, Hernández JA (2007) Oxidative stress induction by *Prunus necrotic ringspot virus* infection in apricot seeds. Physiol Plant 131:302–310

Asada K (1999) The water–water cycle in chloroplasts: scavenging of active oxygens and dissipation of excess photons. Annu Rev Plant Physiol Plant Mol Biol 50:601–639

Ball L, Accotto G, Bechtold U, Creissen G, Funck D, Jimenez A, Kular B, Leyland N, Mejia-Carranza J, Reynolds H, Karpinski S, Mullineaux PM (2004) Evidence for a direct link between glutathione biosynthesis and stress defense gene expression in Arabidopsis. Plant Cell 16:2448–2462

Becana M, Dalton DA, Moran JF, Iturbe-Ormaetxe I, Matamoros MA, Rubio MC (2000) Reactive oxygen species and antioxidants in legume nodules. Physiol Plant 109:372–381

Bernal-Vicente A, Pascual JA, Tittarelli F, Hernández JA, Díaz-Vivancos P (2015) Trichoderma supplementation of compost stimulates the antioxidant defense system in melon plants. J Sci Food Agric 95:2208–2214

Bindschedler LV, Dewdney J, Blee KA, Stone JM, Asai T, Plotnikov J, Denoux C, Hayes T, Gerrish C, Davies DR, Ausubel FM, Bolwell GP (2006) Peroxidase-dependent apoplastic oxidative burst in Arabidopsis required for pathogen resistance. Plant J 47:851–863

Clemente-Moreno MJ, Diaz-Vivancos P, Barba-Espín G, Hernández JA (2010) Benzothiadiazole and L-2-oxothiazolidine-4-carboxylic acid reduced the severity of Sharka symptoms in pea leaves: effect on the antioxidative metabolism at subcellular level. Plant Biol 12:88–97

Clemente-Moreno MJ, Diaz-Vivancos P, Piqueras A, Hernández JA (2012) Plant growth stimulation in *Prunus* species plantlets by BTH or OTC treatments under *in vitro* conditions. J Plant Physiol 169:1074–1083

Clemente-Moreno MJ, Diaz-Vivancos P, Rubio M, Fernandez N, Hernández JA (2013) Chloroplast protection in plum pox virus-infected peach plants by L-2-oxo-4-thiazolidine-carboxylic acid treatments: effect in the proteome. Plant Cell Environ 36:640–654

Dalton DA, Russell SA, Hanus FJ, Pascoe GA, Evans HJ (1986) Enzymatic reactions of ascorbate and glutathione that prevent peroxide damage in soybean root nodules. Proc Natl Acad Sci U S A 83:3811–3815

Dalton DA, Post CJ, Langeberg L (1991) Effects of ambient oxygen and of fixed nitrogen on concentrations of glutathione, ascorbate, and associated enzymes in soybean root nodules. Plant Physiol 96(3):812–818

Diaz-Vivancos P, Clemente-Moreno MJ, Rubio M, Olmos E, Garcia JA, Martinez-Gomez P, Hernandez JA (2008) Alteration in the chloroplastic metabolism leads to ROS accumulation in pea plants in response to *plum pox virus*. J Exp Bot 59:2147–2160

Diaz-Vivancos P, Dong Y, Ziegler K, Markovic J, Pallardo FV, Pellny TK, Verrier PJ, Foyer CH (2010a) Recruitment of glutathione into the nucleus during cell proliferation adjusts whole cell redox homeostasis in *Arabidopsis thaliana* and lowers the oxidative defence shield. Plant J 64:825–838

Diaz-Vivancos P, Wolff T, Markovic J, Pallardo FV, Foyer CH (2010b) A nuclear glutathione cycle within the cell cycle. Biochem J 431:169–178

Diaz-Vivancos P, de Simone A, Kiddle G, Foyer CH (2015) Glutathione – linking cell proliferation to oxidative stress. Free Radic Biol Med 89:1154–1164

Dixon DP, Davis BG, Edwards R (2002) Functional divergence in the glutathione transferase superfamily in plants. J Biol Chem 277:30859–30869

Faize M, Burgos L, Faize L, Petri C, Barba-Espin G, Díaz-Vivancos P, Clemente-Moreno MJ, Alburquerque N, Hernandez JA (2012) Modulation of tobacco bacterial disease resistance using cytosolic ascorbate peroxidase and Cu, Zn-superoxide dismutase. Plant Pathol 61:858–866

Foyer CH, Noctor G (2009) Redox regulation in photosynthetic organisms: signaling, acclimation, and practical implications. Antioxid Redox Signal 11(4):861–905
Foyer CH, Noctor G (2013) Redox signaling in plants. Antioxid Redox Signal 18(16):2087–2090
Foyer CH, Theodoulou FL, Delrot S (2001) The functions of inter- and intracellular glutathione transport systems in plants. Trends Plant Sci 6(10):486–492
Foyer CH, Noctor G (2011) Ascorbate and glutathione: the heart of the redox hub. Plant Physiol 155(1): 2–18
Frendo P, Gallesi D, Turnbull R, Van de Sype G, Hérouart D, Puppo A (1999) Localization of glutathione and homoglutathione in *Medicago truncatula* is correlated to a differential expression of genes involved in their synthesis. Plant J 17:215–219
Frendo P, Harrison J, Norman C, Hernandez-Jimenez MJ, Van de Sype G, Gilabert A, Puppo A (2005) Glutathione and homoglutathione play a critical role in the nodulation process of *Medicago truncatula*. Mol Plant-Microbe Interact 18:254–259
Gao R, Ng FK, Liu P, Wong SM (2012) Hibiscus chlorotic ringspot virus coat protein upregulates sulfur metabolism genes for enhanced pathogen defense. Mol Plant-Microbe Interact 25:1574–1583
Garcia-Limones C, Hervás A, Navas-Cortés JA, Jiménez-Diaz RM, Tena M (2002) Induction of an antioxidant enzyme system and other oxidative stress markers associated with compatible and incompatible interactions between chickpea (*Cicer arietinum* L.) and *Fusarium oxysporum* f.sp. ciceris. Physiol Mol Plant Pathol 61:325–337
Ghanta S, Bhattacharyya D, Sinha R, Banerjee A, Chattopadhyay S (2011) *Nicotiana tabacum* overexpressing γ-ECS exhibits biotic stress tolerance likely through NPR1- dependent salicylic acid-mediated pathway. Planta 233:895–910
Glazebrook J, Rogers EE, Ausubel FM (1996) Isolation of Arabidopsis mutants with enhanced disease susceptibility by direct screening. Genetics 143:973–982
Go YM, Jones DP (2010) Redox control system in the nucleus: mechanisms and functions. Antioxid Redox Signal 12:489–509
Gong H, Jiao Y, Hu WW, Pua EC (2005) Expression of glutathione-S-transferase and its role in plant growth and development in vivo and shoot morphogenesis in vitro. Plant Mol Biol 57:53–66
Green RM, Graham M, O'Donovan MR, Chipman JK, Hodges NJ (2006) Subcellular compartmentalization of glutathione: correlations with parameters of oxidative stress related to genotoxicity. Mutagenesis 21:383–390
Großkinsky DK, Koffler BE, Roitsch T, Maier R, Zechmann B (2012) Compartment-specific antioxidative defense in Arabidopsis against virulent and avirulent *Pseudomonas syringae*. Phytopathology 102(7):662–673
Gullner G, Tóbiás I, Fodor J, Kőmíves T (1999) Elevation of glutathione level and activation of glutathione-related enzymes affect virus infection in tobacco. Free Radic Res 31:S155–S161
Gutscher M, Pauleau AL, Marty L, Brach T, Wabnitz GH, Samstag Y, Meyer AJ, Dick TP (2008) Real-time imaging of the intracellular glutathione redox potential. Nat Methods 5:553–559
Han Y, Chaouch S, Mhamdi A, Queval G, Zechmann B, Noctor G (2013) Functional analysis of Arabidopsis mutants points to novel roles for glutathione in coupling H_2O_2 to activation of salicylic acid accumulation and signaling. Antioxid Redox Signal 18:2106–2121
Hernández JA, Ferrer MA, Jiménez A, Ros-Barceló A, Sevilla F (2001) Antioxidant systems and $O_2^{\cdot-}/H_2O_2$ production in the apoplast of *Pisum sativum* L. leaves: its relation with NaCl-induced necrotic lesions in minor veins. Plant Physiol 127:817–831
Hernández JA, Rubio M, Olmos E, Ros-Barceló A, Martínez-Gómez P (2004) Oxidative stress induced by long-term Plum pox virus infection in peach (*Prunus persica* L. cv GF305). Physiol Plant 122:486–495
Hernández JA, Diaz-Vivancos P, Rubio M, Olmos E, Ros-Barceló A, Martínez-Gómez P (2006) Long-term PPV infection produces an oxidative stress in a susceptible apricot cultivar but not in a resistant cultivar. Physiol Plant 126:140–152

Hernández JA, Gullner G, Clemente-Moreno MJ, Künstler A, Juhász C, Diaz-Vivancos P, Király L (2016) Oxidative stress and antioxidative responses in plant-virus interactions. Physiol Mol Plant Pathol 94:134–148

Horemans N, Foyer CH, Asard H (2000) Transport and action of ascorbate at the plant plasma membrane. Trends Plant Sci 5(6):263–267

Hakmaoui A, Pérez-Bueno ML, García-Fontana B, Camejo D, Jiménez A, Sevilla F, Barón M (2012) Analysis of the antioxidant response of *Nicotiana benthamiana* to infection with two strains of Pepper mild mottle virus. J Exp Bot 63: 5487–5496

Jiménez A, Hernández JA, del Río LA, Sevilla F (1997) Evidence for the presence of the ascorbate–glutathione cycle in mitochondria and peroxisomes of pea leaves. Plant Physiol 114:275–284

Jones JDG, Dangl JL (2006) The plant immune system. Nature 444:323–329

Joo JH, Wang S, Chen JG, Jones AM, Fedoroff NV (2005) Different signaling and cell death roles of heterotrimeric G protein alpha and beta subunits in the Arabidopsis oxidative stress response to ozone. Plant Cell 17(3):957–970

Király L, Künstler A, Höller K, Fattinger M, Juhász C, Müller M, Gullner G, Zechmann B (2012) Sulfate supply influences compartment specific glutathione metabolism and confers enhanced resistance to tobacco mosaic virus during a hypersensitive response. Plant Physiol Biochem 59:44–54

Kopriva S, Rennenberg H (2004) Control of sulphate assimilation and glutathione synthesis: interaction with N and C metabolism. J ExpBot 55:1831–1842

Krueger S, Niehl A, Lopez Martin MC, Steinhauser D, Donath A, Hildebrandt T, Romero LC, Hoefgen R, Gotor C, Hesse H (2009) Analysis of cytosolic and plastidic serine acetyltransferase mutants and subcellular metabolite distributions suggests interplay of the cellular compartments for cysteine biosynthesis in Arabidopsis. Plant Cell Environ 32:349–367

Kuźniak E (2010) The ascorbate–gluathione cycle and related redox signals in plant–pathogen interactions. In: Anjum NA et al (eds) Ascorbate-glutathione pathway and stress tolerance in plants. Springer, Dordrecht, pp 115–136

Kuźniak E, Skłodowska M (1999) The effect of *Botrytis cinerea* infection on ascorbate-glutathione cycle in tomato leaves. Plant Sci 148:69–76

Kuźniak E, Sklodowska A (2001) Ascorbate, glutathione and related enzymes in chloroplasts of tomato leaves infected by *Botrytis cinerea*. Plant Sci 160:723–731

Kużniak E, Sklodowska M (2004) Differential implication of glutathione, glutathione-metabolizing enzymes and ascorbate in tomato resistance to *Pseudomonas syringae*. J Phytopathol 152:529–536

Kuźniak E, Skłodowska M (2005) Compartment-specific role of the ascorbate-glutathione cycle in the response of tomato leaf cells to *Botrytis cinerea* infection. J Exp Bot 56(413):921–933

Lamb C, Dixon RA (1997) The oxidative burst in plant disease resistance. Annu Rev Plant Physiol Plant Mol Biol 48:251–275

Matern S, Peskan-Berghoefer T, Gromes R, Vazquez Kiesel R, Rausch T (2015) Imposed glutathione-mediated redox switch modulates the tobacco wound-induced protein kinase and salicylic acid-induced protein kinase activation state and impacts on defence against *Pseudomonas syringae*. J Exp Bot 66:1935–1950

May MJ, Parker JE, Daniels MJ, Leaver CJ, Cobbett CS (1996) An Arabidopsis mutant depleted in glutathione shows unaltered responses to fungal and bacterial pathogens. Mol Plant-Microbe Interact 9:349–356

Meyer AJ, Fricker MD (2000) Direct measurement of glutathione in epidermal cells of intact Arabidopsis roots by two-photon laser scanning microscopy. J Microsc 198:174–181

Meyer AJ, Brach T, Marty L, Kreye S, Rouhier N, Jacquot JP, Hell R (2007) Redox-sensitive GFP in *Arabidopsis thaliana* is a quantitative biosensor for the redox potential of the cellular glutathione redox buffer. Plant J 52:973–986

Mittler R (2002) Oxidative stress, antioxidants and stress tolerance. Trends Plant Sci 7:405–410

Mou Z, Fan W, Dong X (2003) Inducers of plant systemic acquired resistance regulates NPR1 function through redox changes. Cell 113:935–944

Mukaihara T, Hatanaka T, Nakano M, Oda K (2016) *Ralstonia solanacearum* type III effector RipAY is a glutathione-degrading enzyme that is activated by plant cytosolic thioredoxins and suppresses plant immunity. mBio 7(2):e00359-16

Müller M, Zellnig G, Urbanek A, Zechmann B (2005) Recent developments in methods intracellulary localizing glutathione within plant tissues and cells (a minireview). Phyton 45:45–55

Noctor G, Foyer CH (1998) Ascorbate and glutathione: keeping active oxygen under control. Annu Rev Plant Physiol Plant Mol Biol 49:249–279

Ohkama-Ohtsu N, Radwan S, Peterson A, Zhao P, Badr AF, Xiang C, Oliver DJ (2007) Characterization of the extracellular gamma-glutamyltranspeptidases, GGT1 and GGT2, in Arabidopsis. Plant J 49(5):865–877

Parisy V, Poinssot B, Owsianowski L, Buchala A, Glazebrook J, Mauch F (2007) Identification of PAD2 as a γ-glutamylcysteine synthetase highlights the importance of glutathione in disease resistance of Arabidopsis. Plant J 49:159–172

Porcel R, Barea JM, Ruiz-Lozano JM (2003) Antioxidant activities in mycorrhizal soybean plants under drought stress and their possible relationship to the process of nodule senescence. New Phytol 157:135–143

Schnaubelt D, Queval G, Dong Y, Diaz-Vivancos P, Makgopa ME, Howell G, De Simone A, Bai J, Hannah MA, Foyer CH (2015) Low glutathione regulates gene expression and the redox potentials of the nucleus and cytosol in *Arabidopsis thaliana*. Plant Cell Environ 38:266–279

Simon UK, Polanschütz LM, Koffler BE, Zechmann B (2013) High resolution imaging of temporal and spatial changes of subcellular ascorbate, glutathione and H_2O_2 distribution during *Botrytis cinerea* infection in Arabidopsis. PLoS ONE 8(6):e65811

Sklodowska M, Gajewka E, Kuźniak E, Wielanek M, Mikiciński A, Sobixzewski P (2011) Antioxidant profile and polyphenol oxidase activities in apple leaves after *Erwinia amylovora* infection and pretreatment with a benzothiadiazole-type resistance inducer (BTH). J Phytopathol 159:495–504

Song XS, Wang YJ, Mao WH, Shi K, Zhou YH (2009) Effect of cucumber mosaic virus infection on electron transport and antioxidant system in chloroplasts and mitochondria of cucumber and tomato leaves. Physiol Plant 135:246–257

Suzuki N, Miller G, Morales J, Shulaev V, Torres MA, Mittler R (2011) Respiratory burst oxidases: the engines of ROS signaling. Curr Opin Plant Biol 14:691–699

Terrer C, Vicca S, Hungate BA, Phillips RP, Prentice IC (2016) Mycorrhizal association as a primary control of the CO_2 fertilization effect. Science 353(6294):72–74

Tiedemann AV (1997) Evidence for a primary role of active oxygen species in induction of host cell death during infection of bean leaves with *Botrytis cinerea*. Physiol Mol Plant Pathol 50:151–166

Tolin S, Arrigoni G, Trentin AR, Veljovic-Jovanovic S, Pivato M, Zechmann B, Masi A (2013) Biochemical and quantitative proteomics investigations in *Arabidopsis* ggt1 mutant leaves reveal a role for the gamma-glutamylcyclein plant's adaptation to environment. Proteomics 13:2031–2045

Torres MA, Dangl JL, Jones JDG (2002) Arabidopsis gp91phox homologues AtrbohD and AtrbohF are required for accumulation of reactive oxygen intermediates in the plant defense response. Proc Natl Acad Sci U S A 99:517–522

Vahisalu T, Puzõrjova I, Brosché M, Valk E, Lepiku M, Moldau H, Pechter P, Wang YS, Lindgren O, Salojärvi J, Loog M, Kangasjärvi J, Kollist H (2010) Ozone-triggered rapid stomatal response involves the production of reactive oxygen species, and is controlled by SLAC1 and OST1. Plant J 62(3):442–453

Vanacker H, Carver TL, Foyer CH (1998) Pathogen-induced changes in the antioxidant status of the apoplast in barley leaves. Plant Physiol 117(3):1103–1114

Vanacker H, Carver TL, Foyer CH (2000) Early H_2O_2 accumulation in mesophyll cells leads to induction of glutathione during the hyper-sensitive response in the barley-powdery mildew interaction. Plant Physiol 123(4):1289–1300

Vernoux T, Wilson RC, Seeley KA, Reichheld JP, Muroy S, Brown S, Maughen SC, Cobbett CS, Van Montagu M, Inzé D (2000) The *ROOT MERISTEMLESS1/CADMIUM SENSITIVE 2* gene defines a glutathione dependent pathway involved in the initiation and maintenance of cell division during postembryonic root development. Plant Cell 12:97–110

Vianello A, Zancani M, Peresson C, Petrussa E, Casolo V, Krajňáková J, Patui S, Braidot E, Macrì F (2007) Plant mitochondrial pathway leading to programmed cell death. Physiol Plant 129:242–252

Wingate VPM, Lawton MA, Lamb CJ (1988) Glutathione causes a massive and selective induction of plant defense genes. Plant Physiol 87:206–210

Wojtaszek P (1997) Oxidative burst: an early plant response to pathogen infection. Biochem J 322(Pt 3):681–692

Zechmann B, Mauch F, Sticher L, Müller M (2008) Subcellular immunocytochemical analysis detects the highest concentrations of glutathione in mitochondria and not in plastids. J Exp Bot 59(14):4017–4027

Zechmann B, Müller M (2010) Subcellular compartmentation of glutathione in dicotyledonous plants. Protoplasma 246(1-4):15–24

Zechmann B, Zellnig G, Müller M (2005) Changes in the subcellular distribution of glutathione during virus infection in *Cucurbita pepo* (L.) Plant Biol 7:49–57

Zechmann B, Zellnig G, Müller M (2007a) Virus-induced changes in the subcellular distribution of glutathione precursors in *Cucurbita pepo* (L.) Plant Biol 9:427–434

Zechmann B, Zellnig G, Urbanek-Krajnc A, Müller M (2007b) Artificial elevation of glutathione affects symptom development in ZYMV-infected *Cucurbita pepo* L. plants. ArchVirol 157:747–762

Chapter 15
The Signaling Roles of Glutathione in Plant Disease Resistance

Gábor Gullner, Bernd Zechmann, András Künstler, and Lóránt Király

Abstract Early studies showed that glutathione (GSH) as an antioxidant has a role in modulating plant tolerance to biotic stresses by suppressing localized necrotic symptoms following virus infections. The role of GSH in reducing severity of pathogen-induced symptoms in plants was confirmed by employing pharmacological and transgenic approaches. However, later studies have shown that GSH also has a key role in restricting pathogen levels. In fact, it seems that GSH is a pivotal factor responsible for signaling processes related to different types of plant disease resistance, including systemic acquired resistance. The signaling role of GSH in these processes is interconnected with reactive oxygen species and salicylic acid. GSH also regulates the function of plant defense-associated transcription factors and the transcriptional coregulator NPR1 by modulating their redox state. Another layer of regulation is provided by the nitric oxide donor S-nitrosoglutathione that promotes S-nitrosylation of defense-related transcription factors and transcriptional coregulators. Importantly, the role of GSH in mediating plant disease resistance-related signaling processes is independent of its antioxidant function. Changes in GSH levels and redox state triggered during plant biotic stress are not simply passive responses to oxidative damage, since GSH status regulates important elements of cellular signaling that leads to activation of defense responses.

Keywords Plant-pathogen interactions • Plant disease resistance • Antioxidants • Reactive oxygen species • Signaling • Glutathione • Salicylic acid • Ethylene • NPR1 • NO

G. Gullner • A. Künstler • L. Király (✉)
Plant Protection Institute, Centre for Agricultural Research, Hungarian Academy of Sciences, H-1525 Budapest, P.O. Box 102, Hungary
e-mail: kiraly.lorant@agrar.mta.hu

B. Zechmann
Center for Microscopy and Imaging, Baylor University, Waco, TX 76798, USA

M.A. Hossain et al. (eds.), *Glutathione in Plant Growth, Development, and Stress Tolerance*, https://doi.org/10.1007/978-3-319-66682-2_15

Contents

1 Introduction

A basic difference between resistant and susceptible plants is the timely recognition of the invading pathogen and the rapid and efficient activation of host defense reactions. In resistant plants, the pathogen is rapidly perceived most often by host resistance (R) proteins. Upon recognition, R-proteins activate several signal transduction pathways via kinase catalyzed phosphorylation cascades, ion fluxes, reactive oxygen species (ROS), and other signaling events (Yang et al. 1997; Hernández et al. 2016). Signals are transmitted to the nucleus where they bring about a rapid and extensive reprogramming of gene expression patterns (Li et al. 2004). This transcriptional reprogramming is largely regulated by various transcription factors (DNA-binding proteins) (Singh et al. 2002). Defense hormones are produced as secondary signals, which in turn also upregulate specific sets of genes (Robert-Seilaniantz et al. 2011). In contrast to resistant plants, only late and weak defense reactions occur in susceptible hosts, allowing the growth and spread of pathogens (Yang et al. 1997).

The ubiquitous and abundant low molecular weight, nonprotein cellular thiol compound glutathione (γ-L-glutamyl-L-cysteinyl-glycine) participates in various antioxidative and detoxification reactions in plant cells due to its highly reactive cysteinyl residue (Foyer and Rennenberg 2000; Noctor et al. 2012). Reduced glutathione (GSH) and its oxidized disulfide form GSSG compose a redox buffer that interacts with numerous cellular components and processes (Schafer and Buettner 2001). For example, the cellular redox state, primarily determined by the GSH/GSSG ratio and total GSH concentration, influences gene expression associated with abiotic and biotic stress responses (Ball et al. 2004; Meyer and Hell 2005; Mullineaux and Rausch 2005). In fact, redox homeostasis acts as a metabolic interface for signals derived from metabolism or the environment modulating thereby the induction of plant stress responses (Foyer and Noctor 2005). The significant role of GSH in the regulation of defense reactions in infected plants has been recognized for a long time (Wingate et al. 1988; Foyer and Rennenberg 2000; Gullner and Kőmíves 2001; Ghanta and Chattopadhyay 2011). In recent years, a large amount of novel information has been gathered about the various functions of GSH during plant responses to pathogen infections.

This chapter is an attempt to recapitulate our current knowledge about the signaling roles of GSH in plant defense reactions against microbial pathogens by analyzing the links between glutathione signaling and plant adaptation (resistance) to pathogen infections. Recent research indicates that the role of GSH in mediating plant disease resistance-related signaling processes is independent of its antioxidant function (Ghanta et al. 2011a; Han et al. 2013a; Kovacs et al. 2015). Therefore, changes in GSH levels and redox state triggered during plant biotic stress (pathogen infections) are not simply passive responses to oxidative damage, since GSH status regulates important elements of cellular signaling leading to activation of defense responses.

2 Glutathione in Plant-Pathogen Interactions: Its Role as an Antioxidant

2.1 Glutathione Controlling Localized Cell and Tissue Death During the Hypersensitive Type of Plant Disease Resistance

During plant disease resistance localized, programmed cell and tissue death (necrosis) is often associated with pathogen arrest at infection sites, a process called the hypersensitive response (HR) (Klement 1982; Goodman and Novacky 1994; Greenberg and Yao 2004). It was already known by 1960 that treatment of leaves with antioxidant compounds like ascorbic acid (AsA), glutathione, etc. markedly decreases the number of virus-elicited HR-type necrotic lesions, while virus titers often do not change significantly (Farkas et al. 1960). A paraquat-tolerant tobacco (*Nicotiana tabacum* cv. Samsun) biotype exhibiting enhanced tolerance to various abiotic and biotic stresses not only displayed elevated GSH levels following exposure to, e.g., herbicides, but enhanced glutathione S-transferase (GST) activity was correlated with reduced development of HR-type necrotic lesions caused by *Tobacco necrosis virus* (TNV) (Barna et al. 1993; Gullner et al. 1991, 1995). In plants, GST isoenzymes usually participate in detoxification reactions by catalyzing conjugation between electrophilic, toxic compounds and GSH (Wagner et al. 2002; Dixon and Edwards 2009; Dixon et al. 2011). However, certain plant GSTs of the theta and tau classes exhibit a marked glutathione peroxidase (GPX)-like (i.e. antioxidant) activity, since they catalyze the breakdown of lipid hydroperoxides derived from lipid peroxidation processes that occur e.g. in dying plant cells (Bartling et al. 1993; Dixon and Edwards 2009; Dixon et al. 2009). Accordingly, GSTs, along with GSH, may have an instrumental role in regulating HR-type necrotization during plant resistance to viral infections, as first suggested by the work of Fodor et al. (1997). These authors demonstrated that the appearance of visible HR after *Tobacco mosaic virus* (TMV) inoculation was preceded by a transient decline in total activities of antioxidant enzymes like ascorbate peroxidase (APX), glutathione reductase (GR), and GST. However, after the onset of localized necrosis, both the activities of these antioxidant enzymes and GSH levels were substantially increased (Fodor et al. 1997). Elevated expression of tobacco GST genes of the tau and theta classes (*NtGSTU1* and *NtGSTT2*) has been also associated with TMV-induced HR (Künstler et al. 2007; Király et al. 2012; Juhász and Gullner 2014). In fact, a further induction of *NtGSTU1* expression and accumulation of GSH and its precursor, cysteine correlated with enhanced HR-type resistance (i.e., significantly less necrotic lesions and reduced TMV replication) in plants with a sufficient sulfate supply (Király et al. 2012). Similarly, sufficient sulfate supply induced TMV resistance (delayed mosaic symptoms and diminished virus accumulation) in genetically susceptible tobacco associated with elevated cysteine and GSH and upregulation of two key genes involved in cysteine and GSH biosynthesis (*NtAPR* and *NtGSH1*) (Höller et al. 2010). Importantly, these results imply that GSH, cysteine, and certain GST isoenzymes are not only antioxidants but may also have a role in the signaling of plant resistance to virus infections.

Similar investigations in a plant-bacterium pathosystem were the first to demonstrate the specific association of GSH-related (antioxidant) enzymes with cell death and resistance during bacteria-induced HR. Inoculation of soybean cell cultures with avirulent *Pseudomonas syringae* pv. *glycinea* caused cell death and resistance in bacteria-infected plant cells. On the other hand, a hydrogen peroxide (H_2O_2)-dependent accumulation of transcripts encoding for e.g. GST and GPX occurred in adjacent healthy cells, implying the operation of a plant defense-related signal transduction process (Levine et al. 1994; Tenhaken et al. 1995). Importantly, elevated GST gene expression was likely specific to disease resistance, similar to the differential induction of distinct GST genes during powdery mildew resistance and abiotic stress (xenobiotics) first observed in wheat (Mauch and Dudler 1993). Furthermore, differential mRNA accumulation of two sunflower GPX genes was observed in plants infected with a virulent (causing disease) or an avirulent (causing resistance) race of the oomycete pathogen *Plasmopara halstedii* (Herbette et al. 2003). The two GPX genes were induced also in response to stimulation with signaling molecules including stress-related phytohormones, ROS, nitric oxide (NO), and protein phosphatase or kinase inhibitors. The above mentioned data suggest that plant GST and GPX enzymes, along with GSH, are involved not only in abiotic stress responses but also in the regulation/signaling of cell death and/or pathogen resistance during an HR.

2.2 Glutathione-Mediated Antioxidant Effects: Conferring Susceptibility to (Hemi)biotrophic and Resistance to Necrotrophic Pathogens

A large portion of plant pathogenic microbes are biotrophic, i.e. they need live host tissues for a successful infection to occur. In fact, viruses can be considered as obligate biotrophs, since their pathogenesis fully depends on live host cells with an active metabolism. In contrast, necrotrophic pathogens must kill invaded plant host tissues in order to obtain nutrients. However, a group of plant pathogenic fungi are hemibiotrophic, their pathogenesis comprising an early biotrophic phase followed by a late necrotrophic phase (Perfect and Green 2001; Barna et al. 2012; Spanu and Panstruga 2017).

It has been shown more than 20 years ago that in barley (*Hordeum vulgare*) infected with its biotrophic powdery mildew pathogen (*Blumeria graminis* f. sp. *hordei*), susceptibility is associated with the activation of several antioxidative processes, including increased levels of nonprotein thiols (GSH and γ-glutamylcysteinylserine) and activities of GST (El-Zahaby et al. 1995). These results were later confirmed when leaves of powdery mildew-susceptible barley (cv. Ingrid) and near isogenic, resistant lines (*Mla12*, *Mlg*, and *mlo5*) were inoculated with race A6 of barley powdery mildew (Harrach et al. 2008). Activities of antioxidant enzymes (e.g., APX and superoxide dismutase [SOD]), GST and GR were significantly induced in susceptible barley leaves at 5–7 days after inoculation (DAI), when pathogen-induced symptoms develop. Similar but less pronounced pathogen-induced changes were detected in inoculated leaves of resistant lines, while total glutathione (GSH + GSSG) contents increased significantly only in susceptible barley at 7 DAI. However, it is noteworthy to mention that glutathione was not assayed at early time points after inoculation, where it could potentially play a role in modulating/signaling resistance responses to powdery mildew (Harrach et al. 2008). GST activities were also found to be correlated with susceptibility during interactions of another biotrophic pathogen (*Uromyces phaseoli*) with its bean (*Phaseolus vulgaris*) host (Gullner and Kőmíves 2006). Elevated GST activities appeared in leaves of the very susceptible bean cultivar ′Békési fehér′ concomitant with the appearance of rust symptoms (i.e. intensive pathogen multiplication and spread), while no GST induction occurred at the same time point in the weakly susceptible cultivar ′Diana′. It seems that during interactions of plants with biotrophic fungi (powdery mildews, rusts), GSTs and GSH may contribute significantly to the development of susceptibility, at least in advanced stages of pathogenesis, when visible disease symptoms develop. Interestingly, certain GST genes/isoenzymes and GSH might also contribute to susceptibility to obligate biotrophic pathogens, i.e. viruses. It has been shown that in *Nicotiana benthamiana*, a significant decrease in *Bamboo mosaic virus* (BaMV) RNA accumulation occurs when the expression of the *NbGSTU4* gene is reduced by gene silencing but viral RNA accumulation increases when *NbGSTU4* is transiently overexpressed (Chen et al. 2013). NbGSTU4 could interact with the 3′ untranslated region (UTR) of BaMV RNA in vitro in the presence of GSH and addition of GSH to the in vitro replication complex caused an enhancement of minus strand viral RNA synthesis. These results suggest that certain plant GSTs may play a role in binding viral RNA and delivering GSH to the replication complex to create reduced conditions for efficient viral RNA synthesis.

In case of hemibiotrophic pathogens, the role of glutathione and GSH-associated enzymes might be more complex, since the pathogenesis of these fungi has an early biotrophic phase followed by conversion to necrotrophy. A tobacco GST gene isolated from roots infected with the hemibiotroph *Phytophthora parasitica* var. *nicotianae* was shown to be involved in susceptibility to this pathogen (Hernández et al. 2009). A significant increase in resistance of *N. tabacum* to infection by *P. parasitica* var.

nicotianae was found in GST-silenced plants, as compared to unsilenced controls. These results imply that in this case, certain GSTs modulate plant defense responses in the initial biotrophic phase, likely by sustaining reduced conditions favoring pathogen spread and multiplication. Accordingly, the influence of *in planta* GSTs and/or GSH in the later necrotrophic infection phase could result in conferring disease resistance due to the same (antioxidant) effects. In fact, this has been demonstrated by Dean et al. (2005) who found the transcription of two GST genes (*NbGSTU1* and *NbGSTU3*) markedly induced in *N. benthamiana* following infection by the hemibiotrophic pathogens *Colletotrichum destructivum* and *C. orbiculare*. Interestingly, the resistance of *N. benthamiana* toward *C. orbiculare* infection was significantly impaired when the expression of *NbGSTU1* was inhibited by gene silencing: 130% more lesions and 67% more colonization were observed as compared to control plants in samples taken at the time of conversion from biotrophy to necrotrophy (4 days after inoculation) (Dean et al. 2005). This observation clearly showed that at least one GST isoenzyme has a pivotal function in disease resistance to a hemibiotrophic fungal pathogen. In addition, the role of the antioxidative AsA-GSH cycle in maintaining resistance to a hemibiotrophic bacterial pathogen, *P. syringae* pv. *tomato*, has been also demonstrated in tomato (*Solanum lycopersicum*) (Kuźniak and Skłodowska 2004). A decreasing GSH content, GPX activity, and GSH redox ratio and the accumulation of GSSG were observed in an inoculated susceptible cultivar. On the other hand, the GSH pool homeostasis was maintained throughout bacterial pathogenesis in a resistant tomato cultivar. Moreover, there was a significantly higher constitutive and pathogen-induced GST activity in this incompatible plant-bacterium interaction. These results showed that the maintenance of GSH pool homeostasis and GST induction contribute to resistance towards a hemibiotrophic bacterial pathogen (Kuźniak and Skłodowska 2004).

As mentioned previously, necrotrophic pathogens must kill invaded plant tissues for a successful pathogenesis implying a pathogen-driven suppression of host antioxidant capacity. Accordingly, infection by e.g. the gray mould pathogen *Botrytis cinerea* induces a marked oxidative stress in different plant species. Plants with elevated antioxidant capacity seem to have an increased resistance to this necrotrophic pathogen, since overexpression of a gene encoding a phospholipid hydroperoxide GPX enzyme in tobacco conferred protection against *B. cinerea* (Chen et al. 2004). Furthermore, activity changes of the AsA-GSH cycle following *B. cinerea* infection were studied in tomato chloroplasts, mitochondria, and peroxisomes. The infection caused a general shift of the cellular redox balance toward a more oxidative state, including a decline of the GSH pool, affecting all cellular compartments, in particular the mitochondria and peroxisomes. *B. cinerea* was able to destroy the protective antioxidant barrier of plants and induce senescence that favored a successful pathogenesis (Kuźniak and Skłodowska 2005b). Also, it has been shown that in barley susceptible to root rot caused by the necrotrophic fungal pathogen *Fusarium culmorum*, infected roots display elevated levels of lipid hydroperoxides, reduced levels and decreased ratios of reduced to oxidized forms of glutathione and ascorbate, along with reduced activities of antioxidant enzymes like SOD, APX, and GR (Harrach et al. 2013). In contrast, roots treated with the mutualistic root-colonizing *Piriformospora indica* prior to inoculation with *F. culmorum* showed disease resistance, along with GSH and ascorbate levels, antioxidant enzyme activities, and lipid peroxidation rates similar to untreated controls. These results suggest that resistance of plant roots to a necrotrophic pathogen like *F. culmorum* is caused, at least partly, by maintaining an activated plant antioxidant capacity, including GSH levels and GR activities, even 2 weeks after infection (Harrach et al. 2013). However, glutathione-related plant defense-signaling mechanisms resulting in disease resistance might be also activated at a very early stage of infection by necrotrophic pathogens. In wheat, resistance to seedling blight caused by *Fusarium asiaticum* is clearly associated with an early, dramatic, and transient induction (3–6 h after inoculation) of the GST gene *TaGSTF5* (Li et al. 2010).

From the above, it seems obvious that glutathione and glutathione-associated enzymes (e.g. GST, GPX, GR) are not only antioxidants that reduce or eliminate the mostly necrotic symptoms (cell and tissue death) associated with plant infections but often significantly contribute to disease resistance. In fact, evidence from earlier and recent research indicates that glutathione-driven *in planta* redox reactions may participate in signaling processes leading to successful pathogen defense (e.g., Mauch and Dudler 1993; Levine et al. 1994; Fodor et al. 1997; Herbette et al. 2003; Höller et al. 2010; Li et al. 2010; Király et al. 2012; Harrach et al. 2013).

3 Glutathione as a Signaling Agent Contributing to Plant Disease Resistance

3.1 Modulation of GSH Contents in Infected Plant Tissues

The modulation of GSH in plants infected with diverse pathogens has been reported in a number of papers (May et al. 1996a; Fodor et al. 1997; Foyer and Rennenberg 2000; Gullner and Kőmíves 2001; Zechmann et al. 2005; Höller et al. 2010; Király et al. 2012). A significant accumulation of GSH has been often observed in incompatible plant-pathogen interactions (in resistant plants), but this accumulation was usually preceded by a transient decline of GSH levels (Vanacker et al. 1998, 2000; Mou et al. 2003). Several lines of evidence showed that the artificial increase of cellular GSH contents can markedly improve the disease resistance of plants. The synthetic cysteine precursor L-2-oxothiazolidine-4-carboxylic acid (OTC) can significantly elevate GSH levels in plant tissues (Hausladen and Kunert 1990). Interestingly, OTC pretreatment considerably decreased both the number of necrotic lesions and virus contents in TMV-infected tobacco leaf discs concomitantly with elevated GSH levels (Gullner et al. 1999). Similar antiviral effects of OTC treatments were explored also in other plant-pathogen interactions. Styrian oil pumpkin seedlings were pretreated with OTC and inoculated with *Zucchini yellow mosaic virus* (ZYMV). OTC treatment and subsequently elevated GSH contents led to a strong decrease in ZYMV contents and a suppression of disease symptoms

(Zechmann et al. 2007). In contrast, pretreatment of pea (*Pisum sativum*) plants with OTC afforded partial protection against *Plum pox virus* (PPV) infection, but it did not significantly reduce virus contents (Clemente-Moreno et al. 2010). In *Prunus* (peach and plum) plantlets, OTC treatments resulted in a significant increase in plant growth. However, OTC did not reduce virus contents in PPV-inoculated plantlets, although GSH contents were elevated (Clemente-Moreno et al. 2012).

To further explore the effects of elevated GSH contents in plant defense, transgenic *N. tabacum* were created (*NtGB* lines) that overexpressed the tomato γ-glutamylcysteine synthetase (γ-glutamylcysteine ligase, *GSH1*) gene (Ghanta et al. 2011b). This gene encodes the GSH1 protein, which catalyzes the first, rate-determining step of de novo GSH biosynthesis (Noctor et al. 2012). The transgenic *NtGB* lines, in which GSH1 accumulated in the cytosol, displayed elevated GSH levels and strongly increased transcript abundance of genes encoding the transcriptional coregulator NPR1, the pathogenesis-related 1 (PR-1) protein, a mitogen-activated protein kinase kinase (MAPKK), a thioredoxin (TRX-h), and S-nitrosoglutathione reductase 1 (GSNOR1). On the other hand, the expression levels of NPR1-independent genes like *PR-2* and *PR-5* and that of isochorismate synthase 1 (*ICS1*) did not change significantly or were downregulated (Ghanta et al. 2011a, b). In further experiments, transgenic *N. tabacum* plants overexpressing the *GSH1* gene encoding a chloroplast-targeted GSH1 were also constructed (*NtGp* line). SA-related genes like *PR-1a* and *SAR8.2* and also genes encoding transcription factors like *WRKY1*, *WRKY3*, and ethylene-responsive factor 4 (*ERF4*) were markedly activated in the *NtGp* line. The proteomic profiling of *NtGp* line revealed an increased accumulation of defense-related proteins like serine/threonine protein kinase, heat shock 70 protein (HSP70), and proteins participating in ethylene biosynthesis (Ghanta et al. 2014).

The identification of various GSH-deficient mutants of *Arabidopsis thaliana* also demonstrated that adequate levels of GSH are important for the establishment of disease resistance. GSH-deficient *Arabidopsis* mutants usually demonstrated enhanced susceptibility or hypersusceptibility to diverse bacterial, fungal, or oomycete pathogens and a diminished accumulation of resistance-related compounds such as camalexin and indole glucosinolates (Ball et al. 2004; Parisy et al. 2007; Dubreuil-Maurizi and Poinssot 2012). The decreased GSH levels in these lines are due to the mutation in the gene encoding the GSH1 enzyme. GSH deficiency of these mutants affected oxidative stress-related events, early signaling events, defense gene expression, and the HR (Dubreuil-Maurizi and Poinssot 2012). Interestingly, GSH deficiency markedly altered the levels of the key signaling compounds H_2O_2 and SA in the *pad2-1 Arabidopsis* mutant (Dubreuil-Maurizi et al. 2011). In these mutant plants, an impaired H_2O_2 production was observed during pathogen infection, which correlated with a reduced HR. In addition, a lack of pathogen-triggered expression of the *ICS1* gene, which encodes the SA-biosynthetic enzyme isochorismate synthase, was identified as the cause of SA deficiency (Dubreuil-Maurizi et al. 2011).The proteomic analysis of the *pad2-1* mutant infected with *P. syringae* pv. *tomato* revealed lower accumulations of a leucine-rich repeat receptor kinase (LRR-RK) and a nucleotide-binding site-leucine-rich repeat resistance protein (NBS-LRR) as compared to wild type. Transcriptional and posttranscriptional regulators like MYB-P1 and a glycine-rich repeat RNA-binding protein (GRP) as well as several other stress-related proteins were also differentially regulated in *pad2-1*, as compared to wild-type plants (*Col-0*), in response to infection (Datta and Chattopadhyay 2015). It is notable also, however, that one GSH-deficient mutant displayed unchanged resistance to *Hyaloperonospora parasitica* and *P. syringae* pv. *tomato* (May et al. 1996b).

3.2 Subcellular Changes of GSH Metabolism in Infected Plants

In recent years, novel techniques have been developed to detect GSH levels in various subcellular compartments of infected plant cells that offered new possibilities to gain a deeper insight into the signaling roles of GSH (Zechmann 2014). GSH was found in nearly every compartment of the cell, including the nucleus. Subcellular changes in GSH contents are important for the fine-tuning of plant defense mechanisms (Noctor and Foyer 2016). During *P. syringae* pv. *tomato* infection in *A. thaliana*, peroxisomes were identified as hotspots of GSH accumulation at the beginning of the infection, whereas the collapse of the antioxidative system in peroxisomes in the long term correlated with ROS accumulation and the progress of disease symptoms (Großkinsky et al. 2012). Similar results have been also obtained during *B. cinerea* infection in tomato where the breakdown of the antioxidative system in peroxisomes could be correlated with pathogen-induced leaf senescence (Kuźniak and Skłodowska 2005a, b). Thus, it seems that high GSH levels in peroxisomes are essential for proper plant defense against pathogens. It seems that a drop of GSH contents in peroxisomes leads to the accumulation of ROS in plant tissues and a progression of symptom development.

A similar situation was observed in mitochondria where the glutathione status seems to play an important role in the development of localized necrotic lesions (HR). During an incompatible TMV infection in *N. tabacum* plants, the development of necrotic lesions was accompanied by a decrease of glutathione contents in mitochondria only, which was not found in all other cell compartments (Király et al. 2012). A similar situation could be observed in *B. cinerea*-infected *Arabidopsis* plants where the development of necrosis at the infection site could be correlated with a strong decrease of GSH in mitochondria only, whereas glutathione contents in other cell compartments remained at control levels (Simon et al. 2013). In *B. cinerea*-infected tomato plants, glutathione contents in mitochondria were most severely affected besides peroxisomes. Accumulation of GSSG accompanied by a strong drop in total glutathione contents was observed early in this cell compartment together with pathogen-induced senescence (Kuźniak and Skłodowska 2005a, b). Elevated levels of ROS in mitochondria are involved in the induction of programmed cell death during biotic stress (reviewed by Amirsadeghi et al. 2007; Vianello et al.

2007). Thus, decreased amounts of glutathione in mitochondria following pathogen infections could favor the accumulation of ROS in this cell compartment contributing to the induction of programmed cell death events. Nevertheless, it still remains unclear if glutathione degradation and the accumulation of GSSG in mitochondria observed during pathogen infections are actively governed by the plant in order to allow the rapid accumulation of ROS, which are necessary for the induction of cell death and/or resistance or indirectly caused by the accumulation of ROS in mitochondria due to disturbances in the electron transport chain during biotic stress.

Apoplastic glutathione contents have been found to play important roles during fungal infections (Vanacker et al. 1998, 2000), whereas such effects could not be observed during viral and bacterial infections (Höller et al. 2010; Király et al. 2012; Großkinsky et al. 2012). In the apoplast of oat and barley plants, a strong increase of glutathione was detected which was associated with race and non-race specific resistance to powdery mildew (*B. graminis*) (Vanacker et al. 1998, 2000). In opposite, decreased amounts of apoplastic glutathione were observed after infection of susceptible plants. Interestingly, glutathione contents in the apoplast were not detected or were very low during *B. graminis* infections and could not control H_2O_2 accumulation in the HR type of resistance (Vanacker et al. 1998, 2000). Thus, it seems that in the apoplast, glutathione and/or its redox state (which becomes more oxidized during the HR) may serve important roles as a defense-signaling agent rather than as an antioxidant antagonizing ROS accumulation during pathogen infections.

During pathogen infections, glutathione contents have been found to accumulate in the nuclei. In the nuclei of ZYMV-infected *Cucurbita pepo* plants and TMV-infected *N. tabacum* plants, a strong accumulation of glutathione was detected in younger leaves (Zechmann et al. 2005, 2007; Király et al. 2012). A similar effect has been observed in *Arabidopsis* plants infected with *P. syringae* pv. *tomato* (Großkinsky et al. 2012) and *B. cinerea* at early stages of infection (Simon et al. 2013). The accumulation of glutathione in nuclei during pathogen attack points towards an important role of glutathione in the protection of plants against biotic stress. A pronounced accumulation of GSH was observed in the nucleus at points in the growth cycle at which a high percentage of the cells were in the G_1 phase. Recruitment of GSH into the nucleus led to a high abundance of GSH in the nucleus and severe depletion of the cytoplasmic GSH pool (Diaz-Vivancos et al. 2010a). The abundance of transcripts encoding stress and defense proteins was decreased when GSH was sequestered in the nucleus (Diaz-Vivancos et al. 2010b). The accumulation of GSH could also serve to protect DNA and redox-sensitive nuclear proteins from oxidation (Green et al. 2006; Diaz-Vivancos et al. 2010a, b). Thus, it is very well likely that the accumulation of glutathione in nuclei during biotic stress conditions is used to activate GSH synthesis. Such effects seem plausible as the accumulation of glutathione in nuclei was accompanied by a strong accumulation of glutathione in chloroplasts and the cytosol, considered to be the centers for glutathione synthesis (Wachter et al. 2005), in younger leaves of *C. pepo* during ZYMV infection (Zechmann et al. 2005), in sulfur-depleted *N. tabacum* plants during TMV infection (Király et al. 2012), and in *Arabidopsis* plants infected with *P. syringae* pv.

tomato (Großkinsky et al. 2012) and *B. cinerea* (Simon et al. 2013), respectively. Thus, we can conclude that the fine-tuning of subcellular distribution of glutathione between different cell compartments fulfills important roles in plant defense by keeping ROS under control in organelles where the accumulation of ROS can trigger cell death (e.g., mitochondria), by signaling plant defense (e.g., apoplast), and by protecting critical proteins from oxidation (e.g., nuclei).

3.3 Effects of GSH on the Transcription of Defense Genes

A pioneering study of Wingate et al. (1988) demonstrated that exogenous GSH in physiological concentrations markedly changed the pattern of gene expression and protein synthesis in suspension-cultured cells of bean. GSH strongly and transiently activated the transcription of genes encoding chalcone synthase (CHS) and phenylalanine ammonia lyase (PAL), which participate in the biosynthesis of phenylpropanoids and flavonoids including isoflavonoid phytoalexins, respectively. In addition, exogenous GSH also upregulated the expression of two genes encoding cell wall hydroxyproline-rich glycoproteins. All these effects of GSH closely resembled that of a fungal elicitor (i.e. elicitor of plant defense) prepared from *Colletotrichum lindemuthianum*. Thus, GSH was supposed to be a secondary signal mediating the effect of external stimuli such as fungal elicitors (Wingate et al. 1988). Furthermore, GSH-responsive cis-acting elements were identified in the promoter of a bean CHS gene in electroporated soybean protoplasts (Dron et al. 1988). The expression of a chimeric construct containing the promoter of the bean CHS15 gene linked to a reporter gene was induced by GSH also in electroporated alfalfa protoplasts (Choudhary et al. 1990).

GSH was shown to activate the transcription of a *CHS* gene also by triggering protein phosphorylation. A soybean bZIP-type transcription factor (G/HBF-1) was rapidly phosphorylated in elicited soybean cells, almost exclusively on serine residues. A cytosolic protein-serine kinase was identified that was rapidly and transiently stimulated in cells elicited with either GSH or an avirulent bacterium. Phosphorylation of G/HBF-1 in vitro enhanced its binding to the promoter of a *CHS* gene. Stimulation of G/HBF-1 kinase activity and G/HBF-1 phosphorylation were supposed to be the terminal events in a signaling pathway for activation of early transcription-dependent plant defense responses (Dröge-Laser et al. 1997). Similarly, elicitation of bean cell suspensions with GSH increased the specific nuclear activities of KAP-1 and KAP-2 protein factors that recognize an H-box nucleotide motif (Yu et al. 1993). A recent study explored the impact of elevated GSH contents on the levels of mitogen-activated protein kinases (MAPKs) and on subsequent defense reactions. *N. tabacum* and its transgenic lines producing high GSH levels were transformed with a cytosol-targeted fluorescent redox sensor protein to monitor the cytosolic redox state. Surprisingly, high GSH mutants displayed an oxidative shift and an activation of a wound-induced protein kinase (WIPK) and a SA-induced protein kinase (SIPK). This activation was accompanied by constitutively increased expression of several defense genes. Bacterial infection with *P. syringae* pathovars further amplified the cytosolic redox shift and the defense response (Matern et al. 2015).

The *lesions simulating disease* (*lsd*) mutants of *Arabidopsis* spontaneously develop hypersensitive response-like lesions in the absence of pathogens. The application of L-buthionine-[S,R]-sulfoximine (BSO), a specific inhibitor of GSH biosynthesis, suppressed conditionally induced runaway cell death and expression of the resistance marker *PR-1* gene, suggesting that GSH regulates these processes. The application of GSH or GSSG to *lsd1* mutants led to the accumulation of PR-1 protein. The conditional PR-1 accumulation in the *lsd1* mutant was regulated not by the redox state but by the endogenous levels of GSH (Senda and Ogawa 2004).

The transcripts of a chloroplastic *GR* gene from *Haynaldia villosa* accumulated in response to infection with *B. graminis* f.sp. *tritici* (wheat powdery mildew) in transgenic wheat (*Triticum aestivum*) leaves. Overexpression of this *GR* gene in a susceptible wheat cultivar enhanced resistance to powdery mildew and induced transcript accumulation of the pathogenesis-related (*PR*) genes *PR-1a* and *PR-5* through an increased foliar GSH/GSSG ratio. These results showed that a high ratio of GSH/GSSG was required for wheat defense against powdery mildews and chloroplastic GR enzymes might serve as a redox mediator for NPR1 activation (Chen et al. 2007).

Following the early studies of Wingate et al. (1988), the effects of exogenous GSH were also investigated by high-throughput "omics" methods. Changes in gene expression patterns of *A. thaliana* seedlings following GSH treatment were studied by cDNA microarray analysis (Hacham et al. 2014). Interestingly, many genes involved in biotic stress signaling and in jasmonic acid (JA) biosynthesis were upregulated by exogenous GSH. Thus, genes encoding the PR proteins PR-1, PR-5, and PR-2, several TIR-NBS-type resistance (R) proteins, receptor-like proteins, receptor-like kinases, WRKY transcription factors, as well as numerous protein kinases were strongly activated. On the other hand, GSH downregulated various genes involved in plant growth and development, like those involved in cell wall biosynthesis and extension and genes associated with auxin and cytokinin responses (Hacham et al. 2014). In an integrated transcriptomic and proteomic analysis, the application of exogenous GSH led to the activation of several biotic stress- and ethylene (ET)-related genes as well as to the accumulation of stress and defense-related proteins. Among others, genes encoding essential proteins involved in SA signaling or ET biosynthesis as well as MYB and WRKY transcription factors were markedly activated (Sinha et al. 2015). Furthermore, Cheng et al. (2015) demonstrated by polysomal profiling that global translation was also enhanced after GSH treatment in *A. thaliana*. The translatome analysis revealed that the biosynthesis of abscisic acid (ABA), auxin and JA as well as several signaling molecules is markedly activated during GSH treatments (Cheng et al. 2015).

3.4 Interactions Between GSH and Plant Defense Hormones

3.4.1 Connections Between ROS, Salicylic Acid, and Glutathione

It is known that a tight correlation exists among salicylic acid (SA), H_2O_2, and GSH contents in plants. SA has been extensively studied as a key signal in defense reactions (Enyedi et al. 1992; Vlot et al. 2009). The cross dependence among these metabolites markedly influences the signaling processes leading to the activation of defense responses (Mateo et al. 2006; Herrera-Vásquez et al. 2015). The metabolic links among these signaling compounds are presented in Fig. 15.1. The well-documented relationship between ROS and SA has been recognized for a long time. Initially, H_2O_2 was proposed as a signal downstream of SA, and H_2O_2 was supposed to upregulate the *PR-1* gene associated with systemic acquired resistance (SAR) (Chen et al. 1993). However, later studies proved that H_2O_2 was unable to induce the *PR-1* gene in transgenic *NahG* tobacco, in which SA is degraded to catechol (Bi et al. 1995; Neuenschwander et al. 1995; Chamnongpol et al. 1998). In addition, exogenous H_2O_2 was found to induce the accumulation of benzoic acid and SA in tobacco leaves (Léon et al. 1995). The debate on whether SA is located upstream or downstream of H_2O_2 in defense-signaling pathways was settled with the proposition that both molecules act together in a self-amplifying system (Van Camp et al. 1998). Applications of H_2O_2 and SA have shown that these two compounds induce each other and thus they form a feed-forward loop (León et al. 1995; Kauss and Jeblick 1995; Bi et al. 1995; Rao et al. 1997). The self-amplification loop of H_2O_2 and SA probably amplifies the H_2O_2 signal required for different types of plant disease resistance responses, e.g., the HR and SAR, the latter conferring immunity to a broad spectrum of pathogens (Alvarez et al. 1998; Van Camp et al. 1998).

The signaling role of GSH is strongly linked to both H_2O_2 and SA. GSH is present in plant cells in millimolar concentrations, and it is therefore regarded as a key determinant of the cellular redox status (Foyer and Noctor 2005; Mullineaux and Rausch 2005). GSH can react directly with ROS to detoxify them (Winterbourn 2013) or can act as the substrate for ROS-scavenging GPXs (Navrot et al. 2006) and GSTs (Edwards and Dixon 2005; Rahantaniaina et al. 2013). Moreover, GSH is used to regenerate oxidized ascorbate and oxidized thiol groups of proteins (Noctor et al. 2002). The regeneration of GSH from GSSG is catalyzed by the GR enzyme in an NADPH-consuming reaction, which maintains the high GSH/GSSG ratio in plant cells (Noctor et al. 2002; Rahantaniaina et al. 2013).

Exogenously added SA led to increasing GSH and decreasing GSSG levels in *P. sativum* seedlings. These changes might protect plant cells against oxidative damage caused by increased H_2O_2 levels upon SA treatment (Srivastava and Dwivedi 1998). Similar changes of GSH and GSSG levels were observed in TMV-infected Xanthi-nc tobacco leaves (Fodor et al. 1997). On the other hand, in a tobacco line (*NahG*) unable to accumulate SA, the GSH/GSSG ratio dramatically decreased within the first 4 days after TMV inoculation (Király et al. 2002). Elevation of free

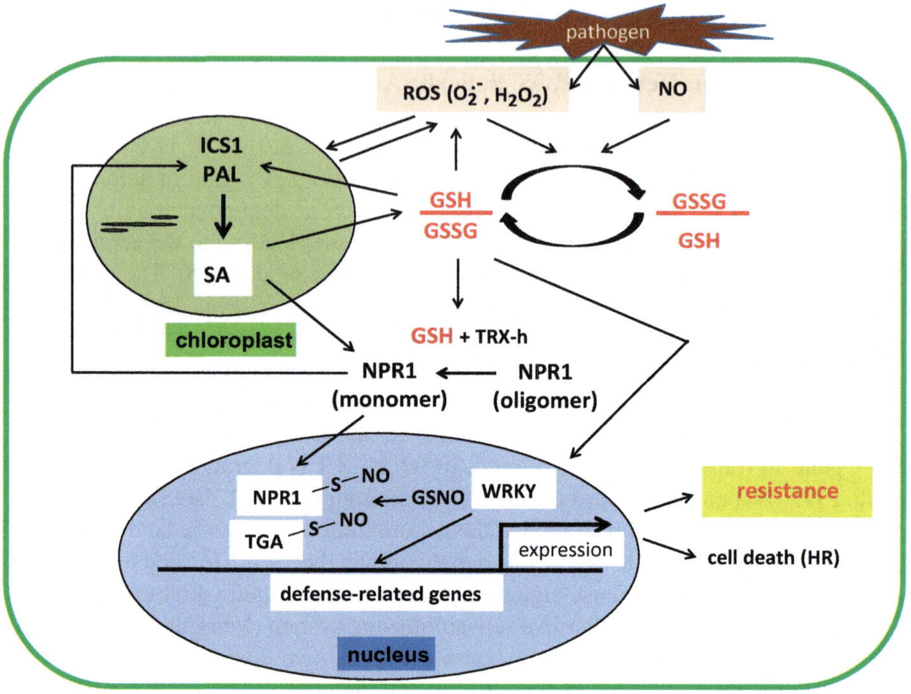

Fig. 15.1 Schematic representation of the main mechanisms by which glutathione (*GSH*) participates in signaling processes leading to the activation of defense-related genes in infected plants. The accumulation of reactive oxygen species (*ROS*) and nitrogen monoxide (*NO*) in infected plants leads to markedly elevated GSH levels preceded by the transient depletion of the GSH pool. Both ROS and GSH are interconnected with salicylic acid (*SA*) levels probably through the modulation of expression of the SA-biosynthetic isochorismate synthase 1 (*ICS1*) and phenylalanine ammonia lyase (*PAL*) genes in chloroplasts. In the cytosol, elevated GSH levels can induce the transformation of inactive NPR1 oligomers to active NPR1 monomers through thioredoxin-h (TRX-h) enzymes. The monomeric NPR1 is translocated to the nucleus where it interacts with TGA transcription factors. Both NPR1 and TGA proteins can be nitrosylated by S-nitrosoglutathione (*GSNO*). The interaction of activated NPR1 and TGA factors results in the enhanced transcription of defense-related genes. Alternatively elevated GSH levels can activate WRKY transcription factors causing same/similar patterns of defense gene induction. The ultimate outcome of these processes is the development of enhanced plant disease resistance often accompanied by localized (hypersensitive) cell death

SA levels in *A. thaliana*, both genetically and by exogenous feeding, enhanced the specific activity of serine acetyltransferase (a precursor of cysteine and GSH biosynthesis) leading to elevated GSH levels (Freeman et al. 2005). In a SA-deficient *NahG* rice mutant, the GSH pool was constitutively diminished as compared to control plants. *NahG* seedlings showed an increased susceptibility to oxidative stress. Conversely, treatment with an activator of the SA-mediated defense-signaling pathway, probenazole, increased the GSH pool size (Kusumi et al. 2006).

Furthermore, *A. thaliana* mutants with constitutive accumulation of SA had strongly increased H_2O_2 and GSH contents, but the mutants were still able to maintain the same glutathione redox state as wild-type plants (Mateo et al. 2006). Interestingly, deficiency of SA and lower H_2O_2 levels did not lead to significantly decreased GSH levels, but in *NahG* plants, which constitutively degrade SA, the glutathione pool was more oxidized. Surprisingly, exogenously supplied SA could induce GSH contents and vice versa. Injection of physiological levels of SA into *Arabidopsis* leaves caused an increase of both GSH and GSSG levels. In the inverse experiment, elevated free and glucosylated SA levels were detected in response to GSH application (Mateo et al. 2006).The role of GSH in elevating SA levels was also confirmed in a transgenic tobacco with enhanced levels of GSH (*NtGB* lines) (Ghanta et al. 2011b). These plants were found to synthesize more SA and being capable of enhanced expression of genes that belong to the NPR1-dependent SA-mediated defense pathway (see earlier). *NtGB* tobaccos exhibited an enhanced resistance to the hemibiotrophic *P. syringae* pv. *tabaci,* while resistance to necrotrophic fungi (*A. alternata* and *B. cinerea*) was only slightly increased (Ghanta et al. 2011a, b). These results suggest that the influence of constitutively high levels of GSH on plant disease resistance could be associated with SA-mediated defense signaling (limiting [hemi] biotrophic pathogens) rather than an antioxidant function suppressing necrotrophic infections (see, e.g., Glazebrook 2005; Barna et al. 2012). In fact, it has been shown through an elegant analysis of *Arabidopsis* mutants by Han et al. (2013a) that GSH acts independently of its antioxidant function allowing increased intracellular H_2O_2 to activate SA accumulation and SA-mediated defense signaling. Mutant plants overproducing H_2O_2 and deficient in GSH synthesis (*cat2 cad2*) did not show increased oxidation of AsA or NADPH or an enhanced accumulation of peroxides but displayed a compromised accumulation of SA and associated defense responses. In other words, the antioxidant properties of GSH could be uncoupled from its role in transmitting defense (disease resistance) signals.

3.4.2 Role of GSH in Jasmonic Acid and Ethylene Signaling

The well-recognized antagonistic effect of SA on JA/ET-mediated transcriptional responses (Pieterse et al. 2009; Caarls et al. 2015) is mediated by class II TGA transcription factors (Ndamukong et al. 2007; Zander et al. 2010). The suppression of JA signaling correlated exactly with the temporal window in which SA-triggered redox perturbations can be detected. Interestingly, the application of the GSH biosynthesis inhibitor BSO strongly mitigated the suppression of defense gene expression by SA, suggesting that GSH-mediated redox modulation plays an important role in the attenuation of the JA signaling pathway (Koornneef et al. 2008). In addition, ectopic expression of the pathogen-inducible glutaredoxin *GRX480* negatively regulated the expression of JA/ET-induced defense genes through an unknown mechanism that required clade II transcription factors TGA2, TGA5, and/or TGA6. GRX480-mediated repression of these defense responses depended on the

GSH-binding site, suggesting that redox modification of either TGA factors or as yet unknown target proteins is important for the suppressive effect (Zander et al. 2012).

In *Arabidopsis* cell cultures, JA treatment but not H_2O_2 markedly increased the mRNA levels of *GSH1* and glutathione synthetase (*GSH2*), which encodes the second enzyme of GSH biosynthesis. On the other hand, JA did not alter GSH contents in unstressed plants (Xiang and Oliver 1998; Noctor et al. 2012).

The involvement of intracellular GSH contents in methyl jasmonate (MeJA) signaling was demonstrated in *A. thaliana* guard cells. The *chlorina1-1* (*ch1-1*) mutation decreased GSH contents in guard cells and narrowed stomatal aperture. External GSH monoethyl ester increased intracellular GSH levels and diminished the narrow stomatal aperture phenotype. GSH did not affect MeJA-induced ROS production or cytosolic Ca^{2+} oscillations, suggesting that GSH modulates MeJA signaling downstream of ROS and Ca^{2+} effects (Akter et al. 2010).

While most attention on GSH signaling functions has been focused on the thiol-regulated protein NPR1, a comparison of JA-linked gene expression in the *A. thaliana cat2 cad2* and *cat2 npr1* double mutants provided substantial evidence that GSH acts also through determining JA-related gene expression. Interestingly, basal expression of JA-related genes correlated with leaf GSH content, and the upregulation of the JA pathway triggered by intracellular oxidation required an accompanying GSH accumulation. Genetically blocking this accumulation in a *cat2 cad2* line largely annulled the H_2O_2-induced expression of JA-linked genes, and this effect can be rescued by exogenous GSH (Han et al. 2013b).

In transgenic *N. tabacum* plants overexpressing a tomato gene encoding a chloroplast-targeted GSH1, the marked upregulation of genes encoding a 1-ami nocyclopropane-1-carboxylate oxidase (ACO) and the ethylene response factor 4 (ERF4) was observed. In addition, proteomic profiling of this transgenic line revealed the elevated accumulation of 1-aminocyclopropane-1-carboxylate synthase (ACS) as compared to wild-type plants. Since ACS and ACO are key enzymes of ethylene biosynthesis, these data demonstrated the active involvement of GSH in the synergistic multiple crosstalk with SA and ethylene to combat biotic stress (Ghanta et al. 2014). In further experiments, transgenic *A. thaliana* plants overexpressing the *GSH1* gene were constructed that showed elevated GSH contents and improved resistance to the necrotrophic pathogen *B. cinerea*. Intriguingly, these transgenic plants exhibited a strong upregulation of *ACS2*, *ACS6*, and *ACO1* at the transcript as well as protein levels, while these genes were downregulated in the GSH-depleted *pad2-1* mutant. The GSH-induced *ACS2* and *ACS6* transcription occurred in a WRKY33-dependent manner. On the other hand, GSH improved the messenger RNA stability of *ACO1*. Furthermore, the ACO1 protein was posttranslationally modified by S-glutathionylation, while ACS2 and ACS6 were not. These results clearly proved that GSH-mediated resistance to necrotrophic plant pathogens may occur via an ethylene-mediated pathway (Datta et al. 2015).

3.5 *Role of GSH in the Posttranslational Modifications of Defense Proteins*

Cellular redox status is tightly controlled in plant cells. The GSH/GSSG redox couple plays a principal role in the buffering of cellular redox state. The reduction state of this redox couple and the GSH concentration are important parameters in mediating dynamic signaling in response to infections. Modifications of the GSH/GSSG redox equilibrium are integral parts of plant responses to infections, and these redox modifications play an important signaling role in defense reactions (Fobert and Després 2005; Noctor 2006; Spoel and Loake 2011).

Thiol-disulfide exchange reactions are accelerated through enzymatic catalysis usually by TRX and glutaredoxin (GRX) enzymes (Meyer et al. 2012). In plants, TRXs and GRXs are encoded by large multigene families, and these ubiquitous proteins show very diverse functions by regulating the structure or activity of many proteins (Geigenberger et al. 2017; Rouhier et al. 2008). TRXs catalyze the reversible reduction of protein disulfide bridges by transferring the reducing equivalents from thioredoxin reductase proteins and ultimately from NADPH (Meyer et al. 2012). Several members of the TRX-h subfamily have emerged as important factors in plant defense reactions (Laloi et al. 2004; Tada et al. 2008; Sun et al. 2010). In *A. thaliana*, the expression of *TRXh5* and a *GRX* was induced during infection with five different RNA viruses (Whitham et al. 2003). The *AtTRXh5* gene was upregulated also during an incompatible interaction with the bacterial pathogen *P. syringae* pv. *tomato*. An electrophoretic mobility shift assay revealed the binding of a WRKY transcription factor to a W-box nucleotide motif in the promoter region of *AtTRXh5* (Laloi et al. 2004). In tobacco, the expression of the *TRXh3* gene was induced by TMV and *Cucumber mosaic virus*, and the overexpression of *TRXh3* conferred increased resistance against both viruses (Sun et al. 2010). GRXs are also thiol-disulfide oxidoreductases constituting an alternative reducing system besides TRXs. GRXs catalyze the reduction of protein-GSH mixed disulfide bonds thereby reversing protein S-glutathionylation. The oxidized form of GRXs is usually reduced by the GSH/GSSG redox couple, and the resulting GSSG is recycled to GSH by GR enzymes by using NADPH (Rouhier et al. 2008). Thus, shifts in the GSH/GSSG redox potential can be sensed by GRXs, which reversibly transfer electrons between the glutathione redox buffer and thiol groups of target proteins (Meyer 2008). At present, the role of GRXs in plant defense reactions is poorly understood (Whitham et al. 2003; Ndamukong et al. 2007; Wang et al. 2009; La Camera et al. 2011; Zander et al. 2012).

GSH has a key role in the redox regulation of the SA-responsive transcriptional coactivator protein NPR1 in *A. thaliana* (Mou et al. 2003; Dong 2004; Tada et al. 2008). NPR1 is an essential regulator of plant SAR, which confers durable resistance (immunity) to a broad spectrum of pathogens. Conserved cysteine residues of NPR1 form intermolecular disulfide bridges between NPR1 molecules that establish a high molecular weight NPR1 oligomer that resides in the cytosol (Mou et al. 2003). Upon pathogen infection, SA accumulation triggers cycles of reduction and oxidation reactions of specific cysteine residues that result in disulfide reduction and S-nitrosylation, respectively (Tada et al. 2008). As a result of these processes, monomeric NPR1 is released and translocates into the nucleus, where it activates defense gene transcription. Mutations of the Cys82 or Cys216 residues in NPR1 led to constitutive monomerization, nuclear localization of the mutant proteins, and defense gene expression. The GSH/GSSG ratio required for in vitro NPR1 reduction was similar to that reached in plants after SAR induction (Mou et al. 2003). Some years later, a thioredoxin (TRX-h5) was identified as an enzyme catalyzing the reduction of the NPR1 oligomer showing that redox signals are conveyed through cytosolic TRXs (Tada et al. 2008). Conversely, S-nitrosylation of NPR1 facilitates the formation of disulfide bridges (Tada et al. 2008). In the nucleus, NPR1 proteins cannot bind directly to DNA, but they form transactivation complexes with TGA transcription factors and thereby regulate their DNA-binding activity. TGA1 and TGA4 were shown to contain disulfide bridges that preclude interactions with NPR1. Interestingly, reduction of these disulfide bridges stimulated the NPR1-TGA1/4 interactions. Furthermore, both NPR1 and TGA proteins can be nitrosylated at specific Cys residues by S-nitrosoglutathione (GSNO) (Després et al. 2003; Tada et al. 2008). The interaction of activated NPR1 and TGA factors results in the enhanced transcription of defense genes. Alternatively elevated GSH levels can activate WRKY transcription factors causing same/similar patterns of defense gene induction. The ultimate outcome of these processes is the development of enhanced plant disease resistance often accompanied by localized (hypersensitive) cell death (Fig. 15.1).

3.6 Glutathione and Nitric Oxide in Plant Defense Signaling

Besides ROS, NO has emerged as a key regulator of plant defense responses. S-nitrosylation, the addition of a NO to protein cysteine residues in order to form an S-nitrosothiol (SNO), is a major redox-based posttranslational modification during plant immune signaling functions (Yun et al. 2012). Total cellular levels of protein S-nitrosylation (SNO) are controlled predominantly by S-nitrosoglutathione reductase 1 (GSNOR1) which turns over the natural NO donor S-nitrosoglutathione (GSNO) (Yun et al. 2012, 2016). GSNO was shown to promote the nuclear accumulation of the NPR1 protein accompanied by an elevated SA concentration and the activation of PR genes, leading to induced resistance of *A. thaliana* against *P. syringae* pv. *tomato* infection (Kovacs et al. 2015). Moreover, NO induced a rapid change in the glutathione status, resulting in increased concentrations of GSH, which is required for SA accumulation and activation of the NPR1-dependent defense response. These data imply a crosstalk between NO and GSH and further demonstrate that GSH is not only an important cellular redox buffer but also a signaling molecule in plant defense (Kovacs et al. 2015).

The significance of GSNOR in regulating S-nitrosylation during plant disease resistance responses was first demonstrated by Feechan et al. (2005) who found that

GSNOR1 of *A. thaliana* positively regulates the signaling network controlled by SA. Increased GSNOR1 activity reduced SNO formation and conferred resistance to normally virulent pathogens. However, loss of AtGSNOR1 function increased SNO and GSNO levels, disabling plant defense responses, including basal and non-host disease resistance (Feechan et al. 2005; Yun et al. 2016). The role of GSNOR in promoting disease resistance was also demonstrated during infections by a biotrophic and hemibiotrophic fungal pathogen. Using the *Medicago truncatula*/*Aphanomyces euteiches* pathosystem, Thalineau et al. (2016) showed that GSNOR activity significantly contributes to resistance against the biotrophic *A. euteiches*. Pathogen levels in GSNOR-overexpressing roots were much lower than in control transformed roots. Also, in potato resistant to the hemibiotrophic oomycete *Phytophthora infestans* (incompatible interaction), a transient but significant increase in GSNOR activity occurred at 3 h after inoculation (hpi), paralleled by a transient decrease in SNO levels (Abramowski et al. 2015), suggesting a role of these processes in the signaling of resistance. Indeed, an early NO and superoxide ($O_2^{\bullet -}$) generation in resistant plants led to H_2O_2 production at 3 hpi, a time point when *P. infestans* pathogenesis is in the initial biotrophic phase (Abramowski et al. 2015). Therefore, these results imply that elevated GSNOR activity coupled to a decline in SNO levels may induce ROS and/or suppress antioxidants in order to confer SA-mediated resistance to (hemi)biotrophic pathogens. In contrast, a loss of GSNOR1 function results in increased SNO and GSNO levels (Feechan et al. 2005; Yun et al. 2016) that could induce activities of antioxidant enzymes like APX and GR (Begara-Morales et al. 2016; Kovacs et al. 2016) conferring thereby tolerance to ROS-elicited abiotic stress. In fact, inhibition of GSNOR activity by H_2O_2 or the superoxide-generating herbicide paraquat results in enhanced levels of SNOs followed by GSH accumulation. Furthermore, transcript levels of redox-regulated genes (e.g., peroxidases, peroxiredoxins, TRXs) and activities of glutathione-dependent – antioxidant – enzymes (GR and GST) are increased in GSNOR-deficient plants, likely contributing to their enhanced resistance against oxidative stress (Kovacs et al. 2016).

Although a GSNOR activity-driven decline in SNO may contribute to plant disease resistance, the role of S-nitrosylation in pathogen defense could be more complex. For example, *S*-nitrosylation inhibits the peroxidase activity of a peroxiredoxin (thiol-based peroxidase), PrxIIE (Romero-Puertas et al. 2007). Interestingly, PrxIIE also possesses peroxynitrite ($ONOO^-$) reductase activity, protecting against oxidative and nitrosative stresses mediated by $ONOO^-$, a key signaling agent of disease resistance. In plants, PrxIIE is *S*-nitrosylated during the hypersensitive type of resistance (HR) (Romero-Puertas et al. 2008), and this modification inhibits both its peroxidase and peroxynitrite reductase activities, which may contribute to both localized plant cell death and resistance during HR (Romero-Puertas et al. 2007, 2008).

4 Conclusions and Future Perspectives

During the last 30 years, the crucial role of glutathione in signaling processes of infected plants and its contribution to the development of disease resistance have been clearly recognized, but the underlying biochemical mechanisms are still largely unknown. Although recent research could uncouple the role of GSH in mediating disease resistance-related signaling processes from its antioxidant function, the exact roles of GSH are still very difficult to identify due to the multiplicity and interdependence of GSH-related biochemical pathways. The modulation of glutathione concentrations and the GSH/GSSG ratio in infected plant cells elicits transcriptional reprogramming and induces posttranslational protein modifications. Importantly, GSH metabolism is interconnected with the levels of key plant defense hormones such as SA, JA, and ethylene. However, sophisticated approaches will be required to dissect the cross-connections between glutathione, SA, and H_2O_2. Recent advances in studying redox signaling during plant-pathogen interactions have identified a key role for diverse oxidative cysteine modifications. Although marked changes in the plant cellular redox status have been observed during resistance responses in different plants, the associated signaling mechanisms are still poorly understood. The development of novel redox-sensitive, fluorescent biosensor proteins should significantly advance our understanding of glutathione homeostasis. In this regard, the exploration of changes in GSH metabolism in various subcellular compartments of infected plant cells may provide a further tool to unravel GSH signaling mechanisms during plant disease resistance.

Acknowledgments Research in the laboratory of the authors is supported by grants of the Hungarian National Research, Development and Innovation Office (NKFIH K111995 and PD108455).

References

Abramowski D, Arasimowicz-Jelonek M, Izbiańska K, Billert H, Floryszak-Wieczorek J (2015) Nitric oxide modulates redox-mediated defense in potato challenged with *Phytophthora infestans*. Eur J Plant Pathol 143:237–260

Akter N, Sobahan MA, Hossain MA, Uraji M, Nakamura Y, Mori IC, Murata Y (2010) The involvement of intracellular glutathione in methyl jasmonate signaling in *Arabidopsis* guard cells. Biosci Biotechnol Biochem 74:2504–2506

Alvarez ME, Pennell RI, Meijer PJ, Ishikawa A, Dixon RA, Lamb C (1998) Reactive oxygen intermediates mediate a systemic signal network in the establishment of plant immunity. Cell 92:773–784

Amirsadeghi S, Robson CA, Vanlerberghe GC (2007) The role of the mitochondrion in plant responses to biotic stress. Physiol Plant 129:253–266

Ball L, Accotto GP, Bechtold U, Creissen G, Funck D, Jimenez A, Kular B, Leyland N, Mejia-Carranza J, Reynolds H, Karpinski S, Mullineaux PM (2004) Evidence for a direct link between glutathione biosynthesis and stress defense gene expression in Arabidopsis. Plant Cell 16:2448–2462

Barna B, Ádám A, Király Z (1993) Juvenility and resistance of a superoxide-tolerant plant to diseases and other stresses. Naturwissenschaften 80:420–422

Barna B, Fodor J, Harrach B, Pogány M, Király Z (2012) The Janus face of reactive oxygen species in resistance and susceptibility of plants to necrotrophic and biotrophic pathogens. Plant Physiol Biochem 59:37–43

Bartling D, Radzio R, Steiner U, Weiler EW (1993) A glutathione-S-transferase with glutathione peroxidase activity from *Arabidopsis thaliana*. Molecular cloning and functional characterization. Eur J Biochem 216:579–586

Begara-Morales JC, Sánchez-Calvo B, Chaki M, Valderrama R, Mata-Pérez C, Padilla MN, Corpas FJ, Barroso JB (2016) Antioxidant systems are regulated by nitric oxide-mediated post-translational modifications (NO-PTMs). Front Plant Sci 7:152

Bi YM, Kenton P, Mur L, Darby R, Draper J (1995) Hydrogen peroxide does not function downstream of salicylic acid in the induction of PR protein expression. Plant J 8:235–245

Caarls L, Pieterse CMJ, Van Wees SC (2015) How salicylic acid takes transcriptional control over jasmonic acid signaling. Front Plant Sci 6:170

Chamnongpol S, Willekens H, Moeder W, Langebartels C, Sandermann H Jr, Van Montagu M, Inzé D, Van Camp W (1998) Defense activation and enhanced pathogen tolerance induced by H_2O_2 in transgenic tobacco. Proc Natl Acad Sci U S A 95:5818–5823

Chen Z, Silva H, Klessig DF (1993) Active oxygen species in the induction of plant systemic acquired resistance by salicylic acid. Science 262:1883–1886

Chen S, Vaghchhipawala Z, Li W, Asard H, Dickman MB (2004) Tomato phospholipid hydroperoxide glutathione peroxidase inhibits cell death induced by Bax and oxidative stresses in yeast and plants. Plant Physiol 135:1630–1641

Chen YP, Xing LP, Wu GJ, Wang HZ, Wang XE, Cao AZ, Chen PD (2007) Plastidial glutathione reductase from *Haynaldia villosa* is an enhancer of powdery mildew resistance in wheat (*Triticum aestivum*). Plant Cell Physiol 48:1702–1712

Chen I-H, Chiu M-H, Cheng S-F, Hsu Y-H, Tsai CH (2013) The glutathione transferase of *Nicotiana benthamiana* NbGSTU4 plays a role in regulating the early replication of *Bamboo mosaic virus*. New Phytol 199:749–757

Cheng MC, Ko K, Chang WL, Kuo WC, Chen GH, Lin TP (2015) Increased glutathione contributes to stress tolerance and global translational changes in Arabidopsis. Plant J 83:926–939

Choudhary AD, Kessmann H, Lamb CJ, Dixon RA (1990) Stress responses in alfalfa (*Medicago sativa* L.). IV. Expression of defense gene constructs in electroporated suspension cell protoplasts. Plant Cell Rep 9:42–46

Clemente-Moreno MJ, Diaz Vivancos P, Barba-Espín G, Hernández JA (2010) Benzothiadiazole and L-2-oxothiazolidine-4-carboxylic acid reduce the severity of Sharka symptoms in pea leaves: effect on antioxidative metabolism at the subcellular level. Plant Biol 12:88–97

Clemente-Moreno MJ, Diaz Vivancos P, Piqueras A, Hernández JA (2012) Plant growth stimulation in Prunus species plantlets by BTH or OTC treatments under in vitro conditions. J Plant Physiol 169:1074–1083

Datta R, Chattopadhyay S (2015) Changes in the proteome of *pad2-1*, a glutathione depleted Arabidopsis mutant, during *Pseudomonas syringae* infection. J Proteome 126:82–93

Datta R, Kumar D, Sultana A, Hazra S, Bhattacharyya D, Chattopadhyay S (2015) Glutathione regulates 1-aminocyclopropane-1-carboxylate synthase transcription via WRKY33 and 1-ami nocyclopropane-1-carboxylate oxidase by modulating messenger RNA stability to induce ethylene synthesis during stress. Plant Physiol 169:2963–2981

Dean JD, Goodwin PH, Hsiang T (2005) Induction of glutathione S-transferase genes of *Nicotiana benthamiana* following infection by *Colletotrichum destructivum* and *C. orbiculare* and involvement of one in resistance. J Exp Bot 56:1525–1533

Després C, Chubak C, Rochon A, Clark R, Bethune T, Desveaux D, Fobert PR (2003) The Arabidopsis NPR1 disease resistance protein is a novel cofactor that confers redox regulation of DNA binding activity to the basic domain/leucine zipper transcription factor TGA1. Plant Cell 15:2181–2191

Diaz Vivancos P, Dong Y, Ziegler K, Markovic J, Pallardo FV, Pellny TK, Verrier PJ, Foyer CH (2010a) Recruitment of glutathione into the nucleus during cell proliferation adjusts whole-cell redox homeostasis in *Arabidopsis thaliana* and lowers the oxidative defence shield. Plant J 64:825–838

Diaz Vivancos P, Wolff T, Markovic J, Pallardó FV, Foyer CH (2010b) A nuclear glutathione cycle within the cell cycle. Biochem J 431:169–178

Dixon DP, Edwards R (2009) Selective binding of glutathione conjugates of fatty acid derivatives by plant glutathione transferases. J Biol Chem 284:21249–21256

Dixon DP, Hawkins T, Hussey PJ, Edwards R (2009) Enzyme activities and subcellular localization of members of the *Arabidopsis* glutathione transferase superfamily. J Exp Bot 60:1207–1218

Dixon DP, Sellars JD, Edwards R (2011) The Arabidopsis phi class glutathione transferase AtGSTF2: binding and regulation by biologically active heterocyclic ligands. Biochem J 438:63–70

Dong X (2004) NPR1, all things considered. Curr Opin Plant Biol 7:547–552

Dröge-Laser W, Kaiser A, Lindsay WP, Halkier BA, Loake GJ, Doerner P, Dixon RA, Lamb C (1997) Rapid stimulation of a soybean protein-serine kinase that phosphorylates a novel bZIP DNA-binding protein, G/HBF-1, during the induction of early transcription-dependent defenses. EMBO J 16:726–738

Dron M, Clouse SD, Dixon RA, Lawton MA, Lamb CJ (1988) Glutathione and fungal elicitor regulation of a plant defense gene promoter in electroporated protoplasts. Proc Natl Acad Sci U S A 85:6738–6742

Dubreuil-Maurizi C, Poinssot B (2012) Role of glutathione in plant signaling under biotic stress. Plant Signal Behav 7:210–212

Dubreuil-Maurizi C, Vitecek J, Marty L, Branciard L, Frettinger P, Wendehenne D, Meyer AJ, Mauch F, Poinssot B (2011) Glutathione deficiency of the Arabidopsis mutant *pad2-1* affects oxidative stress-related events, defense gene expression, and the hypersensitive response. Plant Physiol 157:2000–2012

Edwards R, Dixon DP (2005) Plant glutathione transferases. Methods Enzymol 401:169–186

El-Zahaby HM, Gullner G, Király Z (1995) Effects of powdery mildew infection of barley on the ascorbate-glutathione cycle and other antioxidants in different host-pathogen interactions. Phytopathology 85:1225–1230

Enyedi AJ, Yalpani N, Silverman P, Raskin I (1992) Localization, conjugation, and function of salicylic acid in tobacco during the hypersensitive reaction to tobacco mosaic virus. Proc Natl Acad Sci U S A 89:2480–2484

Farkas G, Király Z, Solymosi F (1960) Role of oxidative metabolism in the localization of plant viruses. Virology 12:408–421

Feechan A, Kwon E, Yun BW, Wang Y, Pallas JA, Loake GJ (2005) A central role for S-nitrosothiols in plant disease resistance. Proc Natl Acad Sci U S A 102:8054–8059

Fobert PR, Després C (2005) Redox control of systemic acquired resistance. Curr Opin Plant Biol 8:378–382

Fodor J, Gullner G, Ádám AL, Barna B, Kőmíves T, Király Z (1997) Local and systemic responses of antioxidants to tobacco mosaic virus infection and to salicylic acid in tobacco. Role in systemic acquired resistance. Plant Physiol 114:1443–1451

Foyer CH, Noctor G (2005) Redox homeostasis and antioxidant signaling: a metabolic interface between stress perception and physiological responses. Plant Cell 17:1866–1875

Foyer CH, Rennenberg H (2000) Regulation of glutathione synthesis and its role in abiotic and biotic stress defence. In: Brunold C, Rennenberg H, De Kok LJ, Stulen I, Davidian JC (eds) Sulfur nutrition and sulfur assimilation in higher plants. Paul Haupt, Bern, pp 127–153

Freeman JL, Garcia D, Kim DG, Hopf A, Salt DE (2005) Constitutively elevated salicylic acid signals glutathione-mediated nickel tolerance in Thlaspi nickel hyperaccumulators. Plant Physiol 137:1082–1091

Geigenberger P, Thormählen I, Daloso DM, Fernie AR (2017) The unprecedented versatility of the plant thioredoxin system. Trends Plant Sci 22:249–262

Ghanta S, Chattopadhyay S (2011) Glutathione as a signaling molecule: another challenge to pathogens. Plant Signal Behav 6:783–788

Ghanta S, Bhattacharyya D, Chattopadhyay S (2011a) Glutathione signaling acts through NPR1-dependent SA-mediated pathway to mitigate biotic stress. Plant Signal Behav 6:607–609
Ghanta S, Bhattacharyya D, Sinha R, Banerjee A, Chattopadhyay S (2011b) *Nicotiana tabacum* overexpressing γ-ECS exhibits biotic stress tolerance likely through NPR1-dependent salicylic acid-mediated pathway. Planta 233:895–910
Ghanta S, Datta R, Bhattacharyya D, Sinha R, Kumar D, Hazra S, Mazumdar AB, Chattopadhyay S (2014) Multistep involvement of glutathione with salicylic acid and ethylene to combat environmental stress. J Plant Physiol 171:940–950
Glazebrook J (2005) Contrasting mechanisms of defense against biotrophic and necrotrophic pathogens. Annu Rev Phytopathol 43:205–227
Goodman RN, Novacky AJ (1994) The hypersensitive reaction in plants to pathogens. APS Press, St. Paul
Green RM, Graham M, O'Donovan MR, Chipman JK, Hodges NJ (2006) Subcellular compartmentalization of glutathione: correlations with parameters of oxidative stress related to genotoxicity. Mutagenesis 21:383–390
Greenberg JT, Yao N (2004) The role and regulation of programmed cell death in plant-pathogen interactions. Cell Microbiol 6:201–211
Großkinsky DK, Koffler BE, Roitsch T, Maier R, Zechmann B (2012) Compartment-specific antioxidative defense in Arabidopsis against virulent and avirulent *Pseudomonas syringae*. Phytopathology 102:662–673
Gullner G, Kőmíves T (2001) The role of glutathione and glutathione-related enzymes in plant-pathogen interactions. In: Grill D, Tausz M, De Kok LJ (eds) Significance of glutathione in plant adaptation to the environment. Kluwer Academic Publishers, Dordrecht, pp 207–239
Gullner G, Kőmíves T (2006) Defense reactions of infected plants: roles of glutathione and glutathione S-transferase enzymes. Acta Phytopathol Entomol Hung 41:3–10
Gullner G, Kőmíves T, Király L (1991) Enhanced inducibility of antioxidant systems in a *Nicotiana tabacum* L. biotype results in acifluorfen resistance. Z. Naturforsch 46c:875–881
Gullner G, Fodor J, Király L (1995) Induction of glutathione-S-transferase activity in tobacco by tobacco necrosis virus infection and by salicylic acid. Pestic Sci 45:290–291
Gullner G, Tóbiás I, Fodor J, Kőmíves T (1999) Elevation of glutathione level and activation of glutathione-related enzymes affect virus infection in tobacco. Free Radic Res 31:S155–S161
Hacham Y, Koussevitzky S, Kirma M, Amir R (2014) Glutathione application affects the transcript profile of genes in Arabidopsis seedling. J Plant Physiol 171:1444–1451
Han Y, Chaouch S, Mhamdi A, Queval G, Zechmann B, Noctor G (2013a) Functional analysis of Arabidopsis mutants points to novel roles for glutathione in coupling H_2O_2 to activation of salicylic acid accumulation and signalling. Antioxid Redox Signal 18:2106–2121
Han Y, Mhamdi A, Chaouch S, Noctor G (2013b) Regulation of basal and oxidative stress-triggered jasmonic acid-related gene expression by glutathione. Plant Cell Environ 36:1135–1146
Harrach BD, Fodor J, Pogány M, Preuss J, Barna B (2008) Antioxidant, ethylene and membrane leakage responses to powdery mildew infection of near-isogenic barley lines with various types of resistance. Eur J Plant Pathol 121:21–33
Harrach BD, Baltruschat H, Barna B, Fodor J, Kogel K-H (2013) The mutualistic fungus *piriformospora indica* protects barley roots from a loss of antioxidant capacity caused by the necrotrophic pathogen *Fusarium culmorum*. Mol Plant-Microbe Interact 26:599–605
Hausladen A, Kunert KJ (1990) Effects of artificially enhanced levels of ascorbate and glutathione on the enzymes monodehydroascorbate reductase, dehydroascorbate reductase, and glutathione reductase in spinach (*Spinacia oleracea*). Physiol Plant 79:384–388
Herbette S, Lenne C, de Labrouhe DT, Drevet JR, Roeckel-Drevet P (2003) Transcripts of sunflower antioxidant scavengers of the SOD and GPX families accumulate differentially in response to downy mildew infection, phytohormones, reactive oxygen species, nitric oxide, protein kinase and phosphatase inhibitors. Physiol Plant 119:418–428
Hernández I, Chacón O, Rodriguez R, Portieles R, López Y, Pujol M, Borrás-Hidalgo O (2009) Black shank resistant tobacco by silencing of glutathione S-transferase. Biochem Biophys Res Commun 387:300–304

Hernández JA, Gullner G, Clemente-Moreno MJ, Künstler A, Juhász C, Diaz Vivancos P, Király L (2016) Oxidative stress and antioxidative responses in plant-virus interactions. Physiol Mol Plant Pathol 94:134–148

Herrera-Vásquez A, Salinas P, Holuigue L (2015) Salicylic acid and reactive oxygen species interplay in the transcriptional control of defense genes expression. Front Plant Sci 6:171

Höller K, Király L, Künstler A, Müller M, Gullner G, Fattinger M, Zechmann B (2010) Enhanced glutathione metabolism is correlated with sulfur induced resistance in tobacco mosaic virus-infected genetically susceptible *Nicotiana tabacum* plants. Mol Plant-Microbe Interact 23:1448–1459

Juhász C, Gullner G (2014) The monoterpenoid (S)-carvone massively up-regulates several classes of glutathione S-transferase genes in tobacco leaf discs. Acta Phytopathol Entomol Hung 49:163–176

Kauss H, Jeblick W (1995) Pretreatment of parsley suspension cultures with salicylic acid enhances spontaneous and elicited production of H_2O_2. Plant Physiol 108:1171–1178

Király Z, Barna B, Kecskés A, Fodor J (2002) Down-regulation of antioxidative capacity in a transgenic tobacco which fails to develop acquired resistance to necrotization caused by tobacco mosaic virus. Free Radic Res 36:981–991

Király L, Künstler A, Fattinger M, Höller K, Juhász C, Müller M, Gullner G, Zechmann B (2012) Sulfate supply influences compartment specific glutathione metabolism and confers enhanced resistance to tobacco mosaic virus during a hypersensitive response. Plant Physiol Biochem 59:44–54

Klement Z (1982) Hypersensitivity. In: Mount MS, Lacy GH (eds) Phytopathogenic prokaryotes II. Academic, New York, pp 149–177

Koornneef A, Leon-Reyes A, Ritsema T, Verhage A, Den Otter FC, Van Loon LC, Pieterse CM (2008) Kinetics of salicylate-mediated suppression of jasmonate signaling reveal a role for redox modulation. Plant Physiol 147:1358–1368

Kovacs I, Durner J, Lindermayr C (2015) Crosstalk between nitric oxide and glutathione is required for NONEXPRESSOR OF PATHOGENESIS-RELATED GENES 1 (NPR1)-dependent defense signaling in *Arabidopsis thaliana*. New Phytol 208:860–872

Kovacs I, Holzmeister C, Wirtz M, Geerlof A, Fröhlich T, Römling G, Kuruthukulangarakoola GT, Linster E, Hell R, Arnold GJ, Durner J, Lindermayr C (2016) ROS-mediated inhibition of S-nitrosoglutathione reductase contributes to the activation of anti-oxidative mechanisms. Front Plant Sci 7:1669

Künstler A, Király L, Pogány M, Tóbiás I, Gullner G (2007) Lipoxygenase and glutathione peroxidase activity in tobacco leaves inoculated with tobacco mosaic virus. Acta Phytopathol Entomol Hung 42:197–207

Kusumi K, Yaeno T, Kojo K, Hirayama M, Hirokawa D, Yara A, Iba K (2006) The role of salicylic acid in the glutathione-mediated protection against photooxidative stress in rice. Physiol Plant 128:651–661

Kuźniak E, Skłodowska M (2004) Differential implication of glutathione, glutathione-metabolizing enzymes and ascorbate in tomato resistance to *Pseudomonas syringae*. J Phytopathol 152:529–536

Kuźniak E, Skłodowska M (2005a) Fungal pathogen-induced changes in the antioxidant systems of leaf peroxisomes from infected tomato plants. Planta 222:192–200

Kuźniak E, Skłodowska M (2005b) Compartment-specific role of the ascorbate-glutathione cycle in the response of tomato leaf cells to *Botrytis cinerea* infection. J Exp Bot 56:921–933

La Camera S, L'Haridon F, Astier J, Zander M, Abou-Mansour E, Page G, Thurow C, Wendehenne D, Gatz C, Métraux JP, Lamotte O (2011) The glutaredoxin ATGRXS13 is required to facilitate *Botrytis cinerea* infection of *Arabidopsis thaliana* plants. Plant J 68:507–519

Laloi C, Mestres-Ortega D, Marco Y, Meyer Y, Reichheld JP (2004) The Arabidopsis cytosolic thioredoxin *h5* gene induction by oxidative stress and its W-box-mediated response to pathogen elicitor. Plant Physiol 134:1006–1016

León J, Lawton MA, Raskin I (1995) Hydrogen peroxide stimulates salicylic acid biosynthesis in tobacco. Plant Physiol 108:1673–1678

Levine A, Tenhaken R, Dixon R, Lamb C (1994) H_2O_2 from the oxidative burst orchestrates the plant hypersensitive disease resistance response. Cell 79:583–593

Li J, Brader G, Palva ET (2004) The WRKY70 transcription factor: a node of convergence for jasmonate-mediated and salicylate-mediated signals in plant defense. Plant Cell 16:319–331

Li X, Zhang JB, Song B, Li HP, Xu HQ, Qu B, Dang FJ, Liao YC (2010) Resistance to Fusarium head blight and seedling blight in wheat is associated with activation of a cytochrome P450 gene. Phytopathology 100:183–191

Mateo A, Funck D, Mühlenbock P, Kular B, Mullineaux PM, Karpinski S (2006) Controlled levels of salicylic acid are required for optimal photosynthesis and redox homeostasis. J Exp Bot 57:1795–1807

Matern S, Peskan-Berghoefer T, Gromes R, Kiesel RV, Rausch T (2015) Imposed glutathione-mediated redox switch modulates the tobacco wound-induced protein kinase and salicylic acid-induced protein kinase activation state and impacts on defence against *Pseudomonas syringae*. J Exp Bot 66:1935–1950

Mauch F, Dudler R (1993) Differential induction of distinct glutathione-S-transferases of wheat by xenobiotics and by pathogen attack. Plant Physiol 102:1193–1201

May MJ, Hammond-Kosack KE, Jones JDG (1996a) Involvement of reactive oxygen species, glutathione metabolism, and lipid peroxidation in the *Cf*-gene-dependent defense response of tomato cotyledons induced by race-specific elicitors of *Cladosporium fulvum*. Plant Physiol 110:1367–1379

May MJ, Parker JE, Daniels MJ, Leaver CJ, Cobbett CS (1996b) An Arabidopsis mutant depleted in glutathione shows unaltered responses to fungal and bacterial pathogens. Mol Plant-Microbe Interact 9:349–356

Meyer AJ (2008) The integration of glutathione homeostasis and redox signaling. J Plant Physiol 165:1390–1403

Meyer AJ, Hell R (2005) Glutathione homeostasis and redox-regulation by sulfhydryl groups. Photosynth Res 86:435–457

Meyer Y, Belin C, Delorme-Hinoux V, Reichheld JP, Riondet C (2012) Thioredoxin and glutaredoxin systems in plants: molecular mechanisms, crosstalks, and functional significance. Antioxid Redox Signal 17:1124–1160

Mou Z, Fan W, Dong X (2003) Inducers of plant systemic acquired resistance regulate NPR1 function through redox changes. Cell 113:935–944

Mullineaux PM, Rausch T (2005) Glutathione, photosynthesis and the redox regulation of stress-responsive gene expression. Photosynth Res 86:459–474

Navrot N, Collin V, Gualberto J, Gelhaye E, Hirasawa M, Rey P, Knaff DB, Issakidis E, Jacquot JP, Rouhier N (2006) Plant glutathione peroxidases are functional peroxiredoxins distributed in several subcellular compartments and regulated during biotic and abiotic stresses. Plant Physiol 142:1364–1379

Ndamukong I, Abdallat AA, Thurow C, Fode B, Zander M, Weigel R, Gatz C (2007) SA-inducible Arabidopsis glutaredoxin interacts with TGA factors and suppresses JA-responsive *PDF1.2* transcription. Plant J 50:128–139

Neuenschwander U, Vernooij B, Friedrich L, Uknes S, Kessmann H, Ryals J (1995) 1s hydrogen peroxide a second messenger of salicylic acid in systemic acquired resistance? Plant J 8:227–233

Noctor G (2006) Metabolic signalling in defence and stress: the central roles of soluble redox couples. Plant Cell Environ 29:409–425

Noctor G, Foyer CH (2016) Intracellular redox compartmentation and ROS-related communication in regulation and signaling. Plant Physiol 171:1581–1592

Noctor G, Gomez L, Vanacker H, Foyer CH (2002) Interactions between biosynthesis, compartmentation and transport in the control of glutathione homeostasis and signalling. J Exp Bot 53:1283–1304

Noctor G, Mhamdi A, Chaouch S, Han Y, Neukermans J, Marquez-Garcia B, Queval G, Foyer CH (2012) Glutathione in plants: an integrated overview. Plant Cell Environ 35:454–484

Parisy V, Poinssot B, Owsianowski L, Buchala A, Glazebrook J, Mauch F (2007) Identification of *PAD2* as a γ-glutamylcysteine synthetase highlights the importance of glutathione in disease resistance of Arabidopsis. Plant J 49:159–172

Perfect ES, Green JR (2001) Infection structures of biotrophic and hemibiotrophic fungal plant pathogens. Mol Plant Pathol 2:101–108

Pieterse CM, Leon-Reyes A, Van der Ent S, Van Wees SC (2009) Networking by small-molecule hormones in plant immunity. Nat Chem Biol 5:308–316

Rahantaniaina MS, Tuzet A, Mhamdi A, Noctor G (2013) Missing links in understanding redox signaling via thiol/disulfide modulation: how is glutathione oxidized in plants? Front Plant Sci 4:477

Rao MV, Paliyath G, Ormrod DP, Murr DP, Watkins CB (1997) Influence of salicylic acid on H_2O_2 production, oxidative stress, and H_2O_2-metabolizing enzymes. Salicylic acid-mediated oxidative damage requires H_2O_2. Plant Physiol 115:137–149

Robert-Seilaniantz A, Grant M, Jones JD (2011) Hormone crosstalk in plant disease and defense: more than just jasmonate-salicylate antagonism. Annu Rev Phytopathol 49:317–343

Romero-Puertas MC, Laxa M, Mattè A, Zaninotto F, Finkemeier I, Jones AME, Perazzolli M, Vandelle E, Dietz KJ, Delledonne M (2007) S-nitrosylation of peroxiredoxin IIE promotes peroxynitrite-mediated tyrosine nitration. Plant Cell 19:4120–4130

Romero-Puertas MC, Campostrini N, Mattè A, Righetti PG, Perazzolli M, Zolla L, Roepstorff P, Delledonne M (2008) Proteomic analysis of S-nitrosylated proteins in *Arabidopsis thaliana* undergoing hypersensitive response. Proteomics 8:1459–1469

Rouhier N, Lemaire SD, Jacquot JP (2008) The role of glutathione in photosynthetic organisms: emerging functions for glutaredoxins and glutathionylation. Annu Rev Plant Biol 59:143–166

Schafer FQ, Buettner GR (2001) Redox environment of the cell as viewed through the redox state of the glutathione disulfide/glutathione couple. Free Radic Biol Med 30:1191–1212

Senda K, Ogawa K (2004) Induction of PR-1 accumulation accompanied by runaway cell death in the lsd1 mutant of Arabidopsis is dependent on glutathione levels, but independent of the redox state of glutathione. Plant Cell Physiol 45:1578–1585

Simon KU, Polanschütz LM, Koffler BE, Zechmann B (2013) High resolution imaging of temporal and spatial changes of subcellular ascorbate, glutathione and H_2O_2 distribution during *Botrytis cinerea* infection in Arabidopsis. PLoS One 8:e65811

Singh K, Foley RC, Oñate-Sánchez L (2002) Transcription factors in plant defense and stress responses. Curr Opin Plant Biol 5:430–436

Sinha R, Kumar D, Datta R, Hazra S, Bhattacharyya D, Mazumdar AB, Mukhopadhyay R, Sultana A, Chattopadhyay S (2015) Integrated transcriptomic and proteomic analysis of *Arabidopsis thaliana* exposed to glutathione unravels its role in plant defense. Plant Cell Tissue Organ Cult 120:975–988

Spanu PD, Panstruga R (2017) Editorial: biotrophic plant-microbe interactions. Front Plant Sci 8:192

Spoel SH, Loake GJ (2011) Redox-based protein modifications: the missing link in plant immune signalling. Curr Opin Plant Biol 14:358–364

Srivastava MK, Dwivedi UN (1998) Salicylic acid modulates glutathione metabolism in pea seedlings. J Plant Physiol 153:409–414

Sun L, Ren H, Liu R, Li B, Wu T, Sun F, Liu H, Wang X, Dong H (2010) An h-type thioredoxin functions in tobacco defense responses to two species of viruses and an abiotic oxidative stress. Mol Plant-Microbe Interact 23:1470–1485

Tada Y, Spoel SH, Pajerowska-Mukhtar K, Mou Z, Song J, Wang C, Zuo J, Dong X (2008) Plant immunity requires conformational changes of NPR1 via S-nitrosylation and thioredoxins. Science 321:952–956

Tenhaken R, Levine A, Brisson LF, Dixon RA, Lamb C (1995) Function of the oxidative burst in hypersensitive disease resistance. Proc Natl Acad Sci U S A 92:4158–4163

Thalineau E, Truong H-N, Berger A, Fournier C, Boscari A, Wendehenne D, Jeandroz S (2016) Cross-regulation between N metabolism and nitric oxide (NO) signaling during plant immunity. Front Plant Sci 7:472

Van Camp W, Van Montagu M, Inzé D (1998) H_2O_2 and NO: redox signals in disease resistance. Trends Plant Sci 3:330–334

Vanacker H, Carver TLW, Foyer CH (1998) Pathogen-induced changes in the antioxidant status of the apoplast in barley leaves. Plant Physiol 117:1103–1114

Vanacker H, Carver TLW, Foyer CH (2000) Early H_2O_2 accumulation in mesophyll cells leads to induction of glutathione during the hypersensitive response in the barley-powdery mildew interaction. Plant Physiol 123:1289–1300

Vianello A, Zancani M, Peresson C, Petrussa E, Casolo V, Krajnakova J, Patui S, Braidot E, Macri F (2007) Plant mitochondrial pathway leading to programmed cell death. Physiol Plant 129:242–252

Vlot AC, Dempsey DA, Klessig DF (2009) Salicylic acid, a multifaceted hormone to combat disease. Annu Rev Phytopathol 47:177–206

Wachter A, Wolf S, Steininger H, Bogs J, Rausch T (2005) Differential targeting of GSH1 and GSH2 is achieved by multiple transcription initiation: implications for the compartmentation of glutathione biosynthesis in the *Brassicaceae*. Plant J 41:15–30

Wagner U, Edwards R, Dixon DP, Mauch F (2002) Probing the diversity of the Arabidopsis glutathione S-transferase gene family. Plant Mol Biol 49:515–532

Wang Z, Xing S, Birkenbihl RP, Zachgo S (2009) Conserved functions of Arabidopsis and rice CC-type glutaredoxins in flower development and pathogen response. Mol Plant 2:323–335

Whitham SA, Quan S, Chang HS, Cooper B, Estes B, Zhu T, Wang X, Hou YM (2003) Diverse RNA viruses elicit the expression of common sets of genes in susceptible *Arabidopsis thaliana* plants. Plant J 33:271–283

Wingate VPM, Lawton MA, Lamb CJ (1988) Glutathione causes a massive and selective induction of plant defense genes. Plant Physiol 87:206–210

Winterbourn CC (2013) The biological chemistry of hydrogen peroxide. Methods Enzymol 528:3–25

Xiang C, Oliver DJ (1998) Glutathione metabolic genes coordinately respond to heavy metals and jasmonic acid in Arabidopsis. Plant Cell 10:1539–1550

Yang Y, Shah J, Klessig DF (1997) Signal perception and transduction in plant defense responses. Genes Dev 11:1621–1639

Yu LM, Lamb CJ, Dixon RA (1993) Purification and biochemical characterization of proteins which bind to the H-box *cis*-element implicated in transcriptional activation of plant defense genes. Plant J 3:805–816

Yun BW, Spoel SH, Loake GJ (2012) Synthesis of and signalling by small, redox active molecules in the plant immune response. Biochim Biophys Acta 1820:770–776

Yun BW, Skelly MJ, Yin M, Yu M, Mun BG, Lee SU, Hussain A, Spoel SH, Loake GJ (2016) Nitric oxide and S-nitrosoglutathione function additively during plant immunity. New Phytol 211:516–526

Zander M, La Camera S, Lamotte O, Métraux JP, Gatz C (2010) *Arabidopsis thaliana* class-II TGA transcription factors are essential activators of jasmonic acid/ethylene-induced defense responses. Plant J 61:200–210

Zander M, Chen S, Imkampe J, Thurow C, Gatz C (2012) Repression of the *Arabidopsis thaliana* jasmonic acid/ethylene-induced defense pathway by TGA-interacting glutaredoxins depends on their C-terminal ALWL motif. Mol Plant 5:831–840

Zechmann B (2014) Compartment-specific importance of glutathione during abiotic and biotic stress. Front Plant Sci 5:566

Zechmann B, Zellnig G, Müller M (2005) Changes in the subcellular distribution of glutathione during virus infection in *Cucurbita pepo* (L.) Plant Biol 7:49–57

Zechmann B, Zellnig G, Urbanek-Krajnc A, Müller M (2007) Artificial elevation of glutathione affects symptom development in ZYMV-infected *Cucurbita pepo* L. plants. Arch Virol 152:747–762

Chapter 16
Glutathione Transporters in Plants

Xuan Zeng, Diyang Qiu, Rui Hu, and Mingyong Zhang

Abstract Glutathione (GSH) is one of the most abundant small-molecule thiols in plant cells. It is mainly synthesized in chloroplasts and cytosol, and must subsequently be transported to other organelles to carry out its biological functions. It is also transported in the phloem for long distances. Several plant glutathione transporters, which have been characterized for inter- and intracellular transport of glutathione from different species, have been described. In this chapter, biochemical characteristics for substrate transportation, expression patterns, and functions of 11 glutathione transporters from four plant species are reviewed.

Keywords Glutathione • Glutathione transport • Glutathione conjugation • Oligopeptide transporter • Chloroquine-resistance transporter

Contents

X. Zeng • D. Qiu • R. Hu • M. Zhang (✉)
Guangdong Provincial Key Laboratory of Applied Botany and Key Laboratory of South China Agricultural Plant Molecular Analysis and Genetic Improvement, South China Botanical Garden, Chinese Academy of Sciences, Guangzhou, China
e-mail: zhangmy@scbg.ac.cn

M.A. Hossain et al. (eds.), *Glutathione in Plant Growth, Development, and Stress Tolerance*, https://doi.org/10.1007/978-3-319-66682-2_16

1 Introduction

Glutathione (GSH), a tripeptide thiol (γ-L-glutamy-L-cysteinyl-glycine), is the most important antioxidant containing sulfur for plant growth and development, and is essential for plant defense against abiotic and biotic stresses. It plays key roles in a multitude of biological processes both in the cytosol and in organelles. Among other functions, glutathione acts as a redox buffer in the cellular response to oxidative stress (Noctor et al. 2012), as a reagent for protein disulfide formation in the endoplasmic reticulum (Aller and Meyer 2013) and for the detoxification of metals and xenobiotics (Koffler et al. 2013), and as a redox environment modulator in the regulation of plant growth and development (Zechmann et al. 2008; Bachhawat et al. 2013; Ball et al. 2004). Glutathione cannot be replaced functionally in plants except by tripeptide homologs, such as homoglutathione (γ-Glu-Cys-β-Ala) (Klapheck et al. 1988). For example, the block of glutathione synthesis in the *Arabidopsis* mutant *rml1* caused defective root cell division, because glutathione is required for the G_1 to S phase transition of the cell cycle. The phenotype of *rml1* has been shown to be relieved by application of glutathione but not ascorbate, another antioxidant (Vernoux et al. 2000). The depletion of glutathione synthesis by buthioninesulfoximine (BSO), an inhibitor of glutathione synthesis, reduced pollen germination rates. These studies demonstrated that glutathione is essential for plants (Gutierrez-Alcala et al. 2000; Zechmann et al. 2011).

2 Subcellular Distribution of Glutathione

Glutathione is synthesized from glutamate, cysteine, and glycine in two ATP-dependent reactions catalyzed by γ-glutamyl cysteine synthetase (GSH1) and glutathione synthetase (GSH2), respectively. In *Arabidopsis*, these two enzymes are found in the cytosol and chloroplasts (Foyer et al. 2001). However, subcellular localization analysis showed that GSH1 is exclusively targeted to plastids, while GSH2 is targeted to both plastids and the cytosol (Wachter et al. 2005). Therefore, the first step of glutathione synthesis, the formation of γ-glutamycysteine (γ-EC), takes place in plastids, and the second step of glutathione synthesis takes place in both chloroplasts and the cytosol (Wachter et al. 2005). Although most glutathione is retained in cytosol, it is also required in other organelles, including the nucleus, mitochondria, chloroplast, endoplasmic reticulum, vacuole, and apoplast. For example, in *Arabidopsis*, the glutathione imported into the nucleus during cell proliferation plays an important role in the whole-cell redox state (Vivancos et al. 2010). Each organelle shows a difference in the absolute concentration of glutathione and the oxidized to reduced glutathione ratios. Glutathione is also effluxed out of the cell, and acts as an antioxidant pool in the immediate environment of a cell, and for the inter-organ transport of glutathione (Bachhawat et al. 2013).

Glutathione is present in many cellular compartments at millimolar concentration levels (Diaz Vivancos et al. 2010; Koffler et al. 2013). It has been measured by different biochemical methods in different compartments after isolation and fractionation. By using these methods, glutathione has been detected at a concentration between 0.5 and 5 mM in the chloroplasts, between 1 and 3.52 mM in the cytosol, and 0.73 mM in the vacuoles (Krueger et al. 2009; Foyer and Halliwell 1976; Noctor et al. 2002).

Recently, the subcellular glutathione concentrations and volumes of subcellular compartments in different leaf areas of *Arabidopsis* have been determined by using quantitative immunogold electron microscopy combined with biochemical methods (Koffler et al. 2013). Vacuoles are the largest volume within a mesophyll cell and the size increases with leaf age (up to 80% in older leaves). Chloroplasts are the second largest organelle (up to 20% in younger leaves), followed by the nucleus (up to 2.3% in younger leaves), mitochondria (up to 1.6% in younger leaves), and peroxisomes (up to 0.3% in younger leaves). The highest concentrations of glutathione were found in mitochondria in a range between 8.7 and 15.1 mM (in younger and older leaves, respectively). The second highest amount of glutathione is in the nucleus (between 5.5 and 9.7 mM in the base and the center of younger leaves, respectively), followed by peroxisomes (between 2.6 mM in the edge of younger leaves and 4.8 mM in the base of older leaves, respectively) and cytosol (2.8 mM in the edge of younger and 4.5 mM in the center of older leaves, respectively). Chloroplasts contain rather low amounts of glutathione (between 1 mM and 1.4 mM). Vacuoles have the lowest concentrations of glutathione (0.01 and 0.14 mM), but show a great difference between different leaf areas. Differences in glutathione content between different leaf areas was only found in vacuoles and mitochondria, revealing that glutathione in the later cell organelle accumulated with leaf aging to concentrations of up to 15 mM and that the concentrations of glutathione in vacuoles are quite low in comparison to the other cell compartments (Koffler et al. 2013). This difference of intracellular glutathione content implies that transport of glutathione is required between organelles. The existence of glutathione transport across membranes such as the plasma membranes, the tonoplast, and the chloroplast envelopes has been established (Schneider et al. 1992; Jamai et al. 1996; Foyer 2001; Noctor et al. 2002; Pasternak et al. 2008). In *Arabidopsis*, three glutathione transporters (CLT1, CLT2, and CLT3) are essential for glutathione transport between chloroplasts and cytosol (Maughan et al. 2010). Rice putative OsCLT1 has been also described (Yang et al. 2016). These CLT transporters are intracellular glutathione transporters. Other transporters of glutathione in plants are homologs from the oligopeptide transporter (OPT) family, including AtOPT4 (Zhang et al. 2016), AtOPT6 (Cagnac et al. 2004; Pike et al. 2009), BjGT1 (Bogs et al. 2003), ZmGT1 (Pang et al. 2010), and OsGT1 (Zhang et al. 2004), which are involved in the intercellular transport of glutathione (Koh et al. 2002; Cagnac et al. 2004; Zhang et al. 2004; Pike et al. 2009; Zhang et al. 2016). The transport of glutathione conjugates and oxidized glutathione into vacuoles in plants is also facilitated by the transporters of the ATP-binding cassette (ABC) family, including AtMRP1 and AtMRP2 (Lu et al. 1998).

3 Long-Distance Transport of Glutathione

It has long been known that glutathione is the major form of the transported reduced sulfur in plants (Rennenberg et al. 1979). At the onset of grape (*Vitis vinifera* L.) ripening, glutathione content increased in grape berries, and declined in leaves, which suggests that fruit imports large amounts of glutathione from the leaves (Adams and Liyanage 1993). In transgenic poplar plants overexpressing γ-ECS, the elevated glutathione biosynthesis also enhanced glutathione concentration in the leaves of three lines (Arisi et al. 1997) and in both xylem and phloem as well as in leaves and roots (Herschbach et al. 2000). Radioactively labeled glutathione (S^{35}-GSH) of a spruce needle indicated that glutathione was exchanged between phloem and xylem in both directions (Schneider et al. 1994; Rennenberg et al. 1979). For example, maize scutella can export glutathione to developing roots and shoots (Rauser et al. 1991), whereas mature *Ricinus* leaves synthesize and export glutathione via phloem (Bonas et al. 1982).

The shoot-specific expression of the key enzyme (GSH1) of glutathione biosynthesis in the *Arabidopsis* glutathione-deficient mutant restored the levels of all thiols [γ-glutamylcysteine (γ-EC), glutathione and phytochelatins (PCs)] in roots (Li et al. 2006). In the phloem and xylem sap of *Brassica napus*, large quantities of PCs and glutathione are present, and are sufficient to form stable complexes with cadmium (Cd). The high ratios of [PCs]/[Cd] and [glutathione]/[Cd] in the phloem sap suggest that PCs and glutathione can function as long-distance carriers of Cd (Mendoza-Cozatl et al. 2008). Glutathione also repressed the sulfur influx and sulfur reduction by ATP sulfurylase (Lappartient et al. 1999). When glutathione biosynthesis was increased, root sulfate uptake was proportionally enhanced (Herschbach et al. 2000). Taking together, these results suggest that glutathione is a major form of long-distance transport for sulfur.

Nevertheless, identification of transporters for the long-distance transport of glutathione is progressing slowly. Only AtOPT4 was shown to transport glutathione in the siliques. Loss of function of *AtOPT4* caused lower glutathione content only in siliques, and *opt2opt4* double mutant showed a severe decrease of glutathione content in siliques (Zhang et al. 2016). Although other glutathione transporters (AtOPT6, BjGT1, and OsGT1) are mainly expressed in vasculature tissue, their involvement in the long-distance transport of glutathione has not been demonstrated up to now.

4 Intercellular Glutathione Transport

The existence of glutathione transporters has been known for many years, and there are several early reports on the biochemical characterization of glutathione transport across the plasma membrane into different organelles in eukaryotes, including yeasts, plants, and animals. The investigation of glutathione uptake in plants has

Table 16.1 Overview of glutathione transporter genes in plants

Transporter	Species	Activity	Roles
CLT1 *CLT2* *CLT3*	*Arabidopsis*	Glutathione uptake in oocytes	The three CLTs are localized on chloroplasts, and efflux glutathione from the chloroplast to the cytoplasm. Mutants (*clt1*, *clt3* and *clt1clt2clt3*) showed lower cytosolic glutathione, and only *clt1clt3* and *clt1clt2clt3* showed increased Cd sensitivity
AtOPT4	*Arabidopsis*	Glutathione uptake in yeast; oligopeptide uptake in oocytes	Glutathione is only lower in siliques of *atopt4*
AtOPT6	*Arabidopsis*	Glutathione uptake in yeast; oligopeptide uptake in oocytes	*AtOPT6* is mainly expressed in dividing areas, and induced by the herbicide primisulfuron
AtMRP1 *AtMRP2*	*Arabidopsis*	GS-conjugates uptake with different preference in yeast	*AtMRP1* is mainly expressed in aerial organs, and *AtMRP2* is mainly expressed in roots
BjGT1	*Brassica juncea*	Glutathione uptake in yeast	*BjGT1* is highly expressed in leaves, and strongly induced by Cd in stems and leaves
OsCLT1	*Oryza sativa*	A CRT-like transporter	Decreased glutathione and γ-EC contents were detected in roots of *osclt1*, and *osclt1* is sensitive to arsenate (As)
OsGT1	*Oryza sativa*	Glutathione and GS conjugates uptake in yeast	*OsGT1* is induced by Cd
ZmGT1	*Zea mays*	Glutathione and GS conjugates uptake in yeast	*ZmGT1* is induced by xenobiotics atrazine

revealed the presence of different transport systems (Table 16.1) reflected by the different kinetic parameters. Early kinetic analysis of glutathione transport into tobacco cells revealed both high-affinity (*K*m = 17 μM) and low-affinity (apparent *K*m = 310 μM) systems in plants (Schneider et al. 1992). In protoplasts of broad bean (*Vicia faba*), glutathione uptake followed single-saturation kinetics with a *K*m of 0.4 mM, and two systems for GSSG uptake were observed (*K*m = 7 μM and 3.7 mM) (Jamai et al. 1996). The uptake of radiolabeled glutathione into bean protoplasts was inhibited by GSSG and GS-X, and GSSG uptake was also inhibited by glutathione and GS-X. Likewise, the uptake of metolachlor-GS was inhibited by GSSG. The plasma membrane of plant cells, therefore, contains a specific transport system for glutathione that imports GSSG and GS-X as well as glutathione (Jamai et al. 1994). Proton flux measurements and electrophysiological data indicated that glutathione and GSSG are taken up via a proton symporter. However, the ion fluxes accompanying GSSG uptake differed from those invoked by glutathione. These data suggest that several glutathione transport mechanisms exist in parallel in plant cells (Horemans et al. 2000). The molecular identification of the glutathione transporters in plants with a high affinity for glutathione is still lacking, while a few transporters with a low affinity for glutathione have been reported.

5 Intracellular Glutathione Transport

Glutathione has been found within most organelles of plant cells (Zechmann 2014). Import of glutathione into the nucleus was identified in *Arabidopsis* during cell proliferation (Vivancos et al. 2010). Biochemical studies of glutathione uptake into wheat chloroplasts using radiolabeled glutathione revealed the existence of both high-affinity and low-affinity glutathione transporter systems (Noctor et al. 2012; Queval et al. 2011). Isolated barley vacuoles could rapidly uptake oxidized glutathione, while the uptake of reduced glutathione by vacuoles was only marginal. The GSSG transport of vacuoles is strictly ATP-dependent and is a saturable process (Km 0.4 to 0.6 mM). The uptake of glutathione S-conjugate into vacuoles is mediated by a specific ATPase (Tommasini et al. 1993; Martinoia et al. 1993). In plants, only transporters for glutathione efflux from chloroplasts into cytosol, the CRT-like transporters (CLTs), have been identified (Maughan et al. 2010). Identification of glutathione transporter for other organelles is progressing slowly.

6 Glutathione Transporters from Plants

In plants, high-affinity and low-affinity systems for glutathione transport were shown in physiological studies (Schneider et al. 1992; Jamai et al. 1996). Below are the details of the characterized glutathione transporters in plants.

6.1 CLT1, CLT2, and CLT3 of Arabidopsis

Three chloroplast glutathione transporters have been identified in *Arabidopsis*. They belong to the chloroquine-resistance transporter (CRT)-like transporter family (named *CLTs*), and were demonstrated to function as glutathione transporters on chloroplasts. *AtCLT1* was firstly cloned with complementation for the severely glutathione-deficient *glutathione1* mutant for BSO-resistance, while *AtCLT2* and *AtCLT3* were identified by homologous analysis. *Xenopus* oocytes expressing *AtCLT1* increased intracellular glutathione content three- to four-fold. *AtCLT1*, *AtCLT2*, and *AtCLT3* could all cause the accumulation of labeled glutathione in *Xenopus* oocytes. The expression of AtCLT proteins allows net transport of glutathione into oocytes, showing that all three these AtCLTs can mediate glutathione uptake (Maughan et al. 2010).

AtCLTs are localized to the plastids via plastid target signal peptide and are responsible for glutathione efflux from the chloroplast to the cytoplasm. Mutations of *AtCLT1* and *AtCLT3* led to lower cytoplasmic glutathione levels, susceptibility to microbial infections, and sensitivity to Cd (Maughan et al. 2010). However, these mutants did not exhibit growth difference to wild type under routine growth conditions. Glutathione levels in the roots of double (*clt1clt3*) and tri-mutants (*clt1clt-2clt3*) were decreased around fourfold compared to wild type (WT). In situ labeling of cytosolic glutathione showed that cytosolic glutathione is decreased in tri-mutant root cells, which may suggest that *AtCLTs* function in maintaining the root glutathione pool (Maughan et al. 2010). Single *clt* mutants showed no sensitivity to Cd, only *clt1clt3* and *clt1clt2clt3* showed increased Cd sensitivity, suggesting that the three genes function redundantly in plant cells.

6.2 *OsCLT1 of Rice*

OsCLT1 is a causal gene for an arsenate (As)-sensitive mutant of rice (Yang et al. 2016). It encodes a CRT-like transporter localized to the envelope membrane of plastids (Yang et al. 2016). The glutathione and γ-EC contents in the roots of *osclt1* and RNA-interference lines were decreased markedly compared with the WT plants. The concentrations of phytochelatin (c-EC)2-G (PC2) in *osclt1* roots were only 32% and 12% of that in wild type after arsenate [As(V)] and arsenite [As(III)] treatments, respectively. *OsCLT1* mutation resulted in lower As accumulation in roots but higher As accumulation in shoots when exposed to As(V). Under As(III) treatment, *osclt1* accumulated a lower As concentration in roots but a similar As concentration in shoots to wild type. Further analysis showed that the reduction of As(V) to As(III) was decreased in *osclt1*, which was also hypersensitive to Cd. These data indicate that *OsCLT1* plays an important role in glutathione homeostasis, probably by mediating the export of γ-EC and glutathione from plastids to cytosol, which in turn affects As and Cd detoxification in rice. However, its glutathione transport activity has not been confirmed experimentally yet.

6.3 *AtOPT4 of Arabidopsis*

AtOPT4 was screened from all of the nine *Arabidopsis* oligopeptide transporter (OPT) genes (*AtOPT1–9*) using a yeast sulfur-amino-acid auxotrophic mutant strain *met15 opt1*, which is defective in methionine (MET) synthesis and glutathione uptake (Zhang et al. 2016). The 9 *AtOPT* genes were cloned into the *pYES2* expression vector and were transformed individually into the yeast *met15 opt1* mutant. Transformed strains were then used for growth complementation assays on media containing glutathione as the sole sulfur source. Only *AtOPT4* was able to rescue the growth of the *met15 opt1* mutant in the presence of glutathione as the sole sulfur source. [^{35}S]GSH uptake by AtOPT4 followed Michaelis–Menten kinetics with an apparent affinity constant (K_m) of 1.4 ± 0.3 mM glutathione and a maximum transport rate (V_{max}) of 2.1 ± 0.3 nmol (mg protein)$^{-1}$ min^{-1}. Their data suggest that AtOPT4 is a low-affinity glutathione transporter. Subcellular analysis showed that AtOPT4-GFP is localized at the plasma membrane of the transiently expressed onion epidermal cell. The glutathione content was not significantly different in roots, stems, and leaves of either *atopt4* mutant or *atopt2/atopt4* mutant compared with the wild-type lines. However, the glutathione content in siliques was consistently lower in *atopt4* mutant (93%) or *atopt2/atopt4* mutant (98%) compared with the wild-type lines. The glutathione content in the siliques of *atopt2/atopt4* mutants was decreased, suggesting a role of AtOPT2 and AtOPT4 in silique glutathione transport.

AtOPT4 is preferentially expressed throughout the vasculature of rosette leaves, stems, and roots (Stacey et al. 2006), suggesting that *AtOPT4* may function broadly in mediating the long-distance transport of glutathione to various tissues in plants. When *atopt4* and *atopt2* mutant seedlings were germinated and grown on media supplemented with 40 mM Cd or 5 mM As(III), no visible growth defects were observed. No significant differences were observed in root growth between the wild-type and the *atopt2/atopt4* double mutant lines when germinated and grown on 40 mM Cd or 5 mM As(III). *AtOPT4* encodes a low-affinity plasma membrane glutathione transporter, which contributes to glutathione loading/unloading in siliques.

AtOPT4 was also identified as a high-affinity oligopeptide transporter (OPT) (Osawa et al. 2006). When it was expressed in yeast strain BY4730, AtOPT4 mediated oligopeptide KLGL transport at pH 5.0. It could also transport other tetrapeptides (GGFM, YGGFM, and IIGLM). When it was expressed in oocytes, *Sc5′UTR*::*AtOPT4* exhibited pronounced inward currents in response to GGFL, GGFM, and KLGL at pH 5.0. The apparent $K_{0.5}$ of AtOPT4 for KLGL was 15 μM at −80 mV, which was significantly lower than the $K_{0.5}$ of ScOPT1 for GGFL or even glutathione and PC_2. However, Osawa et al. (2006) did not observe glutathione transport activities in 6-hour yeast experiments. Specific growth conditions may be needed for the identification of AtOPT4 as it is a low-affinity glutathione transporter (Zhang et al. 2016).

6.4 AtOPT6 of Arabidopsis

Similar to AtOPT4, AtOPT6 is both a glutathione transporter and an oligopeptide transporter (Cagnac et al. 2004; Pike et al. 2009). It was firstly identified as a dual-affinity glutathione transporter (Cagnac et al. 2004). *AtOPT6* could restore the growth of the *hgt1* yeast mutant (Bourbouloux et al. 2000) on a medium containing reduced glutathione as the sole sulfur source and induced the uptake of [^{3}H]GSH. In yeast, AtOPT6-dependent glutathione uptake was mediated by a high-affinity (K_m = 400 μM) and a low-affinity (K_m = 5 mM) phase. AtOPT6 may also transport Cd, Cd/GSH conjugate, and glutathione-*N*-ethylmaleimide conjugate (Cagnac et al. 2004). However, when it was expressed in *Xenopus laevis* oocytes, AtOPT6 displayed low affinity for glutathione (Pike et al. 2009). AtOPT6 displayed high affinity for penta- and dodeca-peptides in *Xenopus laevis* oocytes (Pike et al. 2009). Oocytes expressing AtOPT6 exhibited significantly more inward current induced by glutathione at pH 5 than at pH 6 or pH 7, indicating that AtOPT6 is a proton-coupled low-affinity glutathione transporter. Glutathione-evoked currents were saturated between 1.25 and 5 mM glutathione with a $K_{0.5}$ > 500 μM. AtOPT6 is also an oligopeptide transporter with exceptionally broad substrate specificity along with glutathione (Pike et al. 2009). AtOPT6 can also transport plant signaling peptides, including the amino acid conjugate jasmonate-isoleucine, the active form of the oxylipin signal jasmonic acid, and the sulfated pentapeptide hormone phytosulfokine. In AtOPT6-expressing oocytes, AtOPT6 could transport hormone phytosulfokines: AtCLE5 (RVSPGGPDPQHH), AtCLE19 (RVIPTGPNPLHN), and HgCLE (RLSPSGPDPHHH) from *Heterodera glycines*, and Mi16D10p13 (GKKPSGPNPGGNN) from *Meloidogyne incognita* .

AtOPT6 is mainly expressed in the dividing areas of the plant. Its expression was strongly induced by the herbicide primisulfuron and, to a lesser extent, by abscisic acid but not by Cd (Cagnac et al. 2004). AtOPT6 may be a prime candidate for being the transporter that would be needed for glutathione uptake into phloem companion cells or to permit the uptake of transported glutathione into cells in other parts of plants. Promoter-GUS experiments showed that *AtOPT6* is expressed throughout the vasculature of Arabidopsis seedlings. In adult plants, it is expressed in the cambial zone of petioles and stem vascular bundles and in the vasculature of leaves and ovaries (Cagnac et al. 2004; Stacey et al. 2006). AtOPT6 located on the plasma membrane of specific cell types are needed to support the function of AtOPT6 in the long-distance movement of glutathione. The transport of longer peptides by AtOPT6 that are involved in plant development or nematode pathogenicity, AtCLE19p12, AtCLE5p12, HgCLEp12, and Mi16D10p13, suggests that AtOPT6 may play a role in plant signaling (Pike et al. 2009).

6.5 *BjGT1 of Brassica juncea*

BjGT1 was cloned and functionally characterized from *Brassica juncea,* a plant that may be used for phytoremediation (Bogs et al. 2003). *BjGT1* cDNA showed homology with the high affinity glutathione transporter *HGT1* from *Saccharomyces cerevisiae* (Bourbouloux et al. 2000). When expressed in the *S. cerevisiae hgt1*Δ strain, *BjGT1* could restore growth of the mutant on the medium with glutathione as the only sulfur source and mediated the uptake of [^{3}H]GSH. The glutathione uptake mediated by BjGT1 in the yeast ABC822 strain was pH-dependent and the optimal pH was at pH 5.0. However, Bogs et al. (2003) did not determine Km of BjGT1, and their data showed that the initial rate of glutathione uptake mediated by BjGT1 did not follow simple Michaelis–Menten kinetics.

BjGT1, which is highly expressed in leaves, only slightly in stems, and not at all in roots, was strongly induced by Cd in stems and leaves. The strong regulation of *BjGT1* by Cd suggests a role of this glutathione transporter during heavy metal exposure (Bogs et al. 2003).

6.6 *OsGT1 of Rice*

OsGT1, a glutathione transporter of rice (*Oryza sativa* L.), was screened from a Cd-induced rice cDNA library with yeast *HGT1* (Zhang et al. 2004). *OsGT1* could restore the growth of yeast *hgt1* mutant on a medium containing glutathione as the sole sulfur source. The yeast strain expressing *OsGT1* mediated [^{3}H]GSH uptake, and this uptake was significantly competed by not only unlabeled GSSG and GS conjugates but also some amino acids and peptides, suggesting that OsGT1 has a wide substrate specificity. Two saturable phases of glutathione uptake mediated by OsGT1 are apparent with *K*m values of about 400 μM and 23 mM. *OsGT1* is only weakly expressed in various parts of rice plants, and is induced by Cd. Its function may be to retrieve GSSG, GS conjugates, and peptides under biotic or abiotic stress (Zhang et al. 2004).

6.7 *ZmGT1 of Maize*

ZmGT1 was cloned from maize (*Zea mays*) using RACE-PCR. The deduced ZmGT1 protein is highly homologous to glutathione transporters from other plants. *ZmGT1* could restore growth of the *hgt1Δ* mutant strain (Bourbouloux et al. 2000) on a medium containing glutathione as the sole sulfur source, and it is also involved in the uptake of the glutathione conjugate GS-N-ethylmaleimide (GS-NEM). [^{14}C] GS-NEM uptake mediated by ZmGT1 in the ABC822 strain varied with the pH of the incubation medium. The uptake rate peaked at pH 5.0, and decreased at either a higher or a lower pH. *ZmGT1* was expressed in all organs of maize seedlings, with a higher level of transcripts being found in leaves. The expression of *ZmGT1* was strongly induced by atrazine with a four- to fivefold increase in the transcript level being detected in leaves after 96-hour treatment. The strong up-regulation of *ZmGT1* by atrazine suggests that this glutathione transporter may be involved in the detoxification of xenobiotics (Pang et al. 2010, 2012).

6.8 AtMRP1 and AtMRP2 of Arabidopsis

The two multispecific ABC transporters AtMRP1 and AtMRP2 from *Arabidopsis* can transport GS conjugates. AtMRP1 is a structural and functional homolog of the human multidrug resistance-associated protein (*HmMRP1*) gene product. It has been identified on the basis of its ability to confer GS conjugate pump activity on transfected cells (Muller et al. 1994). AtMRP2 has the facility for simultaneous high-efficiency parallel transport of GS conjugates and the *Brassica napus* chlorophyll catabolite transporter substrate (*Bn*-NCC-1) (Lu et al. 1998). These two genes exhibited differential patterns of expression in plants. AtMRP2 shows a high capacity for transport of glutathionated herbicides and anthocyanins. Heterogeneously expressed AtMRP2 not only has a GS conjugate transport activity that is several fold greater than that of AtMRP1, but it is also capable of high-affinity transport of *Bn*-NCC-1 (Lu et al. 1998).

7 Conclusion

The wide subcellular distribution of glutathione in plants is of great importance as this multifunctional metabolite is essential for plant development and growth. Thus, glutathione transporters must be present in membranes of all organelles in order to facilitate the import and export of glutathione. Although several inter- and intracellular glutathione transporters have been identified, further efforts need to be invested into the identification of intracellular long-distance glutathione transporters, as well as in the clarification of the biological functions of the known glutathione transporters.

Acknowledgements The research in MYZ's lab is supported by National Key Research and Development Program of China (2017YFD0100100), Guangdong Science and Technology Department of China (2015B020231009), Guangdong "Pearl River Talents Plan"-Postdoctoral Project, and Guangdong Agriculture Department of China (Yuenongji 201742).

References

Adams DO, Liyanage C (1993) Glutathione increases in grape berries at the onset of ripening. Am J Enol Vitic 44(3):333–338

Aller I, Meyer AJ (2013) The oxidative protein folding machinery in plant cells. Protoplasma 250(4):799–816. https://doi.org/10.1007/s00709-012-0463-x

Arisi AC, Noctor G, Foyer C, Jouanin L (1997) Modification of thiol contents in poplars (*Populus tremula* × *P. alba*) overexpressing enzymes involved in glutathione synthesis. Planta 203(3):362–372. https://doi.org/10.1007/s004250050202

Bachhawat AK, Thakur A, Kaur J, Zulkifli M (2013) Glutathione transporters. Biochim Biophys Acta-Gen Subj 1830(5):3154–3164. https://doi.org/10.1016/j.bbagen.2012.11.018

Ball L, Accotto GP, Bechtold U, Creissen G, Funck D, Jimenez A, Kular B, Leyland N, Mejia-Carranza J, Reynolds H, Karpinski S, Mullineaux PM (2004) Evidence for a direct link between glutathione biosynthesis and stress defense gene expression in Arabidopsis. Plant Cell 16(9):2448–2462. https://doi.org/10.1105/tpc.104.022608

Bogs J, Bourbouloux A, Cagnac O, Wachter A, Rausch T, Delrot S (2003) Functional characterization and expression analysis of a glutathione transporter, *BjGT1*, from *Brassica juncea*: evidence for regulation by heavy metal exposure. Plant Cell Environ 26(10):1703–1711. https://doi.org/10.1046/j.1365-3040.2003.01088.x

Bonas U, Schmitz K, Rennenberg H, Bergmann L (1982) Phloem transport of sulfur in *Ricinus*. Planta 155(1):82–88. https://doi.org/10.1007/Bf00402936

Bourbouloux A, Shahi P, Chakladar A, Delrot S, Bachhawat AH (2000) Hgt1p, a high affinity glutathione transporter from the yeast *Saccharomyces cerevisiae*. J Biol Chem 275(18):13259–13265. https://doi.org/10.1074/jbc.275.18.13259

Cagnac O, Bourbouloux A, Chakrabarty D, Zhang MY, Delrot S (2004) AtOPT6 transports glutathione derivatives and is induced by primisulfuron. Plant Physiol 135(3):1378–1387. https://doi.org/10.1104/pp.104.039859

Diaz Vivancos P, Wolff T, Markovic J, Pallardo FV, Foyer CH (2010) A nuclear glutathione cycle within the cell cycle. Biochem J 431:169–178. https://doi.org/10.1042/bj20100409

Foyer CH (2001) Prospects for enhancement of the soluble antioxidants, ascorbate and glutathione. Biofactors 15(2–4):75–78

Foyer CH, Halliwell B (1976) Presence of glutathione and glutathione reductase in chloroplasts – proposed role in ascorbic-acid metabolism. Planta 133(1):21–25. https://doi.org/10.1007/Bf00386001

Foyer CH, Theodoulou FL, Delrot S (2001) The functions of inter- and intracellular glutathione transport systems in plants. Trends Plant Sci 6(10):486–492. https://doi.org/10.1016/S1360-1385(01)02086-6

Gutierrez-Alcala G, Gotor C, Meyer AJ, Fricker M, Vega JM, Romero LC (2000) Glutathione biosynthesis in *Arabidopsis* trichome cells. Proc Natl Acad Sci USA 97(20):11108–11113. https://doi.org/10.1073/pnas.190334497

Herschbach C, van der Zalm E, Schneider A, Jouanin L, De Kok LJ, Rennenberg H (2000) Regulation of sulfur nutrition in wild-type and transgenic poplar over-expressing gamma-glutamylcysteine synthetase in the cytosol as affected by atmospheric H_2S. Plant Physiol 124(1):461–473. https://doi.org/10.1104/pp.124.1.461

Horemans N, Foyer CH, Asard H (2000) Transport and action of ascorbate at the plant plasma membrane. Trends Plant Sci 5(6):263–267. https://doi.org/10.1016/S1360-1385(00)01649-6

Jamai A, Chollet JF, Delrot S (1994) Proton-peptide cotransport in broad bean leaf tissues. Plant Physiol 106(3):1023–1031

Jamai A, Tommasini R, Martinoia E, Delrot S (1996) Characterization of glutathione uptake in broad bean leaf protoplasts. Plant Physiol 111(4):1145–1152

Klapheck S, Zopes H, Levels HG, Bergmann L (1988) Properties and localization of the homoglutathione synthetase from *Phaseolus-Coccineus* leaves. Physiol Plant 74(4):733–739. https://doi.org/10.1111/j.1399-3054.1988.tb02045.x

Koffler BE, Bloem E, Zellnig G, Zechmann B (2013) High resolution imaging of subcellular glutathione concentrations by quantitative immunoelectron microscopy in different leaf areas of *Arabidopsis*. Micron 45:119–128. https://doi.org/10.1016/j.micron.2012.11.006

Koh S, Wiles AM, Sharp JS, Naider FR, Becker JM, Stacey G (2002) An oligopeptide transporter gene family in Arabidopsis. Plant Physiol 128(1):21–29. https://doi.org/10.1104/pp.010332

Krueger S, Niehl A, Martin MC, Steinhauser D, Donath A, Hildebrandt T, Romero LC, Hoefgen R, Gotor C, Hesse H (2009) Analysis of cytosolic and plastidic serine acetyltransferase mutants and subcellular metabolite distributions suggests interplay of the cellular compartments for cysteine biosynthesis in Arabidopsis. Plant Cell Environ 32(4):349–367. https://doi.org/10.1111/j.1365-3040.2009.01928.x

Lappartient AG, Vidmar JJ, Leustek T, Glass ADM, Touraine B (1999) Inter-organ signaling in plants: regulation of ATP sulfurylase and sulfate transporter genes expression in roots mediated by phloem-translocated compound. Plant J 18(1):89–95. https://doi.org/10.1046/j.1365-313X.1999.00416.x

Li Y, Dankher OP, Carreira L, Smith AP, Meagher RB (2006) The shoot-specific expression of gamma-glutamylcysteine synthetase directs the long-distance transport of thiol-peptides to roots conferring tolerance to mercury and arsenic. Plant Physiol 141(1):288–298. https://doi.org/10.1104/pp.105.074815

Lu YP, Li ZS, Drozdowicz YM, Hortensteiner S, Martinoia E, Rea PA (1998) AtMRP2, an Arabidopsis ATP binding cassette transporter able to transport glutathione S-conjugates and chlorophyll catabolites: functional comparisons with AtMRP1. Plant Cell 10(2):267–282. https://doi.org/10.1105/tpc.10.2.267

Martinoia E, Grill E, Tommasini R, Kreuz K, Amrhein N (1993) ATP-dependent glutathione S-conjugate export pump in the vacuolar membrane of plants. Nature 364(6434):247–249. https://doi.org/10.1038/364247a0

Maughan SC, Pasternak M, Cairns N, Kiddle G, Brach T, Jarvis R, Haas F, Nieuwland J, Lim B, Muller C, Salcedo-Sora E, Kruse C, Orsel M, Hell R, Miller AJ, Bray P, Foyer CH, Murray JAH, Meyer AJ, Cobbett CS (2010) Plant homologs of the *Plasmodium falciparum* chloroquine-resistance transporter, PfCRT, are required for glutathione homeostasis and stress responses. Proc Natl Acad Sci U S A 107(5):2331–2336. https://doi.org/10.1073/pnas.0913689107

Mendoza-Cozatl DG, Butko E, Springer F, Torpey JW, Komives EA, Kehr J, Schroeder JI (2008) Identification of high levels of phytochelatins, glutathione and cadmium in the phloem sap of *Brassica napus*. A role for thiol-peptides in the long-distance transport of cadmium and the effect of cadmium on iron translocation. Plant J 54(2):249–259. https://doi.org/10.1111/j.1365-313X.2008.03410.x

Muller M, Meijer C, Zaman GJR, Borst P, Scheper RJ, Mulder NH, Devries EGE, Jansen PLM (1994) Overexpression of the gene encoding the multidrug resistance-associated protein results in increased ATP-dependent glutathione S-conjugate transport. Proc Natl Acad Sci U S A 91(26):13033–13037. https://doi.org/10.1073/pnas.91.26.13033

Noctor G, Gomez L, Vanacker H, Foyer CH (2002) Interactions between biosynthesis, compartmentation and transport in the control of glutathione homeostasis and signalling. J Exp Bot 53(372):1283–1304. https://doi.org/10.1093/jexbot/53.372.1283

Noctor G, Mhamdi A, Chaouch S, Han Y, Neukermans J, Marquez-Garcia B, Queval G, Foyer CH (2012) Glutathione in plants: an integrated overview. Plant Cell Environ 35(2):454–484. https://doi.org/10.1111/j.1365-3040.2011.02400.x

Osawa H, Stacey G, Gassmann W (2006) ScOPT1 and AtOPT4 function as proton-coupled oligopeptide transporters with broad but distinct substrate specificities. Biochem J 393:267–275. https://doi.org/10.1042/Bj20050920

Pang S, Li X-F, Liu Z, Wang C-J (2010) ZmGT1 transports glutathione conjugates and its expression is induced by herbicide atrazine. Prog Biochem Biophys 37(10):1120–1127. https://doi.org/10.3724/sp.j.1206.2010.00188

Pang S, Ran ZJ, Liu ZQ, Song XY, Duan LS, Li XF, Wang CJ (2012) Enantioselective induction of a glutathione-S-transferase, a glutathione transporter and an abc transporter in maize by metolachlor and its (S)-isomer. PloS One 7(10). https://doi.org/10.1371/journal.pone.0048085

Pasternak M, Lim B, Wirtz M, Hell R, Cobbett CS, Meyer AJ (2008) Restricting glutathione biosynthesis to the cytosol is sufficient for normal plant development. Plant J 53(6):999–1012. https://doi.org/10.1111/j.1365-313X.2007.03389.x

Pike S, Patel A, Stacey G, Gassmann W (2009) Arabidopsis OPT6 is an oligopeptide transporter with exceptionally broad substrate specificity. Plant Cell Physiol 50(11):1923–1932. https://doi.org/10.1093/pcp/pcp136

Queval G, Jaillard D, Zechmann B, Noctor G (2011) Increased intracellular H_2O_2 availability preferentially drives glutathione accumulation in vacuoles and chloroplasts. Plant Cell Environ 34(1):21–32. https://doi.org/10.1111/j.1365-3040.2010.02222.x

Rauser W, Schupp R, Rennenberg H (1991) Cysteine, γ-glutamylcysteine, and glutathione levels in maize seedlings. Plant Physiol 97(1):128–138. https://doi.org/10.1104/pp.97.1.128

Rennenberg H, Schmitz K, Bergmann L (1979) Long-distance transport of sulfur in *Nicotiana-tabacum*. Planta 147(1):57–62. https://doi.org/10.1007/Bf00384591

Schneider A, Martini N, Rennenberg H (1992) Reduced glutathione (GSH) transport into cultured tobacco cells. Plant Physiol Biochem 30(1):29–38

Schneider A, Schatten T, Rennenberg H (1994) Exchange between phloem and xylem during long-distance transport of glutathione in spruce trees (*Picea-Abies* [Karst] L). J Exp Bot 45(273):457–462. https://doi.org/10.1093/jxb/45.4.457

Stacey MG, Osawa H, Patel A, Gassmann W, Stacey G (2006) Expression analyses of Arabidopsis oligopeptide transporters during seed germination, vegetative growth and reproduction. Planta 223(2):291–305. https://doi.org/10.1007/s00425-005-0087-x

Tommasini R, Martinoia E, Grill E, Dietz KJ, Amrhein N (1993) Transport of oxidized glutathione into barley vacuoles – evidence for the involvement of the glutathione-S-conjugate ATPase. Z Naturforsch C 48(11–12):867–871

Vernoux T, Wilson RC, Seeley KA, Reichheld JP, Muroy S, Brown S, Maughan SC, Cobbett CS, Van Montagu M, Inze D, May MJ, Sung ZR (2000) The *ROOT MERISTEMLESS1/CADMIUM SENSITIVE2* gene defines a glutathione-dependent pathway involved in initiation and maintenance of cell division during postembryonic root development. Plant Cell 12(1):97–109. https://doi.org/10.1105/tpc.12.1.97

Vivancos PD, Dong YP, Ziegler K, Markovic J, Pallardo FV, Pellny TK, Verrier PJ, Foyer CH (2010) Recruitment of glutathione into the nucleus during cell proliferation adjusts whole-cell redox homeostasis in *Arabidopsis thaliana* and lowers the oxidative defence shield. Plant J 64(5):825–838. https://doi.org/10.1111/j.1365-313X.2010.04371.x

Wachter A, Wolf S, Steininger H, Bogs J, Rausch T (2005) Differential targeting of GSH1 and GSH2 is achieved by multiple transcription initiation: implications for the compartmentation of glutathione biosynthesis in the *Brassicaceae*. Plant J 41(1):15–30. https://doi.org/10.1111/j.1365-313X.2004.02269.x

Yang J, Gao MX, Hu H, Ding XM, Lin HW, Wang L, Xu JM, Mao CZ, Zhao FJ, Wu ZC (2016) OsCLT1, a CRT-like transporter 1, is required for glutathione homeostasis and arsenic tolerance in rice. New Phytol 211(2):658–670. https://doi.org/10.1111/nph.13908

Zechmann B (2014) Compartment-specific importance of glutathione during abiotic and biotic stress. Fron Plant Sci 5. https://doi.org/10.3389/fpls.2014.00566

Zechmann B, Mauch F, Sticher L, Muller M (2008) Subcellular immunocytochemical analysis detects the highest concentrations of glutathione in mitochondria and not in plastids. J Exp Bot 59(14):4017–4027. https://doi.org/10.1093/jxb/ern243

Zechmann B, Koffler BE, Russell SD (2011) Glutathione synthesis is essential for pollen germination *in vitro*. BMC Plant Biol 11. https://doi.org/10.1186/1471-2229-11-54

Zhang MY, Bourbouloux A, Cagnac O, Srikanth CV, Rentsch D, Bachhawat AK, Delrot S (2004) A novel family of transporters mediating the transport of glutathione derivatives in plants. Plant Physiol 134(1):482–491. https://doi.org/10.1104/pp.103.030940

Zhang Z, Xie Q, Jobe TO, Kau AR, Wang C, Li Y, Qiu B, Wang Q, Mendoza-Cozatl DG, Schroeder JI (2016) Identification of AtOPT4 as a plant glutathione transporter. Mol Plant 9(3):481–484. https://doi.org/10.1016/j.molp.2015.07.013

Chapter 17
Importance of Glutathione in the Legume-Rhizobia Symbiosis

Eliana Bianucci, Ana Furlan, and Stella Castro

Abstract Glutathione (GSH) is essential for the proper development of root nodules during the symbiotic association of legume and rhizobia. It is involved in the antioxidant defense, the detoxification of xenobiotics, and the tolerance to abiotic and biotic stresses. The high level of GSH in root nodules and the presence of an active ascorbate-glutathione (AsA-GSH) cycle suggest that GSH participates in the protection of the nitrogen-fixing process against reactive oxygen species (ROS) resulting from the active nodule metabolism. Glutathione-related enzymes also play a critical role in defense against ROS: (a) glutathione peroxidase (GPX) is a H_2O_2 scavenger that uses GSH as a reductant, (b) glutathione reductase (GR) reduces GSSG using NADPH as a source of reducing power and maintaining the GSH/GSSG ratio in cells, (c) glutathione-S-transferase (GST) catalyzes the nucleophilic conjugation of GSH with several electrophilic substrates, and (d) glutaredoxins (GRXs), small redox proteins from the thioredoxin (TRX) superfamily, use GSH as electron donor. In this chapter, the role of GSH and its related enzymes was analyzed in free-living rhizobia and in the symbiosis with the legumes as well as the responses to different abiotic stresses (acid pH, saline, drought, and heavy metals/metalloids).

Keywords Glutathione • GSH-related enzymes • Rhizobia • Legumes • Biological nitrogen fixation • Abiotic stress

Contents

E. Bianucci • A. Furlan • S. Castro (✉)
Departamento de Ciencias Naturales, Facultad de Ciencias Exactas, Físico-Químicas y Naturales, Universidad Nacional de Río Cuarto,
Ruta 36 Km. 601, 5800 Río Cuarto, Córdoba, Argentina
e-mail: scastro@exa.unrc.edu.ar

M.A. Hossain et al. (eds.), *Glutathione in Plant Growth, Development, and Stress Tolerance*, https://doi.org/10.1007/978-3-319-66682-2_17

1 Introduction

Glutathione (γglutamylcysteine-glycine; GSH) is the most abundant intracellular thiol in all living cells and is known to be involved in many biological processes, including protein and DNA synthesis, cell transport, enzyme activity modulation, and cellular metabolism as well as defense against reactive oxygen species (ROS). In plants, GSH is a major water-soluble antioxidant and redox buffer, performing critical functions in cell cycle regulation, development, sulfur transport and storage, stress responses, and heavy metal detoxification (Maughan and Foyer 2006). The legume plants, which are capable of establishing symbiotic associations with rhizobia, have also a GSH homologue called homoglutathione (γglutamylcysteine-βalanine; hGSH) that may partially or completely replace GSH (Matamoros et al. 1999; Moran et al. 2000; Frendo et al. 2001; Matamoros et al. 2003).

The pathway for GSH synthesis is probably shared by all organisms and involves two ATP-dependent steps. In the first step, γglutamylcysteine (γGC) is formed from glutamate (Glu) and cysteine (Cys) by γglutamylcysteine synthetase (γGCS), and in the second step, glycine (Gly) is added to the C-terminal site of γGC by glutathione synthetase (GSHS). It is known that the first enzyme is the rate-limiting step and is subject to feedback inhibition by GSH (Meister and Anderson 1983). In legumes, the synthesis of hGSH shares the same first enzyme γGCS which produces the dipeptide γGC and then requires a specific homoglutathione synthetase (hGSHS) which catalyzes the addition of βalanine to γGC. The biochemical properties of the thiol synthetases (γGCS, GSHS, hGSHS) have been examined in several plants, but little is known about the regulation of the thiol biosynthetic pathway in legume roots and nodules. Interestingly, the hGSHS gene showed high sequence identity with the GSHS gene and probably formed by tandem duplication, as found in *Medicago truncatula* (Frendo et al. 2001) and *Lotus japonicus* (Matamoros et al. 2003).

In spite of the close relationship among GSHS and hGSHS genes, the expression of both depends on the legume species and tissue. Thus, hGSHS was detected in the roots and nodules and GSHS throughout the plant of *Medicago truncatula* (Frendo et al. 1999); meanwhile GSHS was detected in the nodules and hGSHS in leaves and roots of *Lotus japonicus* (Matamoros et al. 2003). These genes were also differentially regulated in response to signaling compounds or stress conditions. In roots of *Medicago truncatula*, the expression of the γGCS and GSHS genes, but not of the hGSHS gene, was induced by nitric oxide (Innocenti et al., 2007). In nodules of bean plants treated with hydrogen peroxide (H_2O_2), γGCS and hGSHS were upregulated; meanwhile treatments with cadmium (Cd), sodium chloride, or jasmonic acid had no effect (Loscos et al. 2008). Clemente et al. (2012) studied the subcellular localization of the γGCS, GSHS, and hGSHS enzymes by electron microscopy and the gene expression in response to different plant hormones. The findings revealed that the pathway for thiol synthesis was compartmentalized in legumes and a differential regulation of the three mRNA levels, hGSHS activity,

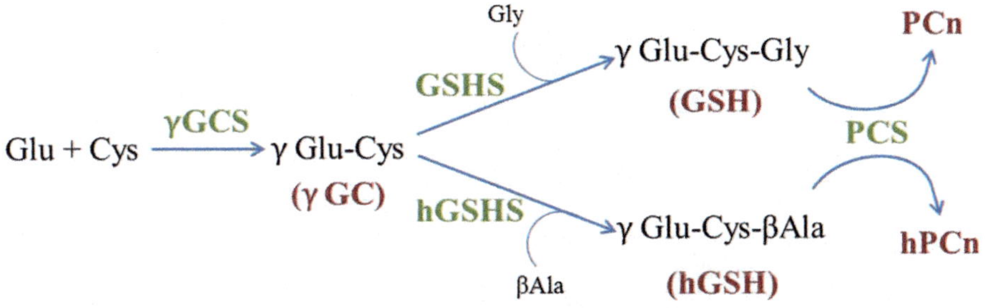

Fig. 17.1 Glutathione, homoglutathione, and phytochelatin synthesis in legume plants

and thiol contents by hormones was observed. These results indicated that GSH and hGSH play different roles in plant development and stress responses. Together, these findings suggest the presence of genes specific for cis-regulatory elements in the GSHS promoter and/or distinct regulatory mechanisms of the GSHS and hGSHS genes. In addition, it provides strength for a different role of GSH and hGSH in plants and especially in nodules showing that these thiols are essential in the nodulation process and biological nitrogen fixation (BNF).

Phytochelatins (PCs) are nonprotein thiols synthesized by phytochelatin synthase (PCS) from GSH and have the general structure, $(\gamma\text{Glu-Cys})_n$-X, where X is commonly glycine (Gly) but may be serine (Ser), glutamine (Gln,) or glutamate (Glu) and n = 2–11 (Zenk 1996; Cobbett and Goldsbrough 2002). The addition of βalanine (βAla) into the PCs general structure leads to hPC formation; however, PCS enzyme has a higher specificity for GSH than for hGSH (Becana et al. 2010). PCS enzyme activity is induced by metal(loid)s, such as cadmium (Cd), mercury (Hg), copper (Cu), arsenic (As), silver (Ag), nickel (Ni), gold (Au), or zinc (Zn), being the principal mechanism of intracellular metal detoxification by complexing and transporting metals into the vacuole (Rauser and Meuwly 1995; Shah and Nongkynrih 2007; Mendoza-Cózatl et al. 2011) (Fig. 17.1).

2 Glutathione in the Legume-Rhizobia Symbiosis

Legumes have ability to establish symbiotic associations with soil bacteria known as rhizobia, and as a result of a molecular dialog between both, in the root surface, a new organ called "nodule" is originated. In the first step, plant signal molecules (such as flavonoid, stachydrine, and aldonic acid) induce the biosynthesis of rhizobial lipochitooligosaccharides, which are termed "Nod factors," through the expression of *nod* genes (Stougaard 2000). These Nod factors alter the hormonal balance

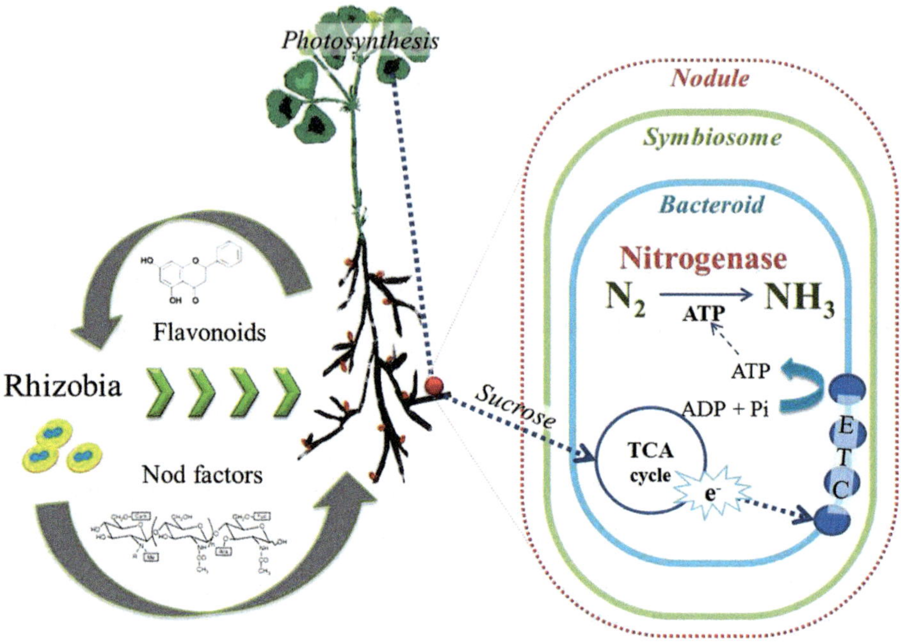

Fig. 17.2 Schematic representation of establishment of legume-rhizobia symbiosis and biological nitrogen-fixation process in nodules

of the plant in order to stimulate mitosis and to allow the development of the symbiosome that hosts the bacteria inside the plant (Ferguson and Mathesius 2003). The plant provides a unique microaerobic low-oxygen environment for the bacteria within the symbiosome that controls the expression of the bacterial nitrogen-fixation genes as well as cytochromes that requires these microaerobic conditions (Long 2001). Inside the nodules, the bacteroids carry out dinitrogen (N_2) fixation, a biological process in which atmospheric dinitrogen is reduced to ammonia by the nitrogenase enzyme complex (Vance 2008). The elevated costs of energy for nitrogen fixation derives ultimately from sucrose which enters the tricarboxylic acid (TCA) cycle and, afterward, the electron transport chain (ETC) that yields ATP. In return, the ammonia produced by the bacteroids is assimilated into organic compounds to fulfill the nitrogen demand of both the bacteria and the plant (Fig. 17.2).

Different researches have showed that GSH produced by both the plant and the bacteria have an important role in the establishment and maintenance of symbiosis (Moran et al. 2000; Matamoros et al. 2003; Frendo et al. 2005; Groten et al. 2005; Harrison et al. 2005). In order to know whether part of the GSH present in the nodule can be synthesized by the bacteria and whether the bacterial GSH pool can modify the nodulation and biological nitrogen-fixation (BNF) processes, studies of the bacterial genes involved in GSH synthesis were performed, and γGCS- and GSHS-defective mutant strains (due to the disruption of *gsh*A and *gsh*B genes, respectively) derived from wild-type rhizobia were constructed. Thus, Ricillo et al. (2000) reported that a mutation in the *gsh*B gene, encoding for GSHS of *Rhizobium tropici* CIAT899, affected its ability to compete for nodule occupancy during the process of bean nodulation. Harrison et al. (2005) found that, in the *Alfalfa-Sinorhizobium meliloti* symbiosis, a mutant strain (SmgshA), unable to synthesize GSH due to a disruption of *gsh*A gene that encodes γGCS enzyme, did not grow or nodulated alfalfa plants. In contrast, a *Sinorhizobium meliloti* mutant strain (SmgshB) with *gsh*B gene deleted, which encodes GSHS enzyme, was able to grow, indicating that γGC (dipeptide intermediate) could partially substitute GSH. However, the SmgshB strain showed a delayed-nodulation phenotype and a 75% reduction in the nitrogen-fixation capacity which was related to an abnormal nodule development with an early senescence process. Both mutant strains showed higher catalase (CAT) activity than the wild-type strain, indicating that they were under oxidative stress. Taken together, these results revealed that the bacterial GSH pool plays an essential role in the growth of *Sinorhizobium meliloti* and during its interaction with the legume.

In the peanut-*Bradyrhizobium* sp. symbiosis, Sobrevals et al. (2006) showed that GSH-deficient mutant strain *Bradyrhizobium* sp. SEMIA 6144S7Z (*gsh*A gene deleted) was capable to form effective nodules as the wild-type strain. However, when peanut plants were co-inoculated with mutant and wild-type strains most of the nodules obtained were occupied by the wild-type strain. Thus, the mutation in the *gsh*A gene appears to affect the bacterial ability to compete during the symbiotic process, possibly due to the lower growth rate of the mutant strain in comparison to the wild-type strain. Muglia et al. (2008) examined the importance of GSH in bean-*Rhizobium tropici* CIAT899 symbiosis, using a GSH-deficient mutant strain (*gsh*B gene deleted). Plants inoculated with the mutant strain presented a delayed nodulation and a reduction in the shoot dry weight, suggesting a decrease of nitrogen-fixation activity. Furthermore, bacterial *gsh*B expression was assayed in wild-type nodules at different steps of nodulation showing an increase in mature and early senescent nodules. Noteworthy, nodules formed by *gsh*B mutant strain presented an early senescent pattern, which was associated with increased levels of ROS such as superoxide accumulation. All these findings showed evidence of the role of bacterial GSH that results to be essential for competitiveness and effectiveness of rhizobia in the BNF process.

2.1 Development of the Root Nodules

In the symbiosis establishment, the plant genome determines the rhizobial entrance mechanism and the type of nodule that is going to form: "indeterminate" and "determinate." The indeterminate nodule has a persistent apical meristem that often yields a cylindrical or branched nodule structure. In contrast, the determinate nodule has no active meristem and thus has a rather different shape and structure to the indeterminate nodule. The bacteria housed in the symbiosomes in both determinate and indeterminate nodules can revert to free-living bacteria as the organs senesce (Müller et al. 2001). The following tissue types can be distinguished in both types of nodule from the periphery to the centre: an external nodule cortex, an endoderm, an internal cortex called the nodule parenchyma, and the central zone housing bacteria (Van de Wiel et al. 1990). Metabolic exchange between the nodule and the other organs of the plant is ensured by the presence of vascular bundles, localized within the nodule parenchyma that is connected to the root vascular system.

Matamoros et al. (1999) demonstrated that GSH/hGSH content and the γGCS, GSHS, and hGSHS activities were high in the meristematic and infected zones, exhibiting a higher hGSHS activity in the cortex of bean nodules. Probably, this specific distribution could be related to a function of this protein in the vascular bundles or in the O_2 diffusion barrier localized in the nodule cortex. These findings were corroborated, using promoter-GUS fusions, to determine the spatio-temporal gene expression of the GSH/hGSH synthesis in *Medicago truncatula.* It was found that the expression of γGCS appeared to be higher in the meristematic and infection zones; meanwhile the hGSHS mRNA was more abundant in the cortex and the GSHS mRNA in the cortex and in the nitrogen-fixing zone of nodules (El Msehli et al. 2011).

Vernoux et al. (2000) reported that GSH/hGSH played a role in the formation and the regulation of the nodule resulting to be very important in the functioning of root tip meristem. In addition, GSH/hGSH induced the expression of plant defense genes indicating that it could be involved in the regulation of the nodule number per plant (Wingate et al. 1988). Thiol concentrations as well as nitrogen-fixing activity were positively correlated during nodule development, and both parameters declined with advancing age (Groten et al. 2005) and during stress-induced senescence (Marino et al. 2007; Naya et al. 2007). These results suggested that both GSH and hGSH were of vital importance for nodule activity, a study that was proved by modulating the thiol contents by pharmacological and genetic approaches. For instance, the transcriptomic analysis of GSH-/hGSH-depleted plants during early nodulation revealed downregulation of genes implicated in meristem formation and upregulation of salicylic acid-related genes after infection with *Sinorhizobium meliloti* (Pucciariello et al. 2009).

The importance of both GSH and hGSH, during the first steps of symbiosis established between *Medicago truncatula* and *Sinorhizobium meliloti*, was examined using both buthionine sulfoximine (BSO), a specific inhibitor of GSH and hGSH synthesis, and transgenic roots expressing GSHS and hGSHS in an antisense orientation. The deficiency in GSH and hGSH synthesis inhibited the formation of root nodules without modification in the infection event number (Cook et al. 1995), indicating that the low level of GSH or hGSH did not alter the first steps of the infection process. By contrast, a strong decrease in the nascent nodule number and in the expression of the early nodulin genes was found in GSH- and hGSH-depleted plants, suggesting that these thiols were involved in the nodule meristem formation. Thus, GSH and hGSH seem to be vital for the proper development of the root nodules resulting from the symbiotic interaction (Frendo et al. 2005).

Bianucci et al. (2008) investigated the effect of the decrease of peanut GSH content in the symbiotic association of peanut-*Bradyrhizobium* sp. SEMIA6144 using BSO. The findings showed that BSO reduced root GSH content without changes in plant growth as well as the typical anatomic structure of the peanut roots in relation to 30-day-old control plants. Thus, the addition of BSO did not affect normal emergence of lateral roots in peanut; however, BSO-treated plants inoculated with wild-type or GSH-deficient mutant strains showed a significant reduction in the nodule number. Taking into account that there was no decrease in the amount of lateral roots in BSO-treated plants, the reduction in the nodule number could be caused by alterations in meristem formation induced by GSH depletion. In addition, a reduction in the nodule dry weight was found with a low GSH content, suggesting a correlation between nodule number and dry weight. As wild-type and GSH-deficient mutant strains of *Bradyrhizobium* sp. were able to nodulate peanut root, these findings infer that the plant GSH and not bacterial GSH appeared to be essential for the proper development of peanut root nodules during symbiotic interaction.

It is well known that ROS accumulate during nodule senescence (Becana and Klucas 1992; Alesandrini et al. 2003) leading to a declination of GSH and hGSH contents. In fact, senescence caused a decrease of the hGSH content in soybean and pea nodules; meanwhile a reduction of GSH content was observed in pea nodules with accumulation of catalytic Fe and oxidation of thiols, lipids, proteins, and DNA (Evans et al. 1999; Matamoros et al. 2003). Matamoros et al. (1999) also reported that the senescent zones of pea nodules have less GSH and hGSH than the meristematic and infected zones. In this sense, pea nodules exhibited a strong correlation between the nitrogen-fixation capacity and the GSH content (Groten et al. 2005). These data strongly suggest a key involvement of thiols in the different stages of nodule development from the first steps to senescence.

2.2 *Biological Nitrogen Fixation*

The high level of GSH/hGSH in root nodules and the presence of an active ascorbate-glutathione (AsA-GSH) cycle suggest that these molecules are involved in the protection of the nitrogen-fixing nodules against ROS, resulting from the high respiration rates in nodule metabolism (Becana et al. 2000). In addition, GSH/hGSH are substrates for antioxidant enzymes, such as glutathione reductases (GR), glutathione-S-transferases (GSTs), and glutaredoxins (GRXs), and therefore both thiols probably take place in the regulation of symbiosis via modulation of enzyme activities (Dalton et al. 2009). In relation to AsA-GSH cycle in nodules, it was more powerful in effective nodules than in ineffective ones, accentuating the protective role that this cycle can play for nitrogen fixation in soybean and alfalfa

(Dalton et al. 1993). Similar results were observed in the leghemoglobin-RNA interference (LbRNAi) lines of *Lotus japonicus*, which were altered in their nitrogen-fixation activity (Ott et al. 2005; Günther et al. 2007). The latter authors found that loss of leghemoglobin resulted in significantly lower H_2O_2 levels in nodules. Transcript levels and catalytic activities of AsA-GSH cycle enzymes involved in H_2O_2 detoxification as well as concentrations of reduced AsA were also altered in LbRNAi nodules. These results suggest that leghemoglobin, which is need for BNF, leads to ROS generation and an enhancing response of the AsA-GSH cycle in nodules.

Exploring the relation between nodule senescence and nitrogenase activity, Dalton et al. (1986) reported that nodule AsA and GSH contents began to decrease early in development, in a similar manner to nitrogenase activity, but the activities of most enzymes of the AsA-GSH cycle were not significantly altered during nodule senescence. Furthermore, the alteration of the nitrogen-fixing efficiency during senescence correlated with a general decrease of the antioxidant defense and to the detection of ROS species in the nodule senescent zone. Additionally, El Msehli et al. (2011) determined the importance of GSH/hGSH in BNF of *Medicago truncatula* using the promoter of nodule cysteine-rich family member 001 (NCR001), which was chosen to build genetic constructs allowing the modification of GSH/hGSH content in the nitrogen-fixing zone. The findings revealed that the increases and decreases in thiol contents were correlated with significant increases and decreases in BNF, respectively. Furthermore, BNF modifications were associated with changes in nodule gene expression levels where γGCS-deficient nodules were smaller than control nodules, indicating that γGCS deficiency disturbed nodule development. Taken together, these data support the idea that cellular redox state plays an important role in the regulation of nodule development and functioning that ultimately benefit the efficiency of the BNF process.

3 Role of Glutathione and Its Related Enzymes in the Defense Mechanism to Abiotic Stress

All types of antioxidants detected in plants have also been found in nodules including (a) enzymatic antioxidants involved in ROS detoxification such as superoxide dismutase (SOD), catalase (CAT), and peroxidases (PRXs) such as glutathione peroxidase (GPX); (b) nonenzymatic antioxidants such as AsA, GSH, and α-tocopherol; (c) the AsA-GSH cycle, which allows the reduction of the two antioxidant molecules by NAD(P)H; (d) glutathione-S-transferase (GST) that catalyzes the nucleophilic conjugation of GSH with several electrophilic substrates; and (e) the enzymes implicated in the disulfide reduction: glutathione reductase (GR), thioredoxins (TRXs), and glutaredoxins (GRXs). A special case is the enzyme GPX which is able to use TRX as a more efficient reducing substrate

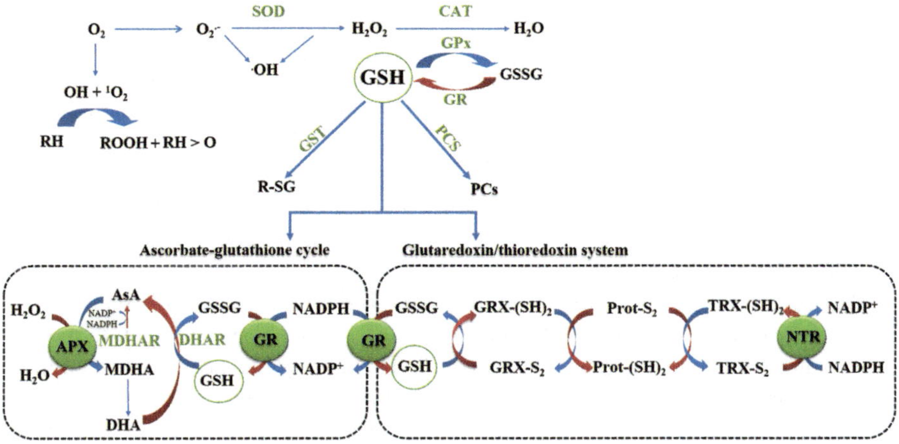

Fig. 17.3 Schematic representation of reactive oxygen species detoxification that involves glutathione. *APX* ascorbate peroxidase, *AsA* ascorbate, *CAT* catalase, *DHA* dehydroascorbate, *DHAR* dehydroascorbate reductase, *GSH* glutathione, *GPX* glutathione peroxidase, *GR* glutathione reductase, *GRX* glutaredoxin, *GSSG* reduced glutathione, *GST* glutathione S-transferase, H_2O_2 hydrogen peroxide, *MDHA* monodehydroascorbate, *MDHAR* monodehydroascorbate reductase, *NTR* NADPH-dependent thioredoxin reductase, $O_2^{\cdot-}$ superoxide anion, $\cdot OH$ hydroxil radical, $^1O_2$ singlet oxygen, *PCs* phytochelatins, *PCS* phytochelatin synthase, *RH* lipids or proteins, *ROOH* hydroperoxides, *R-SG* GSH-conjugated, *SOD* superoxide dismutase

than GSH. Thus, this enzyme could be considered as a peroxiredoxin (PRX) rather than GPX (Navrot et al., 2006), however, it is named in our work as GPX. The bacterial partner differentiated into bacteroid inside the nodule also contains a wide range of antioxidant systems (Fig. 17.3).

This section will focus on the involvement of the GSH and its related enzymes in the response of free-living rhizobia and in the symbiosis with the legumes to different abiotic stresses such as pH acid, saline, drought, and heavy metals/metalloids.

3.1 *Acid pH Stress*

Soil acidity is responsible for significant losses in global legume production, resulting from impaired plant and rhizobia growth, in addition to decreased nodule development and nitrogen fixation (Munns 1986; Evans et al. 1990). Interestingly, legumes tend to acidify soil to a greater extent than many other species. As the soil pH decreases, aluminum (Al^{3+}) is mobilized into the soil solution and may become toxic to plants and microorganisms. Ponsone et al. (2004) demonstrated the effect of the combination of acid pH (5.5) and 50 μM Al^{3+} on two nodulating peanut rhizobia: *Bradyrhizobium* sp. SEMIA6144 and the native isolate NCHAX from Córdoba (Argentina) soils. *Bradyrhizobium* sp. SEMIA6144 revealed an increase of GSH content; however, it was not enough to prevent the significant decrease of bacterial growth rate. In contrast, the isolate NCHAX growth rate and viability were not affected by the stress condition, and GSH content remained unchanged. Probably, GSH content of the native isolate could be enough to prevent the negative effects imposed by this stress allowing to maintain redox state. These findings show that the exposition of each bradyrhizobial strains to different abiotic stress induces diverse responses regarding GSH content.

Results from several studies have indicated that tolerance to acid conditions in rhizobia is often correlated with the ability of strains to maintain internal pH approaching neutrality (Graham et al. 1992). This ability has been suggested to be due to different processes or molecules including GSH. Thus, *Rhizobium tropici gsh*B mutant, which contains only 3% of the GSH present in the wild-type strain, was sensitive to several abiotic stresses, and the addition of GSH restored the responses to wild-type strain levels. This mutant could form effective nodules on bean, but it was outcompeted by the wild-type strain, indicating that GSH was important for stress tolerance as well as the symbiotic process (Ricillo et al. 2000). Moreover, *gsh*B gene expression was found to be induced under acid stress conditions (Muglia et al. 2007). Taking into consideration the acid pH of the peribacteroid space of nitrogen-fixing pea and bean nodules (Krylova et al. 2007), it is possible to suggest that the acidic environment can be involved in *gsh*B gene activation.

In peanut microsymbionts, Sobrevals et al. (2006) revealed that the mutant strain *Bradyrhizobium* sp. SEMIA6144-S7Z, which contains only 4% of the GSH present in the wild-type strain, was unable to grow under different stressing conditions (pH 5.5; 0.3 M NaCl; 0.5 mM H_2O_2) in a minimal medium (MSM). The effect on the mutant strain growth cannot be reverted supplementing exogenous GSH. On the other hand, GSH content of wild-type strain increased when bacterial grew under stressing conditions, demonstrating that GSH had a role in the protection against acid, osmotic, and oxidative stresses. These results support the view that GSH, even though not critical for in vitro growth, confers protection against abiotic stress. An interesting remarkable situation is that the responses to acidity are shared by the already-mentioned abiotic stresses; therefore, these studies are very valuable for the study of legume-rhizobia interaction.

3.2 Saline Stress

Soil salinity is one of the most important abiotic stressors that negatively affects plant growth and agricultural productivity. Nitrogen fixation by legumes is very sensitive to saline stress, which can severely reduce crop productivity and their soil-enriching capacity. Generally, high salinity can disturb essential physiological processes by inducing water deficits, ion imbalance, hyperosmotic stress, nutritional imbalance, metabolic disorders, and even death (Shrivastava and Kumar 2015). To cope with saline stress, plants have evolved complex defense strategies. These include the upregulation of antioxidant enzymes and antioxidants, energy metabolism modifications, and the appearance or disappearance of some proteins (Rubio et al. 2009). These authors revealed that *Lotus japonicus* was more tolerant to saline stress than other legumes, which can be attributed to the capacity of the plant to prevent Na^+ translocation to the shoot and the activation of antioxidant defenses such as SOD and GPX genes.

Studies carried out in transgenic *Medicago truncatula* plants expressing a cyanobacterial flavodoxin, a protein involved in the response to oxidative stress, showed that salt-stressed nodules of transgenic plants had a significantly higher GSH concentration than stressed wild-type nodules. Conversely, the GSSG concentration was increased by salt stress but less so in transgenic plants than in stressed wild-type nodules. Thus, the GSH/GSSG ratio was a parameter indicative of the oxidative balance in nodules, and it decreased notably in both types of nodules when subjected to saline stress (Coba de la Peña et al. 2010). On the other hand, alfalfa plants with active nodules showed higher survival rate associated with reduced lipid peroxidation; higher activities of SOD, CAT, PRX, and APX; as well as higher concentrations of GSH and soluble sugar in comparison to plants with inactive nodules or non-inoculated. These results suggest that alfalfa-*Rhizobium meliloti* symbiosis has a positive effect on saline tolerance by improving the activity of antioxidant enzymes and osmotic adjustment capacity (Wang et al. 2016).

3.3 Drought Stress

Drought stress is among the most severe abiotic stresses to which legumes are exposed in the producing areas. A common response in plants is the accumulation of abscisic acid (ABA) which promotes the stomatal closure, low CO_2 fixation, and accumulation of ROS. Legumes are not extent of this response; therefore, they activate their antioxidant system in an intricate way that is regulated at various levels. In this sense, Sengupta et al. (2012) indicated a stable expression pattern of the gene that encodes γGCS in the roots of *Vigna radiata* during progressive drought stress and recovery. However, the enzyme activity was altered differentially upon exposition to water-deficit conditions and recovery, reflecting the existence of some post-transcriptional or posttranslational regulatory system for the enzyme. The authors performed linear regression analysis between H_2O_2 and lipid peroxidation as well as H_2O_2 and enzyme activity during drought stress and recovery demonstrating interrelationships and putative regulatory mechanisms in the root system of *Vigna radiata* under adverse conditions.

As key antioxidants, AsA and GSH can be used as biochemical markers of general cell redox state (Noctor et al. 2014). For instance, leaf GSH status and intracellular H_2O_2 availability are tightly correlated (Mhamdi et al. 2010). However, these antioxidants can remain relatively unaffected by drought until stress-induced senescence is triggered (Bartoli et al. 2005). At severe drought stress, a decrease in AsA during seed desiccation was accompanied by an accumulation of GSSG with total GSH remaining relatively constant (Colville and Kranner 2010). Measurements using a redox-sensitive GFP revealed that the cytosol becomes more oxidized during drought (Jubany-Mari et al. 2010). Moreover, GSH redox potential is altered suggesting a decrease either in the total cytosolic GSH pool or in the GSH/GSSG ratio (Meyer et al. 2007). The content and redox state of GSH are modified during drought stress in nodules resulting from the interaction between pea and *Rhizobium leguminosarum* biovar *viciae* turning into a more oxidized state. This response is accompanied with a general decrease in antioxidant activity, being GR reduced in a 30% in stressed nodules (Gogorcena et al. 1995) and a decreased nitrogenase activity (Marino et al. 2006). This change in thiol disulfide status is compatible with a role in oxidative stress signaling. Although mild drought does not seem to cause a marked change in the status of key antioxidants measured in whole-tissue extracts, shifts between compartments cannot be discounted. For instance, increases in cytosolic (and therefore nuclear) GSH redox potential during drought may be limited by transfer of GSSG to the vacuole or the apoplast where it can be degraded by γglutamyl transpeptidase-dependent pathways (Noctor et al. 2014).

Studies dealing with drought stress and the responses of the antioxidant system take into account the responses of GSH and its related enzymes in a context that includes the participation of other antioxidant compounds and enzymes which are discussed below. In the soybean-*Bradyrhizobium japonicum* symbiotic interaction, the drought-induced increases in nodule ureides were accompanied by higher levels of nodule protein, pyridine nucleotides, AsA, and GSH. The accumulation of low-molecular-weight antioxidants suggested that the nodule response to the imposition of drought was to exacerbate the defenses against perturbations in cellular redox state, supporting the idea that GSH plays an important role in drought stress tolerance (Marquez-Garcia et al. 2015). These authors also evaluated the transcription of several genes encoding for antioxidant enzymes from thiol metabolism, namely, thioredoxin (TRX1), peroxiredoxins (PRX1, PRX2), and glutaredoxins (GRX1, GRX2), and their upregulation supported the view that oxidative stress in nodules is caused by enhanced ROS formation rather than a decrease in the antioxidant defense. Even though the induction of the components of the antioxidant system appears to be a widespread mechanism associated with ROS accumulation under abiotic stress, different species subjected to drought stress showed contrasting responses, and the GSH-related enzymes do not overcome this aspect. Soybean nodules showed a decrease in GR and APX activities but an increase in SOD and CAT activities (Porcel et al. 2003). However, Naya et al. (2007) found an increase in GR as well as CuZn-SOD (cytosolic), Fe-SOD, and APX transcripts in alfalfa. These authors also studied the rehydration process and reported that increased transcripts were reversed after rehydration for 48 h. The variability of responses underlined the need to study the transcript and activity regulation in more detail, especially in different species.

The induction of the antioxidant system has been correlated with the accumulation of ABA in different species (Zhou et al. 2005; Bright et al. 2006; Zhang et al. 2007; Lu et al. 2009). Being ABA the most important signal upon the exposure of plants to drought stress, it was determined that peanut plants accumulated ABA under stressful conditions and that accumulation was reversed upon rehydration (Furlan et al. 2012). Besides, it was demonstrated that ABA triggered the H_2O_2 production in this legume (Furlan et al. 2013). Also, ABA regulates the expression of many genes, the products of which may function in dehydration tolerance, including the antioxidant system and the expression of AREB1 (ABA response element binding transcription factor) (Furihata et al. 2006). In peanut nodules, AREB1; RPR-10, a hypothetical dehydrin; and Fer1 which codes a ferritin, a protein that sequestrates Fe and diminishes Fenton reactions, were accumulated. It is not discarded that, in peanut nodules, the transcription of several antioxidant enzymes was related to AREB1 expression and therefore with ABA accumulation in response to drought stress. In such way, peanut nodules increased GR and GST together with CuZn-SOD gene expression during drought and reversed upon rehydration. For GR, these changes were associated with increased enzyme activity, whereas GST and SOD activities did not change. Furthermore, total GSH and AsA contents were not affected by drought stress and subsequent rehydration; however, the levels of the oxidized forms increased significantly in response to drought stress (Furlan et al. 2014). Similarly, the aerial part of the legume exhibited an increased specific activity in the enzymes of the AsA-GSH cycle although the levels of GSH were not modified upon the exposure to drought stress and the AsA levels were increased (Furlan et al. 2016).

It is noteworthy that these articles do not exhibit remarkable responses in the activity or transcript level of other GSH-related enzyme, namely, GPX; however, a role in signaling has been attributed for this enzyme. In fact, Miao et al. (2006) demonstrated the dual role of the *Arabidopsis thaliana* GPX3 isoform in abscisic acid and drought stress signaling. This raised the question that up to what extent this response can be achieved by leguminous plants. In this sense, an extensive study of two GPX isoforms abundant in legume nodules from *Lotus japonicus* was conducted. LjGPX1 and LjGPX3 were described as phospholipid hydroperoxidases capable of interacting in vitro with TRXs endogenously present in nodules. The authors demonstrated their localization in nuclei and evaluated the differential expression of LjGPX1 and LjGPX3 in response to nitrosoglutathione (GSNO) and hormones such as ABA and others. Also, the susceptibility to nitrosylation of the catalytic Cys residue provided strong support for GPX signaling roles in addition to their antioxidative properties (Matamoros et al. 2015).

The participation of antioxidant system, especially of GSH and related enzymes, in drought-stressed legume-rhizobia interaction exhibits a variable response, depending on the legume and the severity of the stress imposed. However, a consistent ABA accumulation is observed; therefore, future studies in legume nodules should shed light about the implication of drought-responsive genes associated with ABA accumulation that will determine GSH synthesis, GSH-related enzymes activation, and nitrogen fixation-associated proteins activities such as nitrogenase and leghemoglobin.

3.4 Heavy Metals and Metalloid Stress

Heavy metals and metalloids are widely distributed around the world. Most of them are present as cations associated to different organic and inorganic ligands (Carpena et al. 2006). The increasing level of toxic elements such as cadmium, mercury, lead, and arsenic (Cd, Hg, Pb, As) in the environment are subject to (a) mining and metallurgy activity (Alloway 2012); (b) anthropogenic activities such as the application of phosphate fertilizers, sewage sludge, or pesticides containing metals (Järup 2003; Gratão et al. 2005); and (c) natural sources by volcanic or geothermal activity and rock weathering and erosion (Ziemacki et al. 1989). From all metals and metalloids, Cd and As constitute the most hazardous, being extremely poisoning to human (WHO 2010; IARC 2016). They can be naturally found in soils or groundwater; however, their concentration can be magnified in agricultural soils by sewage sludge supply, atmospheric fallout from industrial processes, artificial irrigation, and use of phosphate fertilizers (Zhu et al. 1999; Mann et al. 2002). Arsenic and cadmium compounds are toxic for cells since they can substitute P and Zn, respectively, in cellular metabolism. The prevalent inorganic chemical species of As are arsenate ($H_2AsO_4^-$, As(V)), which is a phosphate chemical analog, or arsenite ($H_2AsO_3^-$, As(III)), acting as a sulfur-seeking heavy metal ion, as Cd. Arsenate reduction to arsenite can be achieved either enzymatically or nonenzymatically, while Cd is always in the divalent form. It is known that As(III) and Cd cause GSH depletion in cells being the main reason of oxidative stress. These elements can also complex with GSH forming As(III)/Cd-GSH complexes and/or induced PC synthesis (Verbruggen et al. 2009).

One of the most common responses of legume microsymbionts to Cd and As is the cellular oxidative burst that leads to oxidation of lipids, which results in the permeabilization of the plasma membrane (Howlett and Avery 1997; Corticeiro et al. 2006; Bianucci et al. 2012a). Of all the bacterial mechanisms of resistance known so far (Bruins et al. 2000), this chapter makes especially interest of those where GSH and its related enzymes are involved. Bianucci et al. (2011) studied the effect of Cd in *Bradyrhizobium* sp. strains, peanut microsymbionts, finding that the tolerance to metal is related to the intracellular GSH pool. They observed that only in the tolerant strains tested, *Bradyrhizobium* sp. NLH25 and NOD31, metal addition induced GSH synthesis. This behavior was also observed in the tolerant *Rhizobium leguminosarum* strain exposed to metal. Moreover, Cd augmented GSSG levels indicating that cells were exposed to oxidative stress (Corticeiro et al. 2006). These results are in agreement with the role of GSH as a radical scavenger being oxidized by ROS. Regarding GSH-related enzymes activities upon Cd exposure, the mentioned microsymbionts showed distinct response, being the tolerant strains capable to maintain an operating GSH cycle by increasing both GPX and GR activity (Corticeiro et al. 2006; Bianucci et al. 2012a). Interestingly, in the peanut microsymbionts, GST enzyme activity was reduced in both tolerant and sensitive strains. Since GPX activity was enhanced in all strains, this enzyme may not be involved in the tolerance to this metal but has an important role in detoxifying H_2O_2 compounds.

Thus, increment in GPX activity in strains detoxifies ROS, oxidizing GSSG. The increasing GR activity observed in tolerant strains highlights the important role of this enzyme in maintaining high GSH/GSSG ratio. In *Bradyrhizobium japonicum* USDA110 (soybean microsymbiont), Cd addition not only affected bacterial growth and induced lipid peroxidation but also increased GSH content in a significant way. Additionally, all GSH-related enzyme activities increased by metal addition, being GST activity four times higher than control condition (Bianucci et al. 2013a). Similar studies performed by Corticeiro et al. (2013) in two *Rhizobium leguminosarum* strains revealed that only the tolerant strain enhanced GST activity when exposed to Cd. This strain also expressed six isoforms of GST being peak IV involved in the complexation of Cd to GSH, while the sensitive strain only expressed five isoforms of GST. Previous studies performed in these rhizobia by Lima et al. (2006) showed that 75% of Cd was complexed with GSH by the tolerant strain, while only a 23% was observed in the sensitive one. These findings shows that changes in GST activities and the expression of GST isoforms, induced by the tolerant strain in response to Cd, are involved in the tolerance of *Rhizobium leguminosarum*.

Arsenic toxicity was also determined in two peanut microsymbionts classified as tolerant and sensitive to As, *Bradyrhizobium* sp. SEMIA 6144 and *Bradyrhizobium* sp. C-145, respectively (Bianucci et al. 2016). The addition of As(V) only induced GPX activity in the sensitive strain and was consistent with the evidence of oxidative stress symptoms. Furthermore, GST activity was also decreased in this strain by metalloid treatment. The tolerant strain showed an increase on H_2O_2 content, however, it was not enough to induce oxidative stress. Moreover, GPX and GST enzyme activities remained unchanged by As(V) addition. GST comprise a large family of isoenzymes involved in the detoxification of organic peroxides, radicals, and xenobiotics at the expense of GSH and are induced by Cd and $H_2AsO_4^-$ (Allocati et al. 2008; Schröder et al. 2009). Both GST and GPX deplete GSH cytosolic content directly to reduce H_2O_2 and to detoxify oxidation subproducts, promoting oxidizing conditions (Anjum et al. 2012). Thus, in the peanut-tolerant bradyrhizobial strain, the intracellular GSH pool may cover the demand imposed by GPX and GST allowing strain to grow up at high As concentration without oxidative stress symptoms.

Several research have focused on the study of the effects of heavy metals and/or metalloids in plants; however, the impact of these contaminants in the symbiotic interaction established between legumes and rhizobia is scarce even more regarding the GSH role in the symbiosis under metal(loid) exposure. It is known that metal(loid) persistence in soil strata negatively affects soil microorganism viability including rhizobia, causing changes in the composition of microbial communities (Brookes and Mc Grath 1984; Liao et al. 2005; Abou-Shanab et al. 2005; Paudyal et al. 2007; Sobolev and Begonia 2008; Bamborough and Cummings 2008). Furthermore, decrease of crop growth, reduction in nodule development, and deficit in nitrogen fixation are some of the most common symptoms induced by metal(loid) s (Giller et al. 1998; Alexander et al. 1999; Broos et al. 2004; Wani et al. 2006; Younis 2007).

Studies on heavy metal toxicity in the legume-rhizobia symbiosis have gained special interest in the past few years since the knowledge that inoculation of plant with rhizobia alleviates metal(loid) toxicity (Reichman 2007; Wani et al. 2008; Dary et al. 2010). Researches performed by Bianucci et al. (2012b) in peanut plants exposed to Cd showed significant growth arrest and changes in root histological structure with a deposit of unknown material on the epidermal and endodermal cells. Nevertheless, no symptoms of oxidative stress were evidenced. Interestingly peanut leaves' and roots' GSH content was significantly decreased, but the synthesis of different types of PCs (PC2, PC3, and PC4) was detected. Inoculation of these plants exposed to Cd, with the tolerant *Bradyrhizobium* sp. NLH25 strain to metal, showed reduction of nodule number, nitrogen content, and a significant enhance of ROS accumulation in nodules. These responses were accentuated in plants inoculated with the sensitive *Bradyrhizobium* sp. SEMIA61444 strain exhibiting also increased Cd accumulation. Thus, inoculation of peanut with the tolerant strain could limit Cd accumulation in the plant resulting in a better symbiotic interaction (Bianucci et al. 2013b). Similar results were found in roots of lotus plants inoculated with *Mesorhizobium loti* exposed to Cd (Ramos et al. 2007) where no evidence of heavy metal toxicity was observed; meanwhile, thiol analysis showed that hGSH content was unchanged along treatment but a high increase in GSH, PC, and hPC synthesis was revealed. Moreover, three functional genes of PCS enzyme were differentially regulated when the inoculated plants were exposed to metal at different concentrations (Ramos et al. 2008). In soybean plants exposed to Cd, it was shown that metal addition induced oxidative damage only in nodules and a significant increase of PCs (Balestrasse et al. 2001, 2003, 2006). These data allow to suggest that although metal reduced nodulation in soybean plant, the induction of PC synthesis could prevent metal disturbance on cellular redox balance, avoiding oxidative damage to macromolecules in roots. Furthermore, Cd decrease the activities of enzymes involved in nitrogen assimilation pathway by oxidizing proteins without changes in the GSH/GSSG ratio. Nevertheless, exogenous GSH addition did not modify the amount of Cd accumulated in nodules, but it was enough to preventing the oxidation of these proteins. These researches also showed that GSH-related enzyme activities were also altered by Cd addition revealing an improvement of GR activity. All these findings show that GSH has a potential role as an antioxidant that could prevent damage of nodules subjected to Cd stress. Shvaleva et al. (2010) determined Cd tolerance in alfalfa plant, inoculated with a flavodoxin-overexpressing *Sinorhizobium meliloti* mutant, demonstrating an enhanced nitrogenase activity and a decreased GSH content compared to wild-type strain under Cd stress. However, GSH/GSSG ratio of flavodoxin-overexpressing nodules was significantly enhanced in comparison to wild-type nodules suggesting that the mutation could have a positive effect on nodule oxidative balance under this heavy metal exposition.

Bianucci et al. (2016) revealed the inoculation effect of two bradyrhizobial strains on peanut exposed to realistic groundwater concentrations of As(V). They found that inoculation of As(V)-treated plants reduced oxidative stress symptoms, enhanced GSH pool, and decreased proline content. Peanut inoculation with the tolerant strain maintained an operating GSH cycle showing a significant increase on the detoxifying GST enzyme activity. Thus, GSH and related enzymes play a critical role in the symbiotic interaction established between the microsymbionts and the legume.

Bioremediation of soils contaminated with heavy metals has been reviewed by Gómez-Sagasti and Marino (2015) since it is an inexpensive alternative which also renews soils heavily contaminated; however, the use of legumes for this purpose became a problem when the plant is used as animal or human food. Sriprang et al. (2002, 2003) and Ike et al. (2007) developed a novel plant-bacterial remediation system for heavy metals transforming *Mesorhizobium huakuii* subsp. *rengei* B3 with a human metallothionein (MTL4) and/or PCS genes. These rhizobia symbiotically interact with *Astragalus sinicus*, a legume used as a green manure in rice field in Japan and Southern China (Chen et al. 1991). They observed that the free-living B3 mutant was able to increase Cd content in compare to the wild-type strain. Moreover, PCS mutants exposed to Cd catalyzed the synthesis of PCs. It is known that PCs have many advantages compared to MTs, in particular the continuously repeating γGlu-Cys units that allow PCs to have a higher metal-binding capacity than MTs (Mehra and Mulchandani 1995). These studies showed that the symbiosis will be useful in phytoremediation for heavy metal contaminated soils and the biotechnological approach using biothiols in special PCs is a novel goal to achieve it.

4 Conclusion and Future Perspectives

This chapter is focused on the importance of GSH/hGSH in the legume-rhizobia symbiosis whose protective and regulatory functions involve thiol groups, with special attention to the contribution of their redox activities during BNF, from root cell infection to nodule senescence. Additionally, since the tolerance to different stresses of the microsymbionts is often associated to changes in GSH/GSSG ratio, the comprehension of the role of these thiols in the free-living rhizobia results in a significant importance, especially when they are selected to be inoculants for legume to improve crop yield.

The use of enzyme inhibitors and transgenic plants has demonstrated that GSH/hGSH is essential for the proper development of the root nodules resulting from the rhizobia-legume symbiosis. Evidence shows that nodules will not form on roots whether GSH/hGSH synthesis is blocked by addition of BSO, specific inhibitor of GSH and hGSH synthesis, suggesting that like the root meristem, the nodule meristem is unable to develop in the absence of these thiols. However, several important issues need to be resolved. More research will be required to establish whether GSH and hGSH play different roles, especially in redox homeostasis and signaling during nodule development.

The nodule antioxidants are very important for metabolism homeostasis since this organ contains multiple proteins that are sensitive to oxidation, such as nitrogenase, ferredoxins, and leghemoglobins. Therefore, attempts to study the antioxidant system, in particular SOD, CAT, and AsA-GSH cycle enzymes, have been made and have been described in some detail. Even though, there is much less information regarding other antioxidant and redox sensor enzymes such as PRXs, TRXs, GRXs and NTRs from host cells and bacteroids. The comprehension of the functioning of this antioxidant system is particularly relevant since it participates not only in the ROS-detoxification process but also in signaling.

Thus, the challenge for the future is not only to characterize the role of GSH/hGSH and related enzymes and to assess their interactions with other types of redox regulation but also to evaluate the significance of these changes within nodule subjected at different environmental stresses. Furthermore, comparative analyses of the thiol-based signaling mechanisms underpin the response of legume-rhizobia symbiosis to environmental signals that will provide critical information to enhance BNF as well as the tolerance to different stresses. In this aspect, the improvement of the BNF efficiency is expected, in order to achieve environmental benefits which will eventually lead to a reduction in the input of cost and contamination with nitrogen fertilizers.

Acknowledgment This work was supported by Secretaría de Ciencia y Técnica de la Universidad Nacional de Río Cuarto (SECYT-UNRC), Consejo Nacional de Investigaciones Científicas y Técnicas (CONICET), and Ministerio de Ciencia y Tecnología Córdoba. E. Bianucci and A. Furlan are members of the research career from CONICET.

References

Abou-Shanab RA, Ghozlan H, Ghanem K, Moawad H (2005) Behaviour of bacterial populations isolated from rhizosphere of *Diplachne fusca* dominant in industrial sites. World J Microbiol Biotechnol 21:1095–1101

Alesandrini F, Mathis R, Van de Sype G, Hérouart D, Puppo A (2003) Possible roles of a cysteine protease and hydrogen peroxide in soybean nodule development and senescence. New Phytol 158:131–138

Alexander E, Pham D, Steck TR (1999) The viable but non-culturable condition is induced by copper in *Agrobacteriuim* Tumefaciens and *Rhizobium leguminosarum*. Appl Env Microbiol 65:3754–3756

Allocati N, Federici L, Masulli M, Di Llio C (2008) Glutathione transferases in bacteria. FEBS J 276:58–75

Alloway BJ (2012) Sources of heavy metals and metalloids in soils. In: Alloway BJ (ed) Heavy metals in soils. Springer, Heidelberg, pp 15–50

Anjum NA, Ahmad I, Mohmood I, Pacheco M, Duarte AC, Pereira E, Umar S, Ahmad A, Khan NA, Iqbal M (2012) Modulation of glutathione and its related enzymes in plants' responses to toxic metals and metalloids – a review. Environ Exp Bot 75:307–324

Balestrasse KB, Gardey L, Gallego SM, Tomaro ML (2001) Response of antioxidant defence system in soybean nodules and roots subjected to cadmium stress. Aust J Plant Physiol 28:497–504

Balestrasse KB, Benavides MP, Gallego SM, Tomaro ML (2003) Effect of cadmium stress on nitrogen metabolism in nodule and roots of soybean plants. Funct Plant Biol 30:57–64

Balestrasse KB, Gallego SM, Tomaro ML (2006) Oxidation of the enzymes involved in nitrogen assimilation plays an important role in the cadmium-induced toxicity in soybean plants. Plant Soil 284:187–194

Bamborough L, Cummings SP (2008) The impact of increasing heavy metal stress on the diversity and structure of the bacterial and actinobacterial communities of metallophytic grassland soil. Biol Fertil Soils 45:273–280

Bartoli CG, Guaimet JJ, Kiddle G, Pastori GM, Di Cagno R, Theodoulou F, Foyer CH (2005) Ascorbate content of wheat leaves is not determined by maximal l-galactono-1,4-lactone dehydrogenase (GalLDH) activity under drought stress. Plant Cell Environ 28:1073–1081

Becana M, Klucas RV (1992) Transition metals in legume root nodules. Iron dependent free radical production increases during nodule senescence. PNAS 89:8958–8962

Becana M, Dalton DA, Moran JF, Iturbe-Ormaetxe I, Matamoros MA, Rubio MC (2000) Reactive oxygen species and antioxidants in legume nodules. Physiol Plant 109:372–381

Becana M, Matamoros MA, Udvardi M, Dalton DA (2010) Recent insights into antioxidant defenses of legume root nodules. New Phytologist 188:960–976D

Bianucci E, Fabra A, Castro S (2008) Growth of *Bradyrhizobium* sp. SEMIA 6144 in response to methylglyoxal: role of glutathione. Curr Microbiol 56:371–375

Bianucci E, Fabra A, Castro S (2011) Cadmium accumulation and tolerance in *Bradyrhizobium* spp. (peanut microsymbionts). Curr Microbiol 62:96–100

Bianucci E, Fabra A, Castro S (2012a) Involvement of glutathione and enzymatic defense system against cadmium toxicity in *Bradyrhizobium* sp. strains (peanut symbionts). Biometals 25:23–32

Bianucci E, Sobrino-Plata J, Carpena-Ruiz R, Tordable MC, Fabra A, Hernández L, Castro S (2012b) Contribution of phytochelatins to cadmium tolerance in peanut plants. Metallomics 4:1119–1124

Bianucci E, Fullana C, Furlan A, Castro S (2013a) Antioxidant defense system responses and role of nitrate reductase in the redox balance maintenance in *Bradyrhizobium japonicum* strains exposed to cadmium. Enzyme Microb Technol 53:345–350

Bianucci E, Furlan A, Rivadeneira J, Sobrino-Plata J, Carpena-Ruiz R, Tordable MC, Fabra A, Hernández L, Castro S (2013b) Influence of cadmium on the symbiotic interaction established between peanut (*Arachis hypogaea* L.) and sensitive or tolerant bradyrhizobial strains. J Environ Manag 130:126–134

Bianucci E, Furlan A, Isaia A, Peralta JM, Hernández LE, Castro S (2016) Impact of arsenic in bradyrhizobia strains and in the symbiotic interaction with peanut plant. Biocell 40(1):113

Bright J, Desikan R, Tancock JT, Weir IS, Neill SJ (2006) ABA-induced NO generation and stomatal closure in Arabidopsis are dependent on H_2O_2 synthesis. Plant J 45:113–122

Brookes PC, Mc Grath SP (1984) Effect of metal toxicity on the size of the soil microbial biomass. Eur J Soil Sci 10:1365–2389

Broos K, Uyttebroek M, Mertens J, Smolders E (2004) A survey of symbiotic nitrogen fixation by white clover grown on metal contaminated soils. Soil Biol Biochem 36:633–640

Bruins MR, Kapil S, Oehme FW (2000) Microbial resistance to metals in the environment. Ecotoxicol Environ Saf 45:198–207

Cárpena R, Esteban E, Lucena JJ, Peñalosa S, Vázquez P, Zornoza P, Gárate A (2006) Simbiosis y fitorrecuperación de suelos. In: Bedmar EJ, González J, Lluch C, Rodelas MB (eds) Fijación de Nitrógeno: Fundamentos y Aplicaciones. Granada, España, pp 255–268

Chen WX, Li GS, Qi YL, Wang ET, Yuan HL, Li JL (1991) *Rhizobium huakuii* sp. nov. isolated from the root nodules of *Astragalus sinicus*. Int J Syst Bacteriol 41:275–280

Clemente MR, Bustos-Sanmamed P, Loscos J, James EK, Pérez-Rontomé C, Navascués J, Gay M, Becana M (2012) Thiol synthetases of legumes: immunogold localization and differential gene regulation by phytohormones. J Exp Bot 63:3923–3934

Coba de la Peña T, Redondo FJ, Manrique E, Lucas MM, Pueyo JJ (2010) Nitrogen fixation persists under conditions of salt stress in transgenic *Medicago truncatula* plants expressing a cyanobacterial flavodoxin. Plant Biotechnol J 8:954–965

Cobbett C, Goldsbrough P (2002) Phytochelatins and metallothioneins: roles in heavy metal detoxification and homeostasis. Annu Rev Plant Biol 53:159–182

Colville L, Kranner I (2010) Desiccation tolerant plants as model systems to study redox regulation of protein thiols. Plant Growth Regul 62:241–255

Cook D, Dreyer D, Bonnet D, Howell M, Nony E, Van den Bosch K (1995) Transient induction of a peroxidase gene in *Medicago trunculata* precedes infection by *Rhizobium meliloti*. Plant Cell 7:43–55
Corticeiro SC, Lima AI, Figueira EM (2006) The importance of glutathione in oxidative status of *Rhizobium leguminosarum* biovar *viciae* under Cd exposure. Enzym Microb Technol 40:132–137
Corticeiro S, Freitas R, Figueira E (2013) The role of GSTs in the tolerance of *Rhizobium leguminosarum* to cadmium. Biometals 26:879–886
Dalton DA, Russell SA, Hanus FJ, Pascoe GA, Evans HJ (1986) Enzymatic reactions of ascorbate and glutathione that prevent peroxide damage in soybean root nodules. PNAS 83:3811–3815
Dalton DA, Langeberg L, Treneman N (1993) Correlations between the ascorbate-glutathione pathway and effectiveness in legume root nodules. Physiol Plant 87:365–370
Dalton DA, Boniface C, Turner Z, Lindahl A, Kim HJ, Jelinek L, Govindarajulu M, Finger RE, Taylor CG (2009) Physiological roles of glutathione S-transferases in soybean root nodules. Plant Physiol 150:521–530
Dary M, Chamber-Perez MA, Palomares AJ, Pajuelo E (2010) 'In situ' phytostabilisation of heavy metal polluted soils using *Lupinus luteus* inoculated with metal resistant plant-growth promoting rhizobacteria. J Hazard Mater 177:323–330
El Msehli S, Lambert A, Baldacci-Cresp F, Hopkins J, Boncompagni E, Smiti SA, Hérouart D, Frendo P (2011) Crucial role of (homo)glutathione in nitrogen fixation in *Medicago truncatula* nodules. New Phytol 192:496–506
Evans J, Dear B, O'Connor G (1990) Influence of an acid soil on the herbage yield and nodulation of five annual pasture legumes. Aust J Exp Agric 30:55–60
Evans PJ, Gallesi D, Mathieu C, Hernández MJ, de Felipe N, Halliwell B, Puppo A (1999) Oxidative stress occurs during soybean nodule senescence. Planta 208:73–79
Ferguson BJ, Mathesius U (2003) Signaling interactions during nodule development. J Plant Growth Regul 22:47–72
Frendo P, Gallesi D, Turnbull R, Van de Sype G, Hérouart D, Puppo A (1999) Localisation of glutathione and homoglutathione in *Medicago truncatula* is correlated to a differential expression of genes involved in their synthesis. Plant J 17:215–219
Frendo P, Hernández-Jiménez MJ, Mathieu C, Duret L, Gallesi D, Van de Sype G, Hérouart D, Puppo A (2001) A *Medicago truncatula* homoglutathione synthetase is derived from glutathione synthetase by gene duplication. Plant Physiol 126:1706–1715
Frendo P, Harrison J, Norman C, Hernández-Jiménez MJ, Van de Sype G, Gilabert A, Puppo A (2005) Glutathione and homoglutathione play a critical role in the nodulation process of *Medicago truncatula*. Mol Plant-Microbe Interact 18:254–259
Furihata T, Maruyama K, Fujita Y, Umezawa T, Yoshida R, Shinozaki K, Yamaguchi-Shinozaki K (2006) Abscisic acid dependent multisite phosphorylation regulates the activity of a transcription activator AREB1. PNAS 103:1988–1993
Furlan A, Llanes A, Luna V, Castro S (2012) Physiological and biochemical responses to drought stress and subsequent rehydration in the symbiotic association peanut-*Bradyrhizobium* sp. ISRN Agron. https://doi.org/10.5402/2012/318083
Furlan A, Llanes A, Luna V, Castro S (2013) Abscisic acid mediates hydrogen peroxide production in peanut induced by water stress. Biol Plant 57:555–558
Furlan A, Bianucci E, Tordable MC, Castro S, Dietz K (2014) Antioxidant enzyme activities and gene expression patterns in peanut nodules during a drought and rehydration cycle. Funct Plant Biol 41:704–713
Furlan A, Bianucci E, Tordable MC, Kleinert A, Valentine A, Castro S (2016) Dynamic responses of photosynthesis and antioxidant system during a drought and rehydration cycle in peanut plants. Funct Plant Biol 43:337–345
Giller KE, Witter E, McGrath SP (1998) Toxicity of heavy metals to microorganisms and microbial process in agriculture soils – a review. Soil Biol Biochem 30:1389–1414
Gogorcena Y, Iturbe-Ormaetxe I, Escuredo PR, Becana M (1995) Antioxidant defenses against activated oxygen in pea nodules subjected to water stress. Plant Physiol 108:753–759

Gómez-Sagasti MT, Marino D (2015) PGPRs and nitrogen-fixing legumes: a perfect team for efficient Cd phytoremediation? Front Plant Sci 6:81
Graham P (1992) Stress tolerance in *Rhizobium* and *Bradyrhizobium*, and nodulation under adverse soil conditions. Can J Microbiol 38:475–484
Gratão P, Polle A, Lea PJ, Azevedo RA (2005) Making the life of heavy metal-stressed plants a little easier. Funct Plant Biol 32:481–494
Groten K, Vanacker H, Dutilleul C, Bastian F, Bernard S, Carzaniga R, Foyer CH (2005) The roles of redox processes in pea nodule development and senescence. Plant Cell Environ 28:1293–1304
Günther C, Schlereth A, Udvardi M, Ott T (2007) Metabolism of reactive oxygen species is attenuated in leghemoglobin-deficient nodules of *Lotus japonicas*. MPMI 20:1596–1603
Harrison J, Jamet A, Muglia CI, Van de Sype G, Aguilar OM, Puppo A, Frendo P (2005) Glutathione plays a functional role in growth and symbiotic capacity of *Sinorhizobium meliloti*. J Bacteriol 187:168–174
Howlett NG, Avery SV (1997) Induction of lipid peroxidation during heavy metal stress in *Saccharomyces cerevisiae* and influence of plasma membrane fatty acid unsaturation. Appl Environ Microbiol 63:2971–2976
Ike A, Sriprang R, Ono H, Murooka Y, Yamashita M (2007) Bioremediation of cadmium contaminated soil using symbiosis between leguminous plant and recombinant rhizobia with the MTL4 and the PCS genes. Chem 66:1670–1676
Innocenti G, Pucciariello C, Le Gleuher M, Hopkins J, de Stefano M, Delledonne M, Puppo A, Baudouin E, Frendo P (2007) Glutathione synthesis is regulated by nitric oxide in *Medicago truncatula* roots. Planta 225:1597–1602
International Agency of Reasearch on Cancer (IARC) (2016). On line: www.iarc.fr
Järup L (2003) Hazards of heavy metal contamination. Br Med Bull 68:167–182
Jubany-Mari T, Alegre-Batlle L, Jiang K, Feldman LJ (2010) Use of a redox-sensing GFP (c-roGFP1) for real-time monitoring of cytosol redox status in *Arabidopsis thaliana* water-stressed plants. FEBS Lett 584:889–897
Krylova VV, Dubrovo NP, Izmailov SF (2007) The effect of metabolites on the pH gradient and membrane potential of the bean peribacteroid membrane. Appl Biochem Microbiol 43:292–297
Liao M, Luo YK, Zhao X, Huang CY (2005) Toxicity of cadmium to soil microbial biomass and its activity: effect of incubation time on Cd ecological dose in a paddy soil. J Zhejiang Univ Sci B 6:324–330
Lima AIG, Corticeiro SC, Figueira EMAP (2006) Glutathione- mediated cadmium sequestration in *Rhizobium leguminosarum*. Enzym Microb Technol 39:763–769
Long SR (2001) Genes and signals in the *Rhizobium*-legume symbiosis. Plant Physiol 125:69–72
Loscos J, Matamoros MA, Becana M (2008) Ascorbate and homoglutathione metabolism in common bean nodules under stress conditions and during natural senescence. Plant Physiol 146:1282–1292
Lu S, Su W, Li H, Guo Z (2009) Abscisic acid improves drought tolerance of triploid bermudagrass and involves H_2O_2- and NO-induced antioxidant enzyme activities. Plant Physiol Biochem 47:132–138
Mann SS, Rate AW, Gilkes RJ (2002) Cadmium accumulation in agricultural soils in Western Australia. Water Air Soil Pollut 141:281–297
Marino D, Gonzalez EM, Arrese-Igor C (2006) Drought effects on carbon and nitrogen metabolism of pea nodules can be mimicked by paraquat: evidence for the occurrence of two regulation pathways under oxidative stresses. J Exp Bot 57:665–673
Marino D, Frendo P, Ladrera R, Zabalza A, Puppo A, Arrese-Igor C, González EM (2007) Nitrogen fixation control under drought stress. Localized or systemic? Plant Physiol 143:1968–1974
Marquez-Garcia B, Shaw D, Cooper JW, Karpinska B, Dorcas Quain M, Makgopa EM, Kunert K, Foyer CH (2015) Redox markers for drought-induced nodule senescence, a process occurring after drought-induced senescence of the lowest leaves in soybean (*Glycine max*). Ann Bot 116:495–510

Matamoros MA, Moran JF, Iturbe-Ormaetxe I, Rubio MC, Becana M (1999) Glutathione and homoglutathione synthesis in legume root nodules. Plant Physiol 121:879–888

Matamoros MA, Dalton DA, Clemente MR, Rubio MC, Ramos J, Becana M (2003) Biochemistry and molecular biology of antioxidants in the rhizobia-legume symbiosis. Plant Physiol 133:1–11

Matamoros MA, Saiz A, Peñuelas M, Bustos-Sanmamed P, Mulet JM, Barja MV, Rouhier N, Moore M, James EK, Dietz KJ, Becana M (2015) Function of glutathione peroxidases in legume root nodules. J Exp Bot 66:2979–2990

Maughan S, Foyer CH (2006) Engineering and genetic approaches to modulating the glutathione network in plants. Physiol Plant 126:382–397

Mehra RK, Mulchandani P (1995) Glutathione-mediated transfer of Cu(I) into phytochelatins. Biochem J 307:697–705

Meister A, Anderson ME (1983) Glutathione Ann Rev Biochem 52:711–760

Mendoza-Cozatl DG, Jobe TO, Hauser F, Schroeder JL (2011) Long-distance transport, vacuolar sequestration, tolerance, and transcriptional responses induced by cadmium and arsenic. Curr Opin Plant Biol 14:554–562

Meyer AJ, Brach T, Marty L, Kreye S, Rouhier N, Jacquot JP, Hell R (2007) Redox-sensitive GFP in *Arabidopsis thaliana* is a quantitative biosensor for the redox potential of the cellular glutathione redox buffer. Plant J 52:973–986

Mhamdi A, Hager J, Chaouch S, Queval G, Han Y, Taconnat Y, Saindrenan P, Issakidis-Bourguet E, Gouia H, Renou JP, Noctor G (2010) Arabidopsis glutathione reductase 1 is essential for the metabolism of intracellular H_2O_2 and to enable appropriate gene expression through both salicylic acid and jasmonic acid signaling pathways. Plant Physiol 153:1144–1160

Miao Y, Lv D, Wang P, Wang XC, Chen J, Miao C, Song CP (2006) An *Arabidopsis* glutathione peroxidase functions as both a redox transducer and a scavenger in abscisic acid and drought stress responses. Plant Cell 18:2749–2766

Moran JF, Iturbe-Ormaetxe I, Matamoros MA, Rubio MC, Clemente MR, Brewin NJ, Becana M (2000) Glutathione and homoglutathione synthetases of legumes nodules. Cloning, expression, and subcellular localization. Plant Physiol 124:879–888

Muglia CI, Grasso DH, Aguilar OM (2007) *Rhizobium tropici* response to acidity involves activation of glutathione synthesis. Microbiology 153:1286–1296

Muglia C, Comai G, Spegazzini E, Riccillo PM, Aguilar OM (2008) Glutathione produced by *Rhizobium tropici* is important to prevent early senescence in common bean nodules. FEMS Microbiol Lett 286:191–198

Müller J, Wiemken A, Boller T (2001) Redifferentiation of bacteria isolated from *Lotus japonicus* root nodules colonized by *Rhizobium* sp. NGR234. J Exp Bot 2:2181–2186

Munns D (1986) Acid soil tolerance in legumes and rhizobia. Adv Plant Nutr 2:63–91

Navrot N, Collin V, Gualberto J, Gelhaye E, Hirasawa M, Rey P, Knaff DB, Issakidis E, Jacquot JP, Rouhier N (2006) Plant glutathione peroxidases are functional peroxiredoxins distributed in several subcellular compartments and regulated during biotic and abiotic stresses. Plant Physiol 142:1364–1379

Naya L, Ladrera R, Ramos J, Gonzáez EM, Arrese-Igor C, Minchin FR, Becana M (2007) The response of carbon metabolism and antioxidant defenses of alfalfa nodules to drought stress and to the subsequent recovery of plants. Plant Physiol 144:1104–1114

Noctor G, Mhamdi A, Foyer CH (2014) The roles of reactive oxygen metabolism in drought: not so cut and dried. Plant Physiol 164:1636–1648

Ott T, van Dongen JT, Günther C, Krusell L, Desbrosses G, Vigeolas H, Bock V, Czechowski T, Geigenberger P, Udvardi MK (2005) Symbiotic leghemoglobins are crucial for nitrogen fixation in legume root nodules but not for general plant growth and development. Curr Biol 15:531–535

Paudyal SP, Aryal RR, Chauhan SVS, Maheshwari DK (2007) Effect of heavy metals on growth of *Rhizobium* strains and symbiotic efficiency of two species of tropical legumes. Sci World J 5:27–32

Ponsone L, Fabra A, Castro S (2004) Interactive effects of acidity and aluminium on the growth, lipopolysaccharide and glutathione contents in two nodulating peanut rhizobia. Symbiosis 36:193–204

Porcel R, Barea JM, Ruiz-Lozano JM (2003) Antioxidant activities in mycorrhizal soybean plants under drought stress and their possible relationship to the process of nodule senescence. New Phytol 157:135–143

Pucciariello C, Innocenti G, Van de Velde W, Lambert A, Hopkins J, Clément M, Ponchet M, Pauly N, Goormachtig S, Holsters M, Puppo A, Frendo P (2009) (Homo)glutathione depletion modulates host gene expression during the symbiotic interaction between *Medicago truncatula* and *Sinorhizobium meliloti*. Plant Physiol 151:1186–1196

Ramos J, Clemente MR, Naya L, Loscos J, Pérez-Rontomé C, Sato S, Tabata S, Becana M (2007) Phytochelatin synthases of the model legume *Lotus japonicus*. A small multigene family with differential response to cadmium and alternatively spliced variants. Plant Physiol 143:1110–1118

Ramos J, Naya L, Gay M, Abian J, Becana M (2008) Functional characterization of an unusual phytochelatin synthase, LjPCS3, of *Lotus japonicus*. Plant Physiol 148:536–545

Rauser WE, Meuwly P (1995) Retention of cadmium in roots of maize seedlings. Role of complexation by phytochelatins and related thiol peptides. Plant Physiol 109:195–202

Reichman SM (2007) The potential use of legume-*rhizobium* symbiosis for the remediation of arsenic contaminated sites. Soil Biol Biochem 39:2587–2593

Ricillo PM, Muglia CI, Bruijn F, Roe AJ, Booth IR, Aguilar OM (2000) Glutathione is involved in environmental stress responses in *Rhizobium tropici*, including acid tolerance. J Bacteriol 182:1748–1753

Rubio MC, Bustos-Sanmamed P, Clemente MR, Becana M (2009) Effects of salt stress on the expression of antioxidant genes and proteins in the model legume *Lotus japonicus*. New Phytol 181:851–859

Schröder P, Lyubenova L, Huber C (2009) Do heavy metals and metalloids influence the detoxification of organic xenobiotics in plants? Environ Sci Pollut Res 16:795–804

Sengupta D, Ramesh G, Mudalkar S, Kumar KRR, Kirti PB, Reddy AR (2012) Molecular cloning and characterization of γ-Glutamyl cysteine Synthetase (VrγECS) from roots of *Vigna radiata* (L.) Wilczek under progressive drought stress and recovery. Plant Mol Biol Report 30:894–903

Shah K, Nongkynrih JM (2007) Metal hyperaccumulator and bioremediation. Biol Plant 51:618–634

Shrivastava P, Kumar R (2015) Soil salinity: a serious environmental issue and plant growth promoting bacteria as one of the tools for its alleviation. Saudi J Biol Sci 22:123–131

Shvaleva A, Coba T, Peña D, Rincón A, Lucas MM, Pueyo JJ (2010) Flavodoxin overexpression reduces cadmium-induced damage in alfalfa root nodules. Plant Soil 326:109–121

Sobolev D, Begonia MFT (2008) Effects of heavy metal contamination upon soil microbes: lead-induced changes in general and denitrifying microbial communities as evidenced by molecular markers. Int J Environ Res Public Health 5:450–456

Sobrevals L, Müller P, Fabra A, Castro S (2006) Role of glutathione in growth and symbiotic properties of *Bradyrhizobium* sp (peanut microsymbiont) under different environmental stresses. Can J Microbiol 52:606–616

Sriprang R, Hayashi M, Yamashita M, Ono H, Saeki K, Murooka Y (2002) A novel bioremediation system for heavy metals using the symbiosis between leguminous plant and genetically engineered rhizobia. J Biotechnol 99:279–293

Sriprang R, Hayashi M, Ono H, Takagi M, Hirata K, Murooka Y (2003) Enhanced accumulation of Cd^{2+} by a *Mesorhizobium* sp. transformed with a gene from *Arabidopsis thaliana* coding for phytochelatin synthase. Appl Environ Microbiol 69:1791–1796

Stougaard J (2000) Regulators and regulation of legume root nodule development. Plant Physiol 124:531–540

Van de Wiel C, Scheres B, Franssen H, Van Lierop MJ, Van Lammeren A, Van Kammen A, Bisseling T (1990) The early nodulin transcript ENOD2 is located in the nodule parenchyma (inner cortex) of pea and soybean root nodules. EMBO J 9:1–7

Vance CP (2008) Carbon and nitrogen metabolism in legumes nodules. In: Dilworth MJ, James EK, Sprent JI, Newton WE (eds) Nitrogen-fixing leguminous symbioses. Springer, Dordrecht, pp 293–320

Verbruggen N, Hermans C, Schat H (2009) Mechanisms to cope with arsenic or cadmium excess in plants. Curr Opin Plant Biol 12:364–372

Vernoux T, Wilson RC, Seeley KA, Reichheld JP, Muroy S, Brown S, Maughan SC, Cobbett CS, van Montagu M, Inzé D et al. (2000) The *ROOT MERISTEM LESS1/CADMIUM SENSITIVE2* gene defines a glutathione-dependent pathway involved in initiation and maintenance of cell division during postembryonic root development. Plant Cell 12:97–109

Wang Y, Zhang Z, Pan Zhang P, Cao Y, Hu T, Yan (2016) *Rhizobium* symbiosis contribution to short-term salt stress tolerance in alfalfa (*Medicago sativa* L.). Plant Soil 402:247–261

Wani PA, Khan MS, Zaidi A (2006) An evaluation of the effects of heavy metals on the growth, seed yield and grain protein of lentil in pots. Ann Appl Biol 27:23–24

Wani PA, Khan MS, Zaidi A (2008) Chromium-reducing and plant growth-promoting *Mesorhizobium* improves chickpea growth in chromium-amended soil. Biotechnol Lett 30:159–163

Wingate VP, Lawton MA, Lamb CJ (1988) Glutathione causes a massive and selective induction of plant defense genes. Plant Physiol 87:206–210

World Health Organization (WHO) (2010) Online: www.who.int

Younis M (2007) Response of lablab perpureus–rhizobium symbiosis to heavy metals in pot and field experiments. World J Agric Sci 3:111–122

Zenk MH (1996) Heavy metal detoxification in higher plants – a review. Gene 179:21–30

Zhang A, Jiang M, Zhang J, Ding H, Xu S, Hu X, Tan M (2007) Nitric oxide induced by hydrogen peroxide mediates abscisic acid-induced activation of the mitogen-activated protein kinase cascade involved in antioxidant defense in maize leaves. New Phytol 175:36–50

Zhou B, Guo Z, Xing J, Huang B (2005) Nitric oxide is involved in abscisic acid induced antioxidant activities in *Stylosanthes guianensis*. J Exp Bot 56:3223–3228

Zhu YL, Pilon-Smits AH, Jouarin L, Terry N (1999) Overexpression of glutathione synthetase in *Indian mustard* enhances cadmium accumulation and tolerance. Plant Physiol 119:73–80

Ziemacki G, Viviano G, Merli F (1989) Heavy metals: sources and environmental presence. Ann Ist Super Sanita 25:531–536

Chapter 18
Transgenic Plants Over-expressing Glutathione Biosynthetic Genes and Abiotic Stress Tolerance

Ahmed Gaber, David J. Burritt, and Mohammad Anwar Hossain

Abstract Plants cannot survive without glutathione, or a functionally homologous thiol, as glutathione has diverse functions in plant growth and development, many of which cannot be performed by other thiols or antioxidants. The roles of glutathione in plants include the regulation of redox homeostasis, cell signaling and gene expression, and essential roles in key physiological and metabolic processes such as photosynthesis and sulfur assimilation. The cellular pool of reduced glutathione (GSH) can be depleted by oxidation of GSH to glutathione disulfide (GSSG), by reactive oxygen species (ROS), or by reacting with methylglyoxal (MG). The generation of ROS and MG increases in plant cells under abiotic stress, e.g., in plants exposed to heavy metals, salinity, drought, high or low temperatures, herbicides, or air pollutants. There is considerable evidence to suggest that enhanced activities of GSH utilizing and regenerating enzymes are crucial for abiotic stress tolerance in both model and cultivated plant species. Recently, the use of transgenic plants has clearly demonstrated the importance of GSH for stress tolerance, with plants over-expressing GSH biosynthetic genes and genes associated with maintaining GSH levels having increased GSH levels and showing improved tolerance to individual stressors. In addition, modulating the activities of GSH-related enzymes has also been shown to be important for multiple stress tolerance; however, many of the details of the roles GSH plays in multiple stress tolerance are still unresolved.

A. Gaber (✉)
Department of Genetics, Faculty of Agriculture, Cairo University, Giza, Egypt

Deanship of Scientific Research, Taif University, Taif, Saudi Arabia
e-mail: ahmed.gaber@staff.cu.edu.eg; a.gaber@tu.edu.sa

D.J. Burritt
Department of Botany, The University of Otago, Dunedin, New Zealand

M.A. Hossain (✉)
Department of Genetics and Plant Breeding, Bangladesh Agricultural University, Mymensingh 2202, Bangladesh

Laboratory of Plant Nutrition and Fertilizers, Graduate School of Agricultural and Life Sciences, University of Tokyo, Tokyo, Japan
e-mail: anwargpb@bau.edu.bd

M.A. Hossain et al. (eds.), *Glutathione in Plant Growth, Development, and Stress Tolerance*, https://doi.org/10.1007/978-3-319-66682-2_18

The aim of this chapter is to provide a comprehensive overview of the diverse roles of GSH biosynthetic genes in improving abiotic stress tolerance by critically evaluating the research conducted using transgenic plants, expressing GSH-associated genes, grown under abiotic stress.

Keywords Glutathione • Abiotic stress • Transgenic plants • Antioxidant • Glutathione biosynthetic genes

Contents

1 Introduction

All aerobic organisms including plants require molecular oxygen for their survival. As a consequence of cellular respiration, molecular oxygen is reduced to H_2O, and during this process, reactive oxygen species (ROS) including superoxide anion radical ($O_2{\bullet}^-$), hydrogen peroxide (H_2O_2), and hydroxyl radical (•OH) are often produced (Asada 1999; Halliwell and Gutteridge 2007; Krumova and Cosa 2016; Sewelam et al. 2016). In addition, plants grown in the field are constantly exposed to a range of abiotic stresses including heavy metals, salinity, drought, high or low temperature, herbicides, and air pollutants, and the generation of ROS is a common consequence of such abiotic stresses (Shimazaki and Sugahara 1980; Foyer et al. 1994; Prasad 1996; Gaber et al. 2012; Ahmad et al. 2016; Hussain et al. 2016; Akram et al. 2017). As ROS are extremely reactive at high concentration, they can cause severe damage to cell components, e.g., by oxidizing proteins and inactivating enzymes, oxidizing DNA, and initiating the peroxidation of unsaturated fatty acids in cell membranes (Foyer and Harbinson 1994). However, ROS at lower levels perform important roles in the activation of defense gene expression, as part of the protective mechanisms plants use to cope with biotic and abiotic stressors, and so careful regulation of cellular ROS levels is important for plant survival under field conditions (Karpinski et al. 1999; Grant and Loake 2000; Fryer et al. 2003; op den Camp et al. 2003; Cheng et al. 2015; Avery 2011; Krumova and Cosa 2016).

Plant cells have various mechanisms to regulate cellular ROS levels like enzymatic defense systems, including superoxide dismutase (SOD), ascorbate peroxidase (APX), catalase (CAT), glutathione reductase (GR), glutathione

S-transferases (GSTs), glutathione peroxidase (GPX), peroxiredoxins, and thioredoxins and nonenzymatic antioxidants including reduced glutathione (GSH), ascorbate (AsA), vitamin E, lipoic acid, beta-carotene, and flavonoids (Foyer et al. 1994; Gaber et al. 2004, 2006; Halliwell and Gutteridge 2007; Hossain et al. 2014, 2015; Mostofa et al. 2017). Noctor and Foyer (1998), when reviewing the roles of GSH and AsA in plants, concluded that one of the main functions of these molecules is to "keeping reactive oxygen under control." However, more recently studies have shown that GSH and AsA also play very important roles in the regulation of the cellular redox status and in mediating redox and ROS relating signaling in plant cells (Foyer and Noctor 2005a, b, 2009).

Reduced glutathione is the predominant low-molecular-weight thiol found in plant cells and plays an important role in the protection of cells against free radical-mediated damage (Chen et al. 2012, 2015; Noctor et al. 2012; Munné-Bosch et al. 2013). The majority of the cellular GSH is present in the cytosol, with the remaining found in mitochondria, chloroplasts, the nuclear matrix, and peroxisomes (Zechmann 2014). Due to the presence of the cysteine residue, GSH is readily oxidized nonenzymatically to glutathione disulfide (GSSG) by electrophilic substances, e.g., free radicals and reactive oxygen/nitrogen species. Cellular GSH concentrations often decline in response to biotic or abiotic stressors as a result of increased oxidative and/or MG-induced stress (Lu 2000; Hossain et al. 2011; Hoque et al. 2016). The GSH/GSSG ratio, which is frequently used as an indicator of the cellular redox state, is often >10 under normal physiological conditions but declines rapidly in plants under stress (Lu 2000). In both animals and plants, it is well-known that shifting the GSH/GSSG ratio toward a more oxidizing state can activate several signaling pathways/factors including protein kinase B, calcineurin (calcineurin B-like proteins in plants), nuclear factor-κB, c-Jun N-terminal kinase, apoptosis signal-regulated kinase 1, and mitogen-activated protein kinases, thereby reducing cell proliferation and increasing apoptosis (Jones 2000). Recent studies on plants have shown the importance of GSH for abiotic stress tolerance; therefore, the aim of this chapter is to provide an overview of the use GSH biosynthetic gene expression in transgenic plants as a means for improving plant abiotic stress tolerance and possibly crop yields.

2 The Biosynthesis of Glutathione and Relationship of Glutathione to Plant Stress Tolerance

Glutathione is synthesized in two ATP-dependent steps catalyzed by two enzymes, a plastidial γ-glutamylcysteine synthetase (γ-ECS, GSH1; EC 6.3.2.2) and a cytosolic glutathione synthetase (GS, GSH2; EC 6.3.2.3) (May and Leaver 1993; Rawlins et al. 1995; Cobbett et al. 1998; Noctor et al. 2002; Galant et al. 2011). γ-ECS catalyzes the rate-limiting step in GSH biosynthesis in mammals, in yeasts and in plants, and levels of this enzyme can be regulated at the transcriptional and/or translational levels (Xiang and Oliver 1998; Noctor et al. 1996, 2002, 2012; Liedschulte et al. 2010). Studies have shown that increases in GSH levels observed in response to stress are correlated with increased γ-ECS activity (Chen and

Goldsborough 1994; Kocsy et al. 2001). In addition, the over-expression or inhibition of *GSH1* and *GSH2* causes increased or decreased levels of GSH, respectively, in plants and was found to modulate or inhibit stress-responsive pathways required for plant growth, development, and stress tolerance (Cobbett et al. 1998; Xiang and Oliver 1998; Szalai et al. 2009; Ghanta and Chattopadhyay 2011; Noctor et al. 2011, 2012; Cheng et al. 2015; Liu et al. 2015). Increased expression of γ-ECS in transgenic plants and enzymes associated with sulfur assimilation pathway or GR has been shown to cause substantial increases in leaf GSH levels (Harms et al. 2000; Foyer et al. 1995; Noctor et al. 1996). Also, it has been reported that an increased capacity to maintain GSH levels or an increase in the cellular pool of GSH can lead to modified amino acid metabolism and enhanced stress tolerance (Noctor et al. 1998a, b). In addition to the transgenic approach, analysis of mutants deficient in GSH and the treatment of plants with GSH, combined with transcript profiling, has provided valuable information on how GSH and GSSG regulate cell signaling and plant development and ability to tolerate stress. For example, transcript profiling studies have identified the relationships that exist between the regulation of stress-related defensive networks and antioxidant metabolism in plants (Willekens et al. 1997; Rossel et al. 2002; Pneuli et al. 2003). Studies have also shown that GSH can influence cellular levels of the regulatory proteins NPR1 and protein phosphatase 2C (ABI2), which are important in salicylic acid (SA) and abscisic acid (ABA) signaling, respectively (Meinhard et al. 2002; Mou et al. 2003). In a more recent study, using transcriptomic analyses of steady-state and polysome-bound mRNAs in GSH-treated plants, Cheng et al. (2015) reported that GSH had an even greater potential impact on plant growth, development, and stress tolerance than what was apparent from previous total mRNA profiling studies. They demonstrated that the translational changes induced by GSH treatment were associated with changes in numerous hormone and stress signaling pathways and suggested that these changes might contribute to enhance stress tolerance in GSH-treated plants (Cheng et al. 2015). Recently, studies on a range of plant species that used a priming approach, thermal (heat or cold treatments) or chemical (proline, betaine, nitric oxide, selenium, salicylic acid, polyamines, etc.), have shown that priming can increase GSH biosynthesis, elevate cellular GSH pools, and increase the activities of glutathione-utilizing and glutathione-regenerating enzymes, with the end result being plants that have improved abiotic stress tolerance (Hossain et al. 2010, 2011, 2012, 2013a, b and references therein).

3 Transgenic Plants Over-expressing GSH Biosynthetic Genes: Heavy Metal Tolerance

Several studies have shown that cellular GSH levels can regulate the expression of a range of defense genes that confer stress tolerance to plants (Wingsle and Karpinski 1996; Karpinski et al. 1997; Wingate et al. 1988; Loyall et al. 2000; Cheng et al. 2015). A significant role for GSH was found in plants responding to excessive levels of cadmium (Cd) and other heavy metals, as GSH is the precursor of phytochelatins (PCs) ([γ-Glu-Cys]n-Gly) that are synthesized by the enzyme phytochelatin synthase (PCS) (Grill et al. 1987,1989; Cobbett and Goldsbrough 2002; Rea et al. 2004; Hossain et al. 2012). Phytochelatins form complexes with potentially toxic metals, which are then sequestered into the vacuole, reducing their cytotoxicity (Grill et al. 1987, 1989; Cobbett and Goldsbrough 2002; Rea et al. 2004). Transgenic plants over-expressing GSH biosynthetic genes have been generated for various plants species and all showed improved tolerance to metal toxicity as compared to wild-type (WT) plants (Table 18.1).

The above studies clearly indicate that the up-regulation of GSH biosynthesis can improve the tolerance of plants to heavy metals and can, in some cases, enhance metal uptake, utilization, and detoxification. Hence, the development of transgenic plants over-expressing GSH biosynthetic genes and displaying increased heavy metal tolerance could not only be used to allow crop plants to be grown in soils high in heavy metals but could also be used for phytoremediation purposes.

4 Transgenic Plants Over-expressing GSH Biosynthetic Genes: Salt and Osmotic Stress Tolerance

The role of GSH and GSH metabolism in tolerance to salt stress has been studied using salt-tolerant and salt-susceptible genotypes in several plant species (Mittova et al. 2003a, b; El-Shabrawi et al. 2010). In general, salt-tolerant species show greater GSH biosynthesis and cellular GSH levels; higher GSH/GSSG ratio; higher GPX, GST, GR, Gly I, and Gly II activities; and lower levels of oxidative damage when exposed to salt stress (Mittova et al. 2003a, b; El-Shabrawi et al. 2010). Application of GSH to plants has also been shown to improve salt tolerance and yields in crop plants (Hussain et al. 2016; Akram et al. 2017). Several studies using transgenic plants have shown that over-expression of GSH biosynthetic genes improves salt tolerance in a wide range of plant species (Bae et al. 2013; Choe et al. 2013; Li et al. 2015; Park et al. 2017). For example, Choe et al. (2013) showed that transgenic rice (*Oryza sativa*) plants over-expressing *OsECS* had improved salinity stress tolerance, as indicated by a bright green phenotype, and maintained a higher GSH/GSSG ratio as compared to WT plants. These transgenic plants also showed lower ion leakage and higher chlorophyll fluorescence when exposed to MV-induced oxidative stress, and the seeds of these plants showed higher germination rates under saline conditions. In addition, *OsECS* over-expressing rice plants accumulated more biomass and had higher yields when grown in paddy fields in the absence of any stress. Other studies on rice plants in which GSH levels have also been manipulated have also shown improved tolerance to salinity. Bae et al. (2013) reported that transgenic rice plants over-expressing a *Brassica juncea* L. *ECS* (*BrECS*) gene showed improved salt tolerance and higher yields and biomass when grown in the paddy fields. Li et al. (2015) showed that transgenic plants over-expressing a *Pyrus*

Table 18.1 Transgenic plants over-expressing GSH biosynthetic genes and heavy metal stress tolerance

Gene name	Transgenic plant species	Tolerance to	Phenotypic response and factors determining tolerance	References
γ-ECS	*Agrostis palustris*	Cd	Transgenic plants exhibited more effective growth as compared to WT plants under Cd stress due to increased sequestering of Cd^{2+} and PC synthesis	Zhao et al. (2010)
γ-ECS + *ArsC* (arsenate reductase)	*Arabidopsis thaliana*	Arsenic (As)	Transgenic plants over-expressing both genes showed 4- to 17-fold higher shoot fresh weights and accumulated two- to three-fold higher As per gram tissue as compared to WT or plant over-expressing *γ-ECS* or *ArC*	Dhankher et al. (2002)
γ-ECS	*cad2-1* mutant of *Arabidopsis thaliana*	As	Transgenic plants showed 6- to 100-fold higher γ-glutamylcysteine (EC), PC2, and PC3 peptide levels in root tissues over mutants that were equivalent to WT plants. The shoot and root levels of GSH were two- to five-fold above those in WT plants, with or without treatment with As	Li et al. (2006a)
γ-ECS or *GS*	*Arabidopsis thaliana*	Mercury (Hg) and As	Over-expression of *γ-ECS* or *GS* caused a significant increase in GSH and/or PCs, and plants showed resistance to As and Hg, as compared to WT plants. Co-expression of both *ECS* and *GS* resulted in tolerance to Hg, and plants accumulated 35-fold more biomass and three-fold more Hg aboveground than the WT when grown on Hg(II), due to enhanced synthesis of PCs as compared to the WT plants or lines expressing *ECS* or *GS* alone	Li et al. (2006b)
γ-ECS + *PCS* + *serine acetyltransferase*	*Nicotiana tabacum*	Cd	Plants over-expressing transgenes (either separately or in combination) showed increased Cd sequestering in roots but not in shoots compared with WT plants. Importantly, transgenic plants over-expressing all three genes showed the greatest effects (about eight-fold elevation of thiols) as compared to single gene transgenic or to WT plants	Wawrzyński et al. (2006)
γ-ECS	*Arabidopsis thaliana*	As, Hg, and Cd	The level of GSH, PCs were increased 3- to 20-fold in response to As, Hg, and Cd exposure as compared to WT plants. Transgenic plants were highly resistant to As and weakly resistant to Hg but sensitive to Cd indicating different tolerance mechanisms for the metals	Li et al. (2005)

γ-ECS or *GS*	*Brassica juncea*	Heavy metal and metalloids	Transgenic plants over-expressing *γ-ECS* or *GS* showed significantly higher capacities to tolerate and accumulate a variety of metals (particularly As, Cd, and Cr) as well as mixed-metal combinations (As, Cd, Zn/As, Pb, and Zn) due to greater availability of GSH and higher PC synthesis as compared to WT plants	Reisinger et al. (2008)
GS	*Brassica juncea*	Cd	Transgenic plants showed enhanced Cd tolerance at various stages of plant development and accumulated significantly (25%) higher Cd levels than the WT plants. Cd accumulation and tolerance were correlated with the *gshII* expression levels. Transgenic plants also showed higher GSH, thiol, S, Ca, and PC contents as compared to WT plants	Zhu et al. (1999a)
γ-ECS	*Brassica juncea*	Cd	The transgenic plants showed increased tolerance to Cd and had higher concentrations of PCs, GSH, and total nonprotein thiols compared to WT plants. The transgenic plants accumulated more Cd than WT plants: shoot Cd concentrations were 40–90% higher. Over-expression of *γ-ECS* increases biosynthesis of GSH and PCs, which in turn enhances Cd tolerance and accumulation	Zhu et al. (1999b)
γ-ECS	Cottonwood (*Populus deltoides*)	As	Transgenic plans had higher ECS activity and thiol levels and showed enhanced growth in the presence of As as compared to WT plants. Furthermore, roots of transgenic plants accumulated significantly more (two-fold in root tissues and two- to three-fold in shoot tissues) As compared to non-transgenic plants	LeBlanc et al. (2011)
ECS + *PCS*	*Arabidopsis thaliana*	Cd and As	Transgenic plants over-expressing either *ECS* or *PCS* showed higher tolerance to Cd and As as compared to WT plants. Importantly, transgenic plants over-expressing both of the genes showed 2-fold Cd accumulation and higher PCs synthesis as compared to single-gene transgenic lines	Guo et al. (2008)

(continued)

Table 18.1 (continued)

Gene name	Transgenic plant species	Tolerance to	Phenotypic response and factors determining tolerance	References
γ-ECS	*Populus nigra*, *Populus × canescens*	Zinc (Zn)	Transgenic plants over-expressing *γ-ECS* either in the cytosol or chloroplasts showed elevated levels of GSH under Zn stress. Transgenic plants showed higher HM uptake and GST activity as compared to WT plants	Bittsánszky (2005)
γ-ECS	*Populus tremula × P. alba*	Heavy metals and metalloids	Transgenic plants over-expressing *γ-ECS* in the cytosol showed a two-fold increase in foliar GSH concentrations. Biomass accumulation of WT plants decreased in contaminated soil by more than 30-fold, whereas γ-ECS transgenic plants showed only a two-fold decrease compared to control plants grown in uncontaminated soils	Ivanova et al. (2011)

calleryana γ-ECS (*Pc*γ*ECS*) gene showed higher GSH biosynthesis and salt tolerance. Recently, Park et al. (2017) reported that transgenic rice plants over-expressing a GS gene (*OsGS*) showed improved growth and oxidative stress tolerance when planted in paddy fields. The transgenic plants showed improved oxidative stress (induced by MV) tolerance as indicated by lower MDA and H_2O_2 accumulation. Importantly, the transgenic plants showed improved grain yields and increased biomass under variable climatic conditions. The above evidence demonstrates that genetically engineered plants over-expressing *γ-ECS* or *GS* genes show improved salt tolerance as well as increased biomass and yields under salt stress and in the absence of stress.

5 Transgenic Plants Over-expressing GSH Biosynthetic: Drought Stress Tolerance

Drought is one of the most important stressors that impacts crop productivity worldwide (George and Parida 2010; Prabu et al. 2011; Su et al. 2011). It is well known that roots are the first plant organs to respond to drought (Davies and Zhang 1991; Sengupta and Reddy 2011; Sengupta et al. 2011). Drought stress is associated with reduced CO_2 fixation and higher ROS accumulation that can cause oxidative damage (Baena-Gonzalez et al. 2007; Cruz de Carvalho 2008; Miller et al. 2010; Gechev et al. 2012). Therefore, for the survival of plants under drought stress, fully functional ROS-detoxifying systems are essential for normal plant growth and development (Kranner et al., 2002; Hossain et al. 2013a, b). Ahmed et al. (2013) showed that drought-tolerant wild barley showed greater GSH biosynthesis under drought or salt stress alone or in combination and higher levels of antioxidant enzymes. Greater synthesis of GSH under drought stress was also found in drought-tolerant wheat genotypes, as compared to susceptible cultivars (Islam et al. 2015). Exogenous application of GSH improved drought stress tolerance through the up-regulation of ROS and MG detoxification pathways (Nahar et al. 2015). Compartment specific studies of ROS and antioxidant metabolism in GSH-deficient *pad2-1* mutants demonstrated diverse roles for GSH in regulating drought tolerance. Koffler et al. (2014) showed that GSH-deficient *pad2-1* mutant plants exposed to drought stress had significantly lower GSH levels in most cell compartments (51% in mitochondria, 31% in chloroplasts, 34% in nuclei, and 28% in the cytosol), whereas increased GSH levels were found in WT plants under drought. The levels of GSH and AsA decreased significantly in chloroplasts and peroxisomes with a large increase in cellular H_2O_2 levels. Sengupta et al. (2012) investigated the importance of *Vigna radiata* (L.) γ-ECS (*Vr*γECS) under progressive drought stress. Analysis of H_2O_2 levels, lipid peroxidation and *Vr*γECS enzyme activity was linked during drought stress and recovery. Additionally, the delicate inter-relationships, putative regulatory mechanism and functioning in the root system under adverse drought conditions, was associated with these factors (Sengupta et al. 2012). In another study, transgenic tobacco plants over-expressing γ-ECS showed greater drought stress tolerance as indicated by higher germination rate, water retention, water recovery, chlorophyll, and proline content as compared with WT plants. The transgenic tobacco plants also showed higher levels of expression of stress-related genes including *heat shock protein 70(HSP70)*, *GPX*, *thioredoxin peroxidase*, *chalcone synthase*, *1-aminocyclopropane-1-carboxylic acid (ACC) oxidase*, and *heme oxygenase I* (Kumar et al. 2014).

6 Transgenic Plants Over-expressing GSH Biosynthetic Genes: Herbicide Tolerance

In modern agriculture, herbicides are frequently applied to eradicate weeds as they are more labor- and energy-effective than manual or mechanical weed control (De Block et al. 1987). Recent studies have investigated the role of GSH and its related enzymes and herbicide tolerance (Katerova and Miteva 2010; Burns et al. 2017). GSH can directly detoxify herbicides by forming conjugates, a process that can also be catalyzed by the enzyme GST. Once formed, conjugates can be metabolized and excreted or can be stored in vacuoles or dead cells (Katerova and Miteva 2010). Proteomic and genomic studies of GSH-related proteins in *Avena fatua* L. have shown higher DHAR activity in herbicide-resistant genotypes as compared to susceptible genotypes (Burns et al. 2017). Tseng et al. (2013) also showed that a paraquat-tolerant rice mutant showed higher GSH biosynthesis (3.5-fold) as compared to susceptible one. Transgenic poplar hybrids (*Populus tremula* × *Populus alba*) over-expressing a bacterial γ-ECS gene in the cytosol or in chloroplasts displayed increased resistance to chloroacetanilide herbicides and had higher cellular levels of γ-ECS and GSH as well as higher GST activity as compared to WT plants (Gullner et al. 2001). In *Brassica juncea*, over-expressing γ-ECS or GS gene showed a twofold increase in nonprotein thiol levels and enhanced plant resistance to herbicide atrazine (Flocco et al. 2004). These studies demonstrate the potential for that over-expressing genes associated with GSH metabolism has for the production of transgenic herbicide-tolerant crops plants.

7 Conclusions and Future Perspectives

Glutathione has a broad range of functions in plant growth, development, and stress tolerance, and glutathione metabolism is now considered as a prime candidate for the deliberate manipulation of plants to enhance stress tolerance and to improve yields and quality and also for phytoremediation purposes. Transgenic plants over-expressing GSH biosynthetic genes have showed higher stress tolerance, due to greater GSH accumulation, modulation of redox homeostasis, and increased expression of genes associated with stress tolerance in plants. However, there are still numerous questions to be answered with respect to the roles glutathione plays in plant stress tolerance. For instance, the regulation of GSH biosynthesis and its mechanistic interaction with other redox active molecules and the interconnections between GSH and various biochemical pathways concern with abiotic stress tolerance. A complete understanding of the regulatory factors associated with GSH biosynthesis in plants and how this important molecule interacts with other plant metabolic processes will open up a new horizon for stress tolerance and crop improvement through the genetic engineering of GSH biosynthetic genes into plant cells.

References

Ahmad N, Malagoli M, Wirtz M, Hell R (2016) Drought stress in maize causes differential acclimation responses of glutathione and sulfur metabolism in leaves and roots. BMC Plant Biol 16(1):247

Ahmed IM, Cao F, Zhang M, Chen X, Zhang G, Wu F (2013) Difference in yield and physiological features in response to drought and salinity combined stress during anthesis in Tibetan wild and cultivated barleys. PLoS One 8:e77869

Akram S, Siddiqui MN, Hussain BMN, Bari MAA, Mostofa MG, Hossain MA, Tran LSP (2017) Exogenous glutathione modulates salinity tolerance of soybean [*Glycine max* (L.) Merrill] at reproductive stage. J Plant Growth Regul 36:877–888

Asada K (1999) The water-water cycle in chloroplasts: scavenging of active oxygens and dissipation of excess photons. Annu Rev Plant Physiol Plant Mol Biol 50:601–639

Avery SV (2011) Molecular targets of oxidative stress. Biochem J 434:201–210

Bae MJ, Kim YS, Kim IS (2013) Transgenic rice overexpressing the *Brassica juncea* gamma-glutamylcysteine synthetase gene enhances tolerance to abiotic stress and improves grain yield under paddy field conditions. Mol Breed 31:931–945

Baena-Gonzalez E, Rolland F, Thevelein JM, Sheen J (2007) A central integrator of transcription networks in plant stress and energy signalling. Nature 448:938–942

Bittsanszky A, Komives T, Gullner G, Gyulai G, Kiss J, Heszky L, Radimszky L, Rennenberg H (2005) Ability of transgenic poplars with elevated glutathione content to tolerate zinc (2+) stress. Environ Int 31:251–254

Burns EE, Keith BK, Refai MY, Bothner B, Dyer WE (2017) Proteomic and biochemical assays of glutathione-related proteins in susceptible and multiple herbicide resistant *Avena fatua* L. Pestic Biochem Physiol. https://doi.org/10.1016/ j.pestbp.2017.06.007

Chen J, Goldsborough PB (1994) Increased activity of γ-glutamylcysteine synthetase in tomato cells selected for cadmium tolerance. Plant Physiol 106:233–239

Chen JH, Jiang HW, Hsieh EJ, Chen HY, Chien CT, Hsieh HL, Lin TP (2012) Drought and salt stress tolerance of an *Arabidopsis* glutathione S-transferase U17 knockout mutant are attributed to the combined effect of glutathione and abscisic acid. Plant Physiol 158:340–351

Cheng MC, Ko K, Chang WL, Kuo WC, Chen GH, Lin TP (2015) Increased glutathione contributes to stress tolerance and global translational changes in *Arabidopsis*. Plant J 83(5):926–939

Choe YH, Kim YS, Kim IS, Bae MJ, Lee EJ, Kim YH, Park HM, Yoon HS (2013) Homologous expression of gamma-glutamylcysteine synthetase increases grain yield and tolerance of transgenic rice plants to environmental stresses. J Plant Physiol 170:610–618

Cobbett C, Goldsbrough P (2002) Phytochelatins and metallothioneins: roles in heavy metal detoxification and homeostasis. Annu Rev Plant Biol 53:159–182

Cobbett CS, May MJ, Howden R, Rolls B (1998) The glutathione-deficient, cadmium-sensitive mutant, cad2-1, of *Arabidopsis thaliana* is deficient in gamma-glutamylcysteine synthetase. Plant J 16:73–78

Cruz de Carvalho MH (2008) Drought stress and reactive oxygen species: production, scavenging and signaling. Plant Signal Behav 3:156–165

Davies WJ, Zhang J (1991) Roots signals and the regulation of growth and development of plant in dry soil. Annu Rev Plant Physiol Plant Mol Biol 42:55–76

De Block M, Botterman J, Vandewiele M, Dockx J, Thoen C, Gosselé V, RaoMovva N, Thompson C, Van Montagu M, Leemans J (1987) Engineering herbicide resistance in plants by expression of a detoxifying enzyme. EMBO J 6(9):2513–2518

Dhankher OP, Li Y, Rosen BP, Shi J, Salt D, Senecoff JF, Sashti NA, Meagher RB (2002) Engineering tolerance and hyperaccumulation of arsenic in plants by combining arsenate reductase and gamma-glutamylcysteine synthetase expression. Nat Biotechnol 20: 1140–1145

El-Shabrawi H, Kumar B, Kaul T, Reddy MK, Singla-Pareek SL, Sopory SK (2010) Redox homeostasis, antioxidant defense, and methylglyoxal detoxification as markers for salt tolerance in Pokkali rice. Protoplasma 245:85–96

Flocco CG, Lindblom SD, Pilon-Smits EAH (2004) Overexpression of enzymes involved in glutathione synthesis enhances tolerance to organic pollutants in *Brassica juncea*. Int J Phytochem 6:289–304

Foyer CH, Harbinson J (1994) Oxygen metabolism and the regulation of photosynthetic electron transport. In: Foyer CH, Mullineaux PM (eds) Causes of photooxidative stress and amelioration of defense system in plants. CRC Press, Boca Raton, pp 1–42

Foyer CH, Noctor G (2005a) Oxidant and antioxidant signaling in plants: a re-evaluation of the concept of oxidative stress in a physiological context. Plant Cell Environ 29:1056–1071

Foyer CH, Noctor G (2005b) Redox homeostasis and antioxidant signaling: a metabolic interface between stress perception and physiological responses. Plant Cell 17:1866–1875

Foyer CH, Noctor G (2009) Redox regulation in photosynthetic organisms: signaling, acclimation, and practical implications. Antioxid Redox Signal 11:861–905

Foyer CH, DescoUrvieres P, Kunert KJ (1994) Protection against oxygen radicals; an important defense mechanism studied in transgenic plants. Cell Environ 17:507–523

Foyer CH, Souriau N, Perret S, Lelandais M, Kunert KJ, Pruvost C, Jouanin L (1995) Overexpression of glutathione-reductase but not glutathione synthetase leads to increases in antioxidant capacity and resistance to photoinhibition in poplar trees. Plant Physiol 109:1047–1057

Fryer MJ, Ball L, Oxborough K, Karpinski S, Mullineaux PM, Baker NR (2003) Control of ascorbate peroxidase 2 expression by hydrogen peroxide and leaf water status during excess light stress reveals a functional organisation of *Arabidopsis* leaves. Plant J 33:691–705

Gaber A, Yoshimura K, Tamoi M, Takeda T, Nakano Y, Shigeoka S (2004) Induction and functional analysis of two reduced nicotinamide adenine dinucleotide phosphate-dependent glutathione peroxidase-like proteins in *Synechocystis* PCC 6803 during the progression of oxidative stress. Plant Physiol 136(1):2855–2861

Gaber A, Yoshimura K, Yamamoto T, Yabuta Y, Takeda T, Kanaboshi H, Miyasaka H, Nakano Y, Shigeoka S (2006) Glutathione peroxidase-like protein of *Synechocystis* PCC 6803 confers tolerance to oxidative and environmental stresses in transgenic *Arabidopsis*. Physiol Plant 128:251–262

Gaber A, Maruta T, Ogata T, Yoshimura K, Tamoi M, Shigeoka S (2012) Glutathione peroxidase 8 is a novel enzyme for the detoxification of reactive oxygen species in the nucleus of *Arabidopsis*. Plant Cell Physiol 53(9):1596–1606

Galant A, Preuss ML, Cameron JC, Jez JM (2011) Plant glutathione biosynthesis-diversity in biochemical regulation and reaction products. Front Plant Sci 2:45

Gechev TS, Dinakar C, Benina M, Toneva V, Bartels D (2012) Molecular mechanisms of desiccation tolerance in resurrection plants. Cell Mol Life Sci 69:3175–3186

George S, Parida A (2010) Characterization of an oxidative stress inducible nonspecific lipid transfer protein coding cDNA and its promoter from drought tolerant plant *Prosopis juliflora*. Plant Mol Biol Report 28:32–40

Ghanta S, Chattopadhyay S (2011) Glutathione as a signaling molecule: another challenge to pathogens. Plant Signal Behav 6:783–788

Grant JJ, Loake GJ (2000) Role of reactive oxygen intermediates and cognate redox signaling in disease resistance. Plant Physiol 124:21–29

Grill E, Winnacker EL, Zenk MH (1987) Phytochelatins, a class of heavy-metal-binding peptides from plants, are functionally analogous to metallothioneins. Proc Natl Acad Sci U S A 84:439–443

Grill E, Lffler S, Winnacker EL, Zenk MH (1989) Phytochelatins, the heavy-metal binding peptides of plants, are synthesized from glutathione by a specific c-glutamylcysteine dipeptidyl transpeptidase (phytochelatin synthase). Proc Natl Acad Sci U S A 86:6838–6842

Gullner G, Kömives T, Rennenberg H (2001) Enhanced tolerance of transgenic poplar plants overexpressing γ-glutamylcysteine synthetase towards chloroacetanilide herbicides. J Exp Bot 52:971–979

Guo J, Dai X, Xu W, Ma M (2008) Overexpressing GSH1 and AsPCS1 simultaneously increases the tolerance and accumulation of cadmium and arsenic in *Arabidopsis thaliana*. Chemosphere 72:1020–1026

Halliwell B, Gutteridge JMC (2007) Free radicals in biology and medicine, 4th edn. Oxford University Press, Oxford

Harms K, von Ballmoos P, Brunold C, Hofgen R, Hesse H (2000) Expression of a bacterial serine acetyltransferase in transgenic potato plants leads to increased levels of cysteine and glutathione. Plant J 22:335–343

Hoque TS, Hossain MA, Mostofa MG, Burritt DJ, Fujita M, Tran LSP (2016) Methylglyoxal: an emerging signaling molecule in plant abiotic stress responses and tolerance. Front Plant Sci 7:1341

Hossain MA, Hasanuzzaman M, Fujita M (2010) Up-regulation of antioxidant and glyoxalase systems by exogenous glycinebetaine and proline in mung bean confer tolerance to cadmium stress. Physiol Mol Biol Plants 16:259–227

Hossain MA, Teixeira da Silva JA, Fujita M (2011) Glyoxalase system and reactive oxygen species detoxification system in plant abiotic stress response and tolerance: an intimate relationship. In: Shanker A, Venkateswarlu B (eds) Abiotic stress in plants-mechanisms and adaptations. INTECH-Open Access Publisher, Rijeka, pp 235–266

Hossain MA, Piyatida P, Teixeira da Silva JA, Fujita M (2012) Molecular mechanism of heavy metal toxicity and tolerance in plants: central role of glutathione in detoxification of reactive oxygen species and methylglyoxal and in heavy metal chelation. J Bot 2012:872875

Hossain MA, Mostofa MG, Fujita M (2013a) Cross protection by cold-shock to salinity and drought stress-induced oxidative stress in mustard (*Brassica campestris* L.) seedlings. Mol Plant Breed 4:50–70

Hossain MA, Mostofa MG, Fujita M (2013b) Heat-shock positively modulates oxidative protection of salt and drought-stressed mustard (*Brassica campestris* L.) seedlings. J Plant Sci Mol Breed 2:1–14

Hossain MA, Hoque MA, Burritt DJ, Fujita M (2014) Proline protects plants against abiotic oxidative stress: biochemical and molecular mechanisms. In: Ahmad P (ed) Oxidative damage to plants. Elsevier, USA, pp 477–522

Hossain MA, Bhattacharjee S, Armin SM, Qian P, Xin W, Li H-Y, Burritt DJ, Fujita M, Tran LSP (2015) Hydrogen peroxide-priming modulates abiotic oxidative stress tolerance: insights from ROS detoxification and scavenging. Front Plant Sci 6:420

Hussain BMN, Akram S, Raffi SA, Burritt DJ, Hossain MA (2016) Exogenous glutathione improves salinity stress tolerance in rice (*Oryza sativa* L.). Plant Gene Trait 7:1–17

Islam M, Begum MC, Kabir AH, Alam MF (2015) Molecular and biochemical mechanisms associated with differential responses to drought tolerance in wheat (*Triticum aestivum* L.). J Plant Interact 10:195–201

Ivanova LA, Ronzhina DA, Ivanov LA, Stroukova LV, Peuke AD, Rennenberg H (2011) Over-expression of *gsh1* in the cytosol affects the photosynthetic apparatus and improves the performance of transgenic poplars on heavy metal-contaminated soil. Plant Biol 13:649–659

Jones DP (2000) Redox potential of GSH/GSSG couple: assay and biological significance. Methods Enzymol 348:93–112

Karpinski S, Escobar C, Karpinska B, Creissen G, Mullineaux PM (1997) Photosynthetic electron transport regulates the expression of cytosolic ascorbate peroxidase genes in *Arabidopsis* during excess light stress. Plant Cell 9:627–640

Karpinski S, Reynolds H, Karpinska B, Wingsle G, Creissen G, Mullineaux P (1999) Systemic signaling and acclimation in response to excess excitation energy in *Arabidopsis*. Science 284:654–657

Katerova ZI, Miteva LPE (2010) Glutathione and herbicide resistance in plants. In: Anjum NA et al (eds) Ascorbate-glutathione pathway and stress tolerance in plants. Springer, Dordrecht, pp 189–207

Kocsy G, Galiba G, Brunold C (2001) Role of glutathione in adaptation and signaling during chilling and cold acclimation in plants. Physiol Plant 113:158–164

Koffler BE, Luschin-Ebengreuth N, Stabentheiner E, Müller M, Zechmann B (2014) Compartment specific response of antioxidants to drought stress in Arabidopsis. Plant Sci 227:133–144

Kranner I, Beckett RP, Wornik S, Zorn M, Pfeifhofer HW (2002) Revival of a resurrection plant correlates with its antioxidant status. Plant J 31:13–24

Krumova K, Cosa G (2016) Overview of reactive oxygen species. In: Nonell S, Flors C (eds) Singlet oxygen: applications in biosciences and nanosciences, vol 1. Royal Society of Chemistry, Cambridge, pp 1–21

Kumar D, Datta R, Sinha R, Ghosh A, Chattopadhyay S (2014) Proteomic profiling of gamma-ECS overexpressed transgenic *Nicotiana* in response to drought stress. Plant Signal Behav 9:e29246

LeBlanc MS, Lima A, Montello P, Kim T, Meagher RB, Merkle S (2011) Enhanced arsenic tolerance of transgenic eastern cottonwood plants expressing gamma-glutamylcysteine synthetase. Int J Phytoremediation 13:657–673

Li Y, Dhankher OP, Carreira L, Balish RS, Meagher RB (2005) Arsenic and mercury tolerance and cadmium sensitivity in *Arabidopsis* plants expressing bacterial gamma-glutamylcysteine synthetase. Environ Toxicol Chem 24(6):1376–1386

Li Y, Dankher OP, Carreira L, Smith AP, Meagher RB (2006a) The shoot-specific expression of γ-glutamylcysteine synthetase directs the long-distance transport of thiol-peptides to roots conferring tolerance to mercury and arsenic. Plant Physiol 141:288–298

Li Y, Heaton ACP, Carreira L, Meagher RB (2006b) Enhanced tolerance to and accumulation of mercury, but not arsenic, in plants overexpressing two enzymes required for thiol peptide synthesis. Physiol Plant 128:48–57

Li H, Jin-long H, Jing L, Qing-song Y, Chang Y (2015) A γ-glutamylcysteine synthetase gene from *Pyrus calleryana* is responsive to ions and osmotic stresses. Plant Mol Biol Report 33:1088–1097

Liedschulte V, Wachter A, An Z, Rausch T (2010) Exploiting plants for glutathione (GSH) production: uncoupling GSH synthesis from cellular controls results in unprecedented GSH accumulation. Plant Biotechnol J 8:807–820

Liu D, An Z, Mao Z, Ma L, Lu Z (2015) Enhanced heavy metal tolerance and accumulation by transgenic sugar beets expressing *Streptococcus thermophilus* StGCS-GS in the presence of Cd, Zn and Cu alone or in combination. PLoS One 10(6):e0128824

Loyall L, Uchida K, Braun S, Furuya M, Frohnmeyer H (2000) Glutathione and a UV light-induced glutathione-S-transferase are involved in signaling to chalcone synthase in cell cultures. Plant Cell 12:1939–1950

Lu SC (2000) Regulation of glutathione synthesis. Curr Top Cell Regul 36:95–116

May MJ, Leaver CJ (1993) Oxidative stimulation of glutathione synthesis in *Arabidopsis thaliana* suspension-cultures. Plant Physiol 103:621–627

Meinhard M, Rodriguez PL, Grill E (2002) The sensitivity of ABI2 to hydrogen peroxide links the abscisic acid-response regulator to redox signalling. Planta 214:775–782

Miller G, Suzuki N, Ciftci-Yilmaz S, Mittler R (2010) Reactive oxygen species homeostasis and signalling during drought and salinity stresses. Plant Cell Environ 33:453–467

Mittova V, Tal M, Volokita M, Guy M (2003a) Up-regulation of the leaf mitochondrial and peroxisomal antioxidative systems in response to salt-induced oxidative stress in the wild salt-tolerant tomato species *Lycopersicon pennellii*. Plant Cell Environ 26:845–856

Mittova V, Theodoulou FL, Kiddle G, Gomez L, Volokita M, Tal M, Foyer CH, Guy M (2003b) Coordinate induction of glutathione biosynthesis and glutathione-metabolizing enzymes is correlated with salt tolerance in tomato. FEBS Lett 554:417–421

Mostofa MG, Hossain MA, Siddiqui MN, Fujita M, Tran LSP (2017) Phenotypical, physiological and biochemical analyses provide insight into selenium-induced phytotoxicity in rice plants. Chemosphere 178:212–223

Mou Z, Fan W, Dong X (2003) Inducers of plant systemic acquired resistance regulate NPR1 function through redox changes. Cell 113:935–944

Munné-Bosch S, Queval G, Foyer CH (2013) The impact of global change factors on redox signaling underpinning stress tolerance. Plant Physiol 161:5–19

Nahar K, Hasanuzzaman M, Alam MM, Fujita M (2015) Glutathione-induced drought stress tolerance in mung bean: coordinated roles of the antioxidant defence and methylglyoxal detoxification systems. AoB Plants 7:plv069

Noctor G, Foyer CH (1998) Ascorbate and glutathione: keeping active oxygen under control. Annu Rev Plant Physiol Plant Mol Biol 49:249–279

Noctor G, Strohm M, Jouanin L, Kunert KJ, Foyer CH, Rennenberg H (1996) Synthesis of glutathione in leaves of transgenic poplar overexpressing gamma-glutamylcysteine synthetase. Plant Physiol 112:1071–1078

Noctor G, Arisi A, Jouanin L, Foyer C (1998a) Manipulation of glutathione and amino acid biosynthesis in the chloroplast. Plant Physiol 118:471–482

Noctor G, Arisi A, Jouanin L, Kunert K, Rennenberg H, Foyer CH (1998b) Glutathione: biosynthesis, metabolism and relationship to stress tolerance explored in transformed plants. J Exp Bot 49:623–647

Noctor G, Gomez L, Vanacker H, Foyer CH (2002) Interactions between biosynthesis, compartmentation and transport in the control of glutathione homeostasis and signalling. J Exp Bot 53:1283–1304

Noctor G, Queval G, Mhamdi A, Chaouch S, Foyer CH (2011) Glutathione. Arabidopsis Book 9:e0142

Noctor G, Mhamdi A, Chaouch S, Han Y, Neukermans J, Marquez-Garcia B, Queval G, Foyer CH (2012) Glutathione in plants: an integrated overview. Plant Cell Environ 35:454–484

op den Camp RGL, Przbyla D, Ochsenbein C, Laloi C, Kim C, Danon A, Wagner D, Hideg E, Göbel C, Feussner I, Nater M, Apel K (2003) Rapid induction of distinct stress responses after the release of singlet oxygen in *Arabidopsis*. Plant Cell 15:2320–2332

Park SI, Kim YS, Kim JJ, Mok JE, Kim YH, Park HM, Kim IS, Yoon HS (2017) Improved stress tolerance and productivity in transgenic rice plants constitutively expressing the *Oryza sativa* glutathione synthetase *OsGS* under paddy field conditions. J Plant Physiol 215:39–47

Pneuli L, Liang H, Rozenberg M, Mittler R (2003) Growth suppression, altered stomatal responses, and augmented induction of heat shock proteins in cytosolic ascorbate peroxidase (Apx1) deficient *Arabidopsis* plants. Plant J 34:187–203

Prabu G, Kawar PG, Pagariya MC, Prasad DT (2011) Identification of water deficit stress upregulated genes in sugarcane. Plant Mol Biol Report 29:291–304

Prasad TK (1996) Mechanisms of chilling-induced oxidative stress injury and tolerance in developing maize seedlings: changes in antioxidant system, oxidation of proteins and lipids, and protease activities. Plant J 10:1017–1026

Rawlins MR, Leaver CJ, May MJ (1995) Characterization of an *Arabidopsis thaliana* cDNA-encoding glutathione synthetase. FEBS Lett 376:81–86

Rea PA, Vatamaniuk OK, Rigden DJ (2004) Weeds, worms, and more. Papain's long-lost cousin, phytochelatin synthase. Plant Physiol 136:2463–2474

Reisinger S, Schiavon M, Terry N, Pilon-Smits EAH (2008) Heavy metal tolerance and accumulation in Indian mustard (*Brassica juncea* L.) expressing bacterial gamma-glutamylcysteine synthetase or glutathione synthetase. Int J Phytoremediation 10:1–15

Rossel JB, Wilson IM, Pogson BJ (2002) Global changes in gene expression in response to high light in *Arabidopsis*. Plant Physiol 130:1109–1120

Sengupta D, Reddy AR (2011) Water deficit as a regulatory switch for legume root responses. Plant Signal Behav 6:914–917

Sengupta D, Kannan M, Reddy AR (2011) A root proteomics-based insight reveals dynamic regulation of root proteins under progressive drought stress and recovery in *Vigna radiata* (L.) Wilczek. Planta 233:1111–1127

Sengupta D, Ramesh G, Mudalkar S, Kumar KRR, Kirti PB, Reddy AR (2012) Molecular cloning and characterization of γ-glutamyl cysteine synthetase (VrγECS) from roots of *Vigna radiata* (L.) Wilczek under progressive drought stress and recovery. Plant Mol Biol Report 30:894–903
Sewelam N, Kazan K, Schenk PM (2016) Global plant stress signaling: reactive oxygen species at the cross-road. Front Plant Sci 7:187
Shimazaki K, Sugahara K (1980) Inhibition site of the electron transport system in lettuce chloroplasts by fumigation of leaves with SO_2. Plant Cell Physiol 12:303–312
Su Z, Li X, Hao Z, Xie C, Li M, Weng J, Zhang D, Liang X, Wang Z, Gao J, Zhang S (2011) Association analysis of the nced and rab28 genes with phenotypic traits under water stress in maize. Plant Mol Biol Report 29:714–772
Szalai G, Kellös T, Galiba G, Kocsy G (2009) Glutathione as an antioxidant and regulatory molecule in plants under abiotic stress conditions. J Plant Growth Regul 28:66–80
Tseng TY, Ou JF, Wang CY (2013) Role of the ascorbate–glutathione cycle in paraquat tolerance of rice. Weed Sci 61:361–373
Wawrzyński A, Kopera E, Wawrzyńska A, Kamińska J, Bal W, Sirko A (2006) Effects of simultaneous expression of heterologous genes involved in phytochelatin biosynthesis on thiol content and cadmium accumulation in tobacco plants. J Exp Bot 57:2173–2182
Willekens H, Chamnongpol S, Davey M, Schraudner M, Langebartels C, Van Montagu M, Inze D, Van Camp W (1997) Catalase is a sink for H_2O_2 and is indispensable for stress defence in C-3 plants. EMBO J 16:4806–4816
Wingate VMP, Lawton MA, Lamb CJ (1988) Glutathione causes a massive and selective induction of plant defense genes. Plant Physiol 87:206–210
Wingsle G, Karpinski S (1996) Differential redox regulation of glutathione reductase and Cu/Zn superoxide dismutase gene expression in *Pinus sylvestris* L. needles. Planta 198:151–157
Xiang CB, Oliver DJ (1998) Glutathione metabolic genes coordinately respond to heavy metals and jasmonic acid in *Arabidopsis*. Plant Cell 10:1539–1550
Zechmann B (2014) Compartment-specific importance of glutathione during abiotic and biotic stress. Front Plant Sci 5:566
Zhao C, Qiao M, Yu Y, Xia G, Xiang F (2010) The effect of the heterologous expression of *Phragmites australis* γ-glutamylcysteine synthetase on the Cd^{2+} accumulation of *Agrostis palustris*. Plant Cell Environ 33:877–887
Zhu YL, Pilon-Smits EAH, Jouanin L, Terry N (1999a) Overexpression of glutathione synthetase in Indian mustard enhances cadmium accumulation and tolerance. Plant Physiol 119:73–79
Zhu YL, Pilon-Smits EAH, Tarun AS, Weber SU, Jouanin L, Terry N (1999b) Cadmium tolerance and accumulation in Indian mustard is enhanced by overexpressing γ-glutamylcysteine synthetase. Plant Physiol 121:1169–1177

Index

M.A. Hossain et al. (eds.), *Glutathione in Plant Growth, Development, and Stress Tolerance*, https://doi.org/10.1007/978-3-319-66682-2

H

Printed by Printforce, the Netherlands